Entomology in Focus

Volume 10

Insects are fundamentally important in the ecology of terrestrial habitats. What is more, they affect diverse human activities, notably agriculture, as well as human health and wellbeing. Meanwhile, much of modern biology has been developed using insects as subjects of study.

To reflect this, our aim with *Entomology in Focus* is to offer a range of titles that either capture different aspects of the diverse biology of insects or their management, or that offer updates and reviews of particular species or taxonomic groups that are important for agriculture, the environment or public health.

The series results from an agreement between Springer and the Entomological Society of Brazil (SEB) and as such may lean towards tropical entomology. The aim throughout is to provide reference texts that are simple in their conception and organization but that offer up-to-date syntheses of the respective areas, offer suggestions of future directions for research (and for management where relevant) and that don't shy away from offering considered opinions.

Inara R. Leal • Fernanda M. P. Oliveira
Xavier Arnan • Alan N. Andersen
Editors

Ant Responses to Climate Change

From Individuals to Ecosystem Services

Springer

Sociedade Entomológica do Brasil • 1972

Editors
Inara R. Leal
Departamento de Botânica
Universidade Federal de Pernambuco
Recife, Pernambuco, Brazil

Fernanda M. P. Oliveira
Departamento de Botânica
Universidade Federal de Pernambuco
Recife, Pernambuco, Brazil

Xavier Arnan
Universidade de Pernambuco – Campus Garanhuns
Garanhuns, Pernambuco, Brazil

Alan N. Andersen
Research Institute for the Environment and Livelihoods
Charles Darwin University
Brinkin, Northern Territory, NT, Australia

ISSN 2405-853X ISSN 2405-8548 (electronic)
Entomology in Focus
ISBN 978-3-032-21126-2 ISBN 978-3-032-21127-9 (eBook)
https://doi.org/10.1007/978-3-032-21127-9

This Springer imprint is published by the registered company Springer Nature Switzerland AG
The registered company address is: Gewerbestrasse 11, 6330 Cham, Switzerland

Preface

Climate change is one of the most pressing challenges of our time. Its wide-reaching effects on natural systems are already evident across ecosystems worldwide, altering species distributions, phenologies, interactions, and functions. Amid this complex and evolving landscape, ants stand out as a particularly important taxon for understanding and monitoring ecological change. Ubiquitous, highly diverse, and ecologically dominant, ants play fundamental roles in nearly every terrestrial ecosystem on Earth. As such, they are not only susceptible to the impacts of climate change but also key agents influencing ecosystem resilience and transformation.

This book, *Ant Responses to Climate Change—From Individuals to Ecosystem Services*, emerges in response to the rapid growth of scientific interest in the effects of climate change on ant biology and ecology. While the number of studies on this topic has increased substantially in recent years, a comprehensive synthesis has been lacking. Our goal with this volume is to fill that gap by compiling the most up-to-date research and perspectives across a range of biological scales—from the responses of individual ants to shifts in ecosystem-level processes. The book is structured into six sections, each addressing a different level of ecological organization. Part I is an introductory chapter that contextualizes anthropogenic climate change within Earth's climatic history and outlines what is currently known about its impacts on ants. Part II comprises four chapters (Chaps. 2, 3, 4, and 5) that focus on individual-level responses, exploring how climate change influences ant behavior, morphology, thermal tolerance, and desiccation resistance. Part III has two chapters (Chaps. 6 and 7) that address ant populations, discussing how climate change affects their dynamics, distributions, and phenological changes, as well as genetic structure. One of the chapters (Chap. 7) deals with invasive species. Part IV moves to the community level, consisting of six chapters (Chaps. 8, 9, 10, 11, and 12), examining how taxonomic, functional, and phylogenetic diversity shift along environmental gradients, including latitude, elevation, aridity, and vertical stratification from soil to canopy. Part V has six chapters (Chaps. 13, 14, 15, 16, 17, and 18) exploring the effects of climate change on plant-ant interactions and the ecosystem services mediated by ants. This includes mutualistic networks, impacts on plant reproduction, seed dispersal, ant-aphid relationships, and greenhouse gas emissions from leaf-cutting ants. Finally, Part VI presents a synthetic chapter (Chap. 19) that integrates findings across all levels, highlights synergies between climate change and other human disturbances, and outlines key directions for future research.

This volume brings together a team of 49 researchers from 36 institutions, all of whom are leading experts in diverse fields of ant biology. They have generously shared their time, expertise, and insights to create a cohesive and comprehensive resource. The preparation of each chapter required a significant commitment—from synthesizing vast bodies of literature to integrating complex and often interdisciplinary data and perspectives. We are deeply grateful for their dedication and collaboration throughout this process. Their collective efforts make this book not only a timely and essential reference on ant responses to climate change, but also a testament to the power of scientific cooperation in addressing urgent global challenges.

We also would like to thank Joao Pildervasser (formerly a Springer Editor in Brazil) for inviting us to organize this book and Arun Siva Shanmugam (Springer Production Editor) for all the support and patience during the execution of this project. Inara R. Leal (process 305248/2022-7) and Xavier Arnan (process 313589/2023-2) received a productivity grant from the Conselho Nacional de Desenvolvimento Científico e Tecnológico (CNPq). Inara R. Leal, Fernanda M. P. Oliveira, and Xavier Arnan also received financial support from CNPq, Coordenação de Aperfeiçoamento de Pessoal de Nível Superior (CAPES), and Fundação de Amparo à Ciência e Tecnologia do Estado de Pernambuco (FACEPE).

We hope this book inspires new research, fosters interdisciplinary collaboration, and ultimately supports global efforts to preserve biodiversity and ecosystem function in the face of an uncertain climatic future.

Recife, Pernambuco, Brazil	Inara R. Leal
Recife, Pernambuco, Brazil	Fernanda M. P. Oliveira
Garanhuns, Pernambuco, Brazil	Xavier Arnan
Brinkin, Northern Territory, Australia	Alan N. Andersen

Contents

Part VI Conclusion

Editors and Contributors

About the Editors

Inara R. Leal graduated in Biology (1990) from the Federal University of Santa Catarina (Brazil), with an MSc (1994) and PhD (1998) in Ecology from the State University of Campinas (Brazil), and postdoctoral research at the Commonwealth Scientific and Industrial Research Organisation (Australia) in 2010 and 2014. Since 2002, she has been a professor at the Federal University of Pernambuco (Brazil), advising undergraduate students in Biology and Environmental Sciences and post-graduate students in Plant Biology and Animal Biology. Her research focuses on understanding how anthropogenic disturbances and climate change affect plant-animal interactions in the two most important ecosystems of northeastern Brazil: the Atlantic Forest and the Caatinga dry forest—both heavily impacted by human activities. In the Atlantic Forest, she has investigated how habitat loss and fragmentation reshape insect communities (including ants, butterflies, dung beetles, and termites), lead to the proliferation of herbivores—particularly leaf-cutting ants—and simplify other plant-ant interactions. In the Caatinga, her work explores the effects of both acute (e.g., slash-and burn agriculture) and chronic (e.g., livestock raising, firewood collection, and non-timber forest product exploitation) disturbances and climate change (e.g., increasing aridity) on the taxonomic, phylogenetic, and functional organization of plant and animal communities and on ecosystem services provided by ants. She is the author or editor of 13 books, more than 50 book chapters, and over 170 scientific publications, with an h-index of 57 and approximately 12,000 career citations.

Fernanda M. P. Oliveira graduated in Biology (2010) from the State University of Ceará (Brazil), with an MSc in Animal Biology (2012) and a PhD in Plant Biology (2018) from the Federal University of Pernambuco (Brazil), including a research period at the Commonwealth Scientific and Industrial Research Organisation (CSIRO)—Tropical Ecosystems Research Centre (Australia). Since 2024, she has been a professor at the Federal University of Pernambuco (Brazil), advising undergraduate students in Biology and Environmental Sciences and postgraduate students in Plant Biology and Animal Biology. Her research spans multiple levels of biological organization and explores the individual and combined effects of human-induced disturbances and climate change on ant and plant traits, biological communities, mutualistic interactions, and, most notably, the ecosystem functions provided by ants. She also studies how these components recover over time through ecological succession. She is the author/editor of 3 books, 23 book chapters, and 27 scientific publications, with an h-index of 10 and approximately 370 career citations.

Xavier Arnan graduated in Biology (2001), with an MSc in Biology (2004) and a PhD in Ecology (2006) from the Universitat Autònoma de Barcelona (UAB, Spain). He held postdoctoral positions in Spain (UAB and the Centre de Recerca Ecològica i Aplicacions Forestals—CREAF, 2006–2011), Australia (Commonwealth Scientific and Industrial Research Organisation, 2008 and 2009), Germany (Technical University of Darmstadt, 2012–2013), and Brazil (Federal University of Pernambuco, 2014–2017). From May 2017 to January 2020, he served as a Ramón y Cajal staff researcher at CREAF. Since January 2020, he has been a professor at the University of Pernambuco (Brazil). He is a terrestrial ecologist aiming to understand the patterns and responses of biodiversity along disturbance and environmental gradients and the mechanisms underlying them at local, regional, and global scales to understand biodiversity responses and species adaptations to global change from evolutionary and conservation standpoints. His work integrates taxonomic, functional, and phylogenetic approaches, with ants as the primary study group, although he also works with other animals, plants, and their interactions. His studies are mainly conducted at the community level in Mediterranean and tropical ecosystems, using a combination of field observations and experiments, along with laboratory research on model organisms. He is the author of 108 scientific publications, with an h-index of 33 and 3568 career citations.

Alan N. Andersen graduated with a BSc (Hons) from Monash University in 1980, and PhD in ecology at the University of Melbourne in 1985. Currently, he is Professor of Research Excellence and Impact in Charles Darwin University's (CDU) Office of Research and Innovation, and Professor of Terrestrial Invertebrates in CDU's Research Institute for the Environment and Livelihoods (Australia). He was previously a Chief Research Scientist with CSIRO Land & Water, and head of CSIRO's Tropical Ecosystems Research Centre in Darwin from 1995 to 2016. His primary research interests are in the global ecology of ant communities, where he integrates community ecology, historical and contemporary biogeography, and systematics to gain a predictive understanding of ant diversity, behavioral dominance and functional composition in relation to environmental stress and disturbance throughout the world. He applies this understanding to the use of ants as bioindicators of ecological change. He is the author of 5 books and 300 scientific publications, with an h-index of 80 and >23,000 career citations. He is a Fellow of the Australian Academy of Science.

Contributors

Michael F. Allen Center for Conservation Biology, Department of Microbiology and Plant Pathology, University of California, Riverside, CA, USA

Alan N. Andersen Research Institute for the Environment and Livelihoods, Charles Darwin University, Brinkin, Northern Territory, NT, Australia

Elena Ângulo Estación Biológica de Doñana, CSIC, Sevilla, Spain

Xavier Arnan Universidade de Pernambuco – Campus Garanhuns, Garanhuns, Pernambuco, Brazil

Olivia K. Bates Department of Ecology and Evolution, University of Lausanne, Lausanne, Switzerland

Cleo Bertelsmeier Department of Ecology and Evolution, University of Lausanne, Lausanne, Switzerland

Tom R. Bishop School of Biosciences, Cardiff University, Cardiff, UK

Department of Zoology and Entomology, University of Pretoria, Pretoria, South Africa

Solène Blanchard Research Centre for Biodiversity and Environment, University of Toulouse III, Toulouse, France

Jelena Bujan Division for Marine and Environmental Research, Ruđer Bošković Institute, Zagreb, Croatia

Stefan Caddy-Retalic University of Adelaide, North Terrace, Adelaide, SA, Australia

Talita Câmara Universidade de Pernambuco—Campus Garanhuns, Garanhuns, PE, Brazil

Programa de Pós-Graduação em Biologia Vegetal, Universidade Federal de Pernambuco, Recife, PE, Brazil

Programa de Pós-Graduação em Ecologia e Evolução, Universidade Estadual de Feira de Santana, Feira de Santana, BA, Brazil

Byanca N. T. Cavalcante Programa de Pós-Graduação em Biologia Vegetal, Universidade Federal de Pernambuco, Recife, PE, Brazil

Diego Centeno-Alvarado Programa de Pós-Graduação em Biologia Vegetal, Universidade Federal de Pernambuco, Recife, PE, Brazil

Xim Cerdá Estación Biológica de Doñana, CSIC, Sevilla, Spain

Leticia M. Classen-Rodriguez SocioEnvironmental and Economic Development Strategies, Merced, CA, USA

SocioEnvironmental and Education Network, Merced, CA, USA

Sierra Nevada Research Institute, Department of Civil and Environmental Engineering, University of California, Merced, CA, USA

Abdallah Dahbi Laboratory of Ecotoxicology, Bioresources and Coastal Geomorphology, Polydisciplinary Faculty of Safi, Cadi Ayyad University, Marrakesh, Morocco

Amanda V. da Silva Centro de Ciências Naturais e Humanas, Universidade Federal do ABC, Alameda da Universidade, São Bernardo do Campo, SP, Brazil

Carlos L. de la Rosa de la Rosa Consultants, Escondido, CA, USA

Douglas D. de Souza Centro de Ciências Naturais e Humanas, Universidade Federal do ABC, Alameda da Universidade, São Bernardo do Campo, SP, Brazil

Claire Detrain Unit of Social Ecology, University of Brussels, Brussels, Belgium

Luciana Elizalde LIHO—INIBIOMA, Universidad Nacional del Comahue, CONICET, San Carlos de Bariloche, Río Negro, Argentina

Angel S. Fernandez-Bou SocioEnvironmental and Economic Development Strategies, LLC, Merced, California, USA

SocioEnvironmental and Education Network, Merced, California, USA

Sierra Nevada Research Institute, Department of Civil and Environmental Engineering, University of California, Merced, CA, USA

Union of Concerned Scientists, Berkeley, California, USA

Heloise Gibb School of Biological, Earth and Environmental Sciences, UNSW, Kensington, NSW, Australia

School of Life and Environmental Sciences, Deakin University, Burwood, VIC, Australia

Thomas C. Harmon Department of Civil and Environmental Engineering, University of California, Merced, CA, USA

Lily Leahy Department of Ecological, Plant, and Animal Sciences, La Trobe University, Melbourne, VIC, Australia

Inara R. Leal Departamento de Botânica, Universidade Federal de Pernambuco, Recife, Pernambuco, Brazil

Laura C. Leal Departamento de Zoologia, Universidade Federal do Paraná, Curitiba, PR, Brazil

Departamento de Ecologia e Biologia Evolutiva, Universidade Federal de São Paulo, Diadema, SP, Brazil

María Natalia Lescano LIHO—INIBIOMA, Universidad Nacional del Comahue, CONICET, San Carlos de Bariloche, Río Negro, Argentina

Bruno Melati Centro de Ciências Naturais e Humanas, Universidade Federal do ABC, Alameda da Universidade, São Bernardo do Campo, SP, Brazil

Geraldo Nascimento Programa de Pós-Graduação em Etnobiologia e Conservação da Natureza, Universidade Federal Rural de Pernambuco, Recife, Pernambuco, Brazil

Douglas Thierry de Melo Neves Programa de Pós-Graduação em Biologia Vegetal, Universidade Federal de Pernambuco, Recife, PE, Brazil

Anselmo Nogueira Centro de Ciências Naturais e Humanas, Universidade Federal do ABC, Alameda da Universidade, São Bernardo do Campo, SP, Brazil

Fernanda M. P. Oliveira Departamento de Botânica, Universidade Federal de Pernambuco, Recife, Pernambuco, Brazil

Paulo S. Oliveira Departamento de Biologia Animal, Universidade Estadual de Campinas, Campinas, SP, Brazil

Gabriela I. Pirk LIHO—INIBIOMA, Universidad Nacional del Comahue, CONICET, San Carlos de Bariloche, Río Negro, Argentina

Luana S. Prochazka Programa de Pós-Graduação em Biodiversidade Vegetal e Meio Ambiente, Instituto de Pesquisas Ambientais, Avenida Miguel Estefno, São Paulo, SP, Brazil

Javier Retana CREAF and Universitat Autònoma de Barcelona, Bellaterra, Spain

Hannah Riskas Centre for Future Landscapes, La Trobe University, Bundoora, VIC, Australia

Mariano A. Rodriguez-Cabal Grupo de Ecología de Invasiones - INIBIOMA, Universidad Nacional del Comahue, CONICET, San Carlos de Bariloche, Río Negro, Argentina

Rubenstein School of Environment and Natural Resources, University of Vermont, Burlington, VT, USA

Karl A. Roeder North Central Agricultural Research Laboratory, Agricultural Research Service, USDA, Brookings, SD, USA

Philip W. Rundel Department of Ecology and Evolutionary Biology, University of California, Los Angeles, CA, USA

Luitgard Schwendenmann Te Kura Mātai Taiao, School of Environment, Waipapa Taumata Rau, University of Auckland, Auckland, Aotearoa, New Zealand

Marcelo Tabarelli Departamento de Botânica, Universidade Federal de Pernambuco, Recife, PE, Brazil

Louise van Oudenhove Institut Sophia Agrobiotech, INRAE, Sophia Antipolis, France

Heraldo L. Vasconcelos Universidade Federal de Uberlândia, Uberlândia, MG, Brazil

Irene Villalta Institut de Recherche sur la Biologie de l'Insecte, Université de Tours, Tours, France

Robert J. Warren II Buffalo State University, Buffalo, NY, USA

Quentin Willot Environmental and Evolutionary Biology Research Unit, Department of Biology, University of Namur, Namur, Belgium

Alana Zammit Keble College, University of Oxford, Oxford, UK

Part I
Introduction

Chapter 1
Ant Responses to Climate Change: An Introduction

Inara R. Leal, Fernanda M. P. Oliveira, Xavier Arnan, and Alan N. Andersen

1.1 Anthropogenic Climate Change

Throughout its 4.5-billion-year history, our planet has experienced a series of drastic climate changes, each playing a pivotal role in shaping the evolution and diversity of life (Crowley and North 1988; Mayhew et al. 2012; Song et al. 2021). For instance, roughly 252 million years ago, rapid temperature rise caused the extinction of an estimated 70% of land plant and animal species, along with 84% of marine species, an event known as the 'end Permian extinctions' (Sun et al. 2012; Stanley 2016). Analyses of Earth's history and fossil records over the past 520 million years indicate that the end of the Permian is not an isolated event, as rapid climate change (mostly warming but also cooling) is consistently associated with planet-wide die-offs (Song et al. 2021).

Historical climate changes have been related to natural causes such as large-scale volcanic activity, weathering of rocks, asteroid impact, changes in Earth's orbit and tilt, and variations in the incidence of solar radiation (see Zachos et al. 2008 for a review). However, we are now experiencing a new event of climate change, caused by the emission of greenhouse gases (primarily carbon dioxide, nitrous oxide,

I. R. Leal (✉) · F. M. P. Oliveira
Departamento de Botânica, Universidade Federal de Pernambuco, Recife, Pernambuco, Brazil
e-mail: inara.leal@ufpe.br; fmp.oliveira@ufpe.br

X. Arnan
Universidade de Pernambuco – Campus Garanhuns, Garanhuns, Pernambuco, Brazil
e-mail: xavier.arnan@upe.br

A. N. Andersen
Research Institute for the Environment and Livelihoods, Charles Darwin University, Brinkin, Northern Territory, NT, Australia
e-mail: alan.andersen@cdu.edu.au

I. R. Leal et al. (eds.), *Ant Responses to Climate Change*, Entomology in Focus 10,
https://doi.org/10.1007/978-3-032-21127-9_1

methane, chlorofluorocarbons) due to human activity (IPCC 2021). The 'greenhouse effect' is a natural phenomenon, in which greenhouse gases let the sun's light shine onto Earth's surface but trap the heat that reflects back up into the atmosphere, keeping our planet heated. However, since the industrial revolution, in the late 1700s and early 1800s, there has been a progressive increase in the concentration of greenhouse gases related mostly to burning of fossil fuels, agriculture and livestock farming, and deforestation (IPCC 2021). This increased concentration of greenhouse gases in the atmosphere traps more heat than usual and causes global warming in what is referred to as anthropogenic climate change (IPCC 2021).

The global mean near-surface temperature in 2025 was 1.55 °C above the pre-industrial average, making it the warmest year in the 175-year observational record (World Meteorological Organization 2025). The past decade (2010–2020) was the warmest on record, with the years 2016 and 2020 tied as the second warmest years ever recorded (IPCC 2021, 2023). There has also been a global increase in aridity over the past century, with dry areas increasing around 30% from 1950 to 2008 (Dai 2011). A large part of this global drying is related to surface warming attributed to human-induced greenhouse gases, but exceptionally warm events associated with the El Niño Southern Oscillation and the North Atlantic Oscillation also contribute to this pattern (Dai 2011). Additionally, there has been an increase in the frequency, intensity, and duration of extreme weather events, such as heatwaves, droughts, intense rainfall, and hurricanes (Dai 2011; IPCC 2021, 2023). Global sea levels have risen by about 20 centimetres since 1880 due to the thermal expansion of seawater as it warms and the melting of glaciers and ice sheets (Diaz and Rosenberg 2008; IPCC 2021, 2023). Furthermore, as more than 90% of the excess heat from global warming has been absorbed by the oceans (Pinsky et al. 2020), there have been significant increases in ocean temperatures, alterations in ocean currents and circulation, and disruption of nutrient cycles (Diaz and Rosenberg 2008; Pinsky et al. 2020; IPCC 2021, 2023). Warmer waters also hold less oxygen, leading to the expansion of hypoxic or 'dead zones', where oxygen levels are too low to support most marine life (Diaz and Rosenberg 2008; Pinsky et al. 2020). Oceans also absorb about 30% of the carbon dioxide (CO_2) emitted by human activities, leading to ocean acidification and threatening marine life (Doney et al. 2009; Pinsky et al. 2020; IPCC 2021).

All these alterations in Earth's systems have a profound impact on terrestrial and marine life, and a myriad of cascading effects of anthropogenic climate change on natural systems have been reported (see Walther et al. 2002; Parmesan 2006; Tylianakis et al. 2008 for reviews). Species shifting their ranges toward the poles or to higher altitudes in response to rising temperatures were initially reported for many species of butterflies, birds, herbs, and trees, particularly in northern Europe, but studies are now spread across all continents and taxonomic groups (Walther et al. 2002; Root et al. 2003; Parmesan 2006; IPCC 2021). There are also numerous records of changes in the timing of key biological events, such as flowering, breeding, and migration (see Tylianakis et al. 2008 and references therein). While anthropogenic enrichment of atmospheric CO_2 and nitrogen may increase rates of plant growth and flowering, climate-induced phenological mismatches between plants and their pollinators may outweigh these benefits, resulting in a net reduction in fruit set (Memmott et al. 2007; Tylianakis et al. 2008). Higher temperatures, droughts, and increased frequency of wildfires are causing not only alterations in the composition, structure, and biogeography of forests in many regions of the globe but also leading to forest dieback, particularly in boreal and tropical forests (see Walther et al. 2002; Dai 2011; IPCC 2021, 2023, and references therein). Additionally, increasing aridity leads to habitat degradation, reduced forest productivity, and loss of ecosystem functions and services (Thomas et al. 2004; Dai 2011; IPCC 2021, 2023).

In the marine realm, warmer ocean temperatures cause coral bleaching, leading to widespread, mass coral mortality and the collapse of reef ecosystems (IPCC 2021). As on land, marine species are shifting their ranges poleward or to deeper, cooler waters, disrupting existing ecosystems and fisheries (Pinsky et al. 2020; IPCC 2021). Increased CO_2 absorption by oceans lowers pH levels, making it harder for calcifying organisms like corals, molluscs, and plankton to form shells and skeletons (Doney et al. 2009; Pinsky et al. 2020; IPCC 2021). This can affect the entire marine food web. Rising sea levels also threaten coastal ecosystems such as mangroves, salt marshes, and estuaries (Mcleod et al. 2011; Pinsky et al. 2020; IPCC 2021). These habitats are crucial for biodiversity, carbon sequestration, and protecting coastlines from storm surges (Mcleod et al. 2011). The interconnectedness of ecosystems means that changes in one can have ripple effects on others. For instance, the loss of coral reefs can affect fish populations, which in turn affects terrestrial species that rely on them.

All these effects demonstrate the magnitude of climate change and its profound impact on the Earth's systems. This book addresses the responses of arguably the world's most ecologically dominant faunal group, ants.

1.2 Ecological Importance of Ants

Ants are one of the most ubiquitous, widespread, and abundant groups of terrestrial animals globally. In the most recent and comprehensive evaluation of ant biomass, it was estimated that 20 quadrillion of ants live on Earth, representing 12 megatons of dry carbon, equivalent to 20% of human biomass (Schultheiss et al. 2022). Ants are also highly diverse, comprising more than 15,700 named species and subspecies (Bolton 2024), although total diversity could well be double (Ward 2014). With this remarkable numerical dominance, ants mediate many key ecological processes and are crucial for the functioning and maintenance of terrestrial ecosystems (see Folgarait 1998; Del Toro et al. 2012; Elizalde et al. 2020 for reviews). Ants interact with a wide range of organisms in both mutualistic and antagonistic relationships (see Rico-Gray and Oliveira 2007 for a review). They interact mutualistically with plants by protecting them against herbivores and dispersing their seeds, as well as contributing to plant nutrition. On the other hand, ants are dominant seed predators in arid ecosystems (Brown et al. 1979; Morton and Davidson 1988), and leaf-cutting ants are among the most voracious herbivores in Neotropical forests and savannas, harvesting from 14% up to 30% of the leaf crop in these biomes (Urbas et al. 2007; Costa et al. 2008; Siqueira et al. 2018). Ants also form mutualistic relationships with sap-sucking insects, they act as both predators and prey for a wide range of arthropods, and their nests are habitats for a plethora of other organisms (Styrsky and Eubanks 2007; Sanders and van Veen 2011). Finally, ants play a pivotal role in soil development, decomposition, and nutrient cycling with such broad effects on other organisms that ants are often referred to as ecosystem engineers (Frouz and Jilková 2008; Sanders and van Veen 2011; Meyer et al. 2013). For all these reasons, ants are important shapers of biodiversity (Parker and Kronauer 2021).

Due to their ubiquity, astonishing biomass, and ecological dominance, ants have been widely used as bioindicators for assessing the effects of human disturbance (Andersen and Majer 2004). There is extensive literature on the responses of ant communities to disturbances such as agriculture (Pacheco et al. 2013), agroforestry (Philpott et al. 2008), mining (Majer et al. 1984; Hoffmann et al. 2000), fire (Andersen et al. 2006; Andersen 2019; Arnan et al. 2006), logging (Arnan et al. 2009; Ross et al. 2018), and pastoralism (Hoffmann 2010; Arcoverde et al. 2018). Ants also respond to habitat loss and fragmentation caused by changes in land use (Vasconcelos et al. 2006; Leal et al. 2012; Dantas and Fonseca 2024). Even chronic disturbances (i.e. continuous removal of small portions of forest biomass) that do not cause habitat loss and fragmentation, such as the exploitation of forest products and extensive livestock raising, can substantially alter ant communities (Oliveira et al. 2017; Ribeiro-Neto et al. 2016, 2023; Dantas and Fonseca 2024). These studies document a predictable replacement of disturbance 'losers' (disturbance-sensitive taxa) by disturbance 'winners' (disturbance-adapted taxa; see Filgueiras et al. 2021), with disturbances typically favouring open-adapted taxa at the expense of highly specialized, forest-associated functional groups (Andersen 2000; Leal et al. 2012, 2017). This occurs because the most significant effects of human disturbance on ants are related to changes in habitat structure, especially habitat openness, which alters microclimate conditions, resource availability, and competitive interactions (Andersen 2019). The impacts on habitat openness as a key driver of variation in disturbance responses likely extend to other faunal groups as well (Rachmansah et al. 2025).

1.3 Ant Responses to Climate Change

With ecosystems facing increasing threats from climate change, many studies have addressed its impacts on ants and their ecological roles (see a review in Parr and Bishop 2022). Many species have been predicted to shift their distribution ranges in response to warming climates (e.g. Colwell et al. 2008; Bishop et al. 2019). Whereas species from lower latitudes may move in to replace those at higher latitudes, lowland tropical species, which are already near their thermal limits, have no such replacements (Deutsch et al. 2008). Range-restricted, high-elevation species are particularly vulnerable, as they have no suitable habitats to disperse to (Nowrouzi et al. 2016). Areas with high ant species turnover, particularly along rainfall gradients, are expected to be highly sensitive to future climate change (Andersen et al. 2015). Additionally, the ranges of invasive ants are likely to expand as global temperatures rise (Roura-Pascual et al. 2004; Fitzpatrick et al. 2007; Bertelsmeier et al. 2015), negatively impacting native species. These changes in the distribution of individual species in response to climate change may lead to alterations in ant communities, although geographic and phylogenetic contexts influence the overall response (Parr and Bishop (2022)). Generally, communities from tropical regions are negatively impacted by rising temperatures, while those from temperate zones might increase in diversity (Diamond et al. 2012). Lastly, climate change may exacerbate the negative impacts of disturbance, with growing evidence suggesting that climate change and disturbance can interact in complex and sometimes synergistic ways (Gibb et al. 2015; Arnan et al. 2018; Ribeiro-Neto et al. 2023). In this context, ant communities in regions with hot, arid climates are at a higher risk (Gibb et al. 2015). This highlights how ants can be used not only as bioindicators of human disturbance and land management but also of climate change.

1.4 Book Scope and Topics

The number of papers addressing ant responses to climate change is increasing exponentially, and we believe that a comprehensive textbook on the subject is timely. The book consists of 19 chapters, structured into six parts that explore different levels of ecological organization, ranging from individual responses to the impact on ecosystem services. There will always be overlap no matter how different parts are structured, and we recognize that responses at higher levels of organization will be driven by those at lower levels. However, as components of a biological level combine to form a higher hierarchical level, new properties emerge that are absent from the lower level.

This chapter opens the book with an introduction to anthropogenic climate change and a summary of what is presently known about its effects on ant biology. Part II explores the individual-level responses of ants to climate change. It covers behaviour (Cerdá et al., Chap. 2), morphology (Gibb and Riskas, Chap. 3), thermal tolerance (Nascimento et al., Chap. 4), and desiccation resistance (Bujan and Roeder, Chap. 5). In part III, the focus shifts to ant populations, including population dynamics and distributions (Centeno-Alvarado et al., Chap. 6) and implications of climate change for ant invasions (Bertelsmeier and Bates, Chap. 7). Part IV is devoted to community-level changes, with chapters addressing latitudinal gradients (Andersen, Chap. 8), elevation gradients from both taxonomic (Warren II, Chap. 9) and functional and phylogenetic (Bishop and Zammit, Chap. 10) perspectives, aridity gradients (Arnan et al., Chap. 11), and the vertical dimension (i.e. from soil to canopy, Leahy, Chap. 12). Part V addresses plant-ant interactions and associated ecosystem services and includes ant-plant mutualistic networks (Câmara et al.,

Chap. 13), plant reproductive output of plant-herbivore-ant interactions (Leal et al., Chap. 14), ant-mediated ecosystem services in general (Elizalde et al., Chap. 15), ant-mediated seed dispersal in detail (Oliveira et al., Chap. 16), and ant-aphid mutualisms (Detrain and Blanchard, Chap. 17) along with a final chapter on gas emissions from leaf-cutting ants (Fernandez-Bou et al., Chap. 18). Finally, part VI includes an integrative chapter that synthesizes the major impacts of climate change on ants and their ecological roles. It also outlines future research directions, focusing on the synergies between human disturbance and climate change, and identifies key questions we need to address to better understand how ants and the ecosystems they inhabit will respond to the ongoing climate crisis (Leal et al., Chap. 19).

We aim for this book to be a valuable resource for students, researchers, conservationists, policymakers, and anyone interested in the complex effects of climate change on biodiversity and ecological function. By compiling cutting-edge knowledge from various fields, this book provides new insights and aims to inspire new research directions, ultimately contributing to the global effort to address the impact of climate change on one of the most remarkable groups of terrestrial organisms—the ants.

References

Andersen AN (2000) A global ecology of rainforest ants: functional groups in relation to stress and disturbance. In: Agosti D, Majer JD, Alonso L, Schultz T (eds) Ants: standard methods for measuring and monitoring biodiversity. Smithsonian Institution Press, Washington, DC, pp 25–34

Andersen AN (2019) Responses of ant communities to disturbance: five principles for understanding the disturbance dynamics of a globally dominant faunal group. J Anim Ecol 88:350–362. https://doi.org/10.1111/1365-2656.12907

Andersen AN, Majer JD (2004) Ants show the way down-under: invertebrates as bioindicators in land management. Front Ecol Environ 2:291–298. https://doi.org/10.1890/1540-9295(2004)002[0292:ASTWDU]2.0.CO;2

Andersen AN, Hertog T, Woinarski JC (2006) Long term fire exclusion and ant community structure in an Australian tropical savanna: congruence with vegetation succession. J Biogeogr 33:823–832. https://doi.org/10.1111/j.1365-2699.2006.01463.x

Andersen AN, del Toro I, Parr CL (2015) Savanna ant species richness is maintained along a bioclimatic gradient of increasing latitude and decreasing rainfall in northern Australia. J Biogeogr 42:2313–2322. https://doi.org/10.1111/jbi.12599

Arcoverde GB, Andersen AN, Leal IR, Setterfield SA (2018) Habitat-contingent responses to disturbance: impacts of cattle grazing on ant communities vary with habitat complexity. Ecol Appl 28:1808–1817. https://doi.org/10.1002/eap.1770

Arnan X, Rodrigo A, Retana J (2006) Post-fire recovery of Mediterranean ground ant communities follows vegetation and dryness gradients. J Biogeogr 33:1246–1258. https://doi.org/10.1111/j.1365-2699.2006.01506.x

Arnan X, Gracia M, Comas L, Retana J (2009) Forest management conditioning ground ant community structure and composition in temperate conifer forests in the Pyrenees Mountains. For Ecol Manag 258:51–59. https://doi.org/10.1016/j.foreco.2009.03.029

Arnan X, Arcoverde GB, Pie MR, Ribeiro-Neto JD, Leal IR (2018) Increased anthropogenic disturbance and aridity reduce phylogenetic and functional diversity of ant communities in Caatinga dry forest. Sci Total Environ 631-632:429–438. https://doi.org/10.1016/j.scitotenv.2018.03.037

Bertelsmeier C, Luque GM, Hoffmann BD, Courchamp F (2015) Worldwide ant invasions under climate change. Biodivers Conserv 24:117–128. https://doi.org/10.1007/s10531-014-0794-3

Bishop TR, Parr CL, Gibb H, Rensburg BJ, Braschler B, Chown SL, Foord SH, Lamy K, Munyai TC, Okey I, Tshivhandekano PG, Werenkraut V, Robertson MP (2019) Thermoregulatory traits combine with range shifts to alter the future of montane ant assemblages. Glob Change Biol 25:2162–2173. https://doi.org/10.1111/gcb.14622

Bolton B (2024) An online catalog of the ants of the world. https://antcat.org. Accessed 28 Aug 2024

Brown JH, Reichman O, Davidson DW (1979) Granivory in desert ecosystems. Annu Rev Ecol Syst 10:201–227

Colwell RK, Brehm G, Cardelús CL, Gilman AC, Longino JT (2008) Global warming, elevational range shifts, and lowland biotic attrition in the wet tropics. Science 322:258–261. https://doi.org/10.1126/science.1162547

Costa AN, Vasconcelos HL, Vieira-Neto EHM, Bruna EM (2008) Do herbivores exert top-down effects in Neotropical savannas? Estimates of biomass consumption by leaf-cutter ants. J Veg Sci 19:849–854. https://doi.org/10.3170/2008-8-18461

Crowley TJ, North GR (1988) Abrupt climate change and extinction events in earth history. Science 240:996–1002. https://doi.org/10.1126/science.240.4855.996

Dai A (2011) Drought under global warming: a review. WIREs Clim Change 2:45–65. https://doi.org/10.1002/wcc.81

Dantas A, Fonseca CR (2024) Synergistic effect of habitat loss and chronic anthropogenic disturbances on ant species richness. Biodivers Conserv 33:205–219. https://doi.org/10.1007/s10531-023-02743-3

Del Toro I, Ribbons RR, Pelini SL (2012) The little things that run the world revisited: a review of ant-mediated ecosystem services and disservices (Hymenoptera: Formicidae). Myrmecol News 17:133–146. https://doi.org/10.25849/myrmecol.news_017:133

Deutsch CA, Tewksbury JJ, Huey RB, Sheldon KS, Ghalambor CK, Haak DC, Martin PR (2008) Impacts of climate warming on terrestrial ectotherms across latitude. Proc Natl Acad Sci USA 105:6668–6672. https://doi.org/10.1073/pnas.0709472105

Diamond SE, Sorger DM, Hulcr J, Pelini SL, Del Toro I, Hirsch C, Oberg E, Dunn RR (2012) Who likes it hot? A global analysis of the climatic, ecological, and evolutionary determinants of warming tolerance in ants. Glob Change Biol 18:448–456. https://doi.org/10.1111/j.1365-2486.2011.02542.x

Diaz RJ, Rosenberg R (2008) Spreading dead zones and consequences for marine ecosystems. Science 321:926–929. https://doi.org/10.1126/science.115640

Doney SC, Fabry VJ, Feely RA, Kleypas JA (2009) Ocean acidification: the other CO_2 problem. Annu Rev Mar Sci 1:169–192. https://doi.org/10.1146/annurev.marine.010908.163834

Elizalde L, Arbetman M, Arnan X, Eggleton P, Leal IR, Lescano MN, Saez A, Werenkraut V, Pirk GI (2020) The ecosystem services provided by social insects: traits, management tools and knowledge gaps. Biol Rev 95:1418–1441. https://doi.org/10.1016/j.ecoser.2018.12.002

Filgueiras BKC, Peres CA, Melo FPL, Leal IR, Tabarelli M (2021) Winner-loser species replacements in human-modified landscapes. Trends Ecol Evol 36:545–555. https://doi.org/10.1016/j.tree.2021.02.006

Fitzpatrick MC, Weltzin JF, Sanders NJ, Dunn RR (2007) The biogeography of prediction error: why does the introduced range of the fire ant over-predict its native range? Glob Ecol Biogeogr 16:24–33. https://doi.org/10.1111/j.1466-8238.2006.00258.x

Folgarait PJ (1998) Ant biodiversity and its relationship to ecosystem functioning: a review. Biodivers Conserv 7:1221–1244. https://doi.org/10.1023/A:1008891901953

Frouz J, Jilková V (2008) The effect of ants on soil properties and processes (Hymenoptera: Formicidae). Myrmecol News 11:191–199

Gibb H, Sanders NJ, Dunn RR, Watson S, Photakis M, Abril S, Andersen AN, Angulo E, Armbrecht I, Arnan X, Baccaro FB, Bishop TR, Boulay R, Castracani C, Del Toro I, Delsinne T, Diaz M, Donoso DA, Enriquez ML, Fayle TM, Feener DH, Fitzpatrick MC, Gomez C, Grasso DA, Groc S, Heterick B, Hoffmann BD, Lach L, Lattke J, Leponce M, Lessard JP, Longino J, Lucky A, Majer J, Menke SB, Mezger D, Mori A, Munyai TC, Paknia O, Pearce-Duvet J, Pfeiffer M, Philpott SM, Souza JLP, Tista M, Vasconcelos HL, Vonshak M, Parr CL (2015) Climate mediates the effects of disturbance on ant assemblage structure. Proc R Soc London Ser B Biol Sci 282:8. https://doi.org/10.1098/rspb.2015.0418

Hoffmann BD (2010) Using ants for rangeland monitoring: global patterns in the responses of ant communities to grazing. Ecol Indic 10:105–111. https://doi.org/10.1016/j.ecolind.2009.04.016

Hoffmann BD, Griffiths AD, Andersen AN (2000) Response of ant communities to dry sulphur deposition from mining emissions in semi-arid northern Australia, with implications for the use of functional groups. Austral Ecol 25:653–663. https://doi.org/10.1111/j.1442-9993.2000.tb00071.x

IPCC (2021) Intergovernmental Panel on Climate Change 2021. In: Masson-Delmotte V, Zhai P, Pirani A, Connors SL, Péan C, Berger S, Caud N, Chen Y, Goldfarb L, Gomis MI, Huang M, Leitzell K, Lonnoy E, Matthews RBR, Maycock TK, Waterfield T, Yelekçi O, Yu R, Zhou B (eds) Climate change 2021: the physical science basis. Contribution of working group I to the sixth assessment report of the intergovernmental panel on climate change. Cambridge University Press, Cambridge/New York, p 2391. https://doi.org/10.1017/9781009157896

IPCC (2023) Intergovernmental panel on climate change 2023. In: Shukla PR, Skea J, Calvo Buendia A et al (eds) Climate change 2023: synthesis report. Contribution of working groups I, II, and III to the sixth assessment report of the intergovernmental panel on climate change. IPCC

Leal IR, Filgueiras BKC, Gomes JP, Andersen AN (2012) Effects of habitat fragmentation on ant richness and functional composition in Atlantic Forest of northeastern Brazil. Biodivers Conserv 21:1687–1701. https://doi.org/10.1007/s10531-012-0271-9

Leal IR, Ribeiro-Neto JD, Arnan X, Oliveira FMP, Arcoverde GB, Feitosa RM, Andersen AN (2017) Ants of the Caatinga: diversity, biogeography, and functional responses to anthropogenic disturbance and climate change. In: Silva JMC, Leal IR, Tabarelli M (eds) Caatinga: the largest Tropical Dry Forest Region in South America. Springer International Publishing, Cham, pp 65–95

Majer J, Day J, Kabay E, Perriman W (1984) Recolonization by ants in bauxite mines rehabilitated by a number of different methods. J Appl Ecol 21:355–375. https://doi.org/10.2307/2403059

Mayhew PJ, Bell MA, Benton TG, McGowan AJ (2012) Biodiversity tracks temperature over time. Proc Natl Acad Sci USA 109:15141–15145. https://doi.org/10.1073/pnas.1200844109

Mcleod E, Chmura GL, Bouillon S, Salm R, Björk M, Duarte CM, Lovelock CE, Schlesinger WH, Silliman BR (2011) A blueprint for blue carbon: toward an improved understanding of the role of vegetated coastal habitats in sequestering CO_2. Front Ecol Environ 9:552–560. https://doi.org/10.1890/110004

Memmott J, Craze PG, Waser NM, Price MV (2007) Global warming and the disruption of plant-pollinator interactions. Ecol Lett 10:710–717. https://doi.org/10.1111/j.1461-0248.2007.01061.x

Meyer ST, Neubauer M, Sayer EJ, Leal IR, Tabarelli M, Wirth R (2013) Leaf-cutting ants as ecosystem engineers: topsoil and litter perturbations around *Atta cephalotes* nests reduce nutrient availability. Ecol Entomol 38:497–504. https://doi.org/10.1111/een.12043

Morton SR, Davidson DW (1988) Comparative structure of harvester ant communities in arid Australia and North America. Ecol Monogr 58:19–38. https://doi.org/10.2307/1942632

Nowrouzi S, Andersen AN, Macfadyen S, Staunton KM, VanDerWal J, Robson SKA (2016) Ant diversity and distribution along elevation gradients in the Australian wet tropics: the importance of seasonal moisture stability. PLoS One 11:e0153420. https://doi.org/10.1371/journal.pone.0153420

Oliveira FMP, Ribeiro-Neto JD, Andersen AN, Leal IR (2017) Chronic anthropogenic disturbance as a secondary driver of ant community structure: interactions with soil type in Brazilian Caatinga. Environ Conserv 44:115–123. https://doi.org/10.1017/S0376892916000291

Pacheco R, Vasconcelos HL, Groc S, Camacho GP, Frizzo TL (2013) The importance of remnants of natural vegetation for maintaining ant diversity in Brazilian agricultural landscapes. Biodivers Conserv 22:983–997. https://doi.org/10.1007/s10531-013-0463-y

Parker J, Kronauer DJC (2021) How ants shape biodiversity. Curr Biol 31:R1141–R1224. https://doi.org/10.1016/j.cub.2021.08.015

Parmesan C (2006) Ecological and evolutionary responses to recent climate change. Annu Rev Ecol Evol Syst 37:637–669. https://doi.org/10.1146/annurev.ecolsys.37.091305.110100

Parr CL, Bishop TR (2022) The response of ants to climate change. Glob Change Biol 28(10):3188–3205. https://doi.org/10.1111/gcb.16140

Philpott SM, Arendt WJ, Armbrecht I, Bichier P, Diestch TV, Gordon C et al (2008) Biodiversity loss in Latin American coffee landscapes: review of the evidence on ants, birds, and trees. Conserv Biol 22:1093–1105. https://doi.org/10.1111/j.1523-1739.2008.01029.x

Pinsky ML, Selden RL, Kitchel ZJ (2020) Climate-driven shifts in marine species ranges: scaling from organisms to communities. Annu Rev Mar Sci 12:153–179. https://doi.org/10.1146/annurev-marine-010419-010916

Rachmansah A, Christian K, Schlesinger C, Murphy BP, Andersen AN (2025) Faunal responses to habitat disturbance: do the principles explaining responses of ant communities also apply to terrestrial reptiles? Ecol Evol 15:e70939. https://doi.org/10.1002/ece3.70939

Ribeiro-Neto JD, Arnan X, Tabarelli M, Leal IR (2016) Chronic anthropogenic disturbance causes homogenization of plant and ant communities in the Brazilian Caatinga. Biodivers Conserv 25:943–956. https://doi.org/10.1007/s10531-016-1099-5

Ribeiro-Neto JD, Oliveira FM, Arcoverde GB, Tabarelli M, Leal IR, Arnan X (2023) Aridity and chronic anthropogenic disturbances cause a taxonomic, functional, and phylogenetic homogenization of ant communities in a Caatinga dry forest. Biol Conserv 284:110151. https://doi.org/10.1016/j.biocon.2023.110151

Rico-Gray V, Oliveira PS (2007) The ecology and evolution of ant-plant interactions. University of Chicago Press, Chicago. Chicago Scholarship Online, 2013. https://doi.org/10.7208/chicago/9780226713540.001.0001

Root TL, Price JT, Hall KR, Schneider SH, Rosenzweig C, Pounds JA (2003) Fingerprints of global warming on wild animals and plants. Nature 421:57–60. https://doi.org/10.1038/nature01333

Ross SRPJ, Garcia FH, Fischer G, Peters MK (2018) Selective logging intensity in an east African rain forest predicts reductions in ant diversity. Biotropica 50:768–778. https://doi.org/10.1111/btp.12569

Roura-Pascual NR, Suarez AV, Gãmez C, Pons P, Touyama Y, Wild AL, Peterson AT (2004) Geographical potential of Argentine ants (*Linepithema humile* Mayr) in the face of global climate change. Proc Roy Soc London Ser B Biol Sci 271:2527–2535. https://doi.org/10.1098/rspb.2004.2898

Sanders D, van Veen FJF (2011) Ecosystem engineering and predation: the multi-trophic impact of two ant species. J Anim Ecol 80:569–576. https://doi.org/10.1111/j.1365-2656.2010.01796.x

Schultheiss P, Nooten SS, Almeida WR, Brandão CRF, Ribas CR, Fayle TM, Meusemann K, Economo EP (2022) The abundance, biomass, and distribution of ants on Earth. Proc Natl Acad Sci USA 119(40):e2201550119. https://doi.org/10.1073/pnas.2201550119

Siqueira FFS, Ribeiro-Neto JD, Tabarelli M, Wirth R, Andersen AN, Leal IR (2018) Human disturbance promotes herbivory by leaf-cutting ants in the Caatinga dry forest. Biotropica 50:779–788. https://doi.org/10.1111/btp.12599

Song H, Kemp DB, Tian L, Chu D, Song H, Dai X (2021) Thresholds of temperature change for mass extinctions. Nat Commun 12:4694. https://doi.org/10.1038/s41467-021-25019-2

Stanley SM (2016) Estimates of the magnitudes of major marine mass extinctions in earth history. Proc Natl Acad Sci USA 113:E6325–E6334. https://doi.org/10.1073/pnas.1613094113

Styrsky JD, Eubanks MD (2007) Ecological consequences of interactions between ants and honeydew-producing insects. Proc Roy Soc London Ser B Biol Sci 274:151–164. https://doi.org/10.1098/rspb.2006.3703

Sun Y, Joachimski MM, Wignall P, Yan C, Chen Y, Jiang H, Wang L, Lai X (2012) Lethally hot temperatures during the early Triassic greenhouse. Science 338:366–370. https://doi.org/10.1126/science.1224126

Thomas CD, Cameron A, Green RE, Bakkenes M, Beaumont LJ, Collingham YC, Erasmus BFN, Siqueira MF, Grainger A, Hannah L, Hughes L, Huntley B, van Jaarsveld AS, Midgley GF, Miles L, Ortega-Huerta MA, Peterson AT, Phillips OL, Williams SR (2004) Extinction risk from climate change. Nature 427:145–148. https://doi.org/10.1038/nature02121

Tylianakis JM, Didham RK, Bascompte J, Wardle DA (2008) Global change and species interactions in terrestrial ecosystems. Ecol Lett 11:1351–1363. https://doi.org/10.1111/j.1461-0248.2008.01250.x

Urbas P, Vieira MA, Leal IR, Wirth R (2007) Cutting more from cut forests: edge effects on foraging and herbivory of leaf-cutting ants in Brazil. Biotropica 39:489–495. https://doi.org/10.1111/j.1744-7429.2007.00285.x

Vasconcelos HL, Vilhena JMS, Magnusson WE, Albernaz ALKN (2006) Long-term effects of forest fragmentation on Amazonian ant communities. J Biogeogr 33:1348–1356. https://doi.org/10.1111/j.1365-2699.2006.01516.x

Walther GR, Post E, Convey P, Menzel A, Parmesan C, Beebee TJC, Fromentin JM, Hoegh-Guldberg O, Bairlein F (2002) Ecological responses to recent climate change. Nature 416:389–395. https://doi.org/10.1038/416389a

Ward PS (2014) The phylogeny and evolution of ants. Annu Rev Ecol Evol Syst 45:23–43. https://doi.org/10.1146/annurev-ecolsys-120213-091824

World Meteorological Organization (WMO) (2025) State of the Global Climate 2024. Retrieved from https://wmo.int/publication-series/state-of-global-climate-2024

Zachos JC, Dickens GR, Zeebe RE (2008) An early Cenozoic perspective on greenhouse warming and carbon-cycle dynamics. Nature 451:279–283. https://doi.org/10.1038/nature06588

Part II
Individuals

Chapter 2
Behavioural Adaptations of Ants to Stressful Warming Conditions

Xim Cerdá, Javier Retana, Irene Villalta, Abdallah Dahbi, Louise van Oudenhove, and Elena Ângulo

2.1 Introduction

Ants are an essential part of terrestrial ecosystems, whether the metric is species number, total biomass, or contributions to ecosystem services (Wilson 1990; Hölldobler and Wilson 2009; Del Toro et al. 2012, 2015; Elizalde et al. 2020; Rocha et al. 2024). Extensive research has been dedicated to the effects of temperature on ants at the individual, colony, population, and community level (e.g. Cerdá and Retana 1997; Azcárate et al. 2007; Clémencet et al. 2010; van Oudenhove et al. 2012; Bishop et al. 2017, 2019; Kwon 2017, 2018; Molet et al. 2017; Sánchez-García et al. 2022; Seifert 2017; Villalta et al. 2020; Ibarra et al. 2024; Trigos-Peral et al. 2024). However, the climate is changing, and the resulting long-term shifts in weather and temperature patterns are leading to a decline in biodiversity (United Nations https://www.un.org/en/climatechange/what-is-climate-change). The few

X. Cerdá (✉) · E. Ângulo
Estación Biológica de Doñana, CSIC, Sevilla, Spain
e-mail: xim@ebd.csic.es; angulo@ebd.csic.es

J. Retana
CREAF and Universitat Autònoma de Barcelona, Bellaterra, Spain
e-mail: javier.retana@uab.es

I. Villalta
Institut de Recherche sur la Biologie de l'Insecte, Université de Tours, Tours, France
e-mail: irene.villalta@univ-tours.fr

A. Dahbi
Laboratory of Ecotoxicology, Bioresources and Coastal Geomorphology, Polydisciplinary Faculty of Safi, Cadi Ayyad University, Marrakesh, Morocco

L. van Oudenhove
Institut Sophia Agrobiotech, INRAE, Sophia Antipolis, France
e-mail: louise.vanoudenhove@inrae.fr

I. R. Leal et al. (eds.), *Ant Responses to Climate Change*, Entomology in Focus 10,
https://doi.org/10.1007/978-3-032-21127-9_2

studies that have examined how ants are affected by these temperature shifts (although see Parr and Bishop 2022) have employed either a theoretical approach implemented through modelling (e.g. Jenkins et al. 2011; Bewick et al. 2014; Kwon et al. 2016) or an experimental approach in which artificially induced warming occurs under seminatural conditions (Pelini et al. 2011a, b; Diamond et al. 2016, 2017; Bujan et al. 2022; Bujan and Yanoviak 2022).

Ants have developed behavioural, physiological, and morphological adaptations that allow them to survive under stressful thermal conditions. Recently, in a review exploring possible ant responses to climate change, Parr and Bishop (2022) stated that ants could use different behavioural strategies to mitigate the effects of global warming while also highlighting that certain species might be unaffected or even benefit from climate change (Parr and Bishop 2022). When a species faces a new thermal challenge, the first and least costly response is likely to be behavioural rather than physiological, given that physiological responses tend to be more complex and energetically expensive. It seems probable that behavioural responses will continue to play a crucial mechanistic role in allowing species to persist under climate change. In this chapter, we focus on ant behavioural adaptations to high temperatures expressed at the individual level (by workers) that can have consequences at the colony level (via within-colony interactions among workers) or community level (via interactions among workers of the same or different species). However, colony-level behaviour, expressed as growth rates or colony mortality, lies outside the scope of the chapter.

2.2 Ants: Who Is the Individual? Worker Versus Colony

This chapter is part of Part II of this book, which is entitled “Individuals”. However, we ask the following question: can an ant colony be viewed as a single organism, functionally acting as an individual? Snir et al. (2022) stated “colonies of eusocial insects are complex biological systems that consist of obligately interacting individuals working cooperatively towards common goals”. For ants, the colony is the reproductive unit and, consequently, the unit upon which selection operates. Therefore, we could consider that the colony is an “individual”. However, within colonies, the queen is responsible for all reproduction, with males contributing their sperm only to die shortly after mating. She is nonetheless reliant on workers to ensure the colony’s annual or multiannual life cycle. Consequently, the presence of both queens and workers is essential to colony persistence.

Hölldobler and Wilson (2009) developed the concept of superorganism to describe the biological organization of ant colonies. They posited that ant colonies are a step up from organisms in organizational hierarchy, given that they possess the attributes of an organism in that their basic compositional elements are the highly collaborative workers. Often, experimental research on ants focuses on the colony as a whole, ignoring the individual workers. As a result, it may be mistakenly assumed that any individual casualties incurred during risky colony activities, such as foraging and nest defence, are unimportant as lost workers will soon be replaced.

However, a growing number of studies have highlighted that ants display rescue behaviour when nestmates fall into antlion traps (Nowbahari et al. 2009; Hollis and Nowbahari 2013; Taylor et al. 2013; Bar et al. 2023, but see Turza et al. 2020) or are injured during fights with termites (Frank et al. 2017) or play dead to avoid predation (Cassill et al. 2008; Petit et al. 2022). The existence of these responses suggests that individual workers are indeed valuable. Frank et al. (2017) estimated that, in the termite-hunting species *Megaponera foetens*, more than 20% of the workers in the raiding column end up with long-term injuries. Subsequent modelling work suggested that the rescue behaviour directed towards the injured allows the colony to remain 30% larger in size. In *Megaponera analis*, as few as 13 new ants are produced each day, which means that rescuing the injured may be essential to the colony (Frank et al. 2017). The logical next step would be for the injuries to be treated, a behaviour that has already been observed in various ant species (Frank et al. 2024; Beydizada et al. 2024).

Among all the workers in a colony, foragers face the greatest thermal risks (Perez and Aron 2020) and are likely to be the most affected by the rising temperatures resulting from climate change. In scavenger desert ant species, foragers have a very short life expectancy: 4.9 days for *Melophorus bagoti* (Muser et al. 2005), 4.7 days for *Cataglyphis nodus* (Wehner 2020), and 6.3 days for *Cataglyphis bicolor* (Schmid-Hempel and Schmid-Hempel 1984) (to put these values into context, life expectancy of foragers of the New World desert harvester ants is significantly longer: 14 days for *Pogonomyrmex owyheei* (Porter and Jorgensen 1981) and up to 25 days for *Pogonomyrmex barbatus* and *P. rugosus* (Gordon and Hölldobler 1987)).

Nonetheless, *Cataglyphis nodus* workers have been observed caring for injured nestmates (Beydizada et al. 2024). These numerous examples illustrate that the value of individual workers should not be ignored. In the context of climate change, it is unclear that selection would favour the strategy of having vast numbers of workers to compensate for the large percentage killed by thermal stress. Indeed, most thermophilic ant species have relatively small colonies (Lecocq de Pletincx et al. 2021). Therefore, colonies might be more likely to use other strategies that favour worker survival.

This chapter explores the different behavioural strategies used by workers to avoid high temperatures; presumably, these strategies reflect how workers will react to the stressful warming conditions associated with climate change. First, we focus on the individual level, describing the thermal advantages ants may gain by raising their gasters away from the hot ground surface and by exploiting thermal refuges in hot environments. Second, we provide examples of collective responses to temperature, which manifest themselves via foraging activity, brood care, and nest building. Shifts in foraging activity are a collective response to stressful temperatures and arise from information processed at the individual and colony level. Increases in temperature also provoke behavioural changes in brood management, such as relocating brood within the nest. Additionally, while some species modify nest structure to better cope with the heat, others move their nests to cooler locations. Global warming may impose new constraints on individual and collective behaviours, and ants may be able to thermally adapt using a variety of behavioural mechanisms.

2.3 Gaster Raising by Desert Ants

At midday in certain deserts of the Palearctic, Africa, or Australia, observers might happen upon ants zipping around at high speeds with their gasters raised high in the air. They are desert ants belonging to the genera *Cataglyphis*, *Calomyrmex*, *Camponotus*, or *Melophorus*. The reason they raise their gasters in this way is a source of debate. Cerdá and Retana (2000) hypothesized that *Cataglyphis rosenhaueri* workers lift their gasters to protect their vital organs from hot ground surfaces. Indeed, when the substrate reaches temperatures of 50–55 °C, a shift from a horizontal to a vertical position can result in a substantial difference in gaster temperature (see Fig. 2.1). McMeeking et al. (2012) rejected the heat tolerance hypothesis based on the reasoning that the vital organs most susceptible to heat stress are the brain and gonads. They proposed an alternative explanation for gaster raising: it allows ants to run faster because it lowers their inertia moment by as much as 50%.

Without entering into further debate about which body parts contribute most to ant survival and recalling that *Cataglyphis* foragers do not have functional ovaries, it is important to highlight that the ant gaster contains the digestive system, the fat body, and major glands, such as the poison gland and Dufour's gland. Thermal stress may affect the chemical contents of the glands as well as hydrocarbon biosynthesis in the fat body (Soroker and Hefetz 2000). More importantly, it can affect fat body lipids (i.e. triglycerides and fatty acids). In insects, the fat body's main

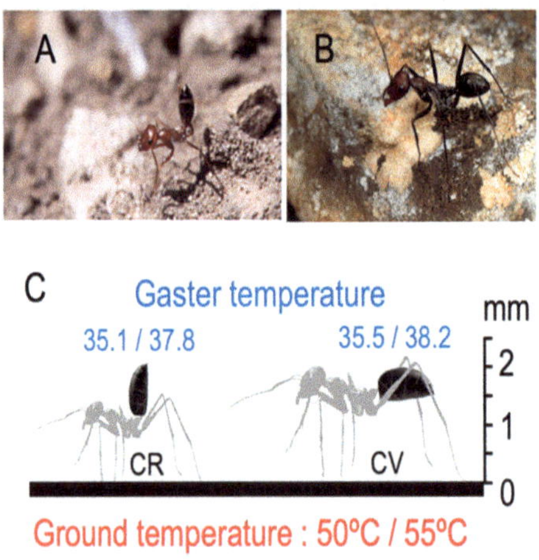

Fig. 2.1 (**a**) Small forager of *Cataglyphis rosenhaueri* running with her gaster in an elevated position (the natural position for this species when temperatures are hot) (photo credit: Xim Cerdá). (**b**) Large forager of *Cataglyphis velox*, a species that cannot raise its gaster (photo credit: Fernando Amor). (**c**) Schematic showing gaster temperatures (in °C) when the gaster is positioned at different heights (in mm) using smaller foragers of *C. rosenhaueri* (CR) and larger foragers of *C. velox* (CV) as examples. The two values above each ant are gaster temperature when the ground temperature is 50 and 55 °C, respectively. (Data from Cerdá and Retana 2000)

function is energy storage (Roma et al. 2006). Although *Cataglyphis* foragers have little developed fat bodies (Cerdá and Retana 1992), their lipid reserves are essential for survival (Dussutour et al. 2016) because they can be mobilized when extra energy is needed (Skowronek et al. 2021). In *C. rosenhaueri*, raised gasters have temperatures of 37.8–38.6 °C, which are lower than the temperatures at which lipids melt (Gibbs et al. 1991), while unraised gasters have temperatures of 44.2–45.2 °C, which equal or exceed the temperatures at which lipids melt (Gibbs et al. 1991; Cerdá and Retana 2000). On the one hand, ants could limit heat-related disruptions to metabolic processes via gaster raising. On the other hand, the speed increase permitted by gaster raising could reduce exposure to high temperatures. Consequently, both potential explanations for gaster raising make sense in that they describe mechanisms that would help foragers cope with extreme temperatures.

Another reason cited by McMeeking et al. (2012) when rejecting the heat tolerance hypothesis (Cerdá and Retana 2000) is that the most thermotolerant *Cataglyphis* species do not employ gaster raising. In his seminal book on desert ants, Dlussky (1981) traced the evolution of *Cataglyphis* propodeum shape, a process that involved the restructuring of the petiole and its associated muscles. These morphological adaptations only occurred in some *Cataglyphis* species. Thus, it remains an open question as to why the morphological changes accompanying this behaviour only occurred in certain *Cataglyphis* species and/or desert ant species. Indeed, Dlussky (1981) himself wondered why the above restructuring did not occur in all desert species.

Gaster raising is not a trait exclusive to desert ants. Ward and Seely (1996) noted “a character may be considered a desert adaptation because many desert species possess it, while non-desert species seldom do”. Many non-desert ant species raise their gasters in a variety of situations as can be seen in the excellent photographs on Alex Wild’s insect photography website (www.alexanderwild.com): the arboreal tree-dwelling *Terataner elegans* in an alarm posture, the Australian spider ant (*Leptomyrmex ruficeps*) in an attack posture spraying defensive chemicals over its head, and various species standing guard near their nests (*Azteca* sp., *Oecophylla smaragdina*, *Forelius nigriventris*) or defending brood or prey (*Crematogaster* spp., *Anonychomyrma* sp.). Recently, Wehner (2020) pointed out that gaster raising also occurs in several other highly agile ant species (e.g. arboreal weaver ants—genus *Oecophylla*) and also stated that the two functions—heat tolerance and locomotor agility—are not mutually exclusive. Indeed, ants may utilize this behaviour for multiple purposes.

2.4 Use of Thermal Refuges by Foragers

Vertebrate ectotherms, such as lizards, rely on external heat sources to regulate their body temperatures and must behaviourally buffer thermal conditions to maintain physiological homeostasis (Kearney et al. 2009). For ants, thermal regulation may be somewhat simpler because of their smaller size and less restrictive thermal requirements. However, most have a thermal range in which they forage (as we will see in Sect. 2.8). They can adjust their foraging activity to thermal conditions, which may mean simply staying inside the nest or looking for thermal shelters if they are outside. In thermally heterogeneous environments, foragers have more opportunities to avoid or take shelter from unfavourable thermal conditions. For vertebrate ectotherms, the priority is to find ways to warm up in environments that are cooler than their thermal optimum. Ants are very different because most species forage in patchy and heterogeneous habitats that are a thermal mosaic (Lőrincz et al. 2024a). Thus, ants can often choose where to forage and when to cool off. Willmer and Unwin (1981) suggested that although small insects heat up more rapidly than large insects, they also cool down more rapidly in any available thermal refuges. Thermal refuges are not only shaded microhabitats but also slightly elevated points, such as the tops of stones, grass stalks, or small sticks. In his research on *Ocymyrmex barbiger* in the Namib Desert, Marsh (1985a) observed that when foragers paused in thermal refuges, their body temperatures always dropped substantially. Under high thermal loads, *O. barbiger* seeks out elevated points, and, while maintaining a firm grip on the stalk or stick with some of its legs, it waves its other legs around in the air to cool down (Marsh 1985a). Depending on wind speed and soil temperature, a pause of 7 s in a thermal refuge can lead to a 5 °C or 6 °C decrease in body temperature (Marsh 1985a). Similarly, foragers of the Australian desert ant *Melophorus bagoti* spend 30–50% of their time in shaded or elevated microhabitats, allowing them to forage continuously despite the high temperatures (Christian and Morton 1992).

In the Spanish desert ants *Cataglyphis velox* and *C. rosenhaueri*, foragers of different sizes (Table 2.1) spent approximately the same percentage of time in thermal refuges (range: 7.3–14.4% of foraging trip duration) (see Fig. 2.2 for an example of ants using dried stalks as a thermal refuge) and had similar pause durations (less than 5 s for workers of all sizes on average). There were only slightly significant differences in pause frequency, which varied from 0.8 pauses/min for large C. *velox* workers to 2.0 pauses/min for very small *C. rosenhaueri* workers and small *C. velox* workers (Cerdá and Retana 2000). *Cataglyphis bombycina* and *Ocymyrmex robustior* foragers spend up to 70–80% of their time outside the nest in thermal refuges (Wehner 2020). Marsh (1985b) highlighted that, as prey do not occur in thermal refuges, any time spent in refuges reduces the amount of time available for foraging. However, this strategy allows ants to remain in or around the foraging area without having to return to the nest to cool down, which would be more costly. During their pauses, foragers release excess heat from their bodies and then resume foraging (Wehner 2020). As Wehner (2020) pointed out, the challenge faced by these thermophilic ants is the sufficient availability of thermal refuges. In their absence, foragers have no way of releasing their heat load and die in an inescapable “thermal trap”.

Table 2.1 Characteristics of foraging trips by different sized workers of the Spanish desert ants *Cataglyphis rosenhaueri* and *Cataglyphis velox* (worker length in mm)

Worker size (length)[a]	Maximum distance (m)	Trip duration (min)	Speed (m/min)	% time spent in thermal refuges	No. pauses per min	Pause duration (s)
Very small (4–5.9 mm) *C. rosenhaueri*	8.7 ± 0.9[b]	11.7 ± 1.4	1.0 ± 0 1.1[b]	14.4 ± 1.8	2.0 ± 0.2[b]	4.6 ± 0.5
Small (6–7.9 mm) *C. rosenhaueri*	10.2 ± 1.1[b,c]	11.6 ± 2.1	1.2 ± 0.1[b]	10.8 ± 2.2	1.6 ± 0.2[b]	4.3 ± 0.4
Small (6–7.9 mm) *C. velox*	11.5 ± 2.4[b,c]	8.2 ± 1.3	1.4 ± 0.2[b,c]	13.4 ± 2.8	2.0 ± 0.4[b]	4.7 ± 0.6
Medium (8–9.9 mm) *C. velox*	21.4 ± 2.6[c]	10.4 ± 1.8	2.1 ± 0.3[b]	11.8 ± 2.3	1.5 ± 0.2[b,c]	4.5 ± 0.8
Large (10–12 mm) *C. velox*	24.6 ± 2.6[c]	9.5 ± 1.5	3.5 ± 0.4[c]	7.3 ± 2.5	0.8 ± 0.2[b]	4.1 ± 1.2
ANOVA (F, *p*)	(5.93, < *0.001*)	(0.22, *0.93*)	(12.6, < *0.0001*)	(1.95, *0.12*)	(2.88, *0.03*)	(0.12, *0.98*)

These foraging trips occurred at midday in the summer in the Cazorla mountains of southeastern Spain (From Cerdá and Retana 2000; see details about methods in the reference)
[a]Differences among worker types for the different variables were examined using one-way ANOVAs and Fisher's protected LSD. Values that were not significantly different based on Fisher's protected LSD post hoc test share a common letter (b, c)

2.5 Collective Behaviour

Ant foraging strategies range from individual-based strategies, where there is no cooperation and communication among nestmates, to well-coordinated group-based strategies, where nestmates forage in swarms, and all the strategies in between, such as tandem running, group recruitment, mass recruitment, and trunk trails. The more complex the strategy, the greater the importance of chemical communication (Beckers et al. 1989; Ruano et al. 2000; Cerdá et al. 2014). In addition, the autonomy of individual foragers appears to decline as strategies become more group based (Anderson and McShea 2001).

In ant species that use individual-based strategies, foragers search for and harvest food items on their own. They rarely signal the presence of food to their nestmates because they cannot use chemical communication. There are some exceptions, such as *C. bicolor* in Tunisia or *Cataglyphis floricola* in Spain, which use a rudimentary form of nonchemical recruitment to alert their nestmates to the presence of heavy prey items or rich food sources near the nest (Wehner 1987; Amor et al. 2010). Regardless of the specifics, temperature can either limit or favour the activity of

Fig. 2.2 Large foragers of *Cataglyphis velox* on a bait at midday in Seville (southern Spain), when the soil temperature was 60 °C. They were shuttling up and down the dried stalks, cooling up top before returning to consume more of the hot bait. After e few minutes of licking the bait and taking short pauses in the thermal refuge, the foragers returned to the nest

species with individual-based foraging. The situation is very different for mass-recruiting species. First, collective behaviour is driven by signal amplification (Gordon 2019); for mass-recruiting species, the signal is trail pheromone, whose effective deposition and detection will be influenced by the characteristics of the ground surface (Detrain et al. 2001; Jeanson et al. 2003). Second, the trail pheromone used by mass-recruiting species decays faster at higher ground surface temperatures, altering communication dynamics among nestmates (van Oudenhove et al. 2011, 2012). As a result, it seems likely that global warming will more dramatically disrupt foraging dynamics in species that use chemical communication than in species that do not. In the latter, foraging can largely continue as usual, with workers making greater use of thermal shelters.

Aphaenogaster senilis illustrates a special case where the interaction between temperature tolerance and group recruitment results in an ecological advantage (Cerdá et al. 2009). The species is relatively thermophilic and uses group recruitment to exploit valuable food resources at temperatures as high as 45 °C. It can continue to forage until temperatures reach 52 °C, but it switches to an individual-based strategy, making shorter trips (i.e. of less than 10 min) and targeting small, easy-to-transport food items. In Doñana National Park (southwestern Spain), *A. senilis* uses its unique combination of traits to successfully exploit numerous food resources at temperatures exceeding 40 °C, conditions that small, dominant, mass-recruiting ant species cannot tolerate. Thus, thanks to its heat tolerance and capacity to recruit nestmates even at high temperatures, *A. senilis* has an ecological advantage over species that use either mass recruitment or individual-based foraging (e.g. *Cataglyphis* species; Cerdá et al. 2009).

2.6 Within the Nest: Moving Brood or Nests

For an ant colony's life cycle to successfully proceed, important decisions must be made around where and when to rear the brood. Brood development occurs within a relatively narrow range of temperatures and is considered to be the most thermally sensitive process within ant colonies (Bujan and Yanoviak 2022). Nest temperature can affect more than brood development; it also influences brood survival and the fate of diploid larvae, determining whether they become queens or workers (Brian 1963, 1973; Kipyatkov and Lopatina 1990; Kipyatkov et al. 2004, 2005; Sanada-Morimura et al. 2006; Boulay et al. 2009; Oms et al. 2017). In some species, workers modify brood rearing temperature by relocating larvae within the nest (Roces and Núñez 1989; Tinaut et al. 1999; Porter and Tschinkel 1993; Penick and Tschinkel 2008).

There are ants with endothermic nests: the colony can maintain a constant temperature within the nest regardless of exterior temperatures. This approach is employed by some wood ant species, which have underground nests associated with above-ground mounds composed of organic matter. This architecture allows for different thermoregulatory strategies: ants can sunbathe outside, accumulating heat that they then release within the nest; the mound can directly absorb the sun's heat; or the nest can be heated via the ants' production of metabolic heat (Gallé 1973; Coenen-Stass et al. 1980; Rosengren et al. 1987; Kadochová and Frouz 2013; Kadochová et al. 2017, 2019). Mound-dwelling wood ants play a dominant role in many boreal and northern temperate forest ecosystems, which may be partially rooted in their ability to maintain nest homeostasis.

Another form of colony-mediated endothermy is illustrated by army ants, whose approach contrasts greatly with that of the mound-dwelling wood ants. Army ants do not reside in nests. Instead, they build temporary bivouacs using the living bodies of the colony's hundreds of thousands of workers. During the colony's stationary phases, the bivouacs serve as starting points for foraging raids (Schneirla et al. 1954). Within the bivouac are empty spaces that function as chambers, where the brood and queen are maintained in relatively stable microclimatic conditions. Indeed, bivouac temperatures remain extremely constant throughout the day (Schneirla et al. 1954). The workers' bodies surround, insulate, and heat the brood and queen. There is extensive evidence that, in both surface-dwelling and subterranean species, the collective metabolic heat in the bivouac helps actively warm brood (Schneirla et al. 1954; Schneirla 1971; Franks 1989; Baudier and O'Donnell 2016).

Ants commonly use nest relocation as a way to ensure that the colony experiences abiotic conditions favourable to growth (McGlynn 2012). Behavioural avoidance is the first line of defence against heat stress (Bujan and Yanoviak 2022). Tropical canopy ants will evacuate their nests in response to heat stress, with different species displaying different degrees of sensitivity. *Camponotus* and *Crematogaster* species consistently removed their brood from experimentally heated nests (83% and 90% of the time, respectively), whereas *Pseudomyrmex* and *Cephalotes* species did not (only 5% and 17% of the time, respectively), likely because brood and workers are tightly packed together in the nests of the latter two genera (Bujan and Yanoviak 2022). In the Mediterranean species *A. senilis*, pupal development is limited by spring and fall temperatures (Boulay et al. 2009). Colonies will relocate to spend longer periods of time at optimal temperatures, moving from vegetated areas, where most nests occur in the spring and fall, to open areas. However, in some habitats, surface temperatures can reach extreme levels in the summer. For example, in Doñana National Park, the temperature in the superficial nest chambers can reach 53 °C when *A. senilis* colonies are located in open areas. Thus, the colonies will return to vegetated areas, where temperatures are considerably lower and more consistent. In *A. senilis*, colonies have a mean residence time of 29 days, which means there is a high frequency of colony relocation, suggesting that these ants are unable to adequately assess site quality before moving their nests. They appear to employ a "trial and error" approach—trying out several successive nest locations until they find the site with the best set of abiotic and biotic conditions, which is where they will settle for longer (Galarza et al. 2012). A similar pattern has been observed in *Proformica longiseta*, a high-elevation ant found in Spain that nests under rocks (Tinaut et al. 1999). In open habitats, it is relatively common for ants to nest under stones, which provide protection from predators and generate favourable thermal conditions (Dean and Turner 1991; Fernández-Escudero et al. 1993; McCaffrey and Galen 2011). However, *P. longiseta* does not seem able to adequately assess rock quality and—like *A. senilis*—does not appear to select nesting sites ahead of time. When a colony nests under a rock whose conditions are optimal for brood development, the colony will remain longer and produce sexuals. However, if it nests under a rock with suboptimal conditions, it will shortly depart and continue its search for a better rock.

Many leaf-cutting ants build nests that are highly structurally complex (e.g. Jonkman 1980; Bollazzi et al. 2012) to maintain optimal thermal conditions for both their brood and the symbiotic fungus that they cultivate within the nest; the growth of this fungus is highly sensitive to temperature conditions. Bollazzi and Roces (2002) explored how workers of *Acromyrmex heyeri* responded to an experimental temperature gradient ranging from 10 °C to 37 °C. When workers started off at the hot end of the gradient, they moved the fungus and brood to locations where the mean temperature was 25 °C and 24 °C, respectively. When workers started off at the cold end of the gradient, they moved the fungus and brood to locations where the mean temperature was lower, 22 °C and 21 °C, respectively. Indeed, these temperatures are essentially those at which maximal growth of the isolated attine ant fungus is achieved (Bollazzi and Roces 2002).

There are, however, at least 50 species of leaf-cutting ants (Schultz 2021), and some build superficial mound-shaped nests. For example, in the aforementioned *A. heyeri*, colonies cultivate fungus on the soil surface and cover their nests with a thatched mound. Other species build multichambered nests that extend down 3 m (Bollazzi et al. 2008) or more (Jonkman 1980). Different nest types are associated with different mechanisms for maintaining environmental conditions within the range that optimizes fungus growth and brood development. In the relatively simple thatched nests of *Acromyrmex lobicornis*, the mound functions as a "thermal buffer", not unlike the mounds of wood ant nests. The thatch helps ensure that temperatures are more stable within the mound than they are outside: fungus garden chamber temperatures fluctuate less than air and soil surface temperatures and are, on average, warmer than air temperatures and cooler than soil surface temperatures (Farji-Brener 2000). In the different *Acromyrmex* species, nesting habits are geographically correlated with soil temperatures: in regions with warmer soils, there are more species with subterranean nests than species with superficial nests (Bollazzi et al. 2008). In addition, species that display plasticity in their nesting habits build subterranean nests in hot soils and superficial nests in cold soils (Bollazzi et al. 2008). Nest architecture plays a determinant role in regulating nest temperature. For instance, in *Acromyrmex ambiguus*, the circulation patterns of dry air currents trigger behavioural responses by workers: the direction of air movement inside the nest's tunnels guides the workers' building patterns, influencing when and where nest structures and tunnel plugs are added, decisions that influence abiotic conditions within the nest (Bollazzi and Roces 2007).

In contrast to some of the species mentioned earlier in this section, leaf-cutting ants appear to rarely resort to nest relocation (but see Wirth et al. 2003). That said, *Atta sexdens* was seen to relocate its nests following a major change in canopy cover; more specifically, the ants moved their nests to a location with more favourable microclimatic conditions (van Gils and Vanderwoude 2012). The impressive nest-building capacities of leaf-cutting ants suggest that these ant species will be able to adapt to future climatic conditions under global warming. However, if they are not, they will always have the option of migrating as long as only there are microhabitats within their present environment that provide suitable conditions free from nonthermal stressors.

2.7 Foraging and Personality

Variability among individual workers is more than statistical noise. In fact, it is at the heart of colony life (Retana and Cerdá 1991): workers often defy the expectations we have for them based on our theoretical or empirical knowledge. What were once viewed as individual idiosyncrasies in ant behaviour studies (Cerdá and Retana 1992) are now being approached from the perspective of personality. More recently, by situating animal personality and behavioural syndromes within an evolutionary context (Dall et al. 2004; Sih et al. 2004), it has become possible to formalize and develop standardized protocols for quantifying individual variability within colonies and personality differences among colonies. In behavioural ecology, animal personality refers to behavioural differences among individuals that are consistent over time and across different situations (Blight et al. 2016). Research on animal personalities has highlighted that a behavioural continuum from boldness to shyness exists in most animal interactions: bold individuals seek out risk and novelty, while shy individuals avoid them (Wilson et al. 1994). Several factors that can affect group-level personality traits must be controlled when studying social species: the level of analysis (individual or group), the colony's age, and past experiences. Although a collective behaviour might result from the average behavioural types of its members (Pinter-Wollman 2012), an analytical approach that focuses on individuals may lead to a mischaracterization of group-level (or colony) personality (Blight et al. 2016), and this is the main reason why it is more correct to focus on the colony personality traits.

Using standardized laboratory colonies of the same age of *Aphaenogaster senilis*, Blight et al. (2016) found that ant colony personality was directly related to resource competition and temperature-related foraging risks. Bold colonies contained workers who more readily explored novel environments, displayed higher levels of aggressiveness, and more efficiently retrieved food when faced with intraspecific competition. However, bold colonies were also more risk prone, and their workers suffered higher mortality rates because they more frequently foraged at temperatures above their critical thermal limit. Therefore, while being bold, active, and aggressive could provide a selective advantage under climate change, colonies are taking an extreme risk by foraging at high temperatures (Blight et al. 2016). In a field study conducted under natural conditions, Kolay et al. (2023) observed that *A. senilis* colonies in a grassland habitat were more active than those in a woodland habitat, even though levels of exploration were similar between the two habitats. However, they did not find any clear evidence of a relationship between colony personality and foraging efficiency. Instead, contrary to what Blight et al. (2016) observed in the laboratory, the bolder, risk-taking colonies—those that were more active and exploratory—did not forage more efficiently than the shy and risk-averse colonies (Kolay et al. 2023).

2.8 Temperature and Foraging Activity: Thermal Performance Curves

To forage or not to forage? The answer to this question comes first from individual workers, who must decide whether or not to leave the nest to forage for the colony (Gordon 1984; Cerdá and Retana 1989; Detrain et al. 1999; Vieira et al. 2024). This decision may be driven by some external abiotic factors such as temperature (Cerda et al. 1989; Crist and MacMahon 1991; Roeder et al. 2022), humidity and rainfall (Friedman et al. 2019; Sundaram et al. 2022; Gordon et al. 2023; Ord 2023), or the position of the sun (Amor et al. 2011). There may also be an influence of biotic factors such as food availability (Crist and MacMahon 1991; Gordon et al. 2013) or nestmate interactions (Pinter-Wollman et al. 2013; Miller and Pinter-Wollman 2023).

To assess the effects of temperature on ant behaviour at the colony or population level, it can be very useful to employ thermal performance curves (TPCs) (Sinclair et al. 2012, 2016; see Chap. 4), which can be estimated from real temperature data and hypothetical temperature data from different climate change scenarios.

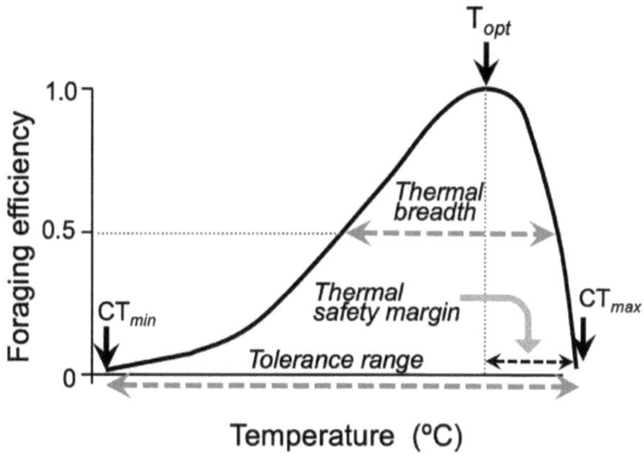

Fig. 2.3 Hypothetical thermal performance curve (TPC) illustrating the relationship between temperature and foraging efficiency, a proxy for fitness in ant colonies. The critical thermal minimum (CT_{min}), critical thermal maximum (CT_{max}), and thermal optimum (T_{opt}) are indicated. Figure modified from Sinclair et al. (2016)

Initially, TPCs were used to describe how body temperature affects an ectotherm's individual performance, which was assumed to be a proxy for fitness and was assessed using physiological measurements directly linked to organismal ecology, such as growth, feeding rate, or locomotion (Huey and Stevenson 1979; Sinclair et al. 2016). A typical TPC (Fig. 2.3) describes how temperature affects performance (e.g. foraging efficiency in the case of ants) across the thermal tolerance range, which lies between the lower and upper critical thermal limits (CT_{min} and CT_{max}, commonly used in TPCs for convenience, but see Chap. 4). Along this hypothetical curve, performance increases slowly with temperature up to a maximum value (T_{opt}), the temperature at which performance peaks, and then drops off quickly. The thermal breadth is the range between the temperature at which performance is 0.5 and the value of CT_{max}. The thermal safety margin is the temperature range between T_{opt} and CT_{max}. The wider the thermal safety margin, the lower the risk the animal faces, and the narrower the thermal safety margin, the greater the thermal risk the animal faces.

In Mediterranean ant communities, each species has specific thermal preferences (e.g. Cros et al. 1997; Sánchez-García et al. 2022). Although foragers avoid lethal temperatures in nature, there are some risk-prone species whose workers forage very close to their lethal limits (Cerdá et al. 1998a). When TPCs are constructed using real-life field data, the shape of the curve will depend on the ant species. Figure 2.4 illustrates the relationship between ground surface temperature (°C) and foraging activity for three Mediterranean ant species: the dominant, mass-recruiting *Tetramorium semilaeve*; the seed harvester *Messor bouvieri*; and the subordinate, thermophilic *Cataglyphis cursor*. *Tetramorium semilaeve* has a low T_{opt} that is quickly reached, and its foraging activity then slowly decreases until its maximum foraging temperature is attained; this species is largely inactive at risky temperatures. *Messor bouvieri* has an intermediate T_{opt}, along with an intermediate-shaped curve, and it takes fewer foraging risks. *Cataglyphis cursor* has a high T_{opt} that is more slowly reached, and its foraging activity then rapidly decreases given the proximity between its T_{opt} and its maximum foraging temperature; this species often forages at risky temperatures (Cerdá et al. 1998a).

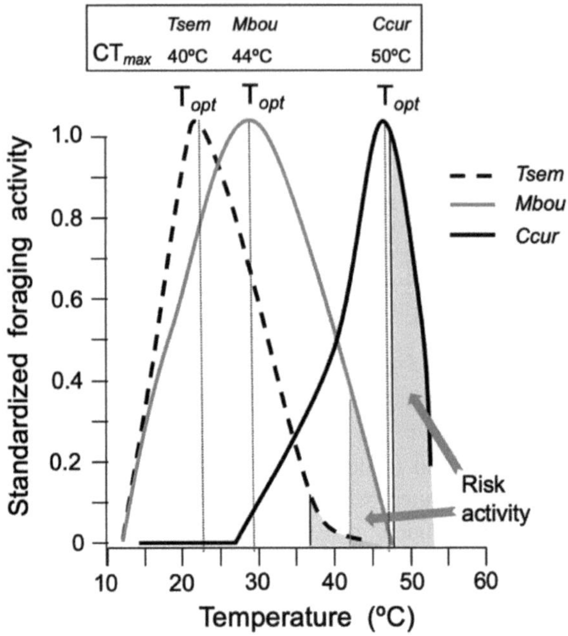

Fig. 2.4 Examples of thermal performance curves for three Mediterranean ant species: *Tetramorium semilaeve* (Tsem), *Messor bouvieri* (Mbou), and *Cataglyphis cursor* (Ccur). These curves describe the relationship between ground surface temperature (°C) and ant foraging activity. The data were collected at all hours of the day across sampling periods in March through November 1990 (Cerdá et al. 1998a). They are organized into 2 °C classes and have been standardized by defining maximum mean activity as 100% activity; the latter corresponds to the temperature at which activity is maximal (T_{opt}). Thermally risky activity is indicated by the shaded areas under the curves and is defined as activity occurring at temperatures ≥ CT_{max}—2 °C. The CT_{max} of each species is indicated at the top of the curves. CT_{max} was calculated in the laboratory as the temperature at which at least 50% of the workers died or lost muscular co-ordination after 10 min of thermal exposure (Cerdá et al. 1997, 1998a), but some workers can be active above their CT_{max} in the field because they used thermal refuges and they were not exposed during 10 min

In Mediterranean communities, high temperatures can present a threat to ant survival, and species have developed different behavioural strategies to protect themselves. Examining the TPCs for three Mediterranean ant species (Fig. 2.4), we can see the spectrum of responses illustrated by *T. semilaeve*, *M. bouvieri*, and *C. cursor*. Dominant species such as *T. semilaeve* are generally heat intolerant and adopt a "minithermy" strategy: they avoid exposure to high temperatures even if they could potentially tolerate them. In contrast, subordinate species such as *C. cursor* are more heat tolerant and adopt a "maxithermy" strategy: they forage at high temperatures, including those close to their CT_{max} (Cerdá et al. 1998a).

It is important to note that species' TPCs vary according to context. For example, in the shrublands of Doñana National Park, three other ant species respond differently to temperature depending on season (spring versus summer) and location (site A versus B) (Fig. 2.5). At site A, the dominant species *Lasius grandis* occupied a

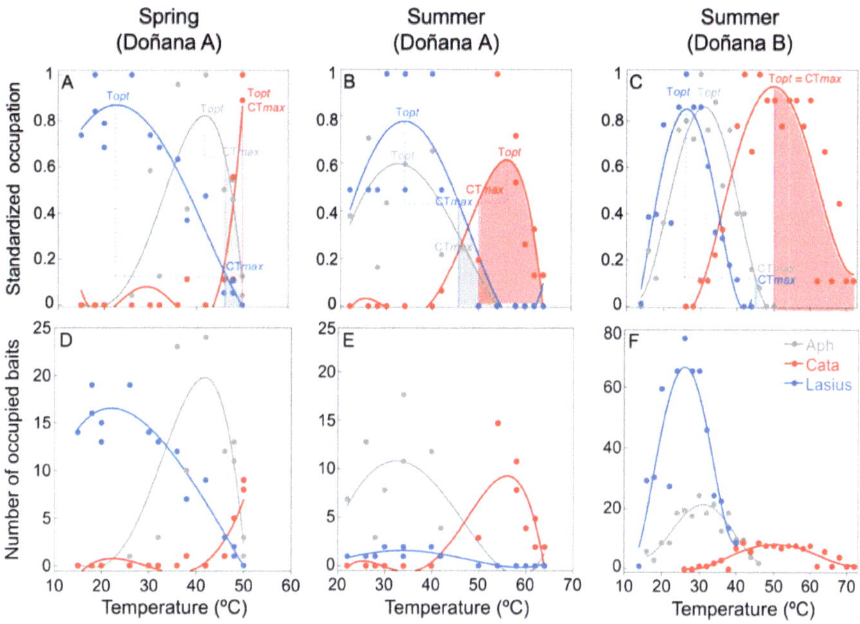

Fig. 2.5 Relationship between ground surface temperature and bait occupation by *Aphaenogaster senilis* (Aph, grey curve), *Cataglyphis floricola* (Cata, red curve), and *Lasius grandis* (Lasius, blue curve) in two shrubland habitats (Doñana **a** and **b**) in the spring and summer in Doñana National Park (southwestern Spain). Shown are standardized bait occupation (**a**–**c**) and the number of occupied baits (**d**–**f**). All the activity curves were fitted using a fourth-degree polynomial, with the exception of *C. floricola* (**d**), for which a third-degree polynomial was used. The temperature associated with maximum activity (T_{opt}) and the critical thermal maximum (CT_{max}) are indicated for each species. Thermally risky activity is indicated by the shaded areas. *L. grandis* and *A. senilis* have the same CT_{max} value. In A and C, the T_{opt} and CT_{max} for *C. floricola* have the same value (from the unpublished data of van Oudenhove de Saint 2010)

large number of baits in the spring (Fig. 2.5d) but not in the summer (Fig. 2.5e). Additionally, it occupied a similar proportion of baits in the summer at site B (Fig. 2.5f) as in the spring at site A. In the summer, the thermophilic species *C. floricola* occupied a similar number of baits at site A (Fig. 2.5e) and site B (note that the scale of the y-axis differs between Fig. 2.5e, f). For *L. grandis*, it seems likely that these seasonal differences are shaped by both environmental constraints on foraging (e.g. temperature) and variation in species abundance or density. It is evident that the same species may have differently shaped TPCs (see Fig. 2.5a–c). It is important to consider context dependence when using species-specific TPCs to gauge the effects of global warming on different species. Indeed, some ant species display remarkable plasticity, which may grant them a competitive advantage under climate change.

Another example of the context-dependent relationship between foraging activity and temperature is displayed by ant species found in different habitats on Surprise Island, a small island in New Caledonia (Cerdá et al. 2012). In the open habitat of the La Plaine site, bait occupation had a negative linear correlation with temperature for the two most abundant species—*Pheidole oceanica* and *Tetramorium simillimum* (Fig. 2.6a). At the coastal shrubland site of Argusia, *Paratrechina longicornis* exhibited this same relationship, but *Monomorium floricola* did not; instead, bait occupation by this species slowly increased up to 38 °C and then slowly decreased (Fig. 2.6b). At Pisonia, a forested site (Fig. 2.6c), the two most abundant species—*Tapinoma melanocephalum* and *Tetramorium simillimum*—displayed a similar non-linear relationship: bait occupation climbed until temperatures reached 30 °C and then declined (maximum environmental temperatures never exceeded 35 °C). In Scaevola, a bushy shrubland, two of the most abundant species—*Tetramorium bicarinatum* and *Pheidole oceanica*—exhibited a negative linear relationship between bait occupation and temperature, but a third species, *M. floricola*, exhibited a positive linear relationship (Fig. 2.6d). Interestingly, *M. floricola* was also present and displayed different dynamics: while in Argusia bait occupation increased up to 38ºC and then decreased, in Scaevola increased until reaching 40ºC. Photographs show *Ph. oceanica* soldier and workers in Scaevola site (Fig. 2.6e) and how a soldier is attacked by *T. bicarinatum* workers (Fig. 2.6f).

2.9 Daily and Seasonal Shifts in Foraging Patterns

As previously mentioned, ant foraging activity is usually influenced by temperature, most often in a species-specific way. In general, ant activity tracks daily temperature fluctuations (Cros et al. 1997; Cerdá et al. 1998b; Calazans et al. 2020). Species differ in their ability to tolerate thermal stress, a fact that shapes their daily and annual activity patterns (Cros et al. 1997). That said, variability in environmental temperatures is rooted in habitat and vegetation structure in addition to daily and seasonal fluctuations (Sánchez-García et al. 2022; Santos et al. 2022; Lőrincz et al. 2024b).

In Mediterranean ant communities, seasonal shifts in foraging activity follow a species-specific pattern. Cros et al. (1997) distinguish between two groups of species: (1) heat-tolerant diurnal species that display small changes in their daily activity patterns across the year and whose maximum activity occurs at temperatures close to their thermal limits ("risk-prone" species as per Cerdá et al. 1998a) and (2) heat-intolerant species that shift their daily activity patterns from diurnal to crepuscular-nocturnal when temperatures become hotter and whose maximum activity occurs at temperatures lower than 30 °C ("risk-averse" species as per Cerdá et al. 1998a). Habitat-specific environmental conditions differentially affect the activity patterns of ant species within these communities. In forests, the canopy cover creates a persistent spatial mosaic of sunny and shaded areas that benefits the heat-intolerant species. In contrast, in grasslands and other open habitats, the wide daily range of diurnal and nocturnal temperatures creates a temporal mosaic of thermal conditions that create windows of opportunity for both the heat-tolerant and heat-intolerant species (Cros et al. 1997). Given this thermal heterogeneity,

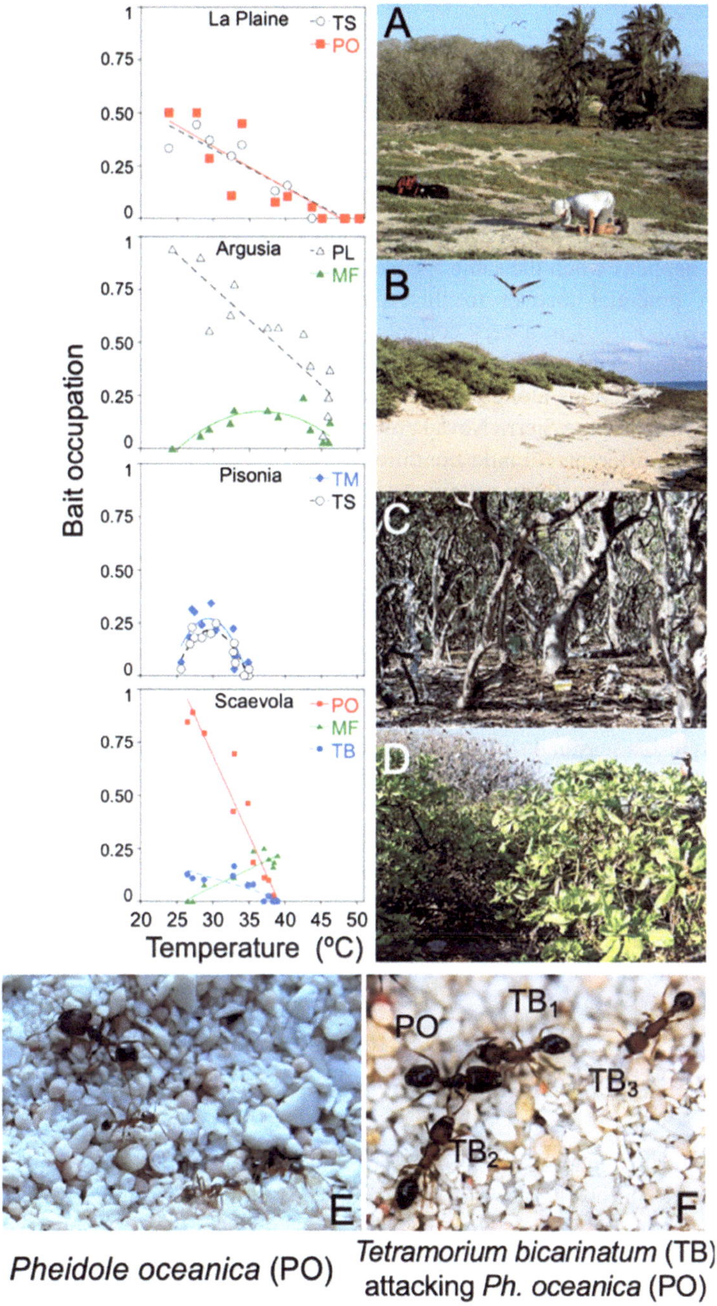

Fig. 2.6 Ant behaviour in different habitats on the small Surprise Island (New Caledonia). Effect of temperature on bait occupation by the most abundant species at each site: (**a**) La Plaine, (**b**) Argusia, (**c**) Pisonia, and (**d**) Scaevola. All the fitted relationships are statistically significant. Data are shown for the most abundant species: *Tetramorium simillimum* (TS), *Pheidole oceanica* (PO), *Paratrechina longicornis* (PL), *Monomorium floricola* (MF), *Tapinoma melanocephalum* (TM), and *Tetramorium bicarinatum* (TB). (**e**) Photograph of four *Pheidole oceanica* foragers—one soldier and three workers. (F) *Pheidole oceanica* soldier being attacked by three *Tetramorium bicarinatum* workers. (Figure modified from Cerdá et al. 2012

different species respond differently to seasonal variation (Cros et al. 1997): some species are diurnal throughout the year (*C. cursor*, *Camponotus foreli*, *A. senilis*); others are diurnal in the spring and crepuscular-nocturnal in the summer (*M. bouvieri*, *Messor capitatus*, *Camponotus cruentatus*); and yet others are more or less crepuscular-nocturnal in both the spring and summer (*Tapinoma nigerrimum*, *Pheidole pallidula*, *T. semilaeve*, *Camponotus sylvaticus*).

What happens when the same species experiences very different climatic conditions? A potential answer to this question is provided by the thermophilic *Aphaenogaster iberica*, a very common species in the Iberian Peninsula that is a strictly diurnal scavenger. Villalta et al. (2020) studied thermal regulation in six *A. iberica* populations along an elevational gradient (100, 600, 1000, 1300, 1700, and 2000 m asl) in the Sierra Nevada mountains of southern Spain. Each population experienced different climatic conditions, and it was hypothesized that foragers would modify their activity patterns to avoid temperatures that exceeded their thermal limits. Colonies behaved as expected—across different elevations and daily temperatures, they shifted their activity so as to avoid overly hot temperatures while simultaneously maintaining food intake levels. One social thermoregulatory mechanism that they employed was to shift daily foraging activity earlier and/or later in the day as mean daily temperatures increased. For this reason, activity patterns were unimodal at higher elevations and bimodal at intermediate and lower elevations, where the midday heat constrained foraging (Fig. 2.7). At intermediate elevations, there was also a seasonal shift: colonies displayed a unimodal activity pattern in the spring when temperatures were mild and a bimodal activity pattern in the summer when temperatures were hotter (Fig. 2.7).

Such shifts allowed the colonies to maintain levels of foraging activity regardless of elevation and daily temperatures in the summer. Moreover, the populations at intermediate elevations (e.g. 1300 m) foraged at temperatures very close to their critical thermal limits, which raises questions about how these ants will cope with climate change.

2.10 Future Predictions? Uncertainty and Challenges

Throughout this chapter, we have discussed different relationships between temperature and ant behaviour with a view to illustrating how climate change could affect these social insects. However, for a variety of reasons, many researchers doubt that realistic predictions can be achieved. Present uncertainty regarding the biological consequences of climate change is perhaps larger than generally acknowledged (Clusella-Trullas et al. 2011). Ants are tiny ectotherms that experience climate on the scale of millimetres to centimetres, and simulations that predict how global warming will shift species distributions using broadscale meteorological patterns are thus unlikely to be accurate. For example, in central European ant communities, 22–31% of the seasonal variance in mean soil temperature is attributable to habitat-specific factors (Seifert 2017). It remains unclear how ants could

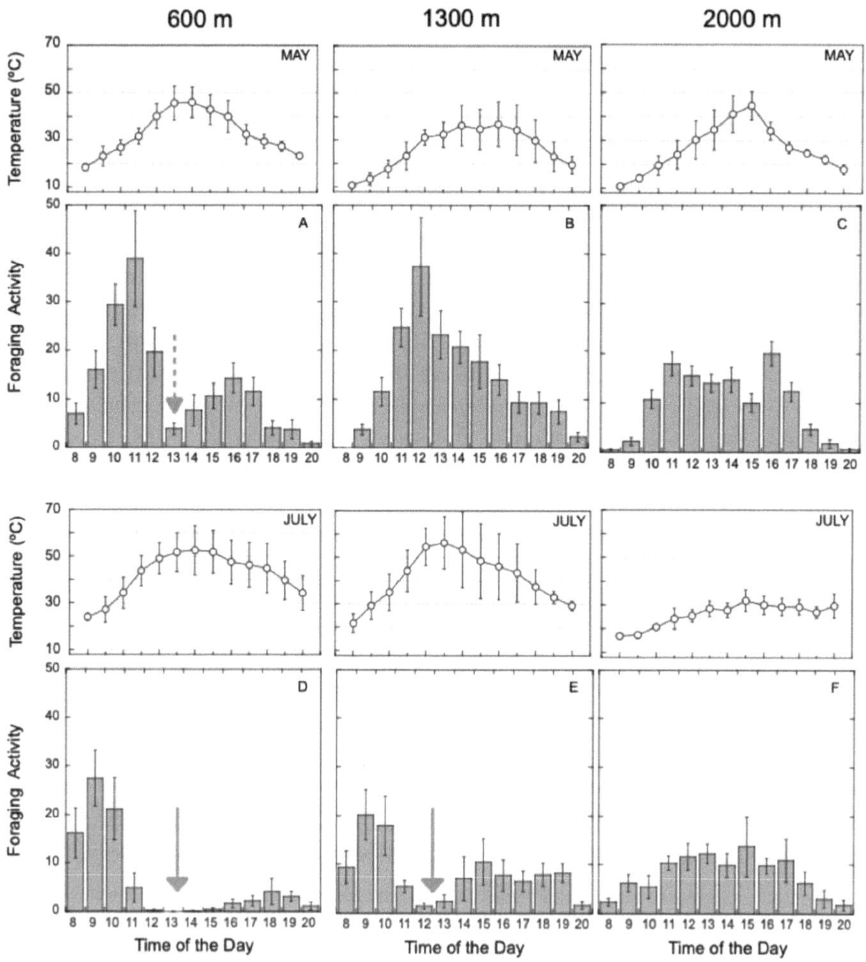

Fig. 2.7 Foraging activity of *Aphaenogaster iberica* along an elevational gradient in the Sierra Nevada mountains of southern Spain. Mean nest entrance temperature (°C) and absolute foraging activity (sum of entries and exits over a 5-min time period; mean ± SE; *n* = 9–13 colonies per month and population) for populations at three elevations (**a** and **d** = 600 m; **b** and **e** = 1300 m; **c** and **f** = 2000 m) in May (**a**–**c**) and July (**d**–**f**). The arrows indicate a dramatic decrease in or cessation of activity. (Figure modified from Villalta et al. 2020)

adapt, acclimate, or alter their behaviour to reduce their exposure to stressful microclimatic conditions. Indeed, it was found that the most common ant species in a North American temperate forest were unable to use either traits or behaviour to buffer the effects of warming (Youngsteadt et al. 2023). In almost all their modelling scenarios, Bewick et al. (2014) found that ant communities might respond to climate warming in unintuitive ways. For example, the most subordinate ant species were predicted to suffer the most as a result of climate change; it was not necessarily the most heat-sensitive species at greatest risk. There has been some criticism about the methods used in this domain of research. For example, Bujan and Yanoviak (2022) noted that congeneric ant species have displayed highly variable responses in experimental warming studies. Additionally, Menzel and Feldmeyer (2021) posited that colony-level resilience to climate change may be highly overestimated when predictions are based on worker survival, given that individual workers may be tolerant of temperatures that are outside the range suitable for colony growth.

Jenkins et al. (2011) suggested that we have an incomplete understanding of ant communities in hot and arid climates: "The best strategy is perhaps to focus more of our efforts in the climates similar to those predicted to expand in the future, even if they are uncomfortable to visit, as they will expand whether we study them or not". *Cataglyphis* desert ants could serve as an interesting study system (Lenoir et al. 2009). The aridification of the Mediterranean Basin triggered geographical spread and rapid species diversification in this genus (Villalta et al. 2018; Lecocq de Pletincx et al. 2024). However, the adaptations displayed by *Cataglyphis* can be viewed as quantitative exaggerations of traits present in most ant species rather than as qualitative innovations. It is difficult to predict the future evolutionary trajectory of these traits, and some may be dead ends that limit a species' capacity to adapt to new environmental conditions (Boulay et al. 2017).

Discussions around the ant species most threatened by climate change tend to focus on ants from tropical ecosystems because they have narrower ranges of thermal tolerance (Janzen 1967; Deutsch et al. 2008; Kingsolver 2009; Diamond et al. 2012) and experience environmental temperatures very close to their thermal limits (Deutsch et al. 2008; Kaspari et al. 2015). However, the same could be said of desert ants: they have narrower ranges of thermal tolerance and experience environmental temperatures so close to their thermal limits that even a slight further increase could have a strong negative impact on foraging efficiency and, therefore, fitness.

In conclusion, thermal adaptation can occur through a variety of mechanisms, from behaviour to natural selection, which are associated with a variety of time scales (Kaspari et al. 2015). Further studies using reciprocal transplant or common garden experiments (with founding queens or small colonies) must be employed to tease apart how local environmental conditions, natural selection, and phenotypic plasticity all affect individual thermal tolerance and/or colony fitness.

Acknowledgements While writing this chapter, Xim Cerdá and Elena Ângulo had the following funding: MCIN/AEI/10.13039/501100011033, ERDF—a way of making Europe; grant reference PID2022-138420NB-I00 (EcoFire); and a grant from the regional government of Andalucía (PROYEXCEL_00688, FaS@InAnt). We thank Jessica Pearce-Duvet for her assistance with the English language editing and Xavier Arnan and Alan N. Andersen for their feedback (which we incorporated as much as possible). The authors would like to dedicate this work to Alain Lenoir, a mentor, advisor, and good friend, as well as an example for us all.

References

Amor F, Ortega P, Cerdá X, Boulay R (2010) Cooperative prey-retrieving in the ant *Cataglyphis floricola*: an unusual short-distance recruitment. Insect Soc 57:91–94. https://doi.org/10.1007/s00040-009-0053-x

Amor F, Ortega P, Cerdá X, Boulay RR (2011) Solar elevation triggers foraging activity in a thermophilic ant. Ethology 117:1031–1039. https://doi.org/10.1111/j.1439-0310.2011.01955.x

Anderson C, McShea DW (2001) Individual versus social complexity, with particular reference to ant colonies. Biol Rev 76:211–237. https://doi.org/10.1017/S1464793101005656

Azcárate FM, Kovacs E, Peco B (2007) Microclimatic conditions regulate surface activity in harvester ants *Messor barbarus*. J Insect Behav 20:315–329. https://doi.org/10.1007/s10905-007-9074-3

Bar A, Gilad T, Massad D, Ferber A, Ben-Ezra D, Segal D, Foitzik S, Scharf I (2023) Foraging is prioritized over nestmate rescue in desert ants and pupae are rescued more than adults. Behav Ecol 34:1087–1096. https://doi.org/10.1093/beheco/arad083

Baudier KM, O'Donnell S (2016) Structure and thermal biology of subterranean army ant bivouacs in tropical montane forests. Insect Soc 63:467–476. https://doi.org/10.1007/s00040-016-0490-2

Beckers R, Goss S, Deneubourg JL, Pasteels JM (1989) Colony size, communication and ant foraging strategy. Psyche 96:239–256. https://doi.org/10.1155/1989/94279

Bewick S, Stuble KL, Lessard JL, Dunn RR, Adler FR, Sanders NJ (2014) Predicting future coexistence in a North American ant community. Ecol Evol 4:1804–1819. https://doi.org/10.1002/ece3.1048

Beydizada NI, Abels A, Schultheiss P, Frank ET (2024) Injury dependent wound care behaviour in the desert ant *Cataglyphis nodus*. Behav Ecol Sociobiol 78:97. https://doi.org/10.1007/s00265-024-03511-1

Bishop TR, Robertson MP, van Rensburg BJ, Parr CL (2017) Coping with the cold: minimum temperatures and thermal tolerances dominate the ecology of mountain ants. Ecol Entomol 42:105–114. https://doi.org/10.1111/een.12364

Bishop TR, Parr CL, Gibb H, van Rensburg BJ, Braschler B, Chown SL, Foord SH, Lamy K, Munyai TC, Okey I, Tshivhandekano PG, Werenkraut V, Robertson MP (2019) Thermoregulatory traits combine with range shifts to alter the future of montane ant assemblages. Glob Change Biol 25:2162–2173. https://doi.org/10.1111/gcb.14622

Blight O, Díaz-Mariblanca GA, Cerdá X, Boulay R (2016) A proactive–reactive syndrome affects group success in an ant species. Behav Ecol 27:118–125. https://doi.org/10.1093/beheco/arv127

Bollazzi M, Roces F (2002) Thermal preference for fungus culturing and brood location by workers of the thatching grass-cutting ant *Acromyrmex heyeri*. Insect Soc 49:153–157. https://doi.org/10.1007/s00040-002-8295-x

Bollazzi M, Roces F (2007) To build or not to build: circulating dry air organizes collective building for climate control in the leaf-cutting ant *Acromyrmex ambiguus*. Anim Behav 74:1349–1355. https://doi.org/10.1016/j.anbehav.2007.02.021

Bollazzi M, Kronenbitter J, Roces F (2008) Soil temperature, digging behaviour, and the adaptive value of nest depth in South American species of *Acromyrmex* leaf-cutting ants. Oecologia 158:165–175. https://doi.org/10.1007/s00442-008-1113-z

Bollazzi M, Forti LC, Roces F (2012) Ventilation of the giant nests of *Atta* leaf-cutting ants: does underground circulating air enter the fungus chambers? Insect Soc 59:487–498. https://doi.org/10.1007/s00040-012-0243-9

Boulay R, Cerdá X, Fertin A, Ichinose K, Lenoir A (2009) Brood development into sexual females depends on the presence of a queen but not on temperature in an ant dispersing by colony fission, *Aphaenogaster senilis*. Ecol Entomol 34:595–602. https://doi.org/10.1111/j.1365-2311.2009.01108.x

Boulay R, Aron S, Cerdá X, Doums D, Graham P, Hefetz A, Monnin T (2017) Social life in arid environments: the case study of *Cataglyphis* ants. Annu Rev Entomol 62:305–321. https://doi.org/10.1146/annurev-ento-031616-034941

Brian MV (1963) Studies of caste differentiation in *Myrmica rubra* L. 6. Factors influencing the course of female development in the early third instar. Insect Soc 10:91–102. https://doi.org/10.1007/BF02336490

Brian MV (1973) Temperature choice and its relevance to brood survival and caste determination in the ant *Myrmica rubra* L. Physiol Zool 46:245–252. https://doi.org/10.1086/physzool.46.4.30155608

Bujan J, Yanoviak SP (2022) Behavioural response to heat stress of twig-nesting canopy ants. Oecologia 198:947–955. https://doi.org/10.1007/s00442-022-05143-6

Bujan J, Nottingham AT, Velasquez E, Meir P, Kaspari M, Yanoviak SP (2022) Tropical ant community responses to experimental soil warming. Biol Lett 18:20210518. https://doi.org/10.1098/rsbl.2021.0518

Calazans EG, da Costa FV, Cristiano MP, Cardoso DC (2020) Daily dynamics of an ant community in a mountaintop ecosystem. Environ Entomol 49:383–390. https://doi.org/10.1093/ee/nvaa011

Cassill DL, Vo K, Becker B (2008) Young fire ant workers feign death and survive aggressive neighbors. Naturwiss 95:617–624. https://doi.org/10.1007/s00114-008-0362-3

Cerdá X, Retana J (1989) Influencia de los factores ambientales sobre la actividad diaria de recolección de la hormiga *Cataglyphis iberica* (Em.) (Hym.: Formicidae). An Biol 15:75–82. https://revistas.um.es/analesbio/article/view/34621

Cerdá X, Retana J (1992) A behavioural study of transporter workers in *Cataglyphis iberica* ant colonies (Hymenoptera Formicidae). Ethol Ecol Evol 4:359–374. https://doi.org/10.1080/08927014.1992.9523128

Cerdá X, Retana J (1997) Links between worker polymorphism and thermal biology in a thermophilic ant species. Oikos 78:467–474. https://doi.org/10.2307/3545608

Cerdá X, Retana J (2000) Alternative strategies by thermophilic ants to cope with extreme heat: individual versus colony level traits. Oikos 89:155–163. https://doi.org/10.1034/j.1600-0706.2000.890117.x

Cerda X, Retana J, Bosch J, Alsina A (1989) Daily foraging activity and food collection of the thermophilic ant *Cataglyphis cursor* (Hymenoptera, Formicidae). Vie Milieu 39:207–212

Cerdá X, Retana J, Cros S (1997) Thermal disruption of transitive hierarchies in Mediterranean ant communities. J Anim Ecol 66:363–374. https://doi.org/10.2307/5982

Cerdá X, Retana J, Cros S (1998a) Critical thermal limits in Mediterranean ant species: trade-off between mortality risk and foraging performance. Funct Ecol 12:45–55. https://doi.org/10.1046/j.1365-2435.1998.00160.x

Cerdá X, Retana J, Manzaneda A (1998b) The role of competition by dominants and temperature in the foraging of subordinate species in Mediterranean ant communities. Oecologia 117:404–412. https://doi.org/10.1007/s004420050674

Cerdá X, Angulo E, Boulay R, Lenoir A (2009) Individual and collective foraging decisions: a field study of worker recruitment in the gypsy ant *Aphaenogaster senilis*. Behav Ecol Sociobiol 63:551–562. https://doi.org/10.1007/s00265-008-0690-5

Cerdá X, Angulo E, Caut S, Courchamp F (2012) Ant community structure on a small Pacific Island: only one native species living with the invaders. Biol Invasions 14:323–339. https://doi.org/10.1007/s10530-011-0065-0

Cerdá X, van Oudenhove L, Bernstein C, Boulay RR (2014) A list of and some comments about the trail pheromones of ants. Nat Prod Commun 9:1115–1122. https://doi.org/10.1177/1934578X140090

Christian KA, Morton SR (1992) Extreme thermophilia in a central Australian ant, *Melophorus bagoti*. Physiol Zool 65:885–905. https://doi.org/10.1086/physzool.65.5.30158548

Clémencet J, Cournault L, Odent A, Doums C (2010) Worker thermal tolerance in the thermophilic ant *Cataglyphis cursor* (Hymenoptera, Formicidae). Insect Soc 57:11–15. https://doi.org/10.1007/s00040-009-0044-y

Clusella-Trullas S, Blackburn TM, Chown SL (2011) Climatic predictors of temperature performance curve parameters in ectotherms imply complex responses to climate change. Am Nat 177:738–751. https://doi.org/10.1086/660021

Coenen-Stass D, Schaarschmidt B, Lamprecht I (1980) Temperature distribution and calorimetric determination of heat production in the nest of the wood ant, *Formica polyctena* (Hymenoptera, Formicidae). Ecology 61:238–244. https://doi.org/10.2307/1935180

Crist TO, MacMahon JA (1991) Foraging patterns of *Pogonomyrmex occidentalis* (Hymenoptera: Formicidae) in a shrub–steppe ecosystem: the roles of temperature, trunk trails, and seed resources. Environ Entomol 20:265–275. https://doi.org/10.1093/ee/20.1.265

Cros S, Cerdá X, Retana J (1997) Spatial and temporal variations in the activity patterns of Mediterranean ant communities. Écoscience 4:269–278. https://doi.org/10.1080/11956860.1997.11682405

Dall SRX, Houston AI, McNamara JM (2004) The behavioural ecology of personality: consistent individual differences from an adaptive perspective. Ecol Lett 7:734–739. https://doi.org/10.1111/j.1461-0248.2004.00618.x

Dean WRJ, Turner JS (1991) Ants nesting under stones in the semi-arid Karoo, South Africa: predator avoidance or temperature benefits? J Arid Environ 21:59–69. https://doi.org/10.1016/S0140-1963(18)30728-6

Del Toro I, Ribbons RR, Pelini SL (2012) The little things that run the world revisited: a review of ant-mediated ecosystem services and disservices (Hymenoptera: Formicidae). Myrmecol News 17:133–146. https://doi.org/10.25849/myrmecol.news_017:133

Del Toro I, Ribbons RR, Ellison AM (2015) Ant-mediated ecosystem functions on a warmer planet: effects on soil movement, decomposition and nutrient cycling. J Anim Ecol 84:1233–1241. https://doi.org/10.1111/1365-2656.12367

Detrain C, Deneubourg JL, Pasteels JM (1999) Decision-making in foraging by social insects. In: Detrain C, Deneubourg JL, Pasteels JM (eds) Information processing in social insects. Birkhäuser, Basel, pp 331–354. https://doi.org/10.1007/978-3-0348-8739-7_18

Detrain C, Natan C, Deneubourg J (2001) The influence of the physical environment on the self-organised foraging patterns of ants. Naturwiss 88:171–174. https://doi.org/10.1007/s001140100217

Deutsch CA, Tewksbury JJ, Huey RB, Sheldon KS, Ghalambor CK, Haak DC, Martin PR (2008) Impacts of climate warming on terrestrial ectotherms across latitude. PNAS 105:6668–6672. https://doi.org/10.1073/pnas.0709472105

Diamond SE, Sorger DM, Hulcr J, Pelini SL, Toro ID, Hirsch C, Oberg E, Dunn RR (2012) Who likes it hot? A global analysis of the climatic, ecological, and evolutionary determinants of warming tolerance in ants. Glob Change Biol 18:448–456. https://doi.org/10.1111/j.1365-2486.2011.02542.x

Diamond SE, Nichols LM, Pelini SL, Penick CA, Barber GW, Cahan SH, Dunn RR, Ellison AM, Sanders NJ, Gotelli NJ (2016) Climatic warming destabilizes forest ant communities. Sci Adv 2:e1600842. https://doi.org/10.1126/sciadv.1600842

Diamond SE, Chick L, Penick CA, Nichols LM, Cahan SH, Dunn RR, Ellison AM, Sanders NJ, Gotelli NJ (2017) Heat tolerance predicts the importance of species interaction effects as the climate changes. Integr Comp Biol 57:112–120. https://doi.org/10.1093/icb/icx008

Dlussky GM (1981) Murav'i Pustyn [Ants of the Deserts, in Russian]. Akademia Nauk, Moscow, CCCP, p 230

Dussutour A, Poissonnier LA, Buhl J, Simpson SJ (2016) Resistance to nutritional stress in ants: when being fat is advantageous. J Exp Biol 219:824–833. https://doi.org/10.1242/jeb.136234

Elizalde L, Arbetman M, Arnan X, Eggleton P, Leal IR, Lescano MN, Saez A, Werenkraut V, Pirk GI (2020) The ecosystem services provided by social insects: traits, management tools and knowledge gaps. Biol Rev 95:1418–1441. https://doi.org/10.1111/brv.12616

Farji-Brener AG (2000) Leaf-cutting ant nests in temperate environments: mounds, mound damages and nest mortality in *Acromyrmex lobicornis*. Stud Neotropical Fauna Environ 35:131–138. https://doi.org/10.1076/0165-0521(200008)35:2;1-9;FT131

Fernández-Escudero I, Tinaut A, Ruano F (1993) Rock selection for nesting in *Proformica longiseta* Collingwood, 1978 (Hymenoptera, Formicidae) in a Mediterranean high mountain. Int J Biometeorol 37:83–88. https://doi.org/10.1007/BF01214386

Frank ET, Schmitt T, Hovestadt T, Mitesser O, Stiegler J, Linsenmair KE (2017) Saving the injured: rescue behavior in the termite-hunting ant *Megaponera analis*. Sci Adv 3:e1602187. https://doi.org/10.1126/sciadv.1602187

Frank ET, Buffat D, Liberti J, Aibekova L, Economo EP, Keller L (2024) Wound dependent leg amputations to combat infections in an ant society. Curr Biol 34:3273–3278. https://doi.org/10.1016/j.cub.2024.06.021

Franks NR (1989) Thermal regulation in army ant bivouacs. Physiol Entomol 14:397–404. https://doi.org/10.1111/j.1365-3032.1989.tb01109.x

Friedman DA, Greene MJ, Gordon DM (2019) The physiology of forager hydration and variation among harvester ant (*Pogonomyrmex barbatus*) colonies in collective foraging behavior. Sci Rep 9:5126. https://doi.org/10.1038/s41598-019-41586-3

Galarza JA, Jovani R, Cerdá X, Rico C, Barroso A, Boulay R (2012) Frequent colony relocations do not result in effective dispersal in the gypsy ant *Aphaenogaster senilis*. Oikos 121:605–613. https://doi.org/10.1111/j.1600-0706.2011.19859.x

Gallé L (1973) Thermoregulation in the nest of *Formica pratensis* Retz. (Hymenoptera, Formicidae). Acta Biol Szeged 19:139–142

Gibbs A, Mousseau TA, Crowe JH (1991) Genetic and acclimatory variation in biophysical properties of insect cuticle lipids. PNAS 88:7257–7260. https://doi.org/10.1073/pnas.88.16.7257

Gordon DM (1984) Species-specific patterns in the social activities of harvester ant colonies (*Pogonomyrmex*). Insect Soc 31:74–86. https://doi.org/10.1007/BF02223693

Gordon DM (2019) The ecology of collective behavior in ants. Annu Rev Entomol 64:35–50. https://doi.org/10.1146/annurev-ento-011118-111923

Gordon DM, Hölldobler B (1987) Worker longevity in harvester ants (*Pogonomyrmex*). Psyche 94:341–346. https://doi.org/10.1155/1987/63930

Gordon DM, Dektar KN, Pinter-Wollman N (2013) Harvester ant colony variation in foraging activity and response to humidity. PLoS One 8:e63363. https://doi.org/10.1371/journal.pone.0063363

Gordon DM, Steiner E, Das B, Walker NS (2023) Harvester ant colonies differ in collective behavioural plasticity to regulate water loss. R Soc Open Sci 10:230726. https://doi.org/10.1098/rsos.230726

Hölldobler B, Wilson EO (2009) The superorganism: the beauty, elegance, and strangeness of insect societies. WW Norton & Co, New York, p 522

Hollis KL, Nowbahari E (2013) A comparative analysis of precision rescue behaviour in sand-dwelling ants. Anim Behav 85:537–544. https://doi.org/10.1016/j.anbehav.2012.12.005

Huey RB, Stevenson RD (1979) Integrating thermal physiology and ecology of ectotherms: a discussion of approaches. Am Zool 19:357–366. https://doi.org/10.1093/icb/19.1.357

Ibarra FG, Jouquet P, Bottinelli N, Bultelle A, Monnin T (2024) Experimental evidence that increased surface temperature affects bioturbation by ants. J Anim Ecol 93:319–332. https://doi.org/10.1111/1365-2656.14040

Janzen DH (1967) Why mountain passes are higher in the tropics. Am Nat 101:233–249. https://doi.org/10.1086/282487

Jeanson R, Ratnieks FLW, Deneubourg JL (2003) Pheromone trail decay rates on different substrates in the Pharaoh's ant, *Monomorium pharaonis*. Physiol Entomol 28:192–198. https://doi.org/10.1046/j.1365-3032.2003.00332.x

Jenkins CN, Sanders NJ, Andersen AN, Arnan X, Bruhl CA, Cerda X, Ellison AM, Fisher BL, Fitzpatrick MC, Gotelli NJ, Gove AD, Guenard B, Lattke JE, Lessard JP, McGlynn TP, Menke SB, Parr CL, Philpott SM, Vasconcelos HL, Weiser MD, Dunn RR (2011) Global diversity in light of climate change: the case of ants. Divers Distrib 17:652–662. https://doi.org/10.1111/j.1472-4642.2011.00770.x

Jonkman JCM (1980) The external and internal structure and growth of nests of the leaf-cutting ant *Atta vollenweideri* Forel, 1893 (Hym.: Formicidae). Part II. The internal nest structure and growth. Z Angew Entomol 89:217–246

Kadochová Š, Frouz J (2013) Thermoregulation strategies in ants in comparison to other social insects, with a focus on red wood ants (*Formica rufa* group). F1000Res 2:280. https://doi.org/10.12688/f1000research.2-280.v2

Kadochová Š, Frouz J, Roces F (2017) Sun basking in red wood ants *Formica polyctena* (Hymenoptera, Formicidae): individual behaviour and temperature-dependent respiration rates. PLoS One 12:e0170570. https://doi.org/10.1371/journal.pone.0170570

Kadochová Š, Frouz J, Tószögyová A (2019) Factors influencing sun basking in red wood ants (*Formica polyctena):* a field experiment on clustering and phototaxis. J Insect Behav 32:164–179. https://doi.org/10.1007/s10905-019-09713-0

Kaspari M, Clay NA, Lucas J, Yanoviak SP, Kay A (2015) Thermal adaptation generates a diversity of thermal limits in a rainforest ant community. Glob Change Biol 21:1092–1102. https://doi.org/10.1111/gcb.12750

Kearney M, Shine R, Porter WP (2009) The potential for behavioral thermoregulation to buffer "cold- blooded" animals against climate warming. PNAS 106:3835–3840. https://doi.org/10.1073/pnas.080891310

Kingsolver JG (2009) The well-temperatured biologist. Am Nat 174:755–768. https://doi.org/10.1086/648310

Kipyatkov VE, Lopatina EB (1990) Seasonal development of *Aphaenogaster sinensis* in the south Primor'e: new type of seasonal cycle in ants. Entomol Rev 69:73–81

Kipyatkov VE, Lopatina EB, Imamgaliev AA, Shirokova LA (2004) Effect of temperature on rearing of the first brood by the founder females of the ant *Lasius niger* (Hymenoptera, Formicidae): latitude-dependent variability of the response norm. J Evol Biochem Physiol 40:165–175

Kipyatkov V, Lopatina E, Imamgaliev A (2005) Duration and thermal reaction norms of development are significantly different in winter and summer brood pupae of the ants *Myrmica rubra* Linnaeus, 1758 and *M. ruginodis* Nylander, 1846 (Hymenoptera: Formicidae). Myrmecol Nachr 7:69–76

Kolay S, Boulay R, Angulo E, d'Ettorre P, Cerdá X (2023) Habitat-dependent variation in consistent behavioural traits does not affect the efficiency of resource acquisition in a thermophilic ant. Behav Ecol Sociobiol 77:1. https://doi.org/10.1007/s00265-022-03274-7

Kwon T-S (2017) Temperature and ant assemblages: biased values of community temperature index. J Asia Pac Entomol 20:1077–1086. https://doi.org/10.1016/j.aspen.2017.04.005

Kwon T-S (2018) High competition between ant species at intermediate temperatures. J Therm Biol 72:59–66. https://doi.org/10.1016/j.jtherbio.2017.11.015

Kwon T-S, Li F, Kim S-S, Chun JH, Park Y-S (2016) Modeling vulnerability and range shifts in ant communities responding to future global warming in temperate forests. PLoS One 11:e0159795. https://doi.org/10.1371/journal.pone.0159795

Lecocq de Pletincx N, Dellicour S, Aron S (2021) The evolution of ant worker polymorphism correlates with multiple social traits. Behav Ecol Sociobiol 75:113. https://doi.org/10.1007/s00265-021-03049-6

Lecocq de Pletincx N, Cerdà X, Kiran K, Karaman C, Taheri A, Aron S (2024) Ecological diversification preceded geographical expansion during the evolutionary radiation of Cataglyphis desert ants. iScience 27:109852. https://doi.org/10.1016/j.isci.2024.109852

Lenoir A, Aron S, Cerdá X, Hefetz A (2009) *Cataglyphis* desert ants: a good model for evolutionary biology in Darwin's anniversary year. Isr J Entomol 39:1–32

Lőrincz Á, Hábenczyus AA, Kelemen A, Ratkai B, Tölgyesi C, Lőrinczi G, Frei K, Bátori Z, Maák IE (2024a) Wood-pastures promote environmental and ecological heterogeneity on a small spatial scale. Sci Total Environ 906:167510. https://doi.org/10.1016/j.scitotenv.2023.167510

Lőrincz Á, Ratkai B, Tölgyesi C, Lőrinczi G, Bán KA, Frei K, Jégh T, Bátori Z, Maák IE (2024b) Ants in space and time: spatiotemporal niche changes facilitate species coexistence in semi-natural ecosystem complexes. Glob Ecol Conserv 54:e03170. https://doi.org/10.1016/j.gecco.2024.e03170

Marsh AC (1985a) Thermal responses and temperature tolerance in a diurnal desert ant, *Ocymyrmex barbiger*. Physiol Zool 58:629–636. https://doi.org/10.1086/physzool.58.6.30156067

Marsh AC (1985b) Microclimatic factors influencing foraging patterns and success of the thermophilic desert ant, *Ocymyrmex barbiger*. Insect Soc 32:286–296. https://doi.org/10.1007/BF02224917

McCaffrey J, Galen C (2011) Between a rock and a hard place: impact of nest selection behavior on the altitudinal range of an alpine ant, *Formica neorufibarbis*. Environ Entomol 40:534–540. https://doi.org/10.1603/EN10304

McGlynn TP (2012) The ecology of nest movement in social insects. Annu Rev Entomol 57:291–308. https://doi.org/10.1146/annurev-ento-120710-100708

McMeeking RM, Arzt E, Wehner R (2012) *Cataglyphis* desert ants improve their mobility by raising the gaster. J Theor Biol 297:17–25. https://doi.org/10.1016/j.jtbi.2011.12.003

Menzel F, Feldmeyer B (2021) How does climate change affect social insects? Curr Opin Insect Sci 46:10–15. https://doi.org/10.1016/j.cois.2021.01.005

Miller JS, Pinter-Wollman N (2023) Social interactions differ in their impact on foraging decisions. Anim Behav 203:183–192. https://doi.org/10.1016/j.anbehav.2023.07.008

Molet M, Péronnet R, Couette S, Canovas C, Doums C (2017) Effect of temperature and social environment on worker size in the ant *Temnothorax nylanderi*. J Therm Biol 67:22–29. https://doi.org/10.1016/j.jtherbio.2017.04.013

Muser B, Sommer S, Wolf H, Wehner R (2005) Foraging ecology of the thermophilic Australian desert ant, *Melophorus bagoti*. Aust J Zool 53:301–311. https://doi.org/10.1071/ZO05023

Nowbahari E, Scohier A, Durand JL, Hollis KL (2009) Ants, *Cataglyphis cursor*, use precisely directed rescue behavior to free entrapped relatives. PLoS One 4:e6573. https://doi.org/10.1371/journal.pone.0006573

Oms CS, Cerdá X, Boulay R (2017) Is phenotypic plasticity a key mechanism for responding to thermal stress in ants? Sci Nat 104:42. https://doi.org/10.1007/s00114-017-1464-6

Ord TJ (2023) Drought-induced relocation of ant colonies and its consequences for the long-term spatial ecology of a population under stress. Funct Ecol 37:2231–2245. https://doi.org/10.1111/1365-2435.14383

Parr CL, Bishop TR (2022) The response of ants to climate change. Glob Change Biol 28:3188–3205. https://doi.org/10.1111/gcb.16140

Pelini SL, Boudreau M, McCoy N, Ellison AM, Gotelli NJ, Sanders NJ, Dunn RR (2011a) Effects of short-term warming on low and high latitude forest ant communities. Ecosphere 2:art62. https://doi.org/10.1890/ES11-00097.1

Pelini SL, Bowles FP, Ellison AM, Gotelli NJ, Sanders NJ, Dunn RR (2011b) Heating up the forest: open-top chamber warming manipulation of arthropod communities at Harvard and Duke Forests. Methods Ecol Evol 2:534–540. https://doi.org/10.1111/j.2041-210X.2011.00100.x

Penick CA, Tschinkel WR (2008) Thermoregulatory brood transport in the fire ant, *Solenopsis invicta*. Insect Soc 55:176–182. https://doi.org/10.1007/s00040-008-0987-4

Perez R, Aron S (2020) Adaptations to thermal stress in social insects: recent advances and future directions. Biol Rev 95:1535–1553. https://doi.org/10.1111/brv.12628

Petit S, Hammond PA, Heterick B, Weyland JJ (2022) *Polyrhachis femorata* (Hymenoptera: Formicidae) habitat and colony defensive immobility strategy. Aust J Zool 70:126–131. https://doi.org/10.1071/ZO22042

Pinter-Wollman N (2012) Personality in social insects: how does worker personality determine colony personality? Curr Zool 58:579–587. https://doi.org/10.1093/czoolo/58.4.580

Pinter-Wollman N, Bala A, Merrell A, Queirolo J, Stumpe MC, Holmes S, Gordon DM (2013) Harvester ants use interactions to regulate forager activation and availability. Anim Behav 86:197–207. https://doi.org/10.1016/j.anbehav.2013.05.012

Porter SD, Jorgensen CD (1981) Foragers of the harvester ant, *Pogonomyrmex owyheei*: a disposable caste? Behav Ecol Sociobiol 9:247–256. https://doi.org/10.1007/BF00299879

Porter SD, Tschinkel WR (1993) Fire ant preferences: behavioral control of growth and metabolism. Behav Ecol Sociobiol 32:321–329. https://doi.org/10.1007/BF00183787

Retana J, Cerdá X (1991) Behavioural variability and development of *Cataglyphis cursor* ant workers (Hymenoptera, Formicidae). Ethology 89:275–286. https://doi.org/10.1111/j.1439-0310.1991.tb00373.x

Roces F, Núñez JA (1989) Brood translocation and circadian variation of temperature preference in the ant *Camponotus mus*. Oecologia 81:33–37. https://doi.org/10.1007/BF00377006

Rocha FP, Bogar TA, Ibañez Weemaels A, Hu J, Liang M, Park C, Lee RH, Khan SA, Han S, Ng YL, Guénard B (2024) Two sides of the same coin? Ants are ecosystem engineers and providers of ecosystem services. Myrmecol News 34:129–157. https://doi.org/10.25849/myrmecol.news_034:129

Roeder DV, Paraskevopoulos AW, Roeder KA (2022) Thermal tolerance regulates foraging behaviour of ants. Ecol Entomol 47:331–338. https://doi.org/10.1111/een.13118

Roma GC, Mathias MIC, Bueno OC (2006) Fat body in some genera of leaf-cutting ants (Hymenoptera: Formicidae). Proteins, lipids and polysaccharides detection. Micron 37:234–242. https://doi.org/10.1016/j.micron.2005.10.012

Rosengren R, Fortelius W, Lindström K, Luther A (1987) Phenology and causation of nest heating and thermoregulation in red wood ants of the *Formica rufa* group studied in coniferous forest habitats in southern Finland. Ann Zool Fenn 2:147–155

Ruano F, Tinaut A, Soler JJ (2000) High surface temperatures select for individual foraging in ants. Behav Ecol 11:396–404. https://doi.org/10.1093/beheco/11.4.396

Sanada-Morimura S, Satoh T, Obara Y (2006) Territorial behavior and temperature preference for nesting sites in a pavement ant *Tetramorium tsushimae*. Insect Soc 53:141–148. https://doi.org/10.1007/s00040-005-0849-2

Sánchez-García D, Cerdá X, Angulo E (2022) Temperature or competition: which has more influence on Mediterranean ant communities? PLoS One 17(4):e0267547. https://doi.org/10.1371/journal.pone.0267547

Santos JT, Brito ELS, Santos GMM (2022) The role of vegetation structure and abiotic factors affecting the temporal dynamics of ant foraging. Sociobiology 69:e7422. https://doi.org/10.13102/sociobiology.v69i2.7422

Schmid-Hempel P, Schmid-Hempel R (1984) Life duration and turnover of foragers in the ant *Cataglyphis bicolor* (Hymenoptera, Formicidae). Insect Soc 31:345–360. https://doi.org/10.1007/BF02223652

Schneirla TC (1971) Army ants. A study in social organization. W. H. Freeman & Co, Francisco, p 349

Schneirla TC, Brown RZ, Brown FC (1954) The bivouac or temporary nest as an adaptive factor in certain terrestrial species of army ants. Ecol Monogr 24:269–296. https://doi.org/10.2307/1948466

Schultz TR (2021) Fungus-farming ants (Attini in part). In: Starr CK (ed) Encyclopedia of social insects. Springer, Cham, pp 404–411. https://doi.org/10.1007/978-3-030-28102-1

Seifert B (2017) The ecology of Central European non-arboreal ants – 37 years of a broad-spectrum analysis under permanent taxonomic control. Soil Organisms 89:1–69. https://soil-organisms.org/index.php/SO/article/view/83

Sih A, Bell AM, Johnson JC, Ziemba RE (2004) Behavioral syndromes: an integrative overview. Q Rev Biol 79:241–277. https://doi.org/10.1086/422893

Sinclair BJ, Williams CM, Terblanche JS (2012) Variation in thermal performance among insect populations. Physiol Biochem Zool 85:594–606. https://doi.org/10.1086/665388

Sinclair BJ, Marshall KE, Sewell MA, Levesque DL, Willett CS, Slotsbo S, Dong Y, Harley CD, Marshall DJ, Helmuth BS, Huey RB (2016) Can we predict ectotherm responses to climate change using thermal performance curves and body temperatures? Ecol Lett 19:1372–1385. https://doi.org/10.1111/ele.12686

Skowronek P, Wójcik Ł, Strachecka A (2021) Fat body – multifunctional insect tissue. Insects 12:547. https://doi.org/10.3390/insects12060547

Snir O, Alwaseem H, Heissel S, Sharma A, Valdés-Rodríguez S, Carroll TS, Jiang CS, Razzauti J, Kronauer DJC (2022) The pupal moulting fluid has evolved social functions in ants. Nature 612:488–494. https://doi.org/10.1038/s41586-022-05480-9

Soroker V, Hefetz A (2000) Hydrocarbon site of synthesis and circulation in the desert ant *Cataglyphis niger*. J Insect Physiol 46:1097–1102. https://doi.org/10.1016/s0022-1910(99)00219-x

Sundaram M, Steiner E, Gordon DM (2022) Rainfall, neighbors, and foraging: the dynamics of a population of red harvester ant colonies 1988–2019. Ecol Monogr 92:e1503. https://doi.org/10.1002/ecm.1503

Taylor K, Visvader A, Nowbahari E, Hollis KL (2013) Precision rescue behavior in North American ants. Evol Psychol 11:665–677. https://doi.org/10.1177/147470491301100312

Tinaut A, Fernández-Escudero I, Ruano F, Cerdá X (1999) The relationship of nest rock dimensions to reproductive success and nest permanence in a high-mountain ant, *Proformica longiseta* (Hymenoptera: Formicidae). Sociobiology 34:99–117

Trigos-Peral G, Maák I, Schmid S, Chudzik P, Czaczkes T, Witek M, Casacci L, Sánchez-García D, Lőrincz Á, Kochanowski M, Heinze J (2024) Urban abiotic stressors drive changes in the foraging activity and colony growth of the black garden ant *Lasius niger*. Sci Total Environ 915:170157. https://doi.org/10.1016/j.scitotenv.2024.170157

Turza F, Zuber G, Bzoma M, Prus M, Filipiak M, Miler K (2020) Ants co-occurring with predatory antlions show unsuccessful rescue behavior towards captured nestmates. J Insect Behav 33:1–6. https://doi.org/10.1007/s10905-020-09739-9

van Gils HAJA, Vanderwoude C (2012) Leafcutter ant (*Atta sexdens*) (Hymenoptera: Formicidae) nest distribution responds to canopy removal and changes in micro-climate in the southern Colombian Amazon. Fla Entomol 95:914–921. https://doi.org/10.1653/024.095.0414

van Oudenhove de Saint Géry L (2010) Trade-offs and foraging activity in Mediterranean ant communities. PhD dissertation, Université Claude Bernard Lyon 1 and Universitat Autònoma de Barcelona. http://hdl.handle.net/10261/89931

van Oudenhove L, Billoir E, Boulay R, Bernstein C, Cerdá X (2011) Temperature limits trail following behaviour through pheromone decay in ants. Naturwiss 98:1009–1017. https://doi.org/10.1007/s00114-011-0852-6

van Oudenhove L, Boulay R, Lenoir A, Bernstein C, Cerda X (2012) Substrate temperature constrains recruitment and trail following behavior in ants. J Chem Ecol 38:802–809. https://doi.org/10.1007/s10886-012-0130-x

Vieira MEL, Chameron S, Châline N (2024) The ethology of foraging in ants: revisiting Tinbergen's four questions. Front Ethol 3:1341120. https://doi.org/10.3389/fetho.2024.1341120

Villalta I, Amor F, Galarza JA, Dupont S, Ortega P, Hefetz A, Dahbi A, Cerdá X, Boulay R (2018) Origin and distribution of desert ants across the Gibraltar Straits. Mol Phylogenet Evol 118:122–134. https://doi.org/10.1016/j.ympev.2017.09.026

Villalta I, Oms CS, Angulo E, Molinas-González CR, Cerdá X, Boulay R (2020) Does social thermal regulation constrain individual thermal tolerance in an ant species? J Anim Ecol 89:2063–2076. https://doi.org/10.1111/1365-2656.13268

Ward D, Seely MK (1996) Adaptation and constraint in the evolution of the physiology and behavior of the Namib desert tenebrionid beetle genus *Onymacris*. Evolution 50:1231–1240

Wehner R (1987) Spatial organization of foraging behavior in individually searching desert ants, *Cataglyphis* (Sahara desert) and *Ocymyrmex* (Namib desert). In: Pasteels JM, Deneubourg JL (eds) From individual to collective behavior in social insects. Birkhauser, Basel, pp 15–42

Wehner R (2020) Desert navigator. The journey of an ant. Belknap Press of Harvard University Press, Cambridge, p 392

Willmer PG, Unwin DM (1981) Field analyses of insect heat budgets: reflectance, size and heating rates. Oecologia 50:250–255. https://doi.org/10.1007/BF00348047

Wilson EO (1990) In: Klinne O (ed) Success and dominance in ecosystems: the case of the social insects. Ecology Institute, Oldendorf/Luhe, p 104

Wilson DS, Clark AB, Coleman K, Dearstyne T (1994) Shyness and boldness in humans and other animals. Trends Ecol Evol 9:442–446. https://doi.org/10.1016/0169-5347(94)90134-1

Wirth R, Beyschlag W, Ryel R, Herz H, Hölldobler B (2003) The herbivory of leaf-cutting ants. A case study on *Atta colombica* in the tropical rainforest of Panama. Springer Verlag, Berlin/Heidelberg/New York, p 233

Youngsteadt E, Prado SG, Keleher KJ, Michelle Kirchner M (2023) Can behaviour and physiology mitigate effects of warming on ectotherms? A test in urban ants. J Anim Ecol 92:568–579. https://doi.org/10.1111/1365-2656.13860

Chapter 3
A Synthesis of Ant Morphological Responses to Climate Change

Heloise Gibb and Hannah Riskas

3.1 Introduction

Within the constraints of ant body form, there is enormous diversity in external morphology. This morphological variation provides a basis for taxonomy and is also critical to our understanding of function (Sosiak and Barden 2021). Many ant morphological traits have known or hypothesised functions and influence the fitness of individuals (Kaspari and Weiser 1999; Gibb et al. 2023). In ants, the focus of research has been on the external morphology of workers, where traits such as body size and colour are linked with the environment at small and large scales (Bishop et al. 2016; Gibb et al. 2018). However, ants are superorganisms, so their morphology is represented not just by the form of individual workers but also by the form of the colony and the relative abundance and form of the different castes. Further, ants are ecosystem engineers and therefore shape their physical environment (Folgarait 1998). As central place foragers, they do this mainly through the structure of the nest, which we consider here as an extended phenotype. Many species are likely to track climates at a large scale due to their dispersal abilities (Parr and Bishop 2022), shift nests to more suitable microclimates or alter activity or mating periods to optimise the match between their traits and the environment. In addition to this filtering of species assemblages, climate change is also likely to drive selection. Both forces

H. Gibb (✉)
School of Biological, Earth and Environmental Sciences, UNSW, Kensington, NSW, Australia

School of Life and Environmental Sciences, Deakin University, Burwood, VIC, Australia
e-mail: h.gibb@deakin.edu.au

H. Riskas
Centre for Future Landscapes, La Trobe University, Bundoora, VIC, Australia
e-mail: h.riskas@latrobe.edu.au

I. R. Leal et al. (eds.), *Ant Responses to Climate Change*, Entomology in Focus 10,
https://doi.org/10.1007/978-3-032-21127-9_3

will affect the morphologies represented in ant assemblages at the scale of the individual, colony and extended phenotype.

Traits often vary in a coordinated fashion, representing dimensions of an ecological strategy (Westoby et al. 2002). Ant morphological traits often covary with other traits related to morphology, physiology or behaviour, contributing to a broader ecological strategy (Gibb et al. 2023). In some cases, this is the result of a trade-off; for example, ant colonies balance the number and size of workers (Kaspari et al. 2000). Due to trait covariances, intra- or interspecific comparisons often cannot clearly attribute causation (Violle et al. 2007). Further, many traits have multiple functions. For example, large ants have slower metabolic rates per unit mass and higher critical thermal maxima than small ants (Chown and Gaston 2010; Nascimento et al. 2022). Laboratory-scale manipulations of traits or climatic factors that may regulate them are relatively rare (e.g. Wittlinger et al. 2006) but can be helpful in understanding the mechanisms through which traits function.

In this chapter, we consider key morphological traits of ants and their known associations with factors likely to shift under climate change, including temperature, precipitation, UV and habitat structure. Although some traits discussed may have effects on the environment (effect traits, Suding et al. 2008), here we consider morphological traits in the context of their response to climate (response traits). First, we discuss caste-specific morphology, including body size, dispersal morphology, sensory and trophic morphology and colour. Next, we consider colony morphology, including colony size and polymorphism. We then discuss the morphology of the extended phenotype of the nest. Finally, we synthesise the literature across these morphological scales to consider how ant assemblages may respond morphologically to climate change.

3.2 Caste-Specific Morphology

3.2.1 Body Size

Body size is a fundamental trait of species, underlying much of their physiology (Chown and Gaston 2010), and linked to metabolic rate, thermoregulation, locomotion, reproduction and lifespan (Peters 1986; Calabi and Porter 1989). It also has important implications for the role of species in ecosystems, such as through predation, competition and resource use (Nylin and Gotthard 1998; Heinrich and Bartholomew 1979; Holter and Scholtz 2005; Andrew et al. 2013) and dispersal ability (e.g. larger queens fly further; Helms 2018). Body size in ants varies 40-fold (Fig. 3.1) and is generally measured as "Weber's length" (the diagonal length of the mesosoma), head length or head width, as the entire body length can be variable due to the expansible gaster (Parr et al. 2017). In social insects, body size is manifested not just in the size of individuals but also in the number of individuals in a colony. In this section, we focus on the body size of individuals, considering colony size

Fig. 3.1 Ants range from 1 mm to 36 mm in body length, from *Carebara minuta* (left) to *Dinoponera gigantea* (right). Small ants also tend to be lighter in colour. (Alex Wild images)

below, but note that there is an energetic trade-off between the size of individuals and the number of individuals produced in the colony (Kaspari et al. 2000; Gibb et al. 2023). The size of individuals is regulated by temperature during development, and size determines responses to environmental temperature and moisture. Temperature and moisture also impact the structure of habitats and food availability, indirectly regulating body size.

The well-established *size-temperature rule* describes the pattern in ectotherms whereby individuals mature earlier and at a smaller size when reared in warmer conditions, despite initially growing faster (Atkinson 1996; Angilletta and Dunham 2003; Kipyatkov and Lopatina 2015). Ants reared at higher temperatures in the laboratory are generally smaller (Molet et al. 2017), although not always (Yilmaz et al. 2019), and a study of *Formica* across North America found a positive size-temperature relationship across workers, queens and males (Brassard et al. 2020). In the wild, ants have a high level of control over their nest temperatures and can relocate larvae within or among nests to optimise their development (Parr and Bishop 2022). However, the size-temperature rule is manifested in intra- and interspecific latitude-size, elevation-size and heat island-size relationships for ants (Cushman et al. 1993; Heinze et al. 2003; Bernadou et al. 2016; Bishop et al. 2016; Merckx et al. 2018; Heinze and Oberstadt 1999; Shik et al. 2019). Further, the very smallest ants are found in the hottest environments (Gibb et al. 2018). Warming climates are therefore expected to increase developmental temperatures, leading to faster maturation of workers at a smaller body size. However, the degree to which ants are able to regulate developmental temperatures in the nest is likely to determine the scale of this impact. A stronger understanding of limits to the regulation of nest temperatures by ants across the variety of nest types (e.g. in soil or twigs, in thatch nests or under rocks) may identify the ant taxa most likely to become smaller under global warming.

Body size is not only determined by temperature but also determines a species' *temperature limits*. Larger ants generally have broader temperature tolerance, equating to higher critical thermal maxima and lower critical thermal minima than smaller ants (see Chap. 4). For example, 82% of the variation in upper lethal temperatures among 18 species of *Cataglyphis* was due to body size (Bujan et al. 2022), and 74%

of the global range of critical thermal maxima of ants can occur at a single site (Kaspari et al. 2015). However, the relationship between size and CT_{max} can be inconsistent (Buxton et al. 2021; Verble-Pearson et al. 2015). Such inconsistency may be due at least in part to variation in the experimental approaches used to measure thermal tolerance; for example, large, long-legged ants may escape the boundary layer in hot plate experiments but not in thermal dry baths (Roeder et al. 2021). The mechanism driving the size-thermal tolerance relationship is thought to be thermal inertia resulting from the greater surface area-to-volume ratio of larger organisms, which reduces cellular damage (Chown and Nicolson 2004). This suggests that smaller individuals resulting from development at higher temperatures might be less resistant to those same warmer temperatures. However, the thermal inertia of larger ants means that they are slower to recover from heat coma and therefore at greater risk of predation if they exceed their thermal upper limits (Leong et al. 2023). Further, laboratory experiments suggest that physiological mechanisms of temperature tolerance, such as heat shock protein synthesis, can compensate for smaller body size in ants reared at higher temperatures (Oms et al. 2017). The negative relationship between body size and cuticle lightness (discussed below) may also ameliorate the effects of accelerated warming in smaller individuals. The relationship between thermal limits and body size suggests that species able to maintain low enough developmental temperatures to rear larger individuals will be favoured because larger size engenders greater temperature inertia.

Body size has also been linked with *CO_2 and food resource and water scarcity* associated with low precipitation. Higher CO_2 during development can reduce body size in insects (Tocco et al. 2021), but ant nests already have high CO_2 (Römer et al. 2018a; Römer et al. 2018b), so it is unclear what impact rising atmospheric CO_2 will have. Body size is linked with precipitation through resistance to desiccation and starvation and habitat structure. Desiccation resistance increases with the declining surface area-to-volume ratio that accompanies increasing body size (Remmert 1981; Hood and Tschinkel 1990; Kaspari 1993; Lighton et al. 1994; Chown and Gaston 2010; Bujan et al. 2016; Arnan and Blüthgen 2015; see Chap. 5). Larger insects also store more water and fat, which can be converted to metabolic water (Hadley 1994), allowing species to better deal with arid environments. On the other hand, small size may favour species in flood-prone environments (Pringle et al. 2019).

Additional fat storage also makes larger individuals more starvation-resistant and therefore better able to persist in seasonal or unpredictable environments (Cushman et al. 1993; Kaspari and Vargo 1995), including some types of deserts (Perez et al. 2023; Araujo et al. 2023). Several studies have reported increases in body size with increasing aridity (Nooten and Guénard 2022; Kaspari 1993). Further, complex habitats (often in wetter environments) obstruct movement of larger ants, favouring reduced body size or sometimes a bimodal body size distribution, where larger ants travel over the litter while small ants travel through it (Kaspari and Weiser 1999; Sarty et al. 2006; Gibb and Parr 2013). Habitat complexity thus limits the range of usable spaces based on body size, influencing the resources a colony can access and likely serving as a key driver of the evolution of body size in ants (Wills et al. 2018). In a global-scale study of ant body size, habitat disturbance (which tends to reduce habitat complexity) had its greatest effects on the very smallest (tenth percentile) and very largest (90th percentile) ants, while the very largest ants were also larger in more arid areas (Gibb et al. 2018, Fig. 3.2a, b). Under climate change, shifts in precipitation and the consequences for food resources and habitat structure will vary among regions (Trenberth 2011). However, shrinking body size is likely to be a net effect of global change (Gibb et al. 2018).

3.2.2 *Dispersal Morphology*

Research on morphology related to locomotion in ants has focussed on worker leg length. Wings, which are only found on the alate reproductive castes, have received less attention, despite wing size being important for dispersal distance (Bruna et al. 2011) and colony founding method (Peeters and Aron 2017), and that body size, wing size and flight musculature may covary (Helms and Kaspari 2015). Body size (discussed above) may also affect locomotion in workers as larger individuals walk further, run faster and are better able to tolerate conditions outside the nest (Hurlbert et al. 2008; Wagner et al. 2021).

Leg length (generally measured as relative femur length) tends to be highly correlated with the length of the antennal scape and eye position (e.g. Silva and Brandão 2010; Gibb et al. 2015, 2024), possibly because these sensory traits align with movement morphology. Oms et al. (2017) showed that relative leg length increases with increasing rearing temperature in *Aphaenogaster senilis* and suggested that this may be adaptive. Leg length often increases with temperature (Schofield et al. 2016; Brassard et al. 2020), and greater leg length is commonly associated with hot, arid climates, where it allows ants to hold their bodies above a hot boundary layer ("stilting", Cerdá and Retana 2000), reducing the operative temperature by up to 4 °C (Sommer and Wehner 2012). Further, in simple habitats, longer legs enable faster running, resulting in greater convective cooling and reduced time outside the nest in high temperatures (Cerdá and Retana 2000; Hurlbert et al. 2008; Sommer and Wehner 2012; Pfeffer et al. 2019). Leg length is often negatively correlated with precipitation, suggesting that long legs might impede movement in the more complex habitats of wetter climates (Oliveira et al. 2022; Nooten and Guénard 2022; Gibb et al. 2024) (but see Ibarra-Isassi et al. 2023). Instead, a bimodal distribution of leg lengths may occur, where relatively short-legged species travel through the leaf litter, while longer-legged species travel over it or are found in the canopy (Kaspari and Weiser 2000; Yanoviak and Kaspari 2000).

Canopy species may elevate their bodies above maximum surface temperatures, depending on substrate colour and wind (Stark et al. 2018). The impact of increasing temperatures on relative appendage length is thus likely to be strongly dependent on precipitation, through its impacts on vegetation structure. Increasing aridity is expected to favour extremely long legs in hot, open environments, where substrates are directly exposed to solar radiation. On the other hand, increasing

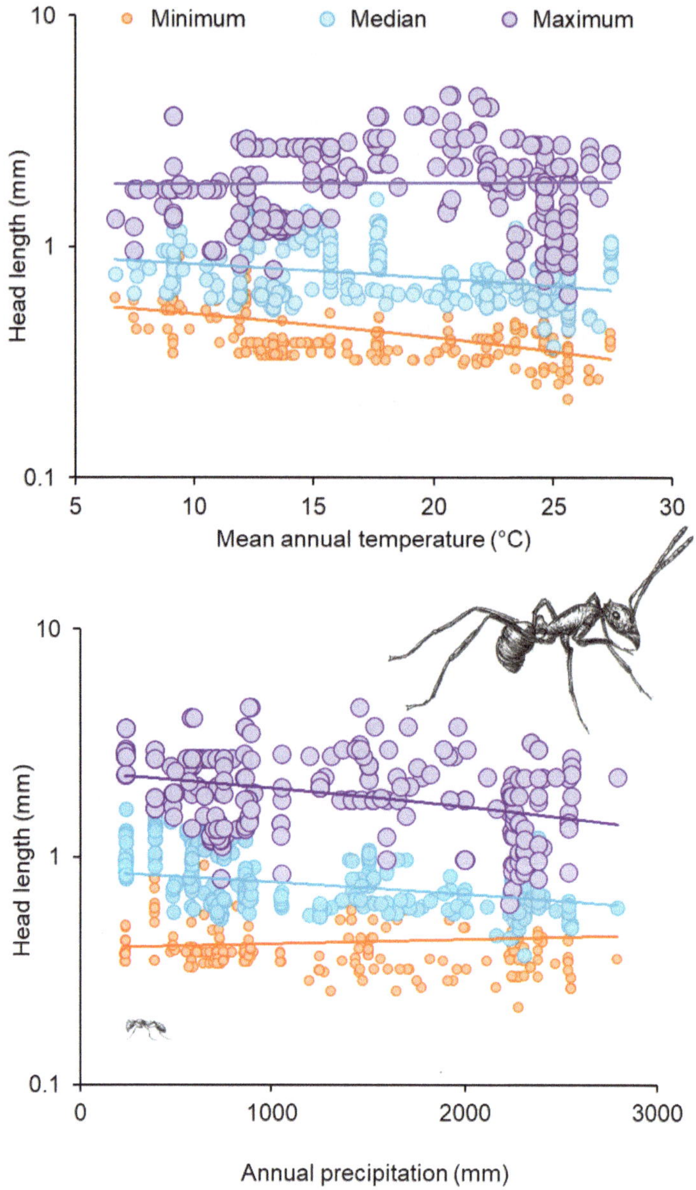

Fig. 3.2 Relationships between the body size of ants (as measured by head length) and mean annual temperature (**a**) and precipitation (**b**). Relationships are considered separately for ants in the bottom tenth percentile (minimum), middle percentile (median) and top tenth percentile (maximum) of the size distribution of local ant assemblages. (Redrawn from Gibb et al. 2018)

precipitation would be expected to result in a bimodal distribution of leg lengths in sheltered, complex forests.

3.2.3 Sensory and Trophic Morphology

Climate change affects habitat structure, which is critical in the evolution of sensory strategies. Eye size and position differ between complex (generally wetter) and simple (usually more arid) habitats, with ants in complex habitats having smaller and more laterally positioned eyes and shorter antennae than those in simple habitats (Gibb and Parr 2010; Oliveira et al. 2022). Visual acuity and long distance vision are likely to be less important in complex habitats where vision is obstructed by habitat (Graham and Philippides 2017). Further, ants commonly navigate using trail pheromones, but these evaporate more rapidly at higher temperatures, favouring visual navigation. In support, Ruano et al. (2000) showed that trail pheromones were used predominantly by species active at lower temperatures.

Climate change can also influence trophic structures, such as mandible and clypeus length. At a local scale, mandibles are longer in more complex ground vegetation (Gibb et al. 2015). Although Ibarra-Isassi et al. (2023) found no relationship between mandible length and precipitation at intermediate scales, a global study showed that specialist predators were most likely to respond negatively to increased habitat disturbance (which commonly results in more open habitats) (Gibb et al. 2018). Clypeus length increases with increasing aridity, suggesting liquid feeding may be more important in arid environments (four species in the Brazilian Caatinga, Oliveira et al. 2022). However, forest flooding has also been associated with longer clypeus length (Pringle et al. 2019), suggesting changes in trophic morphology in response to climate change may be highly variable.

3.2.4 Pigmentation and Structures of the Cuticle

Ant cuticles cover a broad spectrum of colours. While most species have pigmented cuticles that vary from yellows to reds, browns and blacks, some also display structural colours in blue, green and purple iridescence (e.g. *Calomyrmex purpureus*, Fig. 3.3a) or reflective setae in silver or gold (e.g. *Polyrhachis ammon*, Fig. 3.3b) (colour wheel, Parr et al. 2017). Research on the interaction of colour with climate in ants has largely focused on melanism (the variation in dark pigmentation caused by production of the pigment melanin), although iridescence and reflective setae have received some attention (e.g. Andersen 2007; Heterick and Shattuck 2011; Shi et al. 2015; Buxton et al. 2021; Gibb et al. 2024). Melanisation is costly for insects, delaying reproduction (Roff and Fairbairn 2013), and a number of nonexclusive mechanisms have been suggested to explain variation in melanism among species, populations and individuals. Of these, thermal melanism (e.g. Watt 1968; Kettlewell

Fig. 3.3 Structural colour and coloured hairs in melanised species in open habitats in south-eastern Australia: *Calomyrmex purpureus* in semiarid mallee (left, credit HG) and *Polyrhachis ammon* in mesic open woodland of the Sydney region (right, credit Ajay Narendra)

1973), desiccation resistance and UV-B protection (Gunn 1998) relate directly to climate change and have received increasing attention in recent years. Climate change will not only increase temperature and UV-B and alter precipitation, but it is predicted to increase the rate of extreme events, such as bushfires and cyclones, making microclimates hotter, drier and more UV-B exposed. Other possible drivers of melanism, including camouflage (Endler 1984), disease resistance (Wilson et al. 2001), aposematism (Turner et al. 1984) and sexual selection (Wiernasz 1989), might relate indirectly to climate change but have received little attention in ants.

The *thermal melanism hypothesis* states that melanism increases absorption of solar radiation under cool conditions, allowing darker individuals to heat more rapidly than lighter individuals and therefore increase their activity period (Trullas et al. 2007; Clusella-Trullas et al. 2008; Watt 1968). This relationship has been supported in laboratory experiments on insects (Harris et al. 2013; Zverev et al. 2018; Fedorka et al. 2013). Several community-level studies have reported that darker ant species dominate in cooler climates (e.g. elevational gradients, Bishop et al. 2016, globally, Klunk et al. 2022; Idec et al. 2023). Among species, darkness correlates positively with body size, suggesting that melanism may help to offset the thermal inertia of larger-bodied species (Clusella-Trullas et al. 2008; Schweiger and Beierkuhnlein 2016; Bishop et al. 2016; Gibb et al. 2023). Although cuticle melanism is a key determinant of heat gain, most laboratory experiments of thermal limits ignore this trait because direct solar radiation is excluded. Global warming would be expected to lead to increasing success of lighter-coloured species and individuals, and this may be linked with decreasing size or occur independently.

The *photoprotection hypothesis* considers that melanism provides protection against harmful UV-B radiation and that species in high UV-B environments should therefore be darker (Wang et al. 2023; Mosse and Lyakh 1994; Law et al. 2020). Recent studies in tropical rainforest canopies (Law et al. 2020) and along elevational gradients differing in UV-B levels (Gibb et al. 2023) suggest that darker ants dominate assemblages in high UV-B environments. In very high UV environments, photoprotection may become more important than thermal melanism (Bishop et al. 2016). UV-B exposure is expected to increase with climate change, favouring darker individuals. This selection pressure would therefore be opposite to that predicted due to global warming effects on colour through thermal melanism. However, it is possible for species to have melanised cuticles and reflect solar radiation through structural mechanisms such as iridescence and pilosity (e.g. Shi et al. 2015), discussed below.

The presence of dark species, including all those from the highly diverse *Polyrhachis* subgenus *Campomyrma*, in nocturnal ant assemblages (Gibb, pers. obs.; also observed for dung beetles, Hernández 2002) suggests that some benefits of melanism may be unrelated to daytime-relevant theories such as thermal melanism and photoprotection. The *desiccation-resistance hypothesis* posits that increased melanisation increases resistance to desiccation by decreasing cuticular permeability and therefore decreasing water loss (Kalmus 1941; King and Sinclair 2015; see Chap. 5). Laboratory experiments on other taxa support this hypothesis (King and Sinclair 2015; Parkash et al. 2008). In ants, melanism is positively correlated with vapour pressure deficit (tropical canopy ants, Law et al. 2020) and precipitation (in *Pheidole*, Klunk et al. 2022), although other studies have failed to find a link between precipitation and melanism (Idec et al. 2023; Gibb et al. 2024). Melanism is also thought to benefit species in warm, humid habitats by providing camouflage or increased protection against parasites and pathogens (Rensch 1929; Delhey 2019). As camouflage, melanism may also benefit species in burnt habitats (Karlsson et al. 2008). Climate change is expected to have regionally variable impacts on precipitation, so any impacts on ant melanism may be more localised and operate through species filtering, rather than selection.

Iridescence results from the interaction of light with nanostructures in thin films or diffraction gratings in the cuticle (Doucet and Meadows 2009). The thermal properties of iridescence in animals are poorly understood, with suggestions that it decreases solar radiation absorption (Kobelt and Linsenmair 1992), that it aids in warming (Miaoulis and Heilman 1998; Bosi et al. 2008) or that it has no effect (Schultz and Hadley 1987). In plants, structural colour is thought to enhance reflection of shorter wavelengths and therefore protect against UV damage (Vogelmann 1993; Vigneron and Simonis 2010). Cuticular iridescence is common in Australian ants but remains poorly studied. It occurs in a range of Australian genera, such as *Iridomyrmex*, *Rhytidoponera* and *Calomyrmex* (Heterick and Shattuck 2011; Andersen 2007). It is more prevalent in environments that are hot and arid and have high UV-B, although it is rare in the exceptionally diverse and highly thermophilic *Melophorus*.

Setae, produced in the cuticle of ants, provide another mechanism through which ants may insulate themselves from the surrounding microclimate. The prismatic silver hairs of *Cataglyphis bombycina* reflect near- and mid-infrared light and therefore passively cool ants (Shi et al. 2015; Willot et al. 2016). Similar silver or golden hairs are found in a range of ant taxa (e.g. *Polyrhachis*, *Camponotus*, *Dolichoderus*, *Myrmecia*), so this may be a widespread strategy that may be favoured under global warming. However, setae may also protect species from UV-B (Schwind et al. 2023), be hydrophobic (Suter et al. 2004)) and aid in desiccation tolerance (Wolpert 1962, Casey and Hegel 1981). Cross-taxon studies of ants have shown that pilosity is correlated with critical thermal maxima and mass-independent water loss (Buxton et al. 2021) and that robust, pilose ants dominate assemblages in more arid (Gibb et al. 2024) and open environments (Gibb et al. 2015). While predicted changes in precipitation and selection for desiccation tolerance are likely to be regionally variable, increases in aridity, temperature and UV-B are likely to favour pilose species, particularly those with reflective hairs.

Cuticle *microsculpturing* is indicative of thicker cuticles with a lower C:N ratio (Gibb et al. 2015; Buxton et al. 2021). The composition of ant cuticle hydrocarbons changes with developmental temperature (Duarte et al. 2019), and ants trade off investment in cuticle thickness with colony size (Peeters et al. 2017). Defensive structures, such as spines, are also associated with thicker cuticles (Gibb et al. 2023). Greater cuticle sculpturing is associated with tolerance of high temperatures and desiccation resistance (Buxton et al. 2021), while increased cuticle thickness is linked with decreased thermal conductance in insects (Galushko et al. 2005). Desiccation tolerance may be facilitated through shielding the protective barrier provided by the waxy cuticular hydrocarbons of sculptured ants from abrasion (Johnson 2000). It is also proposed to play a role in structural support, communication and insect-microbe interactions (Hellenbrand and Penick 2023). While the cuticle microstructures vary immensely and are evolutionarily labile, the functional differences among microsculpturing types, such as reticulate, striate, punctate and tuberous, are unclear (Hellenbrand and Penick 2023). Under climate change, and particularly in areas of increasing aridity, we might expect species with greater cuticle microsculpturing to be favoured due to increased desiccation resistance.

3.3 Colony-Level Morphology

3.3.1 Polymorphism

Ant colonies vary in the distribution of worker sizes, with some species being very uniform in size (monomorphic), while others vary continuously (polymorphic) or discontinuously (e.g. dimorphic). Worker ants of different sizes exhibit differential efficiencies when performing various tasks within a colony (Hölldobler and Wilson 1990). While genetics strongly influences worker size diversity (Hughes et al. 2003; Huang et al. 2013), a significant portion of size variation within social insect colonies arises from the plastic responses of developing brood to environmental fluctuations (Wheeler 1986; Smith et al. 2008) and external microclimatic conditions (Wheeler 1986; Purcell et al. 2016).

At the colony level, temperature affects caste-associated morphological traits and the allocation to castes, including reproductive females (Trible and Kronauer 2017, 2021a, b). These traits include not only those associated with polymorphism in workers, e.g. larger head and mandibles in soldiers, but also the development of reproductive structures, wings and eyes in reproductive individuals. The role of nutrition, genetics and temperature in the development of reproductive females varies among species, but queens are often only produced at the lower temperatures that allow them to reach a larger size before pupation (Brian

1951, 1979; Du et al. 2007; Penick and Liebig 2012) and require winter chilling in some species (Schwander et al. 2008). Because hymenopterans are haplodiploid, sex is genetically determined, but even the number of males produced by queens decreases with increasing temperature (Aron et al. 1994; Du et al. 2007). This link between temperature and the production of these critical reproductive castes suggests that global warming may impose selection for a switch to the production of reproductives at a smaller minimum size. As discussed above, the implications of smaller body size include reduced dispersal distance and reduced starvation resistance, which may negatively impact the capacity of new queens to build colonies.

Polymorphism in worker size results in individuals differing in thermal and desiccation tolerance (Verble-Pearson et al. 2015), which may increase the activity period of a species across a broader range of temperatures (Cerdá et al. 1998). In laboratory experiments, worker size variation increases with temperature (Molet et al. 2017). Field data show that polymorphism is favoured in hot, arid habitats, both at global (e.g. savannas and deserts, La Richelière et al. 2022) and local (e.g. burnt areas, Arnan et al. 2013, e.g. bare ground and tree trunks, Gibb et al. 2015) scales, and that it is favoured in more seasonal environments (Arnan et al. 2014). In addition to their higher temperatures and increased aridity, these locations are also relatively resource poor, which may also favour polymorphism through increased resource harvesting by size-matching of workers and food resources (Wills et al. 2018; La Richelière et al. 2022). Many parts of the world are likely to become hotter and more arid under climate change, possibly favouring polymorphic species.

3.3.2 Colony Size

Selection on colony size is likely to act at least partially through a trade-off in resource allocation between worker size and worker number, whereby colonies with smaller workers tend to be larger (Bourke 1999; Shik 2008; Gibb et al. 2023). Larger colony size is also linked with greater queen/worker size ratios and worker polymorphism (Bourke 1999; Anderson and McShea 2001; Marting et al. 2018) and smaller brain size (Kamhi et al. 2016; Riveros et al. 2012), suggesting that these traits covary as an ecological strategy (Gibb et al. 2023). The proposition that colony and worker size are negatively correlated is supported by studies within species that show that colony size is larger where temperatures are highest (e.g. simulation in *Solenopsis invicta*, Asano and Cassill 2012, army ants, Purcell et al. 2016). However, Kaspari and Vargo (1995) proposed that large colony size provides a buffer against seasonality and starvation and showed that temperate environments support larger ant colonies than tropical environments. At a local scale, this is supported by evidence that burnt areas, which are drier and more open and presumably lack food resources, support larger ant colonies than unburnt areas (Arnan et al. 2013). Other studies have not supported a relationship between colony size and latitude or altitude in ants or termites (Porter and Hawkins 2001; Geraghty et al. 2007;

Ferguson-Gow et al. 2014; Bernadou et al. 2016). Parr and Bishop (2022) suggested that larger colonies might be more successful under global warming as the impact of losing workers foraging closer to their thermal limits would be less than in small colonies.

3.4 Extended Phenotype

The architecture and size of ant nests are influenced by a combination of local adaptation, plastic responses to environmental factors such as temperature and the social structure of ant colonies (Sankovitz and Purcell 2021; Mikheyev and Tschinkel 2004; Tschinkel 2015). Nest temperatures are critical in determining developmental temperatures and therefore body size. Social insects regulate developmental temperatures through their ability to regulate the environment of the nest by altering shape and orientation to optimise sun exposure and relocate brood within the nest (Jones and Oldroyd 2006; Kadochová and Frouz 2014; Stukalyuk et al. 2020; Villalta et al. 2020; Curbani et al. 2021). Nest temperatures influence the thermal tolerance of workers (Villalta et al. 2020). Mound size is correlated with the number of inhabitants (Tschinkel 1987; Chen and Robinson 2013; Mikheyev and Tschinkel 2004). Surface mounds of the thatch nests of *Formica polyctena* are larger in cooler environments to maximise heat generation through microbial decomposition and worker metabolism (Chen and Robinson 2014; Frouz 2000). Under warmer conditions, nest mounds of cold-climate ants may therefore be smaller. As soil surface temperatures warm, more species occupy subterranean, rather than superficial, nests (e.g. *Acromyrmex* spp., Bollazzi et al. 2008). Further, laboratory experiments show that ants exposed to higher temperatures dig chambers deeper in the soil and close-off upper chambers (e.g. *Lasius niger*, García Ibarra et al. 2024). Warmer climates are therefore likely to lead to reduced nest surface structure and increasingly deep nests.

Fig. 3.4 Flood levy surrounding the entrance of nests of mulga spiny ants (*Polyrhachis macropus*) (left) and mulga spiny ant worker carrying a mulga phyllode to stabilise its levy (right). (Photo credit: Alan Andersen)

In some regions, climate change will lead to increased rates of flooding, which may impact the survival and behaviour of ant colonies. Some soil-nesting ant species alter the morphology (or architecture) of their nests in response to flooding events, with unrelated species in flood-prone environments building structures such as flood levies or raised funnels in response to rainfall (LeBrun et al. 2011). Mulga spiny ants (*Polyrhachis macropus* gp.) provide a spectacular example of this, building levies surrounding nest entrances and covering them with mulga phyllodes to enhance stability (pers. comm. Alan Andersen, Fig. 3.4).

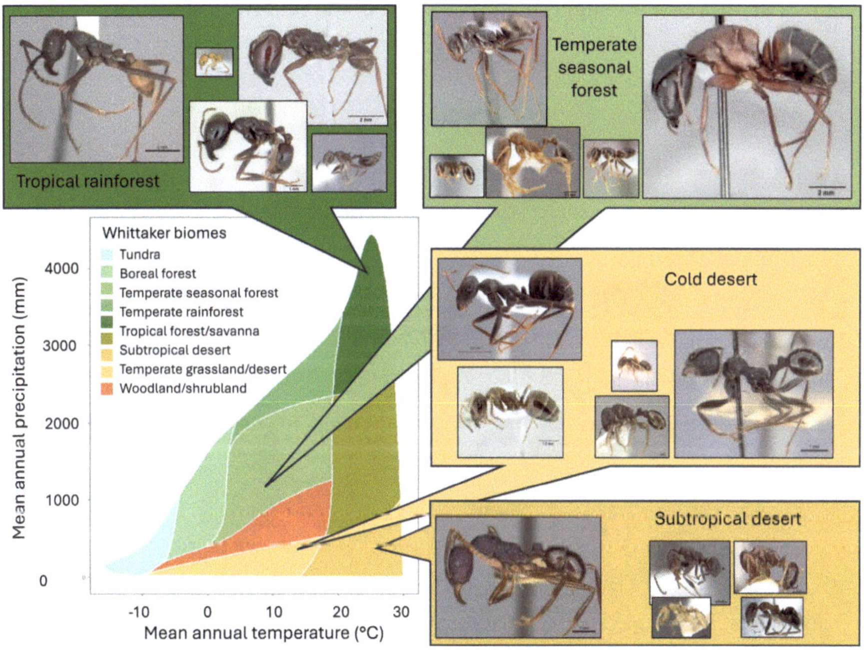

Fig. 3.5 Images of specimens of the most abundant named species in studies from different Whittaker biomes. Relative sizes are representative of actual sizes. Climate is associated with size, colour and a range of other morphological features of ants, but a large proportion of global morphological variation in ants is found in local assemblages, reflecting high variation in local microhabitats and microclimates. Abundances are of pitfall-trapped ants from tropical rainforest in the Brazilian Amazon (clockwise from top left: *Eciton rapax*, *Pheidole fracticeps*, *Labidus coecus*, *Crematogaster tenuicula*, *Labidus praedator*; Oliveira et al. 2009), temperate forest in the USA (*Formica subelongata*, *Camponotus vicinus*, *Tapinoma sessile*, *Aphaenogaster occidentalis*, *Temnothorax nevadensis*; Ratchford et al. 2005), cold desert in Mongolia (*Cataglyphis aenescens*, *Plagiolepis manczshurica*, *Messor aciculatus*, *Tetramorium annectens* [image is similar to *T. sulcinode*], *Proformica mongolica*; Pfeiffer et al. 2003) and subtropical desert in Australia (*Rhytidoponera purpurea*, *Iridomyrmex rufoniger*, *Melophorus fieldi*, *Iridomyrmex chasei*, *Monomorium disetigerum*; Gibb et al. 2019). (Images are from AntWeb, Ant Wiki and iNaturalist)

3.5 Conclusions

Climatic factors influence the morphology of ant castes, colonies and nests in a range of ways, both directly and indirectly, and at small and large scales. Sometimes different elements of climate change select traits in opposing directions. For example, increasing temperature will select for lighter colours, while increasing UV will select for increased melanism, resulting in a final trait value that is a trade-off between these climatic drivers. Similarly, increased colony size might be a consequence of smaller worker body size (in hot, wet environments) or a response to less consistent resource availability, such as in seasonal environments at higher latitudes, or in arid areas with irregular rainfall, where large colonies are often polymorphic. Local assemblages may include three quarters of the variation in traits observed globally (Kaspari et al. 2015) (e.g. Fig. 3.5), probably because microclimates vary enormously at small spatial and temporal scales (Pinker 1980; Kaspari et al. 2015; Aalto et al. 2022). This suggests that changes in the relative abundance of species

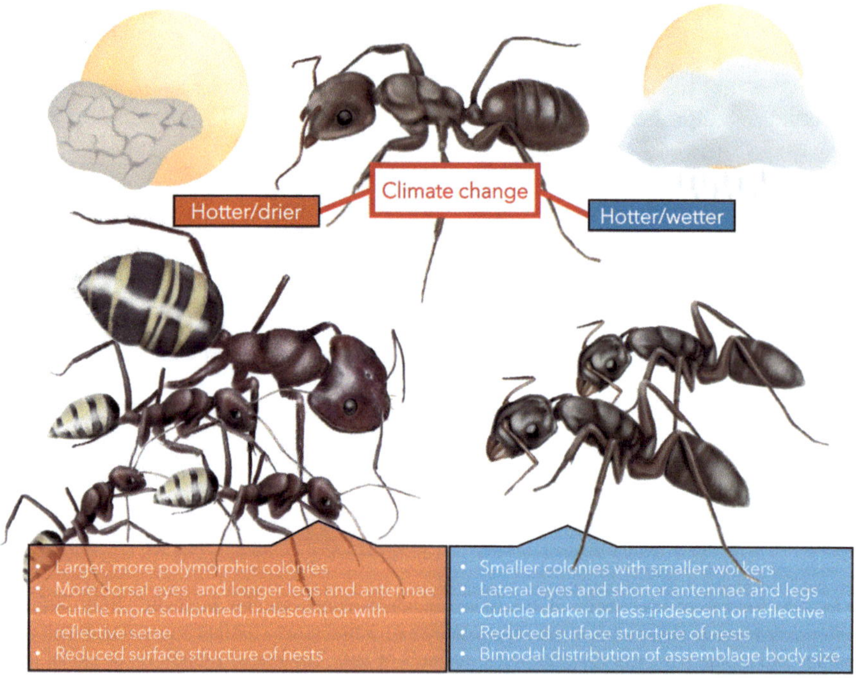

Fig. 3.6 Predictions of change in worker ant morphology in response to climate change. Most parts of the globe are expected to warm, but some may become wetter, while others become drier. Impacts on ant communities are likely to be primarily through species filtering, but selection will also influence species responses. Key species traits expected to shift due to climate change are shown, with illustrations embodying traits typical of increasingly hot and arid and increasingly hot and wet climates. The descriptions refer to morphologies expected to be predominant in assemblages under future climates. (Images by Hannah Riskas)

within local assemblages may dominate early responses to climate change, as indicated by studies that account for relative abundance by considering assemblage-weighted trait means (e.g. Bishop et al. 2016; Law et al. 2020; Gibb et al. 2024). However, dispersal, assemblage filtering and selection will be increasingly important in shaping future ant assemblages. Climate change will increase temperature and UV across the globe but have inconsistent impacts on precipitation, leading to different outcomes for ant morphology (Fig. 3.6).

Globally, databases are increasingly important in our ability to understand relationships between morphological traits and climate (Parr et al. 2017; Gibb et al. 2017, e.g. Bertelsmeier et al. 2013; Guénard et al. 2017). Fine-scale georeferencing and microhabitat information are important in better linking ants with microclimates, and coordination among databases is crucial in avoiding duplication of effort. New studies have commenced large-scale whole-body microCT scanning, providing unprecedented data on the internal morphological structures of ants (Evan Economo, pers. comm.). Given that species traits exist as part of an ecological strategy, research that addresses the response of suites of (ideally continuous) interrelated traits, whether they are morphological, physiological, life history or behavioural, may better address questions around ant responses to climate change (Gibb et al. 2023). While there is some debate about the best traits to measure (Gonçalves-Souza et al. 2023), the covariance of traits likely means that more easily measured "indicator" traits, such as many of the morphological traits considered here, may be adequate to describe ecological strategy and its response to global change.

References

Aalto J, Tyystjärvi V, Niittynen P, Kemppinen J, Rissanen T, Gregow H, Luoto M (2022) Microclimate temperature variations from boreal forests to the tundra. Agric For Meteorol 323:109037. https://doi.org/10.1016/j.agrformet.2022.109037

Andersen AN (2007) Ant diversity in arid Australia: a systematic overview. Mem Am Entomol Soc 80:1–20

Anderson C, McShea DW (2001) Individual versus social complexity, with particular reference to ant colonies. Biol Rev 76:211–237. https://doi.org/10.1017/S1464793101005656

Andrew NR, Hart RA, Jung M-P, Hemmings Z, Terblanche JS (2013) Can temperate insects take the heat? A case study of the physiological and behavioural responses in a common ant, Iridomyrmex purpureus (Formicidae), with potential climate change. J Insect Physiol 59:870–880. https://doi.org/10.1016/j.jinsphys.2013.06.003

Angilletta MJ, Dunham AE (2003) The temperature–size rule in ectotherms: simple evolutionary explanations may not be general. Am Nat 162:332–342. https://doi.org/10.1086/377187

Araujo NDS, Perez R, Willot Q, Defrance M, Aron S (2023) Facing lethal temperatures: heat-shock response in desert and temperate ants. Ecol Evol 13:e10438. https://doi.org/10.1002/ece3.10438

Arnan X, Blüthgen N (2015) Using ecophysiological traits to predict climatic and activity niches: lethal temperature and water loss in Mediterranean ants. Glob Ecol Biogeogr 24:1454–1464. https://doi.org/10.1111/geb.12363

Arnan X, Cerdá X, Rodrigo A, Retana J (2013) Response of ant functional composition to fire. Ecography 36:1182–1192. https://doi.org/10.1111/j.1600-0587.2013.00155.x

Arnan X, Cerdá X, Retana J (2014) Ant functional responses along environmental gradients. J Anim Ecol 83:1398–1408. https://doi.org/10.1111/1365-2656.12227

Aron S, Passera L, Keller L (1994) Queen-worker conflict over sex ratio: a comparison of primary and secondary sex ratios in the Argentine ant, Iridomyrmex humilis. J Evol Biol 7:403–418. https://doi.org/10.1046/j.1420-9101.1994.7040403.x

Asano E, Cassill DL (2012) Modeling temperature-mediated fluctuation in colony size in the fire ant, Solenopsis invicta. J Theor Biol 305:70–77. https://doi.org/10.1016/j.jtbi.2012.03.011

Atkinson D (1996) Ectotherm life history responses to developmental temperature. In: Johnston IA, Bennett AF (eds) Animals and temperature: phenotypic and evolutionary adaptation. Cambridge University Press, Cambridge, pp 183–204

Bernadou A, Römermann C, Gratiashvili N, Heinze J (2016) Body size but not colony size increases with altitude in the holarctic ant, Leptothorax acervorum. Ecol Entomol 41:733–736. https://doi.org/10.1111/een.12338

Bertelsmeier C, Luque GM, Confais A, Courchamp F (2013) Antprofiler: a database of ecological characteristics of ants (Hymenoptera: Formicidae). Myrmecol News 18:73–76. https://doi.org/10.25849/myrmecol.news_018:073

Bishop TR, Robertson MP, Gibb H, van Rensburg BJ, Braschler B, Chown SL, Foord SH, Munyai TC, Okey I, Tshivhandekano PG (2016) Ant assemblages have darker and larger members in cold environments. Glob Ecol Biogeogr 25:1489–1499. https://doi.org/10.1111/geb.12516

Bollazzi M, Kronenbitter J, Roces F (2008) Soil temperature, digging behaviour, and the adaptive value of nest depth in South American species of Acromyrmex leaf-cutting ants. Oecologia 158:165–175. https://doi.org/10.1007/s00442-008-1113-z

Bosi SG, Hayes J, Large MC, Poladian L (2008) Color, iridescence, and thermoregulation in Lepidoptera. Appl Opt 47:5235–5241. https://doi.org/10.1364/AO.47.005235

Bourke AFG (1999) Colony size, social complexity and reproductive conflict in social insects. J Evol Biol 12:245–257. https://doi.org/10.1046/j.1420-9101.1999.00028.x

Brassard F, Francoeur A, Lessard JP (2020) Temperature drives caste-specific morphological clines in ants. J Anim Ecol 89:2517–2530. https://doi.org/10.1111/1365-2656.13330

Brian MV (1951) Caste determination in a Myrmicine ant. Experientia 7:182–183. https://doi.org/10.1007/BF02148904

Brian M (1979) Caste differentiation and division of labor. In: Hermann HR (ed) Social insects, vol 1. Academic, New York, pp 121–222

Bruna EM, Izzo TJ, Inouye BD, Uriarte M, Vasconcelos HL (2011) Asymmetric dispersal and colonization success of Amazonian plant-ant queens. PLoS One 6:e22937. https://doi.org/10.1371/journal.pone.0022937

Bujan J, Yanoviak SP, Kaspari M (2016) Desiccation resistance in tropical insects: causes and mechanisms underlying variability in a Panama ant community. Ecol Evol 6:6282–6291. https://doi.org/10.1002/ece3.2355

Bujan J, Ollier S, Villalta I, Devers S, Cerdá X, Amor F, Dahbi A, Bertelsmeier C, Boulay R (2022) Can thermoregulatory traits and evolutionary history predict climatic niches of thermal specialists? Divers Distrib 28:1081–1092. https://doi.org/10.1111/ddi.13511

Buxton JT, Robert KA, Marshall AT, Dutka TL, Gibb H (2021) A cross-species test of the function of cuticular traits in ants (Hymenoptera: Formicidae). Myrmecol News 31:31–46. https://doi.org/10.25849/myrmecol.news_031:031

Calabi P, Porter SD (1989) Worker longevity in the fire ant, Solenopsis invicta: ergonomic considerations of correlations between temperature, size and metabolic rates. J Insect Physiol 35:643–650

Casey TM, Hegel JR (1981) Caterpillar setae: insulation for an ectotherm. Science 214:1131–1133. https://doi.org/10.1126/science.214.4525.1131

Cerdá X, Retana J (2000) Alternative strategies by thermophilic ants to cope with extreme heat: individual versus colony level traits. Oikos 89:155–163. https://doi.org/10.1034/j.1600-0706.2000.890117.x

Cerdá X, Retana J, Manzaneda A (1998) The role of competition by dominants and temperature in the foraging of subordinate species in Mediterranean ant communities. Oecologia 117(3):404–412. https://doi.org/10.1007/s004420050674

Chen Y-H, Robinson EJH (2013) A comparison of mark–release–recapture methods for estimating colony size in the wood ant Formica lugubris. Insect Soc 60(3):351–359. https://doi.org/10.1007/s00040-013-0300-z

Chen Y-H, Robinson EJ (2014) The relationship between canopy cover and colony size of the wood ant Formica lugubris: implications for the thermal effects on a keystone ant species. PLoS One 9:e116113. https://doi.org/10.1371/journal.pone.0116113

Chown SL, Gaston KJ (2010) Body size variation in insects: a macroecological perspective. Biol Rev 85:139–169. https://doi.org/10.1111/j.1469-185X.2009.00097.x

Chown S, Nicolson S (2004) Insect physiological ecology: mechanisms and patterns. Oxford University Press, Oxford

Clusella-Trullas S, Terblanche JS, Blackburn TM, Chown SL (2008) Testing the thermal melanism hypothesis: a macrophysiological approach. Funct Ecol 22:232–238. https://doi.org/10.1111/j.1365-2435.2007.01377.x

Curbani F, Zocca C, Ferreira RB, Waichert C, Sobrinho TG, Srbek-Araujo AC (2021) Litter surface temperature: a driving factor affecting foraging activity in Dinoponera lucida (Hymenoptera: Formicidae). Sociobiology 68:e6030. https://doi.org/10.13102/sociobiology.v68i1.6030

Cushman JH, Lawton JH, Manly BFJ (1993) Latitudinal patterns in European ant assemblages: variation in species richness and body size. Oecologia 95:30–37. https://doi.org/10.1007/BF00649503

Delhey K (2019) A review of Gloger's rule, an ecogeographical rule of colour: definitions, interpretations and evidence. Biol Rev 94:1294–1316. https://doi.org/10.1111/brv.12503

Doucet SM, Meadows MG (2009) Iridescence: a functional perspective. J R Soc Interface 6:S115–S132. https://doi.org/10.1098/rsif.2008.0395.focus

Du Y, Schrempf A, Heinze J (2007) Environmental determination of the male morph in the ant Cardiocondyla obscurior (Hymenoptera: Formicidae). Eur J Entomol 104:243–246. https://doi.org/10.14411/eje.2007.038

Duarte BF, Michelutti KB, Antonialli-Junior WF, Cardoso CAL (2019) Effect of temperature on survival and cuticular composition of three different ant species. J Therm Biol 80:178–189. https://doi.org/10.1016/j.jtherbio.2019.02.005

Endler JA (1984) Progressive background in moths, and a quantitative measure of crypsis. Biol J Linn Soc 22:187–231. https://doi.org/10.1111/j.1095-8312.1984.tb01677.x

Fedorka KM, Lee V, Winterhalter WE (2013) Thermal environment shapes cuticle melanism and melanin-based immunity in the ground cricket Allonemobius socius. Evol Ecol 27:521–531. https://doi.org/10.1007/s10682-012-9620-0

Ferguson-Gow H, Sumner S, Bourke AF, Jones KE (2014) Colony size predicts division of labour in attine ants. Proc R Soc B Biol Sci 281:20141411. https://doi.org/10.1098/rspb.2014.1411

Folgarait PJ (1998) Ant biodiversity and its relationship to ecosystem functioning: a review. Biodivers Conserv 7:1221–1244. https://doi.org/10.1023/A:1008891901953

Frouz J (2000) The effect of nest moisture on daily temperature regime in the nests of Formica polyctena wood ants. Insect Soc 47:229–235. https://doi.org/10.1007/PL00001708

Galushko D, Ermakov N, Karpovski M, Palevski A, Ishay JS, Bergman DJ (2005) Electrical, thermoelectric and thermophysical properties of hornet cuticle. Semicond Sci Technol 20:286–289. https://doi.org/10.1088/0268-1242/20/3/005

García Ibarra F, Jouquet P, Bottinelli N, Bultelle A, Monnin T (2024) Experimental evidence that increased surface temperature affects bioturbation by ants. J Anim Ecol 93:319–332. https://doi.org/10.1111/1365-2656.14040

Geraghty MJ, Dunn RR, Sanders NJ (2007) Body size, colony size, and range size in ants (Hymenoptera: Formicidae): are patterns along elevational and latitudinal gradients consistent with Bergmann's rule? Myrmecol News 10:51–58

Gibb H, Parr CL (2010) How does habitat complexity affect ant foraging success? A test using functional measures on three continents. Oecologia 164:1061–1073. https://doi.org/10.1007/s00442-010-1703-4

Gibb H, Parr CL (2013) Does structural complexity determine the morphology of assemblages? An experimental test on three continents. PLoS One 8:e69879. https://doi.org/10.1371/journal.pone.0064005

Gibb H, Stoklosa J, Warton DI et al (2015) Does morphology predict trophic position and habitat use of ant species and assemblages? Oecologia 177:519–531. https://doi.org/10.1007/s00442-014-3101-9

Gibb H, Dunn RR, Sanders NJ, Grossman BF, Photakis M, Abril S, Agosti D, Andersen AN, Angulo E, Armbrecht I, Arnan X, Baccaro FB, Bishop TR, Boulay R, Brühl C, Castracani C, Cerda X, Del Toro I, Delsinne T, Diaz M, Donoso DA, Ellison AM, Enriquez ML, Fayle TM, Feener DH Jr, Fisher BL, Fisher RN, Fitzpatrick MC, Gómez C, Gotelli NJ, Gove A, Grasso DA, Groc S, Guenard B, Gunawardene N, Heterick B, Hoffmann B, Janda M, Jenkins C, Kaspari M, Klimes P, Lach L, Laeger T, Lattke J, Leponce M, Lessard J-P, Longino J, Lucky A, Luke SH, Majer J, McGlynn TP, Menke S, Mezger D, Mori A, Moses J, Munyai TC, Pacheco R, Paknia O, Pearce-Duvet J, Pfeiffer M, Philpott SM, Resasco J, Retana J, Silva RR, Sorger MD, Souza J, Suarez A, Tista M, Vasconcelos HL, Vonshak M, Weiser MD, Yates M, Parr CL (2017) A global database of ant species abundances. Ecology 98:883–884. https://doi.org/10.1002/ecy.1682

Gibb H, Sanders NJ, Dunn RR, Arnan X, Vasconcelos HL, Donoso DA, Andersen AN, Silva RR, Bishop TR, Gomez C, Grossman BF, Yusah KM, Luke SH, Pacheco R, Pearce-Duvet J, Retana J, Tista M, Parr CL (2018) Habitat disturbance selects against both small and large species across varying climates. Ecography 41:1184–1193. https://doi.org/10.1111/ecog.03244

Gibb H, Grossman BF, Dickman CR, Decker O, Wardle GM (2019) Long-term responses of desert ant assemblages to climate. J Anim Ecol 88:1549–1563. https://doi.org/10.1111/1365-2656.13052

Gibb H, Bishop TR, Leahy L, Parr CL, Lessard J-P, Sanders NJ, Shik JZ, Ibarra-Isassi J, Narendra A, Dunn RR, Wright IJ (2023) Ecological strategies of (pl)ants: towards a world-wide worker economic spectrum for ants. Funct Ecol 37:13–25. https://doi.org/10.1111/1365-2435.14135

Gibb H, Contos P, Photakis M, Okey I, Dunn RR, Sanders NJ, Jones MM (2024) Morphological strategies in ant communities along Elevational gradients in three mountain ranges. Diversity 16:48. https://doi.org/10.3390/d16010048

Gonçalves-Souza T, Chaves LS, Boldorini GX, Ferreira N, Gusmão RA, Perônico PB, Sanders NJ, Teresa FB (2023) Bringing light onto the Raunkiæran shortfall: a comprehensive review of traits used in functional animal ecology. Ecol Evol 13:e10016. https://doi.org/10.1002/ece3.10016

Graham P, Philippides A (2017) Vision for navigation: what can we learn from ants? Arthropod Struct Dev 46:718–722. https://doi.org/10.1016/j.asd.2017.07.001

Guénard B, Weiser MD, Gomez K, Narula N, Economo EP (2017) The Global Ant Biodiversity Informatics (GABI) database: synthesizing data on the geographic distribution of ant species (Hymenoptera: Formicidae). Myrmecol News 24:83–89. https://doi.org/10.25849/myrmecol.news_024:083

Gunn A (1998) The determination of larval phase coloration in the African armyworm, Spodoptera exempta and its consequences for thermoregulation and protection from UV light. Entomol Exp Appl 86:125–133. https://doi.org/10.1046/j.1570-7458.1998.00273.x

Hadley NF (1994) Water relations of terrestrial arthropods. Cambridge University Press

Harris RM, McQuillan P, Hughes L (2013) Experimental manipulation of melanism demonstrates the plasticity of preferred temperature in an agricultural pest (Phaulacridium vittatum). PLoS One 8:e80243. https://doi.org/10.1371/journal.pone.0080243

Heinrich B, Bartholomew GA (1979) Roles of endothermy and size in inter- and intraspecific competition for elephant dung in an African dung beetle, Scarabaeus laevistriatus. Physiol Zool 52:484–496

Heinze J, Oberstadt B (1999) Worker age, size and social status in queenless colonies of the ant Leptothorax gredleri. Anim Behav 58:751–759. https://doi.org/10.1006/anbe.1999.1188

Heinze J, Foitzik S, Fischer B, Wanke T, Kipyatkov VE (2003) The significance of latitudinal variation in body size in a holarctic ant, Leptothorax acervorum. Ecography 26:349–355. https://doi.org/10.1034/j.1600-0587.2003.03478.x

Hellenbrand JP, Penick CA (2023) Ant cuticle microsculpturing: diversity, classification, and evolution. Myrmecol News 33. https://doi.org/10.25849/myrmecol.news_033:123

Helms JA (2018) The flight ecology of ants (Hymenoptera: Formicidae). Myrmecol News 26:19–30. https://doi.org/10.25849/myrmecol.news_026:019

Helms J, Kaspari M (2015) Reproduction-dispersal tradeoffs in ant queens. Insect Soc 62:171–181. https://doi.org/10.1007/s00040-015-0391-9

Hernández MIM (2002) The night and day of dung beetles (Coleoptera, Scarabaeidae) in the Serra do Japi, Brazil: elytra colour related to daily activity. Rev Bras Entomol 46:597–600. https://doi.org/10.1590/S0085-56262002000400015

Heterick BE, Shattuck S (2011) Revision of the ant genus Iridomyrmex (Hymenoptera: Formicidae). Zootaxa 2845:1–174. https://doi.org/10.11646/zootaxa.2845.1.1

Hölldobler B, Wilson EO (1990) The ants. Harvard University Press, Cambridge, MA

Holter P, Scholtz C (2005) Are ball-rolling (Scarabaeini, Gymnopleurini, Sisyphini) and tunnelling scarabaeine dung beetles equally choosy about the size of ingested dung particles? Ecol Entomol 30:700–705. https://doi.org/10.1111/j.0307-6946.2005.00746.x

Hood WG, Tschinkel WR (1990) Desiccation resistance in arboreal and terrestrial ants. Physiol Entomol 15:23–35. https://doi.org/10.1111/j.1365-3032.1990.tb00489.x

Huang MH, Wheeler DE, Fjerdingstad EJ (2013) Mating system evolution and worker caste diversity in Pheidole ants. Mol Ecol 22:1998–2010. https://doi.org/10.1111/mec.12218

Hughes WO, Sumner S, Van Borm S, Boomsma JJ (2003) Worker caste polymorphism has a genetic basis in Acromyrmex leaf-cutting ants. Proc Natl Acad Sci USA 100:9394–9397. https://doi.org/10.1073/pnas.1633701100

Hurlbert AH, Ballantyne F, Powell S (2008) Shaking a leg and hot to trot: the effects of body size and temperature on running speed in ants. Ecol Entomol 33:144–154. https://doi.org/10.1111/j.1365-2311.2007.00962.x

Ibarra-Isassi J, Handa IT, Lessard JP (2023) Community-wide trait adaptation, but not plasticity, explains ant community structure in extreme environments. Funct Ecol 37:139–149

Idec JH, Bishop TR, Fisher BL (2023) Using computer vision to understand the global biogeography of ant color. Ecography 2023:e06279. https://doi.org/10.1111/ecog.06279

Johnson RA (2000) Water loss in desert ants: caste variation and the effect of cuticle abrasion. Physiol Entomol 25:48–53. https://doi.org/10.1046/j.1365-3032.2000.00170.x

Jones JC, Oldroyd BP (2006) Nest thermoregulation in social insects. Adv Insect Physiol 33:153–191. https://doi.org/10.1016/S0065-2806(06)33003-2

Kadochová Š, Frouz J (2014) Red wood ants Formica polyctena switch off active thermoregulation of the nest in autumn. Insect Soc 61:297–306. https://doi.org/10.1007/s00040-014-0356-4

Kalmus H (1941) The resistance to desiccation of Drosophila mutants affecting body colour. Proc R Soc B 130:185–201. https://doi.org/10.1098/rspb.1941.0011

Kamhi JF, Gronenberg W, Robson SK, Traniello JF (2016) Social complexity influences brain investment and neural operation costs in ants. Proc R Soc B Biol Sci 283:20161949. https://doi.org/10.1098/rspb.2016.1949

Karlsson M, Caesar S, Ahnesjö J, Forsman A (2008) Dynamics of colour polymorphism in a changing environment: fire melanism and then what? Oecologia 154:715–724. https://doi.org/10.1007/s00442-007-0876-y

Kaspari M (1993) Body-size and microclimate use in Neotropical granivorous ants. Oecologia 96:500–507. https://doi.org/10.1007/BF00320507

Kaspari M, Vargo EL (1995) Colony size as a buffer against seasonality – Bergmann's rule in social insects. Am Nat 145:610–632. https://doi.org/10.1086/285758

Kaspari M, Weiser M (1999) The size–grain hypothesis and interspecific scaling in ants. Funct Ecol 13:530–538. https://doi.org/10.1046/j.1365-2435.1999.00343.x

Kaspari M, Weiser MD (2000) Ant activity along moisture gradients in a Neotropical forest. Biotropica 32:703–711. https://doi.org/10.1111/j.1744-7429.2000.tb00518.x

Kaspari M, Alonso L, O'Donnell S (2000) Three energy variables predict ant abundance at a geographical scale. Proc R Soc B Biol Sci 267:485–489. https://doi.org/10.1098/rspb.2000.1026

Kaspari M, Clay NA, Lucas J, Yanoviak SP, Kay A (2015) Thermal adaptation generates a diversity of thermal limits in a rainforest ant community. Glob Chang Biol 21:1092–1102. https://doi.org/10.1111/gcb.12750

Kettlewell B (1973) Evolution of melanism: the study of a recurring necessity. Clarendon Press

King KJ, Sinclair BJ (2015) Water loss in tree weta (Hemideina): adaptation to the montane environment and a test of the melanisation–desiccation resistance hypothesis. J Exp Biol 218:1995–2004. https://doi.org/10.1242/jeb.118711

Kipyatkov VE, Lopatina EB (2015) Comparative study of thermal reaction norms for development in ants. Entomol Sci 18:174–192. https://doi.org/10.1111/ens.12098

Klunk CL, Fratoni RO, Rivadeniera CD, Schaedler LM, Perez DM (2022) Climate and body size have differential roles on melanism evolution across workers in a worldwide ant genus. Oecologia 199:579–587. https://doi.org/10.1007/s00442-022-05211-x

Kobelt F, Linsenmair K (1992) Adaptations of the reed frog Hyperolius viridiflavus (Amphibia: Anura: Hyperoliidae) to its arid environment: VI. The iridophores in the skin as radiation reflectors. J Comp Physiol B 162:314–326. https://doi.org/10.1007/BF00260758

La Richelière F, Muñoz G, Guénard B, Dunn RR, Economo EP, Powell S, Sanders NJ, Weiser MD, Abouheif E, Lessard J-P (2022) Warm and arid regions of the world are hotspots of superorganism complexity. Proc R Soc B 289:20211899. https://doi.org/10.1098/rspb.2021.1899

Law SJ, Bishop TR, Eggleton P, Griffiths H, Ashton L, Parr C (2020) Darker ants dominate the canopy: testing macroecological hypotheses for patterns in colour along a microclimatic gradient. J Anim Ecol 89:347–359. https://doi.org/10.1111/1365-2656.13110

Lebrun E, Moffett M, Holway D (2011) Convergent evolution of levee building behavior among distantly related ant species in a floodplain ant assemblage. Insect Soc 58:263–269. https://doi.org/10.1007/s00040-011-0151-4

Leong CM, Hui TY, Guénard B (2023) The role of body mass in limiting post heat-coma recovery ability in terrestrial ectotherms. Ecol Evol 13:e10218. https://doi.org/10.1002/ece3.10218

Lighton JRB, Quinlan MC, Feener DHJ (1994) Is bigger better? Water balance in the polymorphic desert harvester ant Messor pergandei. Physiol Entomol 19:325–334. https://doi.org/10.1111/j.1365-3032.1994.tb01059.x

Marting PR, Kallman NM, Wcislo WT, Pratt SC (2018) Ant-plant sociometry in the Azteca-Cecropia mutualism. Sci Rep 8:17968. https://doi.org/10.1038/s41598-018-36399-9

Merckx T, Souffreau C, Kaiser A, Baardsen LF, Backeljau T, Bonte D, Brans KI, Cours M, Dahirel M, DeBortoli N (2018) Body-size shifts in aquatic and terrestrial urban communities. Nature 558:113–116. https://doi.org/10.1038/s41586-018-0140-0

Miaoulis IN, Heilman BD (1998) Butterfly thin films serve as solar collectors. Ann Entomol Soc Am 91:122–127. https://doi.org/10.1093/aesa/91.1.122

Mikheyev AS, Tschinkel WR (2004) Nest architecture of the ant Formica pallidefulva: structure, costs and rules of excavation. Insect Soc 51:30–36. https://doi.org/10.1007/s00040-003-0703-3

Molet M, Péronnet R, Couette S, Canovas C, Doums C (2017) Effect of temperature and social environment on worker size in the ant Temnothorax nylanderi. J Therm Biol 67:22–29. https://doi.org/10.1016/j.jtherbio.2017.04.013

Mosse I, Lyakh I (1994) Influence of melanin on mutation load in Drosophila populations after long-term irradiation. Radiat Res 139:357–359. https://doi.org/10.2307/3578834

Nascimento G, Câmara T, Arnan X (2022) Critical thermal limits in ants and their implications under climate change. Biol Rev 97:1287–1305. https://doi.org/10.1111/brv.12843

Nooten SS, Guénard B (2022) Ant communities in disturbed subtropical landscapes: is climate more important than stochastic processes? Oecologia 200:441–454. https://doi.org/10.1007/s00442-022-05276-8
Nylin S, Gotthard K (1998) Plasticity in life-history traits. Annu Rev Entomol 43:63–83. https://doi.org/10.1146/annurev.ento.43.1.63
Oliveira PYD, Souza JLPD, Baccaro FB, Franklin E (2009) Ant species distribution along a topographic gradient in a "terra-firme" forest reserve in Central Amazonia. Pesq Agrop Brasileira 44:852–860. https://doi.org/10.1590/S0100-204X2009000800008
Oliveira FM, Silva LL, Leal IR, Arnan X (2022) Morphology of four common and phylogenetically distant ant species varies along disturbance and aridity gradients in the Caatinga dry forest. Biotropica 54:78–90. https://doi.org/10.1111/btp.13029
Oms CS, Cerdá X, Boulay R (2017) Is phenotypic plasticity a key mechanism for responding to thermal stress in ants? Sci Nat 104:42. https://doi.org/10.1007/s00114-017-1464-6
Parkash R, Rajpurohit S, Ramniwas S (2008) Changes in body melanisation and desiccation resistance in highland vs. lowland populations of Drosophila melanogaster. J Insect Physiol 54:1050–1056. https://doi.org/10.1016/j.jinsphys.2008.04.008
Parr CL, Bishop TR (2022) The response of ants to climate change. Glob Chang Biol 28:3188–3205. https://doi.org/10.1111/gcb.16140
Parr CL, Dunn RR, Sanders NJ, Weiser MD, Photakis M, Bishop TR, Fitzpatrick MC, Arnan X, Baccaro F, Brandao CRF, Chick L, Donoso DA, Fayle TM, Gomez C, Grossman B, Munyai TC, Pacheco R, Retana J, Robinson A, Sagata K, Silva RR, Tista M, Vasconcelos H, Yates M, Gibb H (2017) GlobalAnts: a new database on the geography of ant traits (Hymenoptera: Formicidae). Insect Conserv Diversity 10:5–20. https://doi.org/10.1111/icad.12211
Peeters C, Aron S (2017) Evolutionary reduction of female dispersal in Cataglyphis desert ants. Biol J Linn Soc 122:58–70. https://doi.org/10.1093/biolinnean/blx052
Peeters C, Molet M, Lin C-C, Billen J (2017) Evolution of cheaper workers in ants: a comparative study of exoskeleton thickness. Biol J Linn Soc 121:556–563. https://doi.org/10.1093/biolinnean/blx011
Penick CA, Liebig J (2012) Regulation of queen development through worker aggression in a predatory ant. Behav Ecol 23:992–998. https://doi.org/10.1093/beheco/ars062
Perez R, Benbachir M, Decroo C, Mascolo C, Wattiez R, Aron S (2023) Cataglyphis desert ants use distinct behavioral and physiological adaptations to cope with extreme thermal conditions. J Therm Biol 111:103397. https://doi.org/10.1016/j.jtherbio.2022.103397
Peters RH (1986) The ecological implications of body size. Cambridge University Press
Pfeffer SE, Wahl VL, Wittlinger M, Wolf H (2019) High-speed locomotion in the Saharan silver ant, Cataglyphis bombycina. J Exp Biol 222:jeb198705. https://doi.org/10.1242/jeb.198705
Pfeiffer M, Chimedregzen L, Ulykpan K (2003) Community organization and species richness of ants (Hymenoptera/Formicidae) in Mongolia along an ecological gradient from steppe to Gobi desert. J Biogeogr 30:1921–1935
Pinker R (1980) The microclimate of a dry tropical forest. Agric Meteorol 22:249–265. https://doi.org/10.1016/0002-1571(80)90005-9
Porter EE, Hawkins BA (2001) Latitudinal gradients in colony size for social insects: termites and ants show different patterns. Am Nat 157:97–106. https://doi.org/10.1086/317006
Pringle EG, Santos TF, Gonçalves MS, Hawes JE, Peres CA, Baccaro FB (2019) Arboreal ant abundance tracks primary productivity in an Amazonian whitewater river system. Ecosphere 10:e02902. https://doi.org/10.1002/ecs2.2902
Purcell J, Pirogan D, Avril A, Bouyarden F, Chapuisat M (2016) Environmental influence on the phenotype of ant workers revealed by common garden experiment. Behav Ecol Sociobiol 70:357–367. https://doi.org/10.1007/s00265-015-2055-1
Ratchford JS, Wittman SE, Jules ES, Ellison AM, Gotelli NJ, Sanders NJ (2005) The effects of fire, local environment and time on ant assemblages in fens and forests. Divers Distrib 11:487–497. https://doi.org/10.1111/j.1366-9516.2005.00192.x

Remmert H (1981) Body size of terrestrial arthropods and biomass of their populations in relation to the abiotic parameters of their milieu. Oecologia 50:12–13. https://doi.org/10.1007/BF00378789

Rensch B (1929) Das Prinzip geographischer Rassenkreise und das Problem der Artbildung. Gebrüder Bornträger, Berlin

Riveros AJ, Seid MA, Wcislo WT (2012) Evolution of brain size in class-based societies of fungus-growing ants (Attini). Anim Behav 83:1043–1049. https://doi.org/10.1016/j.anbehav.2012.01.032

Roeder KA, Roeder DV, Bujan J (2021) Ant thermal tolerance: a review of methods, hypotheses, and sources of variation. Ann Entomol Soc Am 114:459–471. https://doi.org/10.1093/aesa/saab018

Roff D, Fairbairn D (2013) The costs of being dark: the genetic basis of melanism and its association with fitness-related traits in the sand cricket. J Evol Biol 26:1406–1416. https://doi.org/10.1111/jeb.12150

Römer D, Bollazzi M, Roces F (2018a) Carbon dioxide sensing in the social context: leaf-cutting ants prefer elevated CO_2 levels to tend their brood. J Insect Physiol 108:40–47. https://doi.org/10.1016/j.jinsphys.2018.05.007

Römer D, Halboth F, Bollazzi M, Roces F (2018b) Underground nest building: the effect of CO_2 on digging rates, soil transport and choice of a digging site in leaf-cutting ants. Insect Soc 65:305–313. https://doi.org/10.1007/s00040-018-0615-x

Ruano F, Tinaut A, Soler JJ (2000) High surface temperatures select for individual foraging in ants. Behav Ecol 11:396–404. https://doi.org/10.1093/beheco/11.4.396

Sankovitz M, Purcell J (2021) Ant nest architecture is shaped by local adaptation and plastic response to temperature. Sci Rep 11:23053. https://doi.org/10.1038/s41598-021-02491-w

Sarty M, Abbott KL, Lester PJ (2006) Habitat complexity facilitates coexistence in a tropical ant community. Oecologia 149:465–473. https://doi.org/10.1007/s00442-006-0453-9

Schofield SF, Bishop TR, Parr CL (2016) Morphological characteristics of ant assemblages (Hymenoptera: Formicidae) differ among contrasting biomes. Myrmecol News 23:129–137

Schultz TD, Hadley NF (1987) Structural colors of tiger beetles and their role in heat transfer through the integument. Physiol Zool 60:737–745. https://doi.org/10.1086/physzool.60.6.30159990

Schwander T, Humbert J-Y, Brent CS, Cahan SH, Chapuis L, Renai E, Keller L (2008) Maternal effect on female caste determination in a social insect. Curr Biol 18:265–269. https://doi.org/10.1016/j.cub.2008.01.024

Schweiger AH, Beierkuhnlein C (2016) Size dependency in colour patterns of Western Palearctic carabids. Ecography 39:846–857. https://doi.org/10.1111/ecog.01570

Schwind B, Wu X, Tiemann M, Fabritius H-O (2023) Broadband Mie scattering effects by structural features of setae from the Saharan silver ant Cataglyphis bombycina. JOSA B 40:B49–B58. https://doi.org/10.1364/JOSAB.474899

Shi NN, Tsai C-C, Camino F, Bernard GD, Yu N, Wehner R (2015) Keeping cool: enhanced optical reflection and radiative heat dissipation in Saharan silver ants. Science 349:298–301. https://doi.org/10.1126/science.aab3564

Shik J (2008) Ant colony size and the scaling of reproductive effort. Funct Ecol 22:674–681. https://doi.org/10.1111/j.1365-2435.2008.01428.x

Shik JZ, Arnan X, Oms CS, Cerdá X, Boulay R (2019) Evidence for locally adaptive metabolic rates among ant populations along an elevational gradient. J Anim Ecol 88:1240–1249. https://doi.org/10.1111/1365-2656.13007

Silva RR, Brandão CRF (2010) Morphological patterns and community organization in leaf-litter ant assemblages. Ecol Monogr 80:107–124. https://doi.org/10.1890/08-1298.1

Smith CR, Toth AL, Suarez AV, Robinson GE (2008) Genetic and genomic analyses of the division of labour in insect societies. Nat Rev Genet 9:735–748. https://doi.org/10.1038/nrg2429

Sommer S, Wehner R (2012) Leg allometry in ants: extreme long-leggedness in thermophilic species. Arthropod Struct Dev 41:71–77. https://doi.org/10.1016/j.asd.2011.08.002

Sosiak CE, Barden P (2021) Multidimensional trait morphology predicts ecology across ant lineages. Funct Ecol 35:139–152. https://doi.org/10.1111/1365-2435.13697
Stark AY, Arstingstall K, Yanoviak SP (2018) Adhesive performance of tropical arboreal ants varies with substrate temperature. J Exp Biol 221:jeb171843. https://doi.org/10.1242/jeb.171843
Stukalyuk S, Radchenko Y, Netsvetov M, Gilev A (2020) Effect of mound size on intranest thermoregulation in the red wood ants Formica rufa and F. polyctena (Hymenoptera, Formicidae). Turk J Zool 44:266–280. https://doi.org/10.3906/zoo-1912-26
Suding KN, Lavorel S, Chapin FS III, Cornelissen JHC, Diaz S, Garnier E, Goldberg D, Hooper DU, Jackson ST, Navas ML (2008) Scaling environmental change through the community-level: a trait-based response-and-effect framework for plants. Glob Chang Biol 14:1125–1140. https://doi.org/10.1111/j.1365-2486.2008.01557.x
Suter RB, Stratton GE, Miller PR (2004) Taxonomic variation among spiders in the ability to repel water: surface adhesion and hair density. J Arachnol 32:11–21. https://doi.org/10.1636/M02-74
Tocco C, Foster J, Venter N, Cowie B, Marlin D, Byrne M (2021) Elevated atmospheric CO_2 adversely affects a dung beetle's development: another potential driver of decline in insect numbers? Glob Chang Biol 27:4592–4600. https://doi.org/10.1111/gcb.15804
Trenberth KE (2011) Changes in precipitation with climate change. Clim Res 47:123–138. https://doi.org/10.3354/cr00953
Trible W, Kronauer DJ (2017) Caste development and evolution in ants: it's all about size. J Exp Biol 220:53–62. https://doi.org/10.1242/jeb.145292
Trible W, Kronauer DJ (2021a) Ant caste evo-devo: size predicts caste (almost) perfectly. Trends Ecol Evol 36:671–673. https://doi.org/10.1016/j.tree.2021.05.004
Trible W, Kronauer DJ (2021b) Hourglass model for developmental evolution of ant castes. Trends Ecol Evol 36:100–103. https://doi.org/10.1016/j.tree.2020.11.010
Trullas SC, van Wyk JH, Spotila JR (2007) Thermal melanism in ectotherms. J Therm Biol 32:235–245. https://doi.org/10.1016/j.jtherbio.2007.01.013
Tschinkel WR (1987) Seasonal life history and nest architecture of a winter-active ant, Prenolepis imparis. Insect Soc 34:143–164. https://doi.org/10.1007/BF02224081
Tschinkel WR (2015) The architecture of subterranean ant nests: beauty and mystery underfoot. J Bioecon 17:271–291. https://doi.org/10.1007/s10818-015-9203-6
Turner JR, Kearney EP, Exton LS (1984) Mimicry and the Monte Carlo predator: the palatability spectrum, and the origins of mimicry. Biol J Linn Soc 23:247–268. https://doi.org/10.1111/j.1095-8312.1984.tb00143.x
Verble-Pearson RM, Gifford ME, Yanoviak SP (2015) Variation in thermal tolerance of North American ants. J Therm Biol 48:65–68. https://doi.org/10.1016/j.jtherbio.2014.12.006
Vigneron J-P, Simonis P (2010) Structural colours. In: Advances in insect physiology. Elsevier, pp 181–218. https://doi.org/10.1016/S0065-2806(10)38004-0
Villalta I, Oms CS, Angulo E, Molinas-González CR, Devers S, Cerdá X, Boulay R (2020) Does social thermal regulation constrain individual thermal tolerance in an ant species? J Anim Ecol 89:2063–2076. https://doi.org/10.1111/1365-2656.13268
Violle C, Navas ML, Vile D, Kazakou E, Fortunel C, Hummel I, Garnier E (2007) Let the concept of trait be functional! Oikos 116:882–892. https://doi.org/10.1111/j.0030-1299.2007.15559.x
Vogelmann TC (1993) Plant tissue optics. Annu Rev Plant Biol 44:231–251. https://doi.org/10.1146/annurev.pp.44.060193.001311
Wagner T, Bachenberg L, Glaser SM, Oikonomou A, Linn M, Grüter C (2021) Large body size variation is associated with low communication success in tandem running ants. Behav Ecol Sociobiol 75:109. https://doi.org/10.1007/s00265-020-02941-x
Wang LY, Franklin AM, Hugall AF, Medina I, Stuart-Fox D (2023) Disentangling thermal from alternative drivers of reflectance in jewel beetles: a macroecological study. Glob Ecol Biogeogr 32:408–420. https://doi.org/10.1111/geb.13632
Watt WB (1968) Adaptive significance of pigment polymorphisms in Colias butterflies. I. Variation of melanin pigment in relation to thermoregulation. Evolution 22:437–458. https://doi.org/10.2307/2406873

Westoby M, Falster DS, Moles AT, Vesk PA, Wright IJ (2002) Plant ecological strategies: some leading dimensions of variation between species. Annu Rev Ecol Evol Syst 33(1):125–159

Wheeler DE (1986) Developmental and physiological determinants of caste in social Hymenoptera: evolutionary implications. Am Nat 128:13–34. https://doi.org/10.1086/284536

Wiernasz DC (1989) Female choice and sexual selection of male wing melanin pattern in Pieris occidentalis (Lepidoptera). Evolution 43:1672–1682. https://doi.org/10.1111/j.1558-5646.1989.tb02617.x

Willot Q, Simonis P, Vigneron J-P, Aron S (2016) Total internal reflection accounts for the bright color of the Saharan silver ant. PLoS One 11:e0152325. https://doi.org/10.1371/journal.pone.0152325

Wills BD, Powell S, Rivera MD, Suarez AV (2018) Correlates and consequences of worker polymorphism in ants. Annu Rev Entomol 63:575–598. https://doi.org/10.1146/annurev-ento-020117-043357

Wilson K, Cotter SC, Reeson AF, Pell JK (2001) Melanism and disease resistance in insects. Ecol Lett 4:637–649. https://doi.org/10.1046/j.1461-0248.2001.00279.x

Wittlinger M, Wehner R, Wolf H (2006) The ant odometer: stepping on stilts and stumps. Science 312:1965–1967. https://doi.org/10.1126/science.1126912

Wolpert A (1962) Heat transfer analysis of factors affecting plant leaf temperature. Significance of leaf hair. Plant Physiology 37:113–122

Yanoviak SP, Kaspari M (2000) Community structure and the habitat templet: ants in the tropical forest canopy and litter. Oikos 89:259–266. https://doi.org/10.1034/j.1600-0706.2000.890206.x

Yilmaz AR, Chick LD, Perez A, Strickler SA, Vaughn S, Martin RA, Diamond SE (2019) Remarkable insensitivity of acorn ant morphology to temperature decouples the evolution of physiological tolerance from body size under urban heat islands. J Therm Biol 85:102426. https://doi.org/10.1016/j.jtherbio.2019.102426

Zverev V, Kozlov MV, Forsman A, Zvereva EL (2018) Ambient temperatures differently influence colour morphs of the leaf beetle Chrysomela lapponica: roles of thermal melanism and developmental plasticity. J Therm Biol 74:100–109. https://doi.org/10.1016/j.jtherbio.2018.03.019

Chapter 4
Ant Responses to Climate Change: Insights from Critical Thermal Limits and Thermal Performance Curves

Geraldo Nascimento, Quentin Willot, and Xavier Arnan

4.1 Introduction

Anthropogenic climate change stands as one of humanity's paramount challenges, exerting a profound impact on global biodiversity (Habibullah et al. 2022). The escalation in average and extreme temperatures presents a particular challenge for ectothermic organisms, whose body temperature is intricately linked to their surrounding climate (Angilletta 2009). For these creatures, the thermal environment plays a pivotal role in governing metabolic processes, behavioural patterns, and reproductive strategies (Angilletta 2009). Fluctuations in temperatures can markedly influence their metabolic rates, locomotor behaviours, reproductive capabilities, and habitat choices (Sinclair et al. 2016; Chou et al. 2019; van Heerwaarden and Sgrò 2021). With climate projections indicating a continual warming trend and an increase in the frequency of unpredictable weather (cold and heat) events, the challenges faced by many ectotherms are significantly magnified (Jørgensen et al. 2022).

In the last decades, as awareness of climate change has increased, researchers globally have focused their efforts on comprehending the thermal biology of

G. Nascimento
Programa de Pós-Graduação em Etnobiologia e Conservação da Natureza, Universidade Federal Rural de Pernambuco, Recife, Pernambuco, Brazil
e-mail: geraldo.nascimento@upe.br

Q. Willot
Environmental and Evolutionary Biology Research Unit, Department of Biology, University of Namur, Namur, Belgium
e-mail: quentin.willot@unamur.be

X. Arnan (✉)
Universidade de Pernambuco – Campus Garanhuns, Garanhuns, Pernambuco, Brazil
e-mail: xavier.arnan@upe.br

I. R. Leal et al. (eds.), *Ant Responses to Climate Change*, Entomology in Focus 10,
https://doi.org/10.1007/978-3-032-21127-9_4

organisms (Nowakowski et al. 2018), especially ectothermic organisms (Sinclair et al. 2016, 2024; Kefford et al. 2022; Terblanche et al. 2024; González-Tokman and Dáttilo 2024). This focus has led to a significant surge in the volume of research published over the past 20 years, particularly relating to critical thermal limits (CTLs) and various thermal performance curves (TPCs) of thermally responsive traits (Sinclair et al. 2016; Jørgensen et al. 2022; Nascimento et al. 2022; Ørsted et al. 2022). CTLs are the lowest (CT_{min}) and highest temperatures (CT_{max}) at which an individual can sustain short-term neuromuscular function before entering chill-induced coma or heat-induced coma, respectively (Andersen et al. 2015; Terblanche et al. 2011), and they determine the operative temperature (i.e. the steady-state body temperature of individuals after accounting for all sources of heat transfer; Sunday et al. 2014) range at which individuals can potentially exhibit activity for a given time interval.

CTLs play a crucial role in shaping species distributions by defining the stressful temperature range organisms can tolerate for a limited amount of time. The thermal adaptation hypothesis suggests that CTLs reflect evolutionary adjustments to local climates, with species in warmer regions having higher CT_{max} and those in colder regions lower CT_{min} (Kaspari et al. 2015). In contrast, the climate variability hypothesis proposes that species from regions with high environmental variability may exhibit broader thermal tolerance ranges, as they are adapted to cope with fluctuating temperatures and other environmental changes (Janzen 1967). Specifically, tropical species tend to have narrower temperature tolerances due to stable, year-round temperatures, while species in temperate regions experience greater temperature fluctuations and thus evolve broader tolerances to survive both hot summers and cold winters. This highlights temperature fluctuation, in addition to average temperatures, as a key driver in the evolution of CTLs. Meanwhile, the thermal asymmetry hypothesis highlights that CT_{min} often imposes stricter constraints on distributions than CT_{max}, as species are more limited in their ability to expand into colder environments (Câmara et al. 2024; see Chaps. 9 and 10).

Together, the thermal adaptation and thermal asymmetry hypotheses explain how thermal physiology underpins species' ranges and predict shifts in distributions under climate change, where warming temperatures may favour poleward expansion but challenge species constrained by cold tolerance. Meanwhile, TPCs are generally expressed as mathematical functions and delineate the effects of body temperature (T_b) on ecological, behavioural, or physiological processes, including traits most critical for fitness such as organism growth, survival, and reproduction (Sinclair et al. 2016) (Fig. 4.1). They are both essential for understanding the impact of temperature on various aspects of organisms' biology (Huey and Stevenson 1979), spanning from biochemical and physiological processes to ecologically significant variables such as predation rates and population growth (Frazier et al. 2006; Sinclair et al. 2016; Rezende and Bozinovic 2019).

Ants are ectothermic animals that play vital roles in terrestrial ecosystems, being essential for many ecological processes (Hölldobler and Wilson 1990). Their presence is linked to a range of ecosystem services, including nutrient cycling, seed dispersal, pest control, and even soil aeration (Del Toro et al. 2012; Elizalde et al.

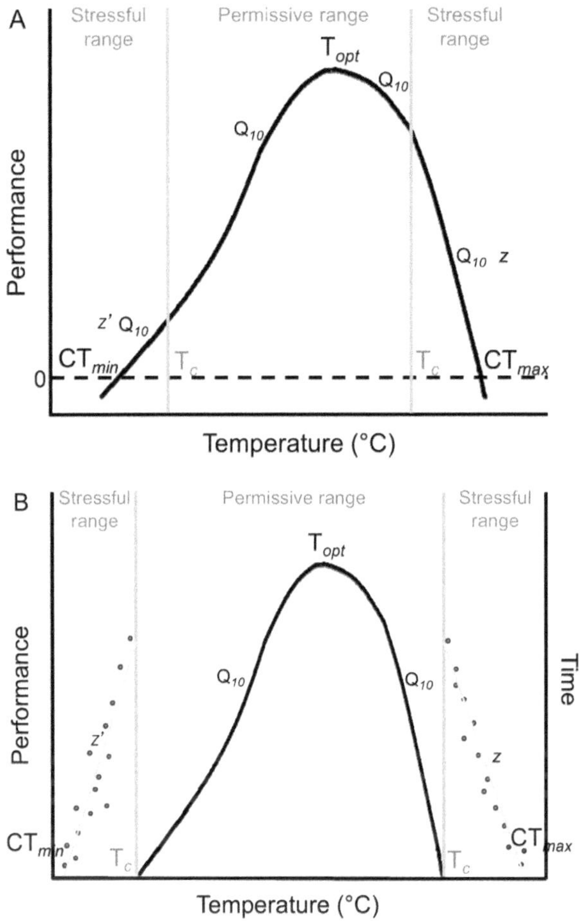

Fig. 4.1 Typical thermal performance curves of ectotherm organisms: (**a**) for traits related to survival (e.g. metabolic, physiological); (**b**) for traits related to fitness (e.g. growth, reproduction—fitness reaches 0 when individuals start experiencing heat-induced damages and their survival becomes a function of time of exposure to these stressful temperatures). The key variables that describe the shape of the curve are indicated, as well as the permissive and stressful thermal ranges transitioned by the critical temperature (T_c). Thermal death time (TDT) curves illustrate the relationship between temperature and survival duration within the stressful temperature range (grey circles and line correspond to the secondary y-axis), alongside acute assessments of CT_{min} (blue circles) and CT_{max} (red circles). Note that the application of the TDT model to estimate CT_{max} (or CT_{min}) and thermal sensitivity (*z* or *z*') only makes sense within any of the stressful ranges. CT_{min} and CT_{max} are also shown for convenience, but they are not part of the TPC per se

2020). Furthermore, ants are considered excellent models for thermal biology studies due to their ecological diversity (Hölldobler and Wilson 1990; Parker and Kronauer 2021), wide geographical distribution (Wilson 1987; Schultheiss et al. 2022), and their ability to adapt to a variety of thermal environments (Bishop et al. 2017; Boulay et al. 2017; Nascimento et al. 2022). They exhibit an impressive range of strategies to cope with temperature variations, from morphological and physiological adjustments to modifications in collective behaviour (Penick and Tschinkel 2008; Wehner and Wehner 2011; Menzel et al. 2018; Willot et al. 2017, 2018; Shik et al. 2019; Perez and Aron 2020; Araujo et al. 2023; see Chaps. 2, 3, and 5). Seminal reviews on ant thermal biology and its relationship with climate change responses have been recently published (Roeder et al. 2021; Nascimento et al. 2022; Parr and Bishop 2022). Most studies to date have focused primarily on the CTLs, while the impact of temperature change on other key traits in temperature located in between these limits has received much fewer attention (Roeder et al. 2021; Nascimento et al. 2022).

While conventional measurements of CTLs can offer valuable insights into the modelling and understanding of present and future distribution of ants (Nascimento et al. 2022), alternative methods capturing additional information regarding species' temperature tolerance and sensitivity may prove extremely effective in providing us useful information to accurately predicts ants' behavioural and performance responses to temperature variations. These alternative approaches include the use of TPCs, which provide a detailed depiction of how biological traits (e.g. reproduction, foraging efficiency, locomotion) vary across a range of temperatures. Unlike CTLs, which are fixed endpoints marking temperatures of failure or mortality, TPCs capture the entire thermal performance spectrum, including critical features such as the optimal temperature (T_{opt})—the point at which performance is maximized—and metrics of sensitivity like the thermal sensitivity (z) and the temperature quotient (Q_{10}). The *z* measures the rate of performance decline within the range of stressful temperatures, while the Q_{10} evaluates the rate of change in biological processes with a 10 °C temperature difference and can be applied both within stressful and permissive temperature ranges (Fig. 4.1). The boundary between these two ranges is defined by the critical temperature (T_c). Furthermore, methods incorporating time--dependent thermal tolerance (TDT) models, which account for the duration of thermal exposure, also expand our understanding by addressing how organisms endure varying thermal environments over time. These approaches can complement and extend CTL-based analyses, particularly when focusing on TPC segments (e.g. ascending and descending parts of the curve) rather than fixed points (Rezende et al. 2014; Jørgensen et al. 2021; Ørsted et al. 2022; Willot et al. 2022).

This review explores how CTLs and TPC variables of key physiological traits inform predictions of ant responses to temperature fluctuations and their ecological relevance. We first describe what these variables represent and how they can be measured and then examine their influence on ant distribution and performance in different environments, as well as their potential for acclimation or evolutionary adaptation. We also assess the ecological redundancy of these variables and suggest future research directions, prioritizing ecological understanding beyond mathematical and physiological aspects. Our ultimate goal is to inspire myrmecologists to conduct studies that accelerate our understanding of ants' responses to climate change using advanced metrics and concepts relating to relevant biological variables and measurement methods related to ant thermal biology.

4.2 Critical Thermal Limits

The CTLs are the most studied and widely used variables describing upper and lower thermal limits in insects and ants, often incorporated on TPCs (see below) to provide relevant survival limits framing the response of key biological traits to temperatures (Roeder et al. 2021; Nascimento et al. 2022), and as such have been often broadly used as a proxy for survival limits (Lutterschmidt and Hutchison 1997; Sinclair et al. 2016) (Table 4.1). The measurement of these thermal limits can vary according to different methodologies, under static and dynamic assays, using devices such as dry baths, water baths, hot plates, or growth chambers (Terblanche et al. 2007; Roeder et al. 2021; Nascimento et al. 2022). To effectively use these measurements for predicting ant responses to climate change, it is crucial to understand precisely what each metric measured from different methods represents, their equivalencies, and their inherent limitations. The less-often-used static assay subjects individuals to a constant and stressful temperature, then measures the time required to reach an endpoint (e.g. loss of muscle coordination), or records survival after a fixed period (Lutterschmidt and Hutchison 1997; Jørgensen et al. 2019; Willot et al. 2022). Meanwhile, the more commonly used dynamic (ramping) assay involves gradually exposing the individual to increasing (CT_{max}) or decreasing (CT_{min}) temperatures and then recording the temperature at which the organism loses its motor coordination (Lutterschmidt and Hutchison 1997). Each of these approaches has distinct features (see Roeder et al. 2021 for a detailed review): the static assay results provide a knockdown time, while the ramping assay results provide a knockdown temperature. Note that ramping assays, on which the large majority of CTL measurements have been based in insects, have been widely recognized to be contingent on methodologies, chiefly starting temperature and ramping rate (Terblanche et al. 2007). Varying ramping rates can, for example, impact recorded knockdown temperatures up to a few degrees and induce large methodological bias across methodologies (Terblanche et al. 2007; Jørgensen et al. 2021; Willot et al. 2022). CTLs in ants using ramping assays have nevertheless been extensively studied in the last two decades, allowing us to advance our knowledge of species' thermal responses (Roeder et al. 2021; Nascimento et al. 2022). These studies have explored how ant CTLs relate to various abiotic and biotic factors, at different levels of biological organization: among individuals, colonies, populations, species, and/or communities (Kaspari et al. 2015; Baudier et al. 2018; Garcia-Robledo et al. 2018). Abiotic factors involve spatial gradients, at both large scales, such as latitudinal (Diamond et al. 2012b; Willot et al. 2022, 2023), altitudinal (Baudier et al. 2018), climatic (Bujan et al. 2020a), or anthropogenic (Diamond et al. 2012b; Diamond et al. 2017) gradients, and small scales, such as nesting microhabitat (Kaspari et al. 2015; Leahy et al. 2022), as well as temporal gradients, such as daily (Garcia-Robledo et al. 2018) or seasonal foraging rhythms (Bujan et al. 2020b). Biotic factors generally include body size (Baudier et al. 2015; Verble-Pearson et al. 2015), caste (Kay and Whitford 1978; Coulin et al. 2019), evolutionary history

Table 4.1 Summary of thermal tolerance traits and key variables that characterize a typical thermal performance curve and that can be used to analyse ant thermal responses, as they are the focus of this study

Variable	Definition	Ecological relevance
CT_{min}	The lowest temperature at which an individual can sustain neuromuscular function before entering chill-induced coma (Andersen et al. 2015)	It specifies the minimum temperature at which an ant can sustain basic neuromuscular function. It is a common metric used to approximate the lower operative temperatures (i.e. the temperature of an ant after accounting for all relevant heat exchange processes; Sunday et al. 2014) at which workers can potentially exhibit activity
CT_{max}	The highest temperature at which an individual can sustain neuromuscular function before entering heat-induced coma (Terblanche et al. 2011)	It indicates the maximum temperature at which an ant can sustain neuromuscular functions and therefore locomotor activities for a limited amount of time before entering heat-induced coma. It is a common metric used to approximate the upper operative temperatures at which workers can potentially exhibit activity
T_{opt}	Temperature at which performance of a trait of interest is maximized (Pawar et al. 2024)	It points out the temperature at which the given trait of an organism performs best (e.g. walking speed, metabolic rate, etc.). T_{opt} is necessary to comprehend how variation in temperature might impact trait performance and ultimately the fitness of individuals. Adaptive genotypical shifts in thermal fitness are mainly dictated by changes in T_{opt}, of key physiological traits rather than in CTLs (which exhibit superior plasticity to short-term thermal fluctuations and extreme heat or cold waves) (Pawar et al. 2024)
z	Coefficient of thermal sensitivity (Rezende et al. 2014)	z denotes the temperature change needed to alter tolerance duration to high temperatures by a factor 10. It provides information on how quickly knockdown times decrease with each additional temperature unit of stressful exposure. This metric relates to the linear slope of the log--transformed relationship between survival time and temperature (see Sect. 4.3). Therefore, it completes absolute heat tolerance metrics such as CT_{max}, by providing additional information on how quickly heat tolerance decreases above the physiological temperatures of individuals
Q_{10}	Quotient that describes the variation in a particular biological rate due to a 10 °C temperature increase (Shik et al. 2019)	It refers to a biological trait's sensitivity to temperature changes regarding a temperature-dependent physiological, metabolic, behavioural, or ecological process. It provides information on how quickly a trait gains or loses performances with each additional temperature unit. This measure can be applied both on the ascending part of the TPCs (activation) and on the descending part (deactivation) and at both the permissive and stressful thermal ranges of the curves

(Diamond and Chick 2018a), foraging activity (Arnan and Blüthgen 2015; Braschler et al. 2020), and nutritional status (Bujan and Kaspari 2017; Freires et al. 2023).

Studies addressing large datasets of ant CTLs have revealed that both CT_{max} and CT_{min} display a strong phylogenetic signal, with a much stronger effect for CT_{max} than for CT_{min} (Diamond and Chick 2018a; Câmara et al. 2024), meaning CT_{max} tends to change more slowly across taxa at an evolutionary scale. These studies also strongly suggest that CT_{max} variations are driven by local microclimate (e.g. in relation to verticality, see Chap. 12), whereas CT_{min} relates to macroclimates (Diamond and Chick 2018a; Bujan et al. 2020a; Leahy et al. 2022; Willot et al. 2023; Câmara et al. 2024). Such asymmetry appears to hold true for insects in general (Sunday et al. 2012, 2019). Overall, the CT_{max} of ants thus appears modulated by specialization in thermal microclimatic niches at evolutionary scales. This is because thermal heterogeneity of the environment at the microclimatic scale probably allows for the emergence of warmer and colder microclimatic niches regardless of the considered location, and it offers the possibility for species to specialize on them. For instance, leaf-litter ants inhabiting colder, damper, and sheltered microenvironments have lower heat tolerance compared with species occurring above them in the canopy that experience a warmer microenvironment (Kaspari et al. 2015; Boulay et al. 2017). Thus, different ant species displaying both low and high heat tolerances can share the same habitat across latitude and climates, which prevents strong correlation between heat tolerance and these factors to be drawn at larger climatic scales (Leahy et al. 2022; Willot et al. 2022; Câmara et al. 2024). Conversely, and perhaps because the thermal heterogeneity of the environment for minimal microclimatic temperatures is less pronounced, cold tolerance of species displays a much stronger correlation with the minimal temperatures of their respective environments (Bujan et al. 2020a; Nascimento et al. 2022; Willot et al. 2023; Diamond and Chick 2018a).

Several studies have also utilized thermal tolerance breadth (CT_{max}–CT_{min}) to predict ant species distributions based on the well-supported 'niche-breadth hypothesis' (Slatyer et al. 2013). According to this hypothesis, species capable of tolerating a wider range of environmental conditions can occupy a broader range of habitats and, consequently, have larger distribution ranges. Diamond and Chick (2018b) characterized thermal tolerance breadth for 41 species from 22 genera, revealing that species with narrow thermal tolerance breadth tended to have more restricted, equatorial ranges, suggesting they may face greater extinction risk under climate change. Similarly, Chick et al. (2020) estimated thermal tolerance breadth for 18 ant species along an elevational gradient in Great Smoky Mountains National Park (USA). They found a moderate positive correlation (50%) between thermal breadth and elevational distribution range. In line with this work, Leahy et al. (2024) reported that arboreal ant species, which experience higher and more variable temperatures, have wider elevational distributions than do ground-active ant species in the Australian wet tropics. Notably, these studies align with Janzen's hypothesis stating that the more variable the climate, the greater the CTLs and thermal breadth and highlight that species distribution patterns may be primarily influenced by physiological constraints imposed by colder temperatures at higher elevations, while biotic interactions shape distributions in warmer climates. All this taken together suggests that CT_{min} might be more informative than CT_{max} for predicting future fitness loss and distributions shifts of ant species under climate change (Coulin et al. 2019; Braschler et al. 2020; Chick et al. 2020; Guo et al. 2020; Willot et al. 2023).

However, when CTL data are used to assess which species may be more vulnerable to climate change, CT_{max} emerges as particularly significant. A recent study with 33 ant communities in the USA involving 28 ant genera, comparing past data with present data in regions where temperatures have increased by an average of 1 °C, shows that genera with higher CT_{max} have benefited the most and expanded their distribution range; conversely, CT_{min} did not explain responses to increased temperature (Roeder et al. 2021). Another study used species-specific CT_{max} to forecast the responses of ant communities to forest-floor warming induced experimentally at the northern and southern boundaries of temperate hardwood forests in eastern North America (Diamond et al. 2012a). Their findings suggest that when microclimatic temperatures and experiences approach a species' physiological limit, CT_{max} becomes a limiting factor of distribution and can accurately forecast future responses, particularly in the context of warming.

Other studies have assessed warming tolerance, defined as the difference between CT_{max} and environmental temperature, to evaluate safety margins for various ant species (Diamond et al. 2012b; Leahy et al. 2022). In their analysis of species most vulnerable to climate warming, Diamond et al. (2012b) estimated warming tolerance for 156 ant species from different latitudes and microhabitats worldwide. They highlighted that ants inhabiting the canopies of hot, tropical forests may face the greatest risk, because their microclimate temperatures already often approach their CT_{max}, and additional warming will prove detrimental. Nevertheless, this idea might be challenged. Warm-adapted species (with high CT_{max}) will still have the option to be active at night or move to a cooler microclimate, giving them the margin to behaviourally adapt. In contrast, cold-adapted ants (with low CT_{max}) will have no choice but to move or disappear, as their niche will vanish. While species with higher CT_{max} might disappear from some areas due to extreme heat, they will gain far more habitats than they lose at larger scales, as many microhabitats presently too cold will become habitable and ultimately their range will expand on average. Under the scope of climate change, many cooler microhabitats sheltering specialized species with low heat tolerance are thus at risk of disappearing, especially in temperate climates. Conversely, the range of warmer microhabitats favouring ants with higher CT_{max} is bound to expand (Thakur et al. 2020), including urban heat islands (Diamond et al. 2017). This dynamic suggests that while some species will struggle to survive (likely cold-adapted species), others will find new opportunities (likely heat-adapted species), leading to significant shifts in ant diversity and distribution. Andrew et al. (2019) demonstrated that warming tolerance in populations of *Iridomyrmex purpureus* declined along an aridity gradient in Australia, which is consistent with the hypothesis that CT_{max} is also related with rainfall (Willot et al. 2022). Increased rainfall generally creates slightly damper and cooler microclimates, providing species with more opportunities to shelter from the heat. In contrast, dry conditions offer fewer refuges from high temperatures, exacerbating thermal stress.

Finally, if we aim to make accurate predictions, it is necessary to understand the potential for adaptation or acclimation in ant CTLs. In this regard, it is crucial to first emphasize that CTLs display a strong phylogenetic signal (Nascimento et al. 2022; Willot et al. 2022), especially CT_{max}. This suggests that there may be strong limits on the evolutionary pace at which heat tolerance can adapt in species, a concern amplified by the looming spectre of fast global warming. In this line, CT_{max} has often been shown to display limited response to acclimation and thus low plasticity in insects (Sørensen et al. 2016; Willot et al. 2021; Weaving et al. 2022). Whether this holds true for ants remains unclear. Low adult acclimation potential for heat tolerance in ants has received support under laboratory conditions (Willot et al. 2022). However, a study comparing populations of *Temnothorax curvispinosus* between rural and urban ('heat islands') areas showed that the urban populations can shift their CTLs to deal with increased temperatures, and these shifts seem to be evolutionarily adapted (Martin et al. 2019). In addition, heat and cold tolerance sometimes exhibit robust, if asymmetrical, responses to acclimation gradients. For example, ants active during different times of day (Garcia-Robledo et al. 2018) or seasons (Bujan et al. 2020b), and thus exposed to different temperature regimes, may exhibit diverging CTLs. Even workers within the same species or colony may vary in their CTLs depending on the temporal context (Esch et al. 2017; Nelson et al. 2018; Bujan et al. 2020b). Some capacity of ants for acclimation to warmer temperatures has also been supported in laboratory settings with select species for CT_{max} (Cokendolpher and Phillips 1990; Jumbam et al. 2008) with a proportionally greater effect of cold acclimation on CT_{min} (Bujan et al. 2021). In summary, when examining the plasticity of CTLs in ants, the literature reveals mixed results, likely due to differing acclimation capabilities across species, but it appears with a relatively greater acclimation plasticity for lower thermal limits than higher ones. Thus, although limited acclimation capabilities and adaption to thermal changes through natural selection are possible, they do not guarantee that species will cope with rising temperatures. On a large scale, it is likely that some species will not be able to adapt and will be outcompeted by those that are better adapted to higher temperatures (Kaspari et al. 2019). Species exposed to above-optimal temperature are likely to alter their behaviour, becoming active at night or for shorter periods during the day and year due to excessive heat. These species tend to be moved to higher altitudes or latitudes (Bishop et al. 2019), while species better adapted to higher temperatures will be able to locally outcompete them for resources.

4.3 Thermal Death Time (TDT) Models

CTLs can be obtained from both static and dynamic methods, which differ in both the rate of temperature change and the duration of organism exposure to thermal stress (Rezende et al. 2014; Jørgensen et al. 2019, 2021). This conundrum has historically given rise to methodological inconsistencies when measuring critical thermal limits of ectotherms, hindering data harmonization across studies, teams, and practices and consequently hindering our ability to predict ant vulnerability to climate change.

To address these challenges, several authors reintroduced a conceptualization that focuses on measuring the interaction between the intensity of thermal stress and the duration of exposure (Bigelow 1921; Rezende et al. 2014). They expanded on the concept of thermal tolerance landscapes in *Drosophila* to build thermal death time (TDT) curves, highlighting how such a parameterization could provide more comprehensive information about heat tolerance than those obtained through isolated static or dynamic assays in ectotherms. TDT models are built based on knockdown time curves, which are compiled from several static experiments at various temperatures (Rezende et al. 2014; Jørgensen et al. 2019, 2021). These curves represent the relationship between the intensity of thermal stress (i.e. temperature) and the time a given temperature can be tolerated and can be derived for any of the stressful ranges (both cold and hot) ectotherms can experience (Figs. 4.1 and 4.2). This is because, mathematically, the relationship between knockdown time and temperature in insects is an exponential function (the time to reach heat-induced knockdown decreases exponentially with increasing temperatures of exposure; Rezende et al. 2014). This exponential function can be log-transformed into a linear relationship of the time to thermal failure vs temperatures (Jørgensen et al. 2019, 2021; Ørsted et al. 2022). This linear regression is mathematically easily described by a slope (or z = −1/slope, the coefficient of thermal sensitivity) and an intercept under the equation y = ax+b (Rezende et al. 2014; Jørgensen et al. 2019, 2021; Ørsted et al. 2022).

The main advantage of TDT parameterization is that it provides approximations of critical temperatures (T_c) that characterize transitions between the permissive range (temperatures which permit completion of the life cycle) and the stressful

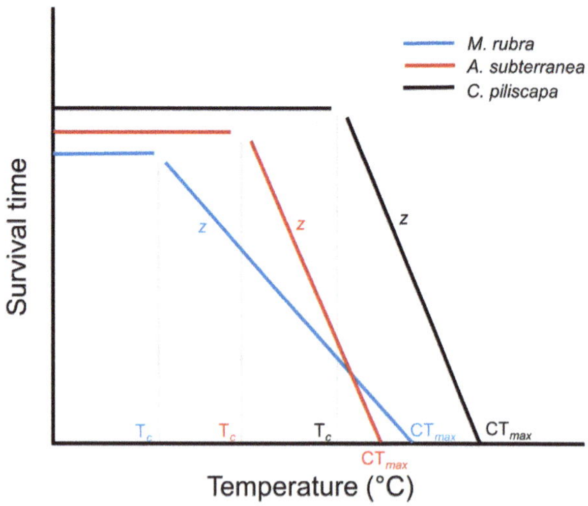

Fig. 4.2 Relationship between the intensity of thermal stress (i.e. temperature) and the median time to knockdown (i.e. thermal death time model) for *Aphaenogaster subterranea*, *Myrmica rubra*, and *Cataglyphis piliscapa*. The critical temperature (T_c), CT_{max}, and *z* that can be estimated from these curves are shown for each species. (This figure is an adaptation of Figure 4.2 from Willot et al. (2022), with authors' permission)

range (temperatures where heat-induced damage accumulates causing mortality—see Sect. 4.4) (Figs. 4.1 and 4.2). The T_c threshold therefore marks the pivotal point where individuals cannot sustain homeostasis, and their fitness becomes null. It represents a critical metric to understand temperature-driven fitness loss in ectotherms beyond endpoints of mortality alone (i.e. CTLs). TDT models also provide values of heat sensitivity (z), which describes how quickly the likelihood of heat-induced mortality progresses once T_c is crossed. Z is particularly important for understanding the vulnerability of species to environmental temperature spikes and fluctuations and provides critical information on how this tolerance dynamically evolves across temperatures (e.g. two species might display the same knockdown time at a given temperature but display difference in tolerance dynamic across a thermal landscape—see *Aphaenogaster subterranea* vs *Myrmica rubra* in Fig. 4.2). Leveraging the information extracted from TDT models for both heat and cold tolerances can offer powerful insight into a species' thermal biology.

Another advantage of TDT models is methodological: they are capable of inter-converting CTL results from static assays (knockdown times) and dynamic assays (knockdown temperatures) while also accounting for ramping rates and starting temperatures in dynamic experiments (see Jørgensen et al. 2021 for accessing TDT model scripts in R). In other words, another powerful output of TDT models is that it can also harmonize CT_{max} values across methodologies, bypassing the main known issue of the classically used dynamic assays (see Sect. 4.2) (e.g. converting CT_{max} results acquired through high ramping rates into their lower ramping rate equivalents and vice versa, allowing accurate CT_{max} comparison across methodologies, authors, and literatures, thus enabling broader generalizations to forecast ant responses to climate change).

To date, only one study has used TDT to measure static and dynamic CTLs, particularly CT_{max}, and the index z of thermal sensitivity in ants (Willot et al. 2022). This study applied the TDT model to 13 ant species from western Europe, and it showed good correlation between predicted and observed values of CT_{max} obtained through static and dynamic methods, indicating accuracy in predictions using TDT variables in ants. All TDT variables (including the thermal sensitivity coefficient z) showed strong phylogenetic signal. However, in line with previous data showing little correlation between CT_{max} and climate in ants (without considering microclimatic gradients), these TDT variables were at most only weakly associated with climatic variables in the locations where the species occurs. Overall, evolutionary history thus appears as a significant predictor of heat tolerance in ants both through CT_{max} and TDT measurements while also being an important predictor of sensitivity to temperature increase (z). This study also supported the notion that ant sensitivity to temperature increase (z) demonstrates a limited potential for acclimation while acknowledging that this requires further investigation.

In conclusion, it appears that TDT can offer extremely valuable insights to our understanding of heat tolerance and heat sensitivity in ants. TDTs have the potential to bring new metrics into the playfield of heat tolerance measurements in ants, such as T_c approximation and z, and yield valuable and novel assessment or reassessment of a species' vulnerability to temperature changes (heat sensitivity). Future studies comparing the z' or z of co-occurring ant species would be highly valuable, particularly in helping to understand the differential responses that groups of co-occurring species with similar CT_{max} or CT_{min} values may exhibit in response to temperature changes under climate change. However, one notable drawback of TDT parameterization is that it takes several static CT_{max} assays at different temperatures to be performed. This makes this approach both more time-consuming and sample demanding than dynamic CT_{max}, making TDTs ill-suited for comparing heat tolerance variables across a large number of populations/species and climatic gradients. Therefore, how this approach can be reconciled with the need of exanimating larger numbers of species under macroecological approaches remains under scrutiny.

4.4 Thermal Performance Curves

CT_{min} and CT_{max} define the boundaries of a species temperature tolerances, but they fail to directly capture fitness change resulting from variation of temperatures still within organism's physiological ranges (e.g. sublethal stress), and they do not incorporate duration of stress exposure into heat-damage models (Terblanche et al. 2007; Ørsted et al. 2022, 2024). Over the past decades, TPCs have been recognized as an essential tool for understanding the physiology and thermal response of organisms (Tüzün and Stoks 2018; Malusare et al. 2023). Often unimodal and asymmetric (left-skewed), TPCs start at their initial point (T_{min}). The performance of the thermally responsive trait under scrutiny usually increases exponentially with temperature until reaching the optimum point (T_{opt} = the point at which performance is maximized) (Fig. 4.1, Table 4.1). Beyond T_{opt}, performance generally decreases exponentially as well as the temperature continues to rise toward an endpoint (T_{max}). Note that T_{min} or T_{max} can easily be confused with the minimum (CT_{min}) and maximum (CT_{max}) critical thermal limits depending on the reference used (e.g. Sinclair et al. 2016; Tüzün and Stoks 2018; Kaspari et al. 2019; Rezende and Bozinovic 2019). Initially, CT_{min} and CT_{max} were proposed to mean minimum and maximum thermal estimates derived from survival or response correction tests (Lutterschmidt and Hutchison 1997) but are often incorporated for reference into TPCs of other physiological traits to overlay TPC information with survival boundaries. However, and to be accurate, these CTLs represent the temperatures at which organisms experience knockdown or the onset of muscle spasms, indicating the thresholds (both warm and cold) at which they lose control over their neuromuscular functions (Lutterschmidt and Hutchison 1997), such as resulting from brief exposures to stressful temperatures (Sinclair et al. 2016; see Sect. 4.3).

Between the endpoints of TPCs, inflection points can be observed on the curve, indicating positive or negative changes in performance related to temperature increases, often captured by Q_{10} values, surrounding T_{opt} or the optimal recorded thermal performance of a given trait (Sinclair et al. 2016). Because associations between performance and temperature across most of TPCs of physiological traits are non-linear and incorporate exponential sections (Figs. 4.1 and 4.3), the impacts

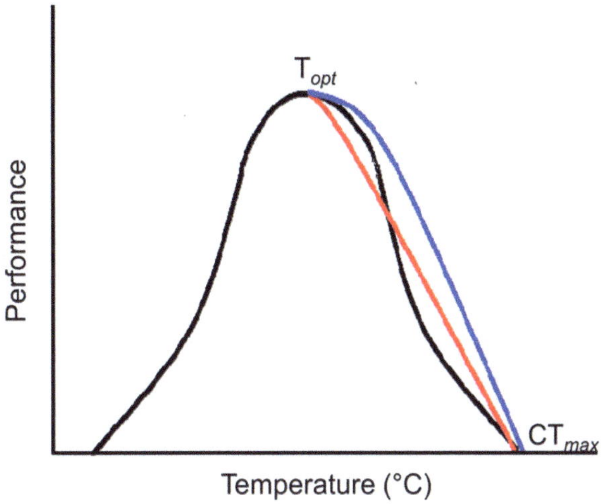

Fig. 4.3 Examples of thermal performance curves displaying differential non-linear reaction norms. At identical optimal temperature (T_{opt}) and critical thermal maximum (CT_{max}), slopes differ in the shape of the descending part of the curve: one is skewed (black curve), one decreases sharply (red curve), and one decreases smoothly (blue curve)

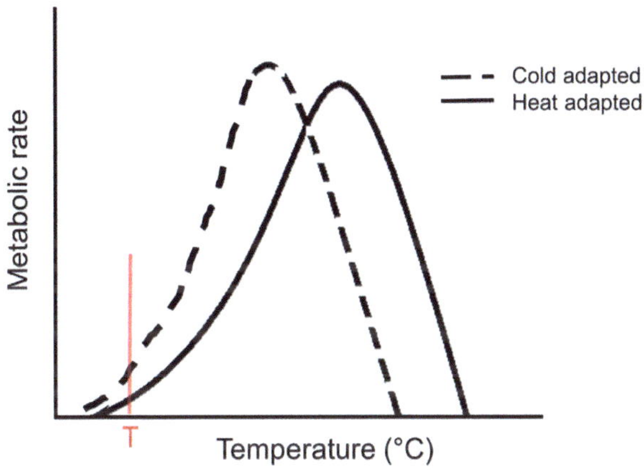

Fig. 4.4 Two thermal performance curves for metabolic rates comparing a cold-adapted species and a heat-adapted species. For example, at the lower temperature range T, the heat-adapted species exhibits low Q_{10} values (shallower TPC slope), indicating low sensitivity to temperature increases at this thermal range. In contrast, the cold-adapted species shows higher Q_{10} values (steeper TPC slope) at T, reflecting greater sensitivity to rising temperatures at this thermal range

of minor temperature fluctuations in different ranges can vary greatly. Whether a given change in temperature has small, negligible, or significant effects on a biological process is therefore function of the shape of its TPC because such traits often display non-linear reaction norms (Sørensen 2020, Fig. 4.3). Thermal sensitivity can be defined as how steeply the change in slope of the organism's performance curve changes with increasing temperature (Rezende et al. 2014; Buckley et al. 2022). For example, in some temperature ranges, a biological process might display low sensitivity to temperature increases (low Q_{10} values), therefore not being significantly impacted by any additional temperature unit (Fig. 4.4). Conversely, the same trait might display increased sensitivity in warmer temperature ranges (higher Q_{10} values), being strongly impacted by any additional temperature unit in that given range (Fig. 4.4) (a concept known as Jensen's inequality—refer to Ruel and Ayres 1999). Note that for cold-adapted species (TPC shifted left towards cooler temperatures, Fig. 4.4), species perform better at cooler temperatures, but they supposedly enter the heat-stressful range at cooler temperatures too, which was also suggested in ants (Willot et al. 2023).

The above complexities underscore the necessity of going beyond the conventional fixed points and linear reaction norms to understand the ultimate impact of temperature shifts on individual fitness and key performance traits and, therefore, to predict the impacts of climate change and temperature shifts on ants. To do so requires measurements that characterize TPCs and assess thermal sensitivity along a non-linear gradient in biologically relevant portions of the curve. Temperature quotient estimates (Q_{10}) can serve this purpose, for example, in the case of estimating the impact of temperatures on metabolic rates (Shik et al. 2019; Willot et al. 2023) (Fig. 4.1, Table 4.1). The thermal sensitivity coefficient (z) (Fig. 4.1, Table 4.1) has been more recently proposed for analysing the impacts of climate change on insects (Rezende et al. 2014; Jørgensen et al. 2022) and more generally the impact of temperature on insect survival (Jørgensen et al. 2019, 2021), including in ants (Willot et al. 2022). z is conceptually related to slope analysis of TPCs: when a trait becomes an exponential function of temperatures in a given range—such as in the case of heat-stress survival time above their physiological damage threshold for insects—a log transformation can be applied to this exponential portion of the TPC to derive a linear function that is easier to manipulate, known in this case as a TDT curve (Rezende et al. 2014; Jørgensen et al. 2019) (see above). While z can only be estimated in the portion of TPCs displaying exponential relationships to stressful temperature (see below), measurements of Q_{10} can be applied across any part of the curve, even considering its ability to estimate both linear and non-linear reaction norms (see Sect. 4.6).

Just as the descending part of the TPC sheds light on how quickly a trait becomes negatively impacted by increasing temperatures, the ascending part provides insight on how quickly a trait becomes positively impacted by temperatures in this range. The thermal sensitivity of key biological traits holds particular significance for ants inhabiting temperate environments with highly variable thermal conditions. For example, when considering metabolic rates, under the metabolic homeostasis hypothesis (MHH), ectotherms inhabiting cooler environments should display higher sensitivity (i.e. high Q_{10} values) of their metabolic rate at lower ranges of temperatures (Cossins and Bowler 1987) to take rapid advantage of mild warming episodes. Conversely, this supposes that cold-adapted species would also reach peak metabolic rates at lower temperatures than in their warm-adapted relatives (Willot et al. 2023). In such environmental settings, where daily favourable temperature ranges for activity are narrow, rapid activation is essential for optimizing energy acquisition and facilitating foraging, growth, and reproduction.

When considering TPCs, a critical point to consider is that they depend on the performance trait considered and whether ants perform in the permissive and/or stressful thermal range for a given trait. For a deeper comprehension of this, we need to define the critical temperature (T_c), which is the temperature (or temperature zone) that defines the transition between the permissive (the temperature range which permits completion of life cycle, i.e. development, maturation, and reproduction, and where thermal injury does not accumulate) and stressful (the temperature range when homeostasis is disrupted such that thermal stress accumulates until it causes heat failure or mortality) temperature range for a population or species (Ørsted et al. 2022) (Fig. 4.1). Hence, TPCs pose a challenge because certain physiological and behavioural traits related to survival must function effectively across both the permissive and stressful temperature ranges (Fig. 4.1a). For traits critical for survival, T_{min} and T_{max} may closely resemble CT_{min} and CT_{max}. However, TPCs for fitness-related traits (e.g. population growth, reproduction, chronic behaviours, etc.) are confined solely to the permissive temperature range (see Ørsted et al. 2024), and in this case their endpoints cannot be strictly considered as the usual measurements of CT_{min} and CT_{max} because the temperature limits of the TPC are not necessarily related to acute failure (Fig. 4.1b). Notably, TPC metrics play crucial roles in our understanding of ant responses to temperature changes. By comprehending the thresholds that define the physiological shift to a stressful range at T_c, and how this relates to fitness, we gain insights into how temperatures and temperature fluctuations might impact the activity and long-term survival of ants in challenging environments.

Despite the importance of considering both the permissive and stressful ranges for the TPCs of key biological trait, most studies investigating the thermal responses of ants have historically focused on survival only (through CT_{max}). Few studies provide comprehensive descriptions of complete TPCs (Roeder et al. 2021; Nascimento et al. 2022). Some studies have offered insights into partial TPCs such as survival time vs temperatures (derived into TDT curves, Willot et al. 2022) or partial TPCs of metabolic rates (Shik et al. 2019; Willot et al. 2023). The majority of other studies that do describe the entirety of TPCs often focus on behavioural traits, such as running speed (e.g. Chick et al. 2021) or foraging activity rates (Andrew et al. 2013; Arnan and Blüthgen 2015; Braschler et al. 2020; Guo et al. 2020). However, these may only partially relate to species' total gain or loss of fitness and survival, and many ant species exhibit locomotor activity at temperatures far from their thermal limits (Cerdá et al. 1998).

Nevertheless, given the significance of the behavioural strategies ants employ to cope with adverse thermal conditions (Wehner and Wehner 2011; Parr and Bishop 2022; see Chap. 2), such behavioural curves could still hold greater relevance than anticipated in forecasting ant responses to thermal change. In fact, and despite the deep contribution of physiological CTLs to our understanding of temperature tolerance in ants, a growing body of literature has more recently proposed to focus on alternative metrics such as behavioural thermal limits derived from local activity measures. For example, recent studies have shown that behavioural limits such as preferred foraging ground temperatures can more accurately predict ant species distributions than CTLs (Braschler et al. 2020; Guo et al. 2020). Constructing behavioural TPCs based on colony activity rates relative to operative temperature could provide a straightforward approach to determine thermal sensitivity of foraging activities and better anticipate ant responses to rising temperatures. Overall, studies focusing on complete TPCs of key traits related to fitness in ants should provide a more comprehensive and detailed insight into how these organisms respond to temperature changes (Sinclair et al. 2016; Little and Seebacher 2021). Conversely, research concentrating on one or a few specific variables or missing the relevant biological range of key trait's TPCs may overlook important information in the broader context (Sinclair et al. 2016), although this is unfortunately the common scenario, at least in the case of ants. Measuring complete ant TPCs is a time-intensive task, often avoided by ecologists seeking simpler metrics for broader species analyses. Understanding the contribution of each TPC variable is crucial to optimize data collection, reconcile approaches, and improve predictions of ant responses to climate change.

Finally, another relevant point to discuss is the phenotypically plastic nature of TPCs (MacLean et al. 2019), either by shifting the T_{opt}, causing the curves to skew to the right or left, or by altering their limits and shape (Fig. 4.3). The modification of the curve's limits through phenotypic plasticity is a key aspect of insect's response to temperature changes and relates overall to the general discussion surrounding CTL plasticity. This aspect will be explored in the following sections.

4.5 Optimal Temperature (T_{opt})

There is a scarcity of available estimates for the T_{opt} of key biological traits in ants, due to the need to measure performance of such traits at various constant temperatures alongside a TPC (Buckley and Kingsolver 2021). However, determining T_{opt} is crucial for understanding the adaptability of organisms in thermally variable environments (Buckley and Kingsolver 2021; Kefford et al. 2022; Pawar et al. 2024). That is because rather than just focusing on short-term responses to extreme temperatures, which are clearly much better evaluated by CTLs and thermal sensitivity measures, through quantifying T_{opt} researchers can better predict how key traits will fare under sustained climate change scenarios (Pawar et al. 2024) and putatively better infer the fitness consequences entailed.

Studies conducted so far with ants have mainly focused on determining the optimal local activity temperature ($T_{opt\text{-}activity}$) at different taxonomic levels and spatial scales (see Guo et al. 2020 for a brief review). Studies on $T_{opt\text{-}activity}$ in ants reveal notable intra- and interspecies variability, reflecting adaptations to different environments and thermal conditions (Guo et al. 2020). At the genus level, $T_{opt\text{-}activity}$ can vary greatly among taxa (Guo et al. 2020). At the community level, $T_{opt\text{-}activity}$ appears to follow average soil and air temperatures in a top mountain ecosystem (Calazans et al. 2020), and the $T_{opt\text{-}activity}$ of 17 Mediterranean ant species differed according to their habitat type (Arnan and Blüthgen 2015). However, within a community, the species $T_{opt\text{-}activity}$ can vary widely, likely in link with the numerous thermal niches species adapt to a local scale (see point on CTLs). For example, in Mediterranean ecosystems, behaviourally dominant ants have significantly lower $T_{opt\text{-}activity}$ than subordinate species (Cerdá et al. 1998; Arnan et al. 2012); this variation in $T_{opt\text{-}activity}$ allows for the coexistence of both behavioural groups each occupying a respective thermal niche in terms of habitat, thereby enhancing local diversity.

As far as we know, no study utilizes $T_{opt\text{-}activity}$ to predict ant responses to climate change. However, there are studies that use T_{opt} of certain traits in conjunction with other variables of the TPC in such analyses. In this regard, Guo et al. (2020) demonstrate that incorporating $T_{opt\text{-}activity}$ into predictive models (behavioural) of ant species distribution greatly enhances the predictive capacity of the models.

Although studies measuring T_{opt} for other traits are rare, research has shown that the running speed of acorn ants (*Temnothorax curvispinosus*), a crucial feature for resource acquisition, shows indications of evolutionary adaptations towards a higher $T_{opt\text{-}running\ speed}$ in urban environments ($T_{opt\text{-}running\ speed}$ = 38.8 °C) compared to rural environments ($T_{opt\text{-}running\ speed}$ = 36.4 °C) (Chick et al. 2021). These findings, although from a single study, may suggest a high adaptive capacity of ants to temperature changes through modification of certain T_{opt} in response to selective pressures, at least from a behavioural standpoint.

In ants, one of the most extensively studied traits in relation to temperature is metabolic rates. However, placing T_{opt} values in a comparative framework for TPCs of metabolic rates in ants is highly complex due to various methodologies used to record these TPCs over the decades. Nonetheless, some literature, including recent studies from Shik et al. (2019) and Willot et al. (2023), suggests that the TPCs of metabolic rates in ants exhibit plasticity. The TPC-metabolic rates, including the optimal temperature for metabolic rate ($T_{opt\text{-}metabolic\ rate}$), tend to shift towards cooler temperatures in response to higher altitudes, higher latitudes, and generally cooler climates. However, it remains unclear whether these shifts are due to local adaptations at the population or species level, acclimation, or both.

Overall, we need more studies on the adaptive potential of T_{opt} of key adaptive traits to gain a comprehensive understanding of how ant populations will respond to long-term sustained climatic warming. In fact, T_{opt} for some key traits appears to be able to evolve rapidly under selection, as they face fewer thermodynamic constraints compared to thermal sensitivity or CT_{max} (Buckley and Kingsolver 2021; Kefford et al. 2022; Pawar et al. 2024). This is especially relevant because unlike immediate, rapid, or short-term responses to environmental temperature changes (i.e. phenotypic plasticity), the adaptive genetic adjustments in thermal fitness are predominantly driven by shifts in the T_{opt} of underlying traits; these adjustments are thus encapsulated by alterations in T_{opt} (Pawar et al. 2024).

4.6 Temperature Quotient (Q_{10})

The Q_{10} measures how a particular physiological/metabolic (also used for behavioural and ecological) rate varies over a temperature range of 10 °C (Ørsted et al. 2022). Simply put, Q_{10} is a measure of the temperature sensitivity of any biological process across a 10 °C range (Lighton 2008). It is calculated by comparing two empirical measurements of biological processes at two temperatures following the formula: $Q_{10} = (R2/R1)(10/(T2-T1))$. The Q_{10} coefficient is thus equal to the rate ratio at a temperature increase of 10° (Lighton 2008; Laspoumaderes et al. 2022). Another way to approach Q_{10} values is through a linear regression transformation, where the physiological (or any) rate is measured at different temperatures in the laboratory, and if the relationship detected is exponential, then the data can be plotted on a log-linear graph and the slope of transformed linear regression estimated (this relates to the *z* parameter in the case of knockdown time vs stressful temperatures; Lighton 2008). It is important to note that Q_{10} values can vary significantly depending on the selected temperature intervals, considering the non-linear shape of the TPC (Shik et al. 2019, Fig. 4.3). For example, the sensitivity of the rate to temperature may be higher in certain intervals. This non-linearity can be captured with Q_{10} and with *z* if the function is exponential (and within the TPC stressful range).

It is important to highlight that Q_{10} research in ants has primarily been used in the field of metabolic rates TPCs, specifically in the case of cold tolerance to determine if there is metabolic compensation as an adaptive response in ants to cope with colder temperatures. This is because according to the MHH in ectotherms (Cossins and Bowler 1987), cold-adapted species are expected to display higher sensitivity of their metabolic rates (higher Q_{10} values) at lower temperatures, allowing them to quickly take advantage of mild warming episodes in their natural environment (Fig. 4.4).

It is worth noting that while several studies have investigated these mechanisms, none specifically within the context of understanding responses to climate change. Initial research on the Q_{10} of key biological traits in ants began with a focus on walking speed in relation to temperature (Shapley 1924). Over the last century, subsequent studies have investigated thermal sensitivity in various other traits, chiefly metabolic rates as a function of latitude, climate, and altitude (Kennington 1957; MacKay 1982; Nielsen 1986; Nielsen et al. 1999; Hurlbert IV et al. 2008; Penick et al. 2017; Shik et al. 2019; Willot et al. 2023) and diet (Schilman and Roces 2006). Interestingly, recent evidences report that ant species from higher latitudes (from a pool of 13 species across a latitudinal gradient in Europe) and altitudes displayed higher Q_{10} values for their metabolic rates at lower temperature as compared to their Mediterranean/lowland relatives (Shik et al. 2019; Willot et al. 2023), i.e. a trend that aligns with the MHH (Cossins and Bowler 1987). Contrary to the body of literature concerning insects in general (Jørgensen et al. 2022), measurements of Q_{10} have been scarcely used to make predictions about ants' responses to climate change. However, Willot et al. (2023) suggested that Q_{10} values of metabolic rates at higher temperatures could be inversely correlated with species' limits of latitudinal distribution (the species tested extending their range at higher latitudes tend to approach peaking metabolic rates with lower Q_{10} values in the 25–30 °C range, suggesting a proximity with T_{opt} for metabolic rate in this range). Moreover, they found that the combination of cold tolerance and metabolic rates at lower temperatures surpasses either variable alone in predicting species' distribution range limits and climatic temperatures. This implies that cold tolerance and metabolic rates provide complementary insights into the evolutionary history to colder climates in ants and, used in tandem, might be good predictors of future ant distribution under climate change scenarios.

Regarding adaptation or acclimation, the scarce literature also emphasizes that ants can achieve metabolic compensation to colder climates (i.e. TPC of metabolic rates shifted left because of cold acclimation), either through phenotypic plasticity, interspecific adaptive shifts, or, more likely, a combination of both mechanisms (Shik et al. 2019; Willot et al. 2023). It appears the TPC for metabolic rates, inherently including T_{opt} for metabolic rate in addition to Q_{10} values describing the slopes of the relationship, tends to shift towards cooler temperatures with increasing altitude (Shik et al. 2019) and higher latitude (Willot et al. 2023) as well. It remains unclear whether these patterns are due to local adaptations at the population or species level, acclimation, or a combination of both. Our understanding of their potential for adaptation or acclimatization of Q_{10} to cooling or warming remains profoundly limited by the scarcity of data to date.

4.7 Integrating CTLs and TPCs

Considering the potential energy costs of maintaining high CT_{max} and low CT_{min} simultaneously (Sinclair et al. 2016), one might expect them to be strongly negatively correlated. However, the evidence presented here challenges this assumption: while evidences point out that CT_{max} is strongly evolutionarily conserved and responds to microclimates, CT_{min} is poorly evolutionarily conserved, is extremely plastic, and responds to macroclimates. It thus seems that the upper and lower thermal limits follow different selective pressures. In this line, research investigating both variables in ants have generally found only a weak positive correlation between them (Diamond and Chick 2018a) or no correlation at all (Kaspari et al. 2015; Baudier and O'Donnell 2018; Andrew et al. 2019; Bujan et al. 2020a; O'Donnell et al. 2020). For example, a study conducted at the genus level of North American and European ants found that some genera have high tolerance to both heat and cold (e.g. arboreal genera such as *Crematogaster*, *Pseudomyrmex*, and *Leptothorax*), while others have low tolerance to both heat and cold (e.g. genera such as *Brachymyrmex*, *Myrmecina*, *Acromyrmex*, and *Atta* and species like *Aphaenogaster subterranea*). On the other hand, some ant genera may have high heat tolerance but low cold tolerance (e.g. *Forelius* and *Pogonomyrmex*), and vice versa (low heat tolerance and high cold tolerance) is also possible (e.g. *Myrmica* and *Prenolepis*) (Bujan et al. 2020a; Roeder et al. 2021; Willot et al. 2022, 2023). While the first two groups would belong to thermal generalists, the last two groups would correspond to thermal specialists, hot and cold specialists, respectively. Whatever the case, it appears that the ability to tolerate high and low temperatures in ants is developed independently and can respond to different respective pressures, and thus they can also differently withstand the challenges of climate change.

When we aim to relate the optimum point of TPCs (T_{opt}) for any given key biological trait, the ant studies are scarce and mainly focusing on local activity and/or metabolic data (Cerdá et al. 1998; Arnan et al. 2012; Shik et al. 2019). Regarding activity and to the best of our knowledge, only one study makes a direct association between the T_{opt} for a trait, namely, locomotor activity in this case, and heat tolerance (Arnan and Blüthgen 2015). This study found that $T_{opt\text{-}locomotor\ activity}$ and CT_{max} from 17 European ant species estimated from local activity data were moderately and positively correlated. We can therefore assume that species with a higher heat tolerance indeed display higher temperature for the T_{opt} of other biological function as well. However, caution in this assumption needs to be exerted, as the distance at which species perform their activities optimally appears not systematically proportional to their thermal limits (Guo et al. 2020). In this line, several studies indirectly support that this relationship between $T_{opt\text{-activity}}$ and CT_{max} would be not generalized and that such a pattern would mostly depend on the regional context and pool of species-specific thermal strategies studied. For instance, various authors have intuitively suggested two strategies, spanning a gradient of intermediate scenarios, while describing thermal behaviour and activity patterns of coexisting ant species in arid regions (Heatwole and Muir 1989; Heatwole and Harrington 1989; Cerdá et al. 1998). This conceptualization distinguishes two extreme foraging strategies concerning temperature:

1. Heat-intolerant ant species exhibit risk-averse behaviour, foraging at temperatures significantly distant from their CTLs. This is the case of species such as *Tapinoma nigerrimum* and *Pheidole pallidula* in Mediterranean ecosystems (Cerdá et al. 1998).
2. Heat-tolerant ant species display risk-prone behaviour, foraging very close to their CTLs and facing a heightened risk of heat-related mortality. This strategy benefits heat-tolerant species by enhancing foraging performance at high temperatures. For instance, this is the case of *Cataglyphis* species in arid and semi-arid regions from North Africa, Europe, and Asia (Cerdá and Retana 2000), of *Ocymyrmex* species from southern and eastern Africa (Marsh 1985), and some *Melophorus* species from central Australia (Christian and Morton 1992).

   Each of these strategies, in terms of mortality risk, aligns with the two predominant risk strategies described in optimal foraging literature: while heat-intolerant species adopt a risk-averse approach, foraging at temperatures far from their CTLs to avoid the risk of heat-induced mortality, heat-tolerant species adopt a risk-prone approach, foraging near their CTLs and accepting a higher risk of heat-related mortality (Stephens and Krebs 1986; see Chap. 3).

To explore such link between CTLs and activity, Guo et al. (2020) constructed TPCs based on behavioural thermal limits of activity and extracted the optimal temperature for foraging activity (i.e. based on activity rhythms, realized niche) for 20 genera distributed worldwide, indirectly linking behavioural thermal minimum, behavioural thermal maximum, and $T_{opt\text{-}activity}$. In link with previous conclusions, they highlight two different strategies. First, for certain genera such as *Dorymyrmex*, *Iridomyrmex*, and *Pogonomyrmex*, $T_{opt\text{-}activity}$ is much closer to behavioural thermal

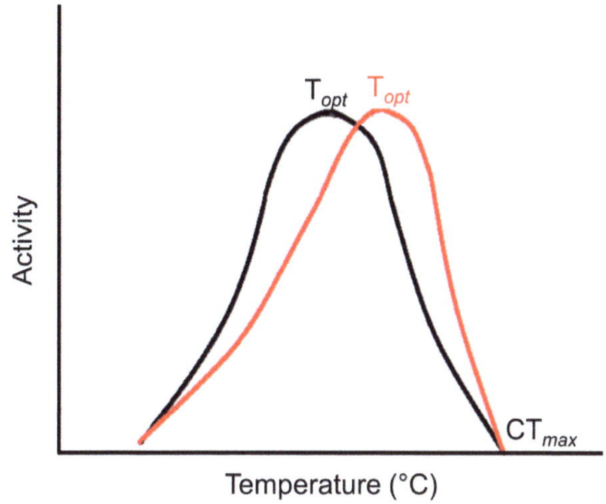

Fig. 4.5 Thermal performance curves depicting the relationship between temperature and activity. The T_{opt} and CT_{max} (not part of the curve per se) are shown. The distance between T_{opt} and CT_{max} is smaller in the red curve, indicating that even slight temperature increases beyond the T_{opt} could significantly impact resource acquisition for this species, thereby affecting its fitness. In contrast, the black curve shows a greater distance between T_{opt} and CT_{max} (and equidistant with CT_{min} and CT_{max}), suggesting that this species would be less impacted by minor temperature increases beyond its T_{opt}

maximum than to behavioural thermal minimum, and the TPC of their foraging activity is heavily skewed to cooler temperatures (Fig. 4.5). In such scenario, even slight temperature increase going beyond $T_{opt\text{-}activity}$ may significantly impact species' performance and fitness by limiting their ability to forage for resources. Second, for other species, $T_{opt\text{-}activity}$ may also be equidistant from behavioural thermal maximum and behavioural thermal minimum, as observed in genera like *Crematogaster*, *Messor*, and *Pheidole*. Here, the TPC of foraging activity tends to be more unimodal rather than skewed to cooler temperatures (Fig. 4.5). This implies that these ants have a broad range of temperatures beyond T_{opt} where they can carry out their activities without experiencing thermal stress, and their activity will be less impacted by marginal temperature increase beyond their optimal activity temperature.

The analysis of the relationship between upper and lower limits of thermal tolerance and the T_{opt} for activity ($T_{opt\text{-}activity}$) inevitably leads us to talk about thermal sensitivity (i.e. a trait's response to changes in temperature). One might think that the greater the distance between $T_{opt\text{-}activity}$ and CT_{min}, the lower the sensitivity to cold increases, and the greater the distance between $T_{opt\text{-}activity}$ and CT_{max}, the lower the sensitivity to temperature increases. However, this is not as trivial, and the shape of the curve can range from linear to non-linear (e.g. TPC for activity can be skewed and decrease sharply or smoothly after T_{opt}; Fig. 4.3), which requires a deeper analysis of these segments of the curve. This can be more accurately assessed using the temperature quotient (Q_{10}), especially for non-linear relationships, or the thermal sensitivity index *z*, providing deeper insights into an organism's responsiveness to temperature variations (see above). In this case, we might expect a strong correlation between the difference between $T_{opt\text{-}activity}$ and one of the thermal limits and the corresponding thermal sensitivity index. However, this has never been analysed before, probably because, if one has determined the T_{opt} of a trait, they have already constructed a full TPC. A full TPC allows for direct calculation of Q_{10} on any given TPC interval, without the need to infer the difference between T_{opt} and CTLs to approximate TPC sensitivity indirectly. Based on the previous paragraph, we would expect there to be no relationship between the CTLs and the thermal sensitivity indices, as it was indeed suggested in other insect taxa as well (Jørgensen et al. 2019). To the best of our knowledge, the only study addressing this point found a lack of clear association between CT_{max} and thermal sensitivity as well (Willot et al. 2022).

Second, considering traits such as metabolic rates and CTLs in ants, only one ant study relates lower thermal limits with the temperature quotient (Q_{10}) of the standard metabolic rates of species at different temperatures (Willot et al. 2023). This data supported an inverted relationship between latitude and $Q_{10\text{-}metabolic\ rate}$ in the 25–30 °C range interval, suggesting that ant species extending their range northward could be closer to T_{opt} in the 25–30 °C (low Q_{10} values) than their southern adapted relatives. However, such direct association between $Q_{10\text{-}metabolic\ rate}$ in the 25–30 °C and cold tolerance (CT_{min})—rather than distributional limits—was much less pronounced (non-significant trend). This means they could provide complementary information: enhanced cold tolerance can be hypothesized to aid survival in extreme cold, while metabolic compensation putatively facilitates resource acquisition in moderately suboptimal conditions (Willot et al. 2023). Regardless, whether Q_{10} values of the metabolic rate TPC truly correlate with CTLs awaits further examination.

In short, exploring potential correlations between variables describing heat tolerance and the TPC of key physiological traits can help us understand the thermal biology of species as a whole and, thus, refine our predictions about the effects of climate change on ants. Just as CT_{max} and CT_{min} appear to respond mostly independently to selective pressure in ants, TPC variables such as T_{min}, T_{max}, T_{opt}, and Q_{10} values could also be uncoupled, as well as their relationship with the corresponding thermal limits. Gaining a global overview of TPC is thus paramount to fully understand how climate change might impact key biological traits and thus infer gain and loss of fitness, under the complexity of a warming climate.

4.8 Final Considerations

Only approximately 3% of ant species have their CTLs documented (Nascimento et al. 2022), often using differing methodologies, which hinders direct comparisons and generalizations. Furthermore, other variables related to thermal biology, such as TPCs and sensitivity metrics, remain even less explored. To improve our understanding of ant thermal responses, it is essential to reconcile existing methodologies and develop a unified framework. This chapter contributes to that goal by synthesizing available data, exploring methodological equivalencies, and proposing pathways to harmonize research approaches.

Despite these challenges, we can derive several general predictions about ant responses to climate change:

1. Species distribution shifts: Thermal tolerance breadth (CT_{max}–CT_{min}) and metrics like warming tolerance suggest that species with narrower tolerance ranges, usually living in equally more thermally stable environment, may be more vulnerable to climate change, while those with broader ranges are likely to expand their habitats into newly suitable areas.
2. Thermal adaptation limits: High phylogenetic signal in CT_{max} suggests strong evolutionary constraints on heat tolerance adaptation at evolutionary scales. Cold tolerance (CT_{min}) and TPCs of metabolic rates, however, show greater plasticity and responsiveness to environmental gradients, which may be linked with the ability of species to expand into colder climates.
3. Behavioural adaptations: Ant species may increasingly rely on behavioural thermoregulation, such as shifting foraging activity to cooler periods, to buffer against thermal stress, highlighting the need for behavioural thermal limits in predictive models.
4. Microclimatic and biotic interactions: Local microclimates and biotic interactions will continue to modulate ant communities, with heat-adapted species likely gaining dominance in warming environments, while cold-adapted specialists face higher extinction risks.
5. Adaptive potential of T_{opt}: To fully comprehend how ant populations and communities will respond to long-term, sustained climatic warming, it is crucial to deepen our understanding of the adaptive potential of T_{opt} for key traits.

Moving forward, integrating CTLs, TPCs, and innovative metrics such as thermal sensitivity coefficients (z) and classic metrics such as Q_{10} will enhance our ability to predict ant responses to climate change. This approach not only unifies disparate methodologies but also fosters more generalizable insights into how ants and their ecosystems will adapt to a warming world. In addition, understanding the potential for phenotypic plasticity of these parameters as well as CTLs will likely be crucial for predicting whether species can rely on plastic responses to withstand the effect of climate change. Addressing these knowledge gaps will provide a robust foundation for predicting and mitigating the impacts of climate change on ants, their communities, and their ecological roles.

References

Andersen JL, Manenti T, Sørensen JG, MacMillan HA, Loeschcke V, Overgaard J (2015) How to assess *Drosophila* cold tolerance: chill coma temperature and lower lethal temperature are the best predictors of cold distribution limits. Funct Ecol 29:55–65. https://doi.org/10.1111/1365-2435.12310

Andrew NR, Hart RA, Jung MP, Hemmings Z, Terblanche JS (2013) Can temperate insects take the heat? A case study of the physiological and behavioural responses in a common ant, Iridomyrmex purpureus (Formicidae), with potential climate change. J Insect Physiol 59:870–880. https://doi.org/10.1016/j.jinsphys.2013.06.003

Andrew NR, Miller C, Hall G, Hemmings Z, Oliver I (2019) Aridity and land use negatively influence a dominant species' upper critical thermal limits. PeerJ 6:e6252. https://doi.org/10.7717/peerj.6252

Angilletta M (2009) Thermal adaptation: a theoretical and empirical synthesis. Oxford Academic. https://doi.org/10.1093/acprof:oso/9780198570875.001.1

Araujo NDS, Perez R, Willot Q, Defrance M, Aron S (2023) Facing lethal temperatures: heat-shock response in desert and temperate ants. Ecol Evol 13:e10438. https://doi.org/10.1101/2022.11.12.516273

Arnan X, Blüthgen N (2015) Using ecophysiological traits to predict climatic and activity niches: lethal temperature and water loss in Mediterranean ants. Glob Ecol Biogeogr 24:1454–1464. https://doi.org/10.1111/geb.12363

Arnan X, Cerdá X, Retana J (2012) Distinctive life traits and distribution along environmental gradients of dominant and subordinate Mediterranean ant species. Oecologia 170:489–500. https://doi.org/10.1007/s00442-012-2315-y

Baudier KM, Mudd AE, Erickson SC, O'Donnell S (2015) Microhabitat and body size effects on heat tolerance: implications for responses to climate change (army ants: Formicidae, Ecitoninae). J Anim Ecol 84:1322–1330. https://doi.org/10.1111/1365-2656.12388

Baudier KM, D'Amelio CL, Malhotra R, O'Connor MP, O'Donnell S (2018) Extreme insolation: climatic variation shapes the evolution of thermal tolerance at multiple scales. Am Nat 192:347–359. https://doi.org/10.1086/698656

Baudier K, O'Donnell S (2018) Complex body size differences in thermal tolerance among army ant workers (Eciton burchellii parvispinum). J Therm Biol 78:277–280. https://doi.org/10.1016/j.jtherbio.2018.10.011

Bigelow WD (1921) The logarithmic nature of thermal death time curves. J Infect Dis 29:528–536

Bishop TR, Robertson MP, Van Rensburg BJ, Parr CL (2017) Coping with the cold: minimum temperatures and thermal tolerances dominate the ecology of mountain ants. Ecol Entomol 42:105–114. https://doi.org/10.1111/een.12364

Bishop TR, Parr CL, Gibb H, van Rensburg BJ, Braschler B, Chown SL, Foord SH, Lamy K, Munyai TC, Okey I, Tshivhandekano PG, Werenkraut V, Robertson MP (2019) Thermoregulatory traits combine with range shifts to alter the future of montane ant assemblages. Glob Chang Biol 25:2162–2173. https://doi.org/10.1111/gcb.14622

Boulay R, Aron S, Cerdá X, Doums C, Graham P, Hefetz A, Monnin T (2017) Social life in arid environments: the case study of *Cataglyphis ants*. Annu Rev Entomol 62:305–321. https://doi.org/10.1146/annurev-ento-031616-034941

Braschler B, Duffy GA, Nortje E, Kritzinger-Klopper S, du Plessis D, Karenyi N, Leihy RI, Chown SL (2020) Realised rather than fundamental thermal niches predict site occupancy: implications for climate change forecasting. J Anim Ecol 89:2863–2875. https://doi.org/10.1111/1365-2656.13358

Buckley LB, Kingsolver JG (2021) Evolution of thermal sensitivity in changing and variable climates. Annu Rev Ecol Evol Syst 52:563–586. https://doi.org/10.1146/annurev-ecolsys-011521-102856

Buckley LB, Huey RB, Kingsolver JG (2022) Asymmetry of thermal sensitivity and the thermal risk of climate change. Glob Ecol Biogeogr 31:2231–2244. https://doi.org/10.1111/geb.13570

Bujan J, Kaspari M (2017) Nutrition modifies critical thermal maximum of a dominant canopy ant. J Insect Physiol 102:1–6. https://doi.org/10.1016/j.jinsphys.2017.08.007

Bujan J, Roeder KA, de Beurs K, Weiser MD, Kaspari M (2020a) Thermal diversity of North American ant communities: cold tolerance but not heat tolerance tracks ecosystem temperature. Glob Ecol Biogeogr 29:1486–1494. https://doi.org/10.1111/geb.13121

Bujan J, Roeder KA, Yanoviak SP, Kaspari M (2020b) Seasonal plasticity of thermal tolerance in ants. Ecology 101:e03051. https://doi.org/10.1002/ecy.3051

Bujan J, Charavel E, Bates OK, Gippet JMW, Darras H, Lebas C, Bertelsmeier C (2021) Increased acclimation ability accompanies a thermal niche shift of a recent invasion. J Anim Ecol 90:483–491. https://doi.org/10.1111/1365-2656.13381

Calazans EG, da Costa FV, Cristiano MP, Cardoso DC (2020) Daily dynamics of an ant community in a mountaintop ecosystem. Environ Entomol 49:383–390. https://doi.org/10.1093/ee/nvaa011

Câmara T, Andersen AN, Nascimento G, Arnan X (2024) Thermal tolerance and species distributions: interactions between latitude, elevation and arboreality in ants. Divers Distrib 30:e13936. https://doi.org/10.1111/ddi.13936

Cerdá X, Retana J (2000) Alternative strategies by thermophilic ants to cope with extreme heat: individual versus colony level traits. Oikos 89:155–163. https://doi.org/10.1034/j.1600-0706.2000.890117.x

Cerdá X, Retana J, Cros S (1998) Critical thermal limits in Mediterranean ant species: trade-off between mortality risk and foraging performance. Funct Ecol 12:45–55. https://doi.org/10.1046/j.1365-2435.1998.00160.x

Chick LD, Lessard JP, Dunn RR, Sanders NJ (2020) The coupled influence of thermal physiology and biotic interactions on the distribution and density of ant species along an elevational gradient. Diversity 12:456. https://doi.org/10.3390/d12120456

Chick LD, Waters JS, Diamond SE (2021) Pedal to the metal: cities power evolutionary divergence by accelerating metabolic rate and locomotor performance. Evol Appl 4:36–52. https://doi.org/10.1111/eva.13083

Chou CC, Perez DM, Johns S, Gardner R, Kerr KA, Head ML, McCullough EL, Backwell PRY (2019) Staying cool: the importance of shade availability for tropical ectotherms. Behav Ecol Sociobiol 73:106. https://doi.org/10.1007/s00265-019-2721-9

Christian KA, Morton SR (1992) Extreme thermophilia in a central Australian ant, *Melophorus bagoti*. Physiol Zool 65:885–905

Cokendolpher JC, Phillips SA (1990) Critical thermal limits and locomotor activity of the red imported fire ant (Hymenoptera: Formicidae). Environ Entomol 19:878–881. https://doi.org/10.1093/ee/19.4.878

Cossins AR, Bowler K (1987) Rate compensations and capacity adaptations. In: Temperature biology of animals. Springer, Dordrecht, pp 155–203

Coulin C, de la Vega GJ, Chifflet L, Calcaterra LA, Schilman PE (2019) Linking thermo-tolerances of the highly invasive ant, *Wasmannia auropunctata*, to its current and potential distribution. Biol Invasions 21:3491–3504. https://doi.org/10.1007/s10530-019-02063-0

Del Toro I, Ribbons RR, Pelini SL (2012) The little things that run the world revisited: a review of ant-mediated ecosystem services and disservices (Hymenoptera: Formicidae). Myrmecol News 17:133–146

Diamond SE, Chick LD (2018a) The Janus of macrophysiology: stronger effects of evolutionary history, but weaker effects of climate on upper thermal limits are reversed for lower thermal limits in ants. Curr Zool 64:223–230. https://doi.org/10.1093/cz/zox072

Diamond SE, Chick LD (2018b) Thermal specialist ant species have restricted, equatorial geographic ranges: implications for climate change vulnerability and risk of extinction. Ecography 41:1507–1509. https://doi.org/10.1111/ecog.03264

Diamond SE, Nichols LM, McCoy N, Hirsch C, Pelini SL, Sanders NJ, Ellison AM, Gotelli NJ, Dunn RR (2012a) A physiological trait-based approach to predicting the responses of species to experimental climate warming. Ecology 93:2305–2312. https://doi.org/10.1890/11-2296.1

Diamond SE, Sorger DM, Hulcr J, Pelini SL, Del Toro I, Hirsch C, Oberg E, Dunn RR (2012b) Who likes it hot? A global analysis of the climatic, ecological, and evolutionary determinants of warming tolerance in ants. Glob Chang Biol 18:448–456. https://doi.org/10.1111/j.1365-2486.2011.02542.x

Diamond SE, Chick L, Perez A, Strickler SA, Martin RA (2017) Rapid evolution of ant thermal tolerance across an urban-rural temperature cline. Biol J Linn Soc 121:248–257. https://doi.org/10.1093/biolinnean/blw047

Elizalde L, Arbetman M, Arnan X, Eggleton P, Leal IR, Lescano MN, Saez A, Werenkraut V, Pirk GI (2020) The ecosystem services provided by social insects: traits, management tools and knowledge gaps. Biol Rev 95:1418–1441. https://doi.org/10.1111/brv.12616

Esch C, Jimenez JP, Peretz C, Uno H, O'Donnell S (2017) Thermal tolerances differ between diurnal and nocturnal foragers in the ant *Ectatomma ruidum*. Insect Soc 64:439–444. https://doi.org/10.1007/s00040-017-0555-x

Frazier M, Huey RB, Berrigan D (2006) Thermodynamics constrains the evolution of insect population growth rates: "warmer is better". Am Nat 168:512–520. https://doi.org/10.1086/506977

Freires ASA, Ferreira CAS, Nascimento G, Arnan X (2023) Carbohydrate-rich diet increases critical thermal maximum in ants. Entomol Exp Appl 171:240–247. https://doi.org/10.1111/eea.13268

Garcia-Robledo C, Chuquillanqui H, Kuprewicz EK, Escobar-Sarria F (2018) Lower thermal tolerance in nocturnal than in diurnal ants: a challenge for nocturnal ectotherms facing global warming. Ecol Entomol 43:162–167. https://doi.org/10.1111/een.12481

González-Tokman D, Dáttilo W (2024) Effects of climate change on insects: physiological, evolutionary, and ecological responses. Oxford University Press. https://doi.org/10.1093/oso/9780192864161.001.0001

Guo F, Guénard B, Economo EP, Deutsch CA, Bonebrake TC (2020) Activity niches outperform thermal physiological limits in predicting global ant distributions. J Biogeogr 47:829–842. https://doi.org/10.1111/jbi.13799

Habibullah MS, Din BH, Tan SH, Zahid H (2022) Impact of climate change on biodiversity loss: global evidence. Environ Sci Pollut Res 29:1073–1086. https://doi.org/10.1007/s11356-021-15702-8

Heatwole H, Harrington S (1989) Heat tolerances of some ants and beetles from the pre-Saharan steppe of Tunisia. J Arid Environ 16:69–77. https://doi.org/10.1016/S0140-1963(18)31048-6

Heatwole H, Muir R (1989) Seasonal and daily activity of ants in the pre-Saharan steppe of Tunisia. J Arid Environ 16:49–67. https://doi.org/10.1016/S0140-1963(18)31047-4

Hölldobler B, Wilson EO (1990) The ants. Harvard University Press, Berlin

Huey RB, Stevenson RD (1979) Integrating thermal physiology and ecology of ectotherms: a discussion of approaches. Am Zool 19:357–366. https://doi.org/10.1093/icb/19.1.357

Hurlbert AH IV, Ballantyne F, Powell S (2008) Shaking a leg and hot to trot: the effects of body size and temperature on running speed in ants. Ecol Entomol 33:144–154. https://doi.org/10.1111/j.1365-2311.2007.00962.x

Janzen DH (1967) Why mountain passes are higher in the tropics. Am Nat 101:233–249. https://doi.org/10.1086/282487

Jørgensen LB, Ørsted M, Malte H, Wang T, Overgaard J (2022) Extreme escalation of heat failure rates in ectotherms with global warming. Nature 611:93–98. https://doi.org/10.1038/s41586-022-05334-4

Jumbam KR, Jackson S, Terblanche JS, McGeoch MA, Chown SL (2008) Acclimation effects on critical and lethal thermal limits of workers of the Argentine ant, *Linepithema humile*. J Insect Physiol 54:1008–1014. https://doi.org/10.1016/j.jinsphys.2008.03.011

Jørgensen LB, Malte H, Overgaard J (2019) How to assess Drosophila heat tolerance: Unifying static and dynamic tolerance assays to predict heat distribution limits. Funct Ecol 33:629–642. https://doi.org/10.1111/1365-2435.13279Digital

Jørgensen LB, Malte H, Ørsted M, Klahn NA, Overgaard J (2021) A unifying model to estimate thermal tolerance limits in ectotherms across static, dynamic and fluctuating exposures to thermal stress. Sci Rep 11:12840. https://doi.org/10.1038/s41598-021-92004-6

Kaspari M, Clay NA, Lucas J, Yanoviak SP, Kay A (2015) Thermal adaptation generates a diversity of thermal limits in a rainforest ant community. Glob Chang Biol 21:1092–1102. https://doi.org/10.1111/gcb.12750

Kaspari M, Bujan J, Roeder KA, de Beurs K, Weiser MD (2019) Species energy and thermal performance theory predict 20-yr changes in ant community abundance and richness. Ecology 100:e02888. https://doi.org/10.1002/ecy.2888

Kay CAR, Whitford WG (1978) Critical thermal limits of desert honey ants: possible ecological implications. Physiol Zool 51:206–213. https://doi.org/10.1086/physzool.51.2.30157867

Kefford BJ, Ghalambor CK, Dewenter B, Poff NL, Hughes J, Reich J, Thompson R (2022) Acute, diel, and annual temperature variability and the thermal biology of ectotherms. Glob Chang Biol 28:6872–6888. https://doi.org/10.1111/gcb.16453

Kennington GS (1957) Influence of altitude and temperature upon rate of oxygen consumption of *Tribolium confusum* Duval and *Camponotus pennsylvanicus* Modoc Wheeler. Ecol Evol Physiol 30:305. https://doi.org/10.1086/physzool.30.4.30152212

Laspoumaderes C, Meunier CL, Magnin A, Berlinghof J, Elser JJ, Balseiro E, Torres G, Modenutti B, Tremblay N, Boersma M (2022) A common temperature dependence of nutritional demands in ectotherms. Ecol Lett 25:2189–2202. https://doi.org/10.1111/ele.14093

Leahy L, Scheffers BR, Williams SE, Andersen AN (2022) Arboreality drives heat tolerance while elevation drives cold tolerance in tropical rainforest ants. Ecology 103:e03549. https://doi.org/10.1002/ecy.3549

Leahy L, Scheffers BR, Andersen AN, Williams SE (2024) Rates of species turnover across elevation vary with vertical stratum in rainforest ant assemblages. Ecography 2024:e06972. https://doi.org/10.1111/ecog.06972

Lighton JRB (2008) Measuring metabolic rates: a manual for scientists. Oxford University Press, New York

Lutterschmidt WI, Hutchison VH (1997) The critical thermal maximum: data to support the onset of spasms as the definitive end point. Can J Zool 75:1553–1560

Little AG, Seebacher F (2021) Physiological performance curves: when are they useful? Front Physiol 12:805102. https://doi.org/10.3389/fphys.2021.805102

MacKay WP (1982) An altitudinal comparison of oxygen consumption rates in three species of *Pogonomyrmex* harvester ants (Hymenoptera: Formicidae). Physiol Zool 55:367–377. https://doi.org/10.1086/physzool.55.4.30155864

MacLean HJ, Sørensen JG, Kristensen TN, Loeschcke V, Beedholm K, Kellermann V, Overgaard J (2019) Evolution and plasticity of thermal performance: an analysis of variation in thermal tolerance and fitness in 22 Drosophila species. Philos Trans R Soc B Biol Sci 374:20180548. https://doi.org/10.1098/rstb.2018.0548

Marsh AC (1985) Thermal responses and temperature tolerance in a diurnal desert ant, *Ocymyrmex barbiger*. Physiol Zool 58:629–636

Martin RA, Chick LD, Yilmaz AR, Diamond SE (2019) Evolution, not transgenerational plasticity, explains the adaptive divergence of acorn ant thermal tolerance across an urban–rural temperature cline. Evol Appl 128:1678–1687. https://doi.org/10.1111/eva.12826

Menzel F, Zumbusch M, Feldmeyer B (2018) How ants acclimate: impact of climatic conditions on the cuticular hydrocarbon profile. Funct Ecol 32:657–666. https://doi.org/10.1111/1365-2435.13008

Malusare SP, Zillio G & Fronhofer EA (2023) Evolution of thermal performance curves: a meta-analysis of selection experiments. J Evol Biol 36:15–28. https://doi.org/10.1111/jeb.14087

Nascimento G, Câmara T, Arnan X (2022) Critical thermal limits in ants and their implications under climate change. Biol Rev 97:1287–1305. https://doi.org/10.1111/brv.12843

Nelson AS, Scott T, Barczyk M, McGlynn TP, Avalos A, Clifton E, Das A, Figueiredo A, Figueroa LL, Janowiecki M, Pahlke S, Rana JD, O'Donnell S (2018) Day/night upper thermal limits differ within *Ectatomma ruidum* ant colonies. Insect Soc 65:183–189. https://doi.org/10.1007/s00040-017-0585-4

Nielsen MG (1986) Respiratory rates of ants from different climatic areas. J Insect Physiol 32:125–131. https://doi.org/10.1016/0022-1910(86)90131-9

Nielsen MG, Elmes GW, Kipyatkov VE (1999) Respiratory Q10 varies between populations of two species of Myrmica ants according to the latitude of their sites. J Insect Physiol 45:559–564. https://doi.org/10.1016/S0022-1910(98)00162-0

Nowakowski AJ, Frishkoff LO, Agha M, Todd BD, Scheffers BR (2018) Changing thermal landscapes: merging climate science and landscape ecology through thermal biology. Curr Landsc Ecol Rep 3:57–72. https://doi.org/10.1007/s40823-018-0034-8

O'Donnell S, Bulova S, Caponera V, Oxman K, Giladi I (2020) Species differ in worker body size effects on critical thermal limits in seed-harvesting desert ants (*Messor ebeninus* and *M. arenarius*). Insect Soc 67:473–479. https://doi.org/10.1007/s00040-020-00782-5

Ørsted M, Jørgensen LB, Overgaard J (2022) Finding the right thermal limit: a framework to reconcile ecological, physiological and methodological aspects of CTmax in ectotherms. J Exp Biol 225:jeb244514. https://doi.org/10.1242/jeb.244514

Ørsted M, Willot Q, Olsen AK, Kongsgaard V, Overgaard J (2024) Thermal limits of survival and reproduction depend on stress duration: a case study of Drosophila suzukii. Ecol Lett 27: e14421. https://doi.org/10.1111/ele.14421Digital

Parker J, Kronauer DJC (2021) How ants shape biodiversity. Curr Biol 31:R1141–R1224

Parr CL, Bishop TR (2022) The response of ants to climate change. Glob Chang Biol 28:3188–3205. https://doi.org/10.1111/gcb.16140

Pawar S, Huxley PJ, Smallwood TR, Nesbit ML, Chan AH, Shocket MS, Johnson LR, Georgios Kontopoulos D, Cator LJ (2024) Variation in temperature of peak trait performance constrains adaptation of arthropod populations to climatic warming. Nat Ecol Evol:1–11

Penick CA, Tschinkel WR (2008) Thermoregulatory brood transport in the fire ant, *Solenopsis invicta*. Insect Soc 55:176–182. https://doi.org/10.1007/s00040-008-0987-4

Penick CA, Diamond SE, Sanders NJ, Dunn RR (2017) Beyond thermal limits: comprehensive metrics of performance identify key axes of thermal adaptation in ants. Funct Ecol 31:1091–1100. https://doi.org/10.1111/1365-2435.12818

Perez R, Aron S (2020) Adaptations to thermal stress in social insects: recent advances and future directions. Biol Rev 95:1535–1553. https://doi.org/10.1111/brv.12628

Rezende EL, Bozinovic F (2019) Thermal performance across levels of biological organization. Philos Trans R Soc B Biol Sci 374:20180549. https://doi.org/10.1098/rstb.2018.0549

Rezende EL, Castañeda LE, Santos M (2014) Tolerance landscapes in thermal ecology. Funct Ecol 28:799–809. https://doi.org/10.1111/1365-2435.12268

Roeder KA, Roeder DV, Bujan J (2021) Ant thermal tolerance: a review of methods, hypotheses, and sources of variation. Ann Entomol Soc Am 114:459–469. https://doi.org/10.1093/aesa/saab018

Ruel JJ, Ayres MP (1999) Jensen's inequality predicts effects of environmental variation. Trends Ecol Evol 14:361–366. https://doi.org/10.1016/S0169-5347(99)01664-X

Schilman PE, Roces F (2006) Foraging energetics of a nectar-feeding ant: metabolic expenditure as a function of food-source profitability. J Exp Biol 209:4091–4101. https://doi.org/10.1242/jeb.02478

Schultheiss P, Nooten SS, Wang R, Wong MKL, Brassard F, Guénard B (2022) The abundance, biomass, and distribution of ants on earth. PNAS 119:e2201550119. https://doi.org/10.1073/pnas.2201550119

Shapley H (1924) Note on the Thermokinetics of Dolichoderine ants. Proc Natl Acad Sci 10:436–439

Shik JZ, Arnan X, Oms CS, Cerdá X, Boulay R (2019) Evidence for locally adaptive metabolic rates among ant populations along an elevational gradient. J Anim Ecol 88:1240–1249. https://doi.org/10.1111/1365-2656.13007

Sinclair BJ, Marshall KE, Sewell MA, Levesque DL, Willett CS, Slotsbo S, Dong Y, Harley CDG, Marshall DJ, Helmuth BS, Huey RB (2016) Can we predict ectotherm responses to climate change using thermal performance curves and body temperatures? Ecol Lett 19:1372–1385. https://doi.org/10.1111/ele.12686

Sinclair BJ, Saruhashi S, Terblanche JS (2024) Integrating water balance mechanisms into predictions of insect responses to climate change. J Exp Biol 15:jeb247167. https://doi.org/10.1242/jeb.247167

Slatyer RA, Hirst M, Sexton JP (2013) Niche breadth predicts geographical range size: a general ecological pattern. Ecol Lett 16:1104–1114. https://doi.org/10.1111/ele.12140

Sørensen JG, Kristensen TN, Overgaard J (2016) Evolutionary and ecological patterns of thermal acclimation capacity in *Drosophila*: is it important for keeping up with climate change? Curr Opin Insect Sci 17:98–104. https://doi.org/10.1016/j.cois.2016.08.003

Sørensen JG, Winther ML, Salachan PV, MacLean HJ (2020) Drawing the line: Linear or non-linear reaction norms in response to adult acclimation on lower thermal limits. J Insect Physiol 124:104075. https://doi.org/10.1016/j.jinsphys.2020.104075

Stephens DW, Krebs JR (1986) Foraging theory. Princeton university press

Sunday J, Bates A, Dulvy N (2012) Thermal tolerance and the global redistribution of animals. Nat Clim Chang 2:686–690. https://doi.org/10.1038/nclimate1539

Sunday JM, Bates AE, Kearney MR, Colwell RK, Dulvy NK, Longino JT, Huey RB (2014) Thermal-safety margins and the necessity of thermoregulatory behavior across latitude and elevation. Proc Natl Acad Sci 111:5610–5615. https://doi.org/10.1073/pnas.1316145111

Sunday J, Bennett JM, Calosi P et al (2019) Thermal tolerance patterns across latitude and elevation. Philos Trans R Soc B 374:20190036. https://doi.org/10.1098/rstb.2019.0036

Terblanche JS, Deere JA, Clusella-Trullas S, Janion-Scheepers C, Chown SL (2007) Critical thermal limits depend on methodological context. Proc R Soc B Biol Sci:2935–2942. https://doi.org/10.1098/rspb.2007.0985

Terblanche JS, Hoffmann AA, Mitchell KA, Rako L, le Roux PC, Chown SL (2011) Ecologically relevant measures of tolerance to potentially lethal temperatures. J Exp Biol 214:3713–3725. https://doi.org/10.1242/jeb.061283

Terblanche JS, Clusella-Trullas S, Lehmann P (2024) How climate change might impact insect movement via physiological mechanisms. One Earth 7:608–622. https://doi.org/10.1016/j.oneear.2024.03.008

Thakur MP, Bakker ES, Veen GF, Harvey JA (2020) Climate extremes, rewilding, and the role of microhabitats. One Earth 2:506–509. https://doi.org/10.1016/j.oneear.2020.05.010

Tüzün N, Stoks R (2018) Evolution of geographic variation in thermal performance curves in the face of climate change and implications for biotic interactions. Curr Opin Insect Sci 29:78–84. https://doi.org/10.1016/j.cois.2018.07.004

van Heerwaarden B, Sgrò CM (2021) Male fertility thermal limits predict vulnerability to climate warming. Nat Commun 12:2214. https://doi.org/10.1038/s41467-021-22546-w

Verble-Pearson RM, Gifford ME, Yanoviak SP (2015) Variation in thermal tolerance of North American ants. J Therm Biol 48:65–68. https://doi.org/10.1016/j.jtherbio.2014.12.006

Weaving H, Terblanche JS, Pottier P, English S (2022) Meta-analysis reveals weak but pervasive plasticity in insect thermal limits. Nat Commun 131:5292. https://doi.org/10.1038/s41467-022-32953-2

Wehner R, Wehner S (2011) Parallel evolution of thermophilia: daily and seasonal foraging patterns of heat-adapted desert ants: *Cataglyphis* and *Ocymyrmex* species. Physiol Entomol 36:271–281. https://doi.org/10.1111/j.1365-3032.2011.00795.x

Willot Q, Gueydan C, Aron S (2017) Proteome stability, heat hardening and heat-shock protein expression profiles in *Cataglyphis* desert ants. J Exp Biol 220:1721–1728. https://doi.org/10.1242/jeb.154161

Willot Q, Mardulyn P, Defrance M, Gueydan C, Aron S (2018) Molecular chaperoning helps safeguarding mitochondrial integrity and motor functions in the Sahara silver ant *Cataglyphis bombycina*. Sci Rep 8:9220. https://doi.org/10.1038/s41598-018-27628-2

Willot Q, Loos B, Terblanche JS (2021) Interactions between developmental and adult acclimation have distinct consequences for heat tolerance and heat stress recovery. J Exp Biol 224:jeb242479. https://doi.org/10.1242/jeb.242479

Willot Q, Ørsted M, Damsgaard C, Overgaard J (2022) Thermal-death-time model as a tool to analyze heat-tolerance, acclimation, and biogeography in ants. Myrmecol News 32:127–138. https://doi.org/10.25849/myrmecol.news_032:127

Willot Q, Ørsted M, Hans M, Overgaard J (2023) Cold comfort: metabolic rate and tolerance to low temperatures predict latitudinal distribution in ants. Proc R Soc B Biol Sci 290:20230985. https://doi.org/10.1098/rspb.2023.0985

Wilson EO (1987) Causes of ecological success: the case of the ants. J Anim Ecol 56:1–9. https://doi.org/10.2307/4795

Chapter 5
Ant Desiccation Resistance in the Light of Climate Change

Jelena Bujan and Karl A. Roeder

5.1 Introduction

Climate change continues to have direct and multifaceted impacts on our planet (Slette et al. 2019; IPCC 2022). Temperatures are predicted to increase along with the frequency of extreme weather events like droughts. Such impacts are particularly concerning for small ectothermic organisms like ants, as their small size and large surface area-to-volume ratios may put them at risk of dehydration. Furthermore, decreased precipitation can limit the distribution of many ant species, including invasives (Heller et al. 2008). However, desiccation resistance (i.e. the ability to withstand dry conditions by preventing or reducing water loss [Hadley 1994]) may enable ants to survive predicted drier conditions. Desiccation resistance and tolerance are often used interchangeably in the ant literature, although the latter strictly refers to the ability to withstand total removal of cellular water without irreversible damage determined after rehydration (Leprince and Buitink 2010) and has not been experimentally examined in ants.

Mary Talbot conducted the first desiccation resistance study on ants approximately 90 years prior to the writing of this chapter (Talbot 1934). Talbot recognised the importance of species' physiology for understanding distributional patterns in ants: *The purpose of this work was to make certain experimental studies with reference to the toleration of ants to extremes of heat and dryness in order to ascertain whether these could be shown to be limiting factors in distribution.* It took another

J. Bujan (✉)
Division for Marine and Environmental Research, Ruđer Bošković Institute, Zagreb, Croatia
e-mail: jelena.bujan@gmail.com

K. A. Roeder
North Central Agricultural Research Laboratory, Agricultural Research Service, USDA, Brookings, SD, USA
e-mail: karl.roeder@usda.gov

I. R. Leal et al. (eds.), *Ant Responses to Climate Change*, Entomology in Focus 10,
https://doi.org/10.1007/978-3-032-21127-9_5

four decades for the next study to emerge (Kay and Whitford 1975), and we have found only 39 studies to date that explore desiccation resistance in ants.

Interest in ant desiccation resistance has historically been similar to that in ant thermal tolerance, but since 2005 there has been an exponential increase in papers on thermal tolerance (Roeder et al. 2021; Nascimento et al. 2022; Parr and Bishop 2022) that has not been matched by those on desiccation resistance (Fig. 5.1).

Desiccation resistance has been measured for 253 ant species and morphospecies from 69 genera in nine subfamilies (out of 16 extant subfamilies). The most studied subfamilies are also the most diverse ones: myrmicines were used in 43% and formicines in 29% of the cases (Fig. 5.2). The three most studied genera are *Camponotus* (N = 29 times, 28 species tested), *Solenopsis* (N = 27, ten species tested), and *Pogonomyrmex* (N = 21, eight species tested). Most studies have focused on a single ant species with only six studies examining more than ten species (Fig. 5.2). Ant

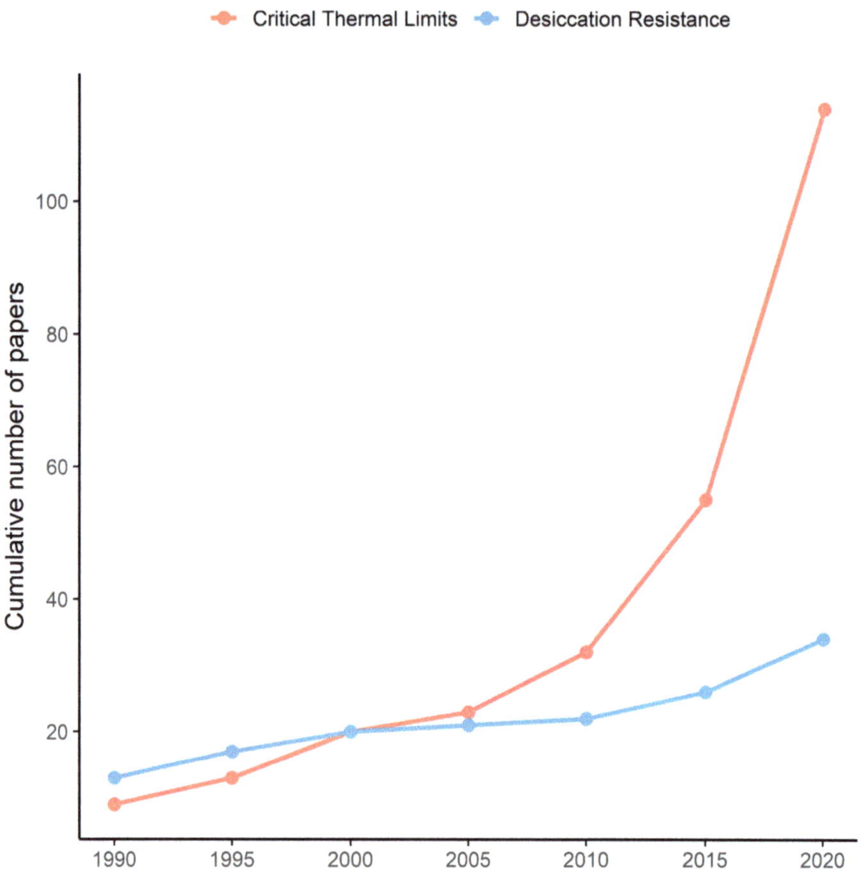

Fig. 5.1 Cumulative number of papers on desiccation resistance compared with thermal tolerance (i.e. critical thermal limits) studies in ants over the last 30 years. The data from 1990 is the sum of all published studies up to that year

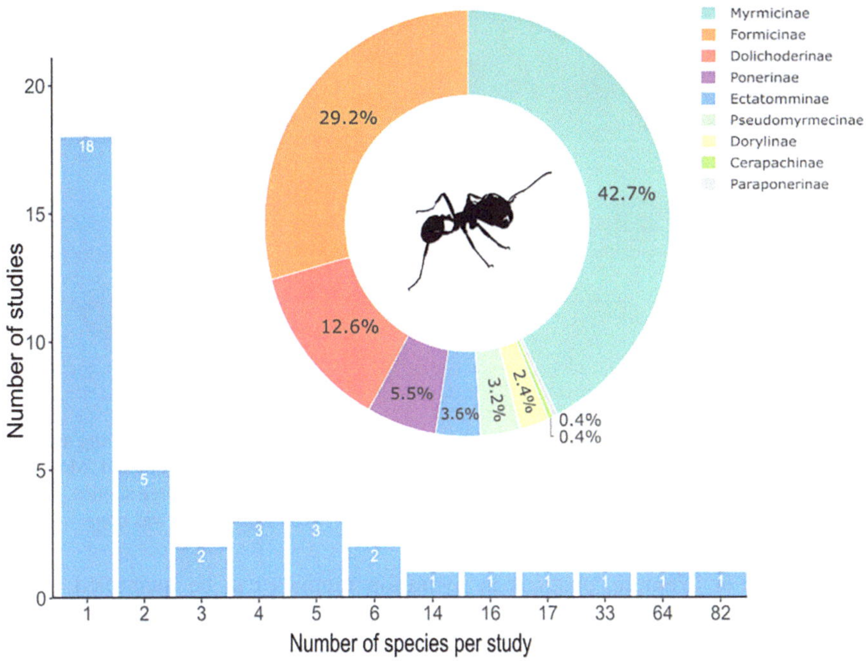

Fig. 5.2 Number of ant species tested for desiccation resistance per study (bar graph). Inset donut plot represents the distribution of studies among subfamilies

desiccation resistance has been tested in nine countries, mostly from the Northern Hemisphere, and biased towards the USA where 28/39 studies were conducted (Fig. 5.3). There is a strong preference towards testing desert ants (especially *Pogonomyrmex* species) and invasive ants, as is understandable given that the former occur in dry environments and the latter can have negative impacts on biodiversity and economies (Simberloff et al. 2013). In this chapter, we discuss commonly tested hypotheses and causes of variability in ant desiccation resistance. We further examine traits that help withstand desiccation and thus enable survival in drier climates including foraging and nest-digging behaviours. Both behaviours could substantially decrease with an increase of droughts and reduced soil moisture. Finally, we discuss future directions in desiccation resistance research and knowledge gaps we need to fill if we want to understand the relationship of desiccation resistance and ant persistence in a drier climate.

5.2 Measuring Desiccation Resistance in Ants

We consider any ant trait measured in response to changes in relative humidity as a measure of desiccation resistance. The most commonly measured responses are water loss and survival. However, there are a variety of ways these traits were

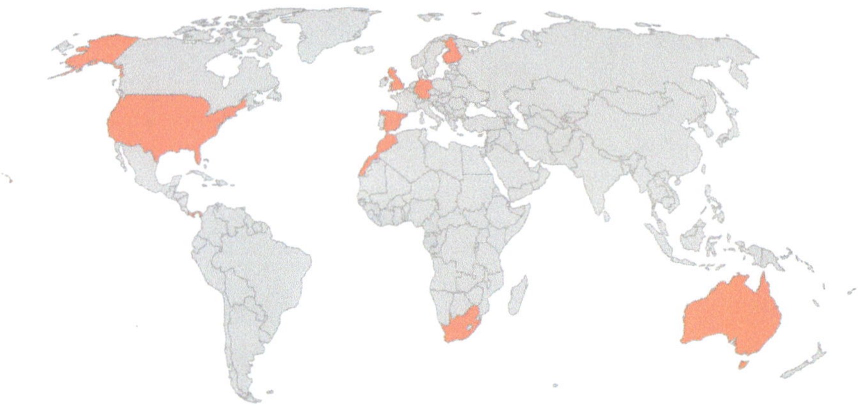

Fig. 5.3 Global distribution of countries represented in each of the 39 studies examining ant desiccation resistance. Ants in most countries were studied only once: Australia, Finland, Morocco, Panama, South Africa, Spain, and the UK. Ants from Germany were tested in three studies, and 28 studies sampled ants in the USA. One study did not report where the sampling was done

measured. For example, changes in hydration regimes can be quantified as an amount of water loss, the rate of water loss, the amount of cuticular water lost, and critical water content (i.e. the point where enough water has been lost that ants lose muscle control). Even the methods for measuring similar water loss traits might affect the results. For example, water loss measured gravimetrically and using air flow respirometry give different results (Schilman et al. 2007).

Survival/mortality are usually quantified as time of death of 50% of individuals (LT50), but sometimes survival is reported as time of death of 90% of individuals (LT90) or 99% (LT99) of individuals. Such diversity of traits and their measurement highlights a major issue across ant desiccation studies—there is no standardised measurement protocol. The question then becomes not only which trait should we measure but also how should we measure it.

Based on ant and other insect physiological studies, we suggest a combination of water loss rates/critical hydration levels and survival time. We acknowledge that measuring water loss in the field remains difficult and, under such circumstances, survival time might be the only feasible trait. Yet, large amounts of data can be quickly collected across colonies/species using vials filled with desiccant set at room temperature. Such approaches are particularly handy for collecting data in the field, which will be invaluable for large-scale comparisons. An increase in standardised studies would provide valuable data on desiccation resistance and enable large-scale predictions about ant range shifts due to climate change based on both thermal tolerance and desiccation resistance.

5.3 Ant Desiccation Resistance Through the Lens of Hypothesis Testing

Ant desiccation resistance is usually explored at the level of a single species (Fig. 5.2) or a couple of congenerics. Researchers have often focused on how environmental factors govern desiccation resistance of species from different castes and habitats (Hood and Tschinkel 1990; Johnson 2000; Ajayi et al. 2020) or on traits important for preventing water loss (see Sect. 5.5.5). Thus, only a few, formally posited hypotheses exist regarding desiccation resistance in ants. There are four main hypotheses that relate directly to climate change, and here we explore their predictions and empirical support.

The desiccation resistance hypothesis posits that organisms exposed and adapted to drier environments will have higher desiccation resistance (see Chap. 3). This hypothesis has had strong support in *Drosophila* flies, where tropical species have lower desiccation resistance compared to species from drier environments at higher latitudes (Stanley and Parsons 1981; Karan et al. 1998; Hoffmann et al. 2003). There has been mixed support for this hypothesis in ants. At the species level, workers of *Solenopsis invicta* from drier, xeric conditions were less prone to desiccation than were those from moist conditions, suggesting this species acclimated to the local conditions in its invasive range (Phillips et al. 1996). Support for this hypothesis at the community level comes from canopy and litter ant comparisons. Canopy ants, exposed to drier conditions than litter ants, can survive longer at low-humidity atmospheres in both the temperate zone (Hood and Tschinkel 1990) and in the tropics (Bujan et al. 2016). This suggests that communities are adapted to the humidity conditions experienced in the microhabitat. However, consistent patterns in desiccation resistance do not always occur. For example, in 16 species of Florida ants, desiccation resistance did not vary with dryness of the environment, even when the foraging time of the ants (day vs. night) was accounted for (Wiescher et al. 2011).

Temporally, we might predict desiccation resistance would follow the same pattern as heat tolerance which is lower in nocturnal ants (Jayatilaka et al. 2011; Garcia-Robledo et al. 2018). Yet results from a comparison of six desert species suggest the night-foraging *Novomessor cockerelli* had the highest desiccation resistance and the lowest desiccation rate (Whitford et al. 1975). More data is desperately needed to confirm if this pattern of nocturnal ants failing to follow abiotic conditions holds true. The seasonality of desiccation resistance has been rarely examined, as we found only two studies that compared seasonal water loss of ants (but see Sect. 5.5.2.3 for thoughts on acclimation and plasticity). Of these two studies, results were mixed as workers of *Formica exsectoides* were found to be more sensitive to water loss during the winter compared to summer (Sigal and Arlian 1982), while no seasonal differences in water loss were found in *F. rufa* (North 1991). Seasonal changes in desiccation resistance might suggest this trait is variable, and such phenotypic plasticity could be beneficial in a drier future climate.

The beneficial acclimation hypothesis posits that the ability to acclimate to novel conditions is adaptive (Leroi et al. 1994; Wilson and Franklin 2002). In the case of desiccation resistance, drier conditions are expected to promote a reversible change towards a phenotype that can reduce water loss. For example, the cuticular hydrocarbon (CHC) profile (see Sect. 5.5.2.3 for details) of two *Temnothorax* species, which differ in their habitat preferences, changed after they were exposed to dry conditions, with workers surviving longer in drier conditions after acclimation (Menzel et al. 2018). Even though the total CHC content did not change with acclimation, dry-acclimated ants had higher percentages of particular CHCs that are responsible for waterproofing (Menzel et al. 2018). Desiccation resistance can also increase in response to acclimation to high temperature: two of three *Lasius* species (*L. niger* and *L. platythorax*) acclimated to 28 °C experienced increased desiccation resistance and changed their CHC profile (Baumgart et al. 2022). This hypothesis may garner the most interest in the light of climate change as it examines the acclimation ability of ants in response to novel climatic conditions, which will be advantageous in a warmer and drier future.

The melanism-desiccation hypothesis posits that the hydrophobic nature of melanin reduces cuticular permeability, with darker cuticles maintaining higher desiccation resistance and lower rates of water loss in insects (Kalmus 1941). Support for this hypothesis has been observed in species of *Drosophila* (Brisson et al. 2005; Parkash et al. 2009) and in other insects like tree weta (King and Sinclair 2015), geometrid moth larvae (Välimäki et al. 2015), and German cockroach (Bai et al. 2022). Yet, recently, it was experimentally observed that there were no detectable differences in desiccation resistance between lighter and darker morphs of *Drosophila melanogaster* (Rajpurohit et al. 2016). In ants, the melanism-desiccation hypothesis has some support, as darker ants were found in the drier canopy, while the wet leaf litter communities contained paler ants (Law et al. 2020), although actual desiccation resistance was not measured. Presently, we do not know if ants are darker in more arid environments compared to ants from humid environments. Based on predictions from the melanism-desiccation hypothesis, we might observe a shift to darker-coloured ants under drier conditions that are predicted with climate change. However, there might be a trade-off with thermal tolerance (see Chap. 3), as insects are darker in cooler environments because melanisation enables insects to warm up faster (Watt 1968). This thermal-melanism hypothesis is supported by observations of mountain ants at high elevations, where darker ants are more common (Bishop et al. 2016). Presently, we do not know which of the two environmental factors—temperature or aridity—is more important in driving ant coloration. The underexplored relationship between ant colour, melanisation, and ecophysiology may become of more interest in the future as novel ways to quantify ant colour increase (Idec et al. 2023).

The body size hypothesis predicts that larger ants will have higher desiccation resistance because of the lower surface area-to-volume ratio. Larger insects can also store more water and fat reserves that can be converted to metabolic water, both of which can further increase desiccation resistance (Hadley 1994). Within a species, large ants generally have higher desiccation resistance—they lose water slower and survive longer in low-relative-humidity atmospheres. This was observed at the caste level where workers lost more water than both male and female reproductive alates (Elzen 1986; North 1991; Johnson 2000). But desiccation resistance can also vary within the worker caste as larger, major workers of *Solenopsis* spp. tolerated 0% RH better than minors, regardless of the temperature (Braulick et al. 1988). The body size hypothesis was also supported at the community level, although the relationship between body size and ant survival was weaker than predicted by theory in temperate ant communities (Hood and Tschinkel 1990) and even weaker in tropical ants (Bujan et al. 2016). Nevertheless, in a drier world, it might be beneficial to produce larger workers to reduce water loss.

5.4 Intraspecific Variability in Desiccation Resistance

Differences in desiccation resistance within species can be driven by abiotic conditions across space (e.g. mesic vs. xeric habitats) and through time (e.g. daily or seasonal fluctuations). Abiotic conditions that a species experience can be quite variable, and this is especially true for introduced ants that occupy a variety of different habitats and climates (Bates et al. 2020). For example, we know that invasive Argentine ant (*Linepithema humile*) is rarely found in drier areas, in part, due to its low desiccation resistance (Schilman et al. 2005; Menke et al. 2007). Ants across their introduced range are often exposed to different climates compared to their native range (Bates et al. 2020) and thus may be able to inform future ant responses to changing climates. Although desiccation resistance of invasive species was studied 19 times, only four invasive species were tested: *Solenopsis invicta*, *Solenopsis geminata*, *Linepithema humile*, and *Tapinoma melanocephalum*. Contrary to predictions, tropical *S. invicta* had higher survival than temperate congenerics in dry conditions (Ajayi et al. 2020). But this species has been within the USA for almost 100 years (Callcott and Collins 1996) and might be acclimatised or adapted to subtropical and temperate conditions by now. Given that invasive ants may exhibit some level of plasticity in physiological traits, we caution extrapolating results from different parts of an invasive ant's range, especially if introduction of the species has occurred many decades prior. At the moment information is lacking about desiccation resistance for most invasive ant species, especially in their native vs. invasive range.

Desiccation resistance can also vary depending on intrinsic factors like developmental stage (e.g. larva, pupa, adult) and caste (e.g. worker, soldier, alate, queen). In *Formica polyctena,* for example, water loss was usually the highest in workers and then decreased from worker to larvae to pupae (Coenen-Staß 1986). Yet larvae may be the most vulnerable developmental stage as their mobility is limited, their cuticle is softer, and they lack protective cocoons present in pupae that are seen in certain ant subfamilies. At the level of caste, water loss was always the highest in workers compared to reproductive alates (Whitford et al. 1975; Elzen 1986; Johnson 2000). Within the worker caste, majors can tolerate desiccation better than minors (Braulick et al. 1988), and young nurse workers survived low RH longer than foragers posing the question about how physiology, morphology, and development may interact to govern desiccation resistance (Menzel et al. 2018; Sprenger et al. 2018).

5.5 Ant Traits Promoting Desiccation Resistance

5.5.1 *Body Size*

Size, be it linear measurements or mass, is perhaps one of the most studied and measured animal traits (Peters 1983; McGill et al. 2006; Roeder et al. 2022). Spanning multiple orders of magnitude even in ants, body size is integral to both how and where an individual can operate. Of the 39 studies examining desiccation resistance, ~79% included some measurement of size. Perhaps this should not be surprising though as trait measurements of desiccation resistance like critical water content and water loss rates are dependent on precise mass measurements. Data generally support the body size hypothesis both intra- and interspecifically, which posits that larger ants are more desiccation resistant. We have discussed a number of mechanisms on how such a phenomenon may occur in Sect. 5.3 of this chapter. However, this does pose an interesting hypothetical question: if body size decreases under a warming climate as predicted by the temperature-size rule (Sheridan and Bickford 2011; Tseng et al. 2018; see Chap. 3), will desiccation resistance follow suit? Such a result could have negative implications for large ground-dwelling species in deserts as they often already reside in low-humidity environments.

5.5.2 *Cuticle Properties*

The majority of water loss in insects happens through the cuticle (Gibbs and Rajpurohit 2010). Research examining the relationship between cuticle properties and desiccation resistance has mostly focused on the impact of soil abrasion on water loss, i.e. the test of the cuticle abrasion hypothesis (Johnson 2000). In recent years, more attention has been paid to morphology of the cuticle (Parr et al. 2017; Buxton et al. 2021; Hellenbrand and Penick 2023) with certain forms being correlated with water loss reduction (Buxton et al. 2021) as well as cuticle waterproofing (Buellesbach et al. 2018; Menzel et al. 2018; Baumgart et al. 2022). The best way to reduce water loss may be achieved via thick, sculptured, or pilose cuticles.

5.5.2.1 Cuticle Thickness and Sculpturing

Thicker cuticles provide a physically larger barrier for water resulting in higher desiccation resistance (Buxton et al. 2021). Comparison of ant cuticles from nine subfamilies (42 ant species) revealed that Formicinae and Dolichoderinae have the thinnest and often smoothest cuticles (Peeters et al. 2017; Hellenbrand and Penick 2023), suggesting species in these subfamilies may also have the lowest desiccation resistance. This idea, although appealing, has yet to be experimentally tested.

The most comprehensive study of ant cuticular sculpturing examined 11,739 species and categorised microsculpturing into five distinct groups from 150 unique terms used by myrmecologists: smooth, striate, reticulate, tuberous, and punctate (Hellenbrand and Penick 2023). Smooth cuticle was the most common, found in ~57% of ants. Tuberous cuticle, on the other hand, was the least common, and it is hypothesised that it has an important role in hosting symbiotic bacteria that are generally present in fungus-growing ants (Currie et al. 2006). Desiccation resistance of several fungus-growing genera (*Apterostigma, Cyphomyrmex, Sericomyrmex, Trachymyrmex*) is some of the lowest from 82 tropical ant species tested (Bujan et al. 2016), making it unlikely that tuberous sculpturing contributes to desiccation resistance.

Other microsculptures (striations, reticulations, and punctations) might provide space for storing cuticular lipids, which would help reduce water loss (Hellenbrand and Penick 2023). Rugosity of the cuticle could also help reduce abrasion caused by the soil, as cuticular microsculptures are generally denser in places most exposed to abrasion (e.g. clypeus, occiput, and pronotum; Johnson et al. 2011). One caveat though is that ants with thicker cuticles often have more microsculpturing, so the contribution of these two traits to desiccation resistance is difficult to uncouple.

5.5.2.2 Pilosity

The pilosity (i.e. hairiness) of insect cuticle is assumed to be beneficial for reducing water loss where dense hairs create a boundary layer at the surface of the cuticle, preventing water loss caused by convection. Pilosity of ant cuticle takes many forms including hairs, setae, and pubescence (Bolton 1994). However, pilosity of ant cuticle is a rarely quantified trait (Parr et al. 2017). Dense pilosity is assumed to increase desiccation resistance, thermoregulation, and defence against predation and to prevent water immersion (Buxton et al. 2021). For thermoregulating insects, dense fur is efficient at preventing heat loss in bees (Church 1960). Such "furry coats" are often observed in species inhabiting cooler climates such as bumblebees, some moths/butterflies, and rain beetles (Heinrich 1993). But using dense hairs for thermoregulation is not limited to cooler climates. The desert ant *Cataglyphis bombycina* uses dense silver, triangular hairs to dissipate heat through enhancing reflectance and emissivity (Shi et al. 2015). Theoretically, the presence of dense pubescence should also be beneficial for preventing water loss; however, this hypothesis has been rarely, if ever, tested in ants. In addition to density, hair length is another important feature of pilosity. Longer-haired bees, for example, are observed in cooler climates and at higher elevations, suggesting that pilosity might be a trait that follows climatic changes at least when it comes to thermoregulation (Peters et al. 2016).

5.5.2.3 Cuticular Hydrocarbon (CHC) Profiles

The outermost (epicuticular) layer of the cuticle includes a waterproofing coating of cuticular lipids that might be one of the most important traits for preventing water loss (Gibbs and Rajpurohit 2010). Cuticular lipids create a waxy barrier on the insect cuticle that prevents desiccation (Gibbs 1998). Unlike cuticle structural adaptations that evolved as a species trait, the benefit of the epicuticular layer is that CHC profiles can respond to changes in temperature and humidity (Menzel et al. 2018). Among cuticular lipids, CHCs are perhaps the best known because they are the most hydrophobic lipid component and a strong candidate for preventing water loss. It does not hurt that CHCs are also abundant and easy to quantify (Gibbs and Rajpurohit 2010). The solid state of CHCs is temperature dependent, and melting of CHCs due to high temperatures makes the cuticle more permeable which increases water loss (Gibbs 2002). Thus, an ideal mixture for waterproofing the cuticle would consist of CHCs that are stable at higher temperatures such as longer-chained hydrocarbons, hydrocarbons with fewer branches, and hydrocarbons with fewer double bonds (Gibbs and Pomonist 1995).

5.5.2.3.1 Ant CHC and Desiccation Resistance

Ant CHCs serve a dual role—they are used for waterproofing the cuticle and for communication and nestmate recognition (Lahav et al. 1999; Menzel et al. 2019; Walsh et al. 2020). Moreover, CHC profiles are often unique to a colony, and not necessarily to the workers within the colony, allowing intercolonial differences to occur within species (Torres et al. 2007; Whyte et al. 2023). Such differences could have genetic underpinnings as certain hydrocarbons in *Monomorium pharaonis* experienced strong directional selection with implications for colony fitness (Walsh et al. 2020). Among ant desiccation resistance studies, 21% (8/39) of them either quantified CHC profiles or tested how the removal of cuticular lipids changed ant water balance. For example, removing cuticular lipids with hexane increased water loss of ants (Hood and Tschinkel 1990; Appel et al. 1991; Chen et al. 2014). Even at relative humidities (RHs) that were much greater than 0%, water loss rates remained high following complete cuticular lipid removal (e.g. *Solenopsis invicta* water loss rate increased 14-fold and *S. richteri* increased 12-fold at 24% RH; Chen et al. 2014). Cuticular hydrocarbons in ants follow the predictions related to their physical properties, so that longer-chained, less branched, and more saturated CHCs are responsible for higher desiccation resistance, as would be predicted based on their biophysical properties (Menzel et al. 2018; Ostwald et al. 2023).

5.5.2.3.2 Acclimation and Plasticity of CHC

One fascinating property of an ant's waterproofing layer is its plasticity. Although sympatric congenerics can differ in CHC profiles (e.g., *Myrmica rubra* and *M. ruginodis*), drought acclimation increased CHC quantity and survival at 0% RH in both species (Sprenger et al. 2018). In the red harvester ant *Pogonomyrmex barbatus*, nest maintenance workers exposed to warm and dry climate had higher abundance of *n*-alkanes compared to workers that were kept in the lab (Wagner et al. 2001). The same pattern was observed in lab conditions where dry-acclimated workers of *Temnothorax* increased the proportion of *n*-alkanes and decreased the proportion of dimethyl alkanes, thus increasing the melting point of their CHC layer (Menzel et al. 2018). Further, temperature acclimation can alter CHC profiles with potential implications for drought survival. For example, warm acclimated *Lasius niger* and *Lasius platythorax* had more *n*-alkanes and monomethyl alkanes and fewer low-melting-temperature CHCs, which resulted in higher desiccation survival (Baumgart et al. 2022). Future tests of the acclimation ability of CHC profiles will lead to mechanistic-level discoveries and give a comprehensive overview of desiccation resistance and reasons for its variability.

5.5.3 Rectal Pads

Rectal pads, or papillae, are structures inside of an insect's rectum that reabsorb water and ions from the faeces (Phillips et al. 1987). They are usually studied as a part of the ant digestive system (Garayoa et al. 1999; Solis et al. 2013), and only one study to our knowledge has examined the relationship between ant rectal pads and desiccation resistance, finding that total rectal pad area was higher in ants from drier, arboreal habitats (Hood and Tschinkel 1990). This suggests that arboreal ants are better at reclaiming faecal water, which might be related to their high reliance on liquid diets of honeydew and extrafloral nectar (Blüthgen et al. 2000). However, there was no direct relationship between the total rectal pad area and desiccation resistance. The importance of rectal pads in preventing water loss in ants is still unknown but may be an interesting avenue of future research.

5.5.4 Water Content

Water balance traits measured in desiccation studies often include total water content, critical water content, and water loss rates. Total water content represents the difference between wet and dry mass values, while critical water content (CWC) is calculated as the difference between moribund and dry mass, divided by wet mass (Schilman et al. 2007; Friedman et al. 2019). However, critical hydration levels have also been reported as simply the mass of body water at the time of coordination loss (Lighton and Feener 1989). Critical water content is often determined using 0% humidity environments where an ant is checked regularly until normal locomotion/orientation stops (Feener and Lighton 1991). Water loss rates on the other hand represent a change in mass over time and can be incredibly sensitive to methodological differences like temperature and RH (if not 0%).

Most measured ants within two Australian communities (Buxton et al. 2021) maintained water content values between 52% and 82% with subfamilies like Formicinae having a higher proportion of water (avg = 0.71) compared to Ectatomminae (avg = 0.55). Similar values were reported from tropical ants in Panama where ants regularly maintained 48–75% water content in their bodies (Bujan et al. 2016). Mass independent water loss rates for these ants averaged 0.06 mg/h, but values varied substantially across subfamilies with Cerapachinae losing water the slowest (avg = 0.02 mg/h) compared to Myrmicinae which lost water the fastest (avg = 0.10 mg/h). Critical water content of ants, on the other hand, appears to be rarely quantified with measurements primarily from desert genera like *Pogonomyrmex* (e.g. 51.7% CWC for *P. rugosus*; Lighton and Feener 1989). This presents an interesting challenge for ant desiccation research as others suggest critical water content may be a trait worth investigating for cross-study comparisons (Burdine and McCluney 2019).

5.5.5 *Other Traits*

Many other traits might be important for desiccation resistance, but they have not been commonly studied in ants in this context. We briefly discuss three of these traits: nutrition, heat shock proteins, and hygrosensors.

5.5.5.1 Nutrition

The amount of consumed resources and the type of ingested food may impact desiccation resistance. For example, when *Drosophila* larvae are fed protein-rich food during development, adult flies have higher desiccation resistance and thermal tolerance compared to larvae fed carbohydrate-rich food (Andersen et al. 2010). Exposure to repeated dehydrating events has also been shown to reduce the nutrient reserves of female mosquitoes causing lower egg production (Benoit et al. 2010). Thus, dehydration in these solitary insects can have a direct effect on reproduction via nutrition. Ant research in this field is rare; however, it has been shown that worker nutritional state can affect desiccation resistance as starved ants will desiccate sooner (Nguyen et al. 2017).

5.5.5.2 Heat Shock Proteins

Heat shock proteins (HSP) are molecular chaperones that protect cells from thermal damage or other stressors including desiccation (Tammariello et al. 1999; Sørensen et al. 2003; Bujan et al. 2022). Consequently, synthesis of HSP at stressful temperatures might also allow ants to withstand stressful humidity levels. Insects predominantly use HSP70 in their response to environmental stress (Zhao and Jones 2012), including chill coma recovery in *Drosophila* (Colinet et al. 2010) and ants (Maisov et al. 2007).

5.5.5.3 Hygrosensors

The ability to sense and avoid unfavourable humidity levels would help ants adjust their foraging, nest building, and brood rearing to appropriate abiotic conditions. Hygrosensors have been studied in a few insect taxa (cockroaches, crickets, fireflies, and *Drosophila*) and were always located in the antennae (Yokohari 1978; Itoh et al. 1984; Iwasaki et al. 1995; Yao et al. 2005). We assume the same antennal position in ants. More than desiccation resistance, this trait would help with desiccation avoidance.

5.6 Foraging and Nest-Digging Behaviour in a Drier Climate

Expected decreases in RH and increases in drought frequency could negatively impact ant foraging and nest-digging behaviour. In a tropical rainforest, vapour pressure deficit differences governed ant foraging such that smaller ants preferred foraging in wetter microclimates, presumably because they are more prone to desiccation (Kaspari 1993). And in the temperate zone, high humidity during the spring (i.e. March and April) promoted the highest foraging activity for the cold-loving winter ant, *Prenolepis imparis* (Talbot 1943). Relative humidity could be an important limiting factor of ant foraging in habitats where ants are not adapted to low RHs. Desert ants, on the other hand, have adaptations which allow foraging during low RH conditions.

Dry habitat conditions might be limiting another type of behaviour—nest digging—which is more difficult in dried soil. Prolonged droughts, which are expected to be more frequent under climate change, may consequently pose a severe threat to reproduction of soil-nesting ants. Additionally, if an already established nest becomes too dry, ants will move to a nest with more acceptable humidity levels (e.g. in *Monomorium antarcticum* [Lamb et al. 2020]). Queens generally have higher desiccation resistance than workers (see Sect. 5.3), and it is the DR of the queen and the queens' RH requirements that could be limiting colony founding. Experimental work on two *Camponotus* species showed that colony founding in *C. modoc* is highly sensitive to decreases in humidity levels, and the queens of this species have lower desiccation resistance than queens of *C. vicinus* (Mankowski and Morrell 2011). However, we do not know whether desiccation resistance of queens is always correlated with colony founding behaviour and how more frequent and severe droughts will affect colony establishment.

5.7 Conclusion

Given that we are in a world that is presently experiencing global insect declines (Goulson 2019; Wagner et al. 2021; Harvey et al. 2023), understanding the physiological traits that enable species to survive and thrive is paramount. Desiccation resistance is one potentially important trait that might limit ant distribution, as it does in the case of invasive Argentine ants (Schilman et al. 2005; Menke et al. 2007), but has been surprisingly little explored, considering the predictions of higher drought frequency and intensity caused by climate change. Ants from the tropics—a place where ant diversity peaks globally—might particularly suffer from increased droughts as tropical insects generally have lower desiccation resistance. Yet desiccation resistance studies in ants are heavily biased towards either desert ecosystems or focus on invasive species. The data at hand (one tropical study) does not allow us to make global predictions on the response of these communities with climate change, but canopy ants in these habitats might be better suited to the drier climates.

Considering the importance of desiccation resistance as one of the traits enabling survival, foraging, and colony founding in a warmer and drier world, we see a pressing need for further research, and we highlight five priority areas of research that we believe will be fruitful avenues of desiccation research over the next decade.

1. *Methodology*: We see a need for standardised sampling of desiccation resistance in future studies. Few studies exist that examine how different methods may impact reported results. We discuss a number of issues throughout the chapter, but we need comparative work to test how different approaches change DR measurements. For example, how does temperature mediate the outcome and what should we be using as a standard temperature? Should we use one relative humidity level as a standard in all experiments? Until some of these methods are agreed upon within the ant desiccation community, results for global studies will be fraught with inconsistencies.
2. *Macrophysiological Patterns*: To test hypotheses at larger spatial and temporal scales, there exists a need for more desiccation resistance data across the ant phylogeny. Desiccation resistance values only exist for ~1% of ant species, and most data have been collected from the southwestern USA within desert biomes. While most studies to date have examined desiccation resistance at a local scale, we have yet to learn about macrophysiological patterns that may be incredibly important in the light of climate change.
3. *Plasticity/Acclimation*: Traits are presently being recognised more as variable and not fixed. However, few standardised data exist that would enable tests of intraspecific variation in desiccation resistance across seasons, days, habitats, or even colonies. Plasticity or acclimation could enable ants to survive drier conditions than we presently predict. Such data will be necessary to test questions like the beneficial acclimation hypothesis (see Sect. 5.3). Further, how does behavioural plasticity in foraging affect our interpretation of desiccation for ant survival under climate change?

4. *Correlative Traits*: A single trait of an organism rarely resides in a vacuum where it alone is the only important factor. Yet, few studies have examined if other physiological or morphological traits are synergistic (i.e. cross-tolerance; Sinclair et al. 2013) or if trade-offs exist. Testing desiccation resistance along with thermal traits or CHC profiles would be useful for studies on the adaptation of ants to climate change. Finally, worker condition (e.g. age, health, nutrition, fat content, etc.) could be incredibly important as starved ants may desiccate at a quicker rate than satiated ones.
5. *Drought*: Given that climate change is predicted to alter both temperature and precipitation patterns, it is surprising how interest has focused little on desiccation resistance compared to thermal tolerance. Drought will likely lead to direct changes in species pools of ants at local and perhaps regional scales via the inability to tolerate certain abiotic conditions. But drought might also indirectly impact an ecosystem in ways that we do not fully understand like reductions in resource availability and nest site locations. Disentangling these complex problems will require multiple lines of research to better understand the myriad drivers of changes in ant diversity over time.

References

Ajayi OS, Appel AG, Chen L, Fadamiro HY (2020) Comparative cutaneous water loss and desiccation tolerance of four *Solenopsis* spp. (Hymenoptera: Formicidae) in the Southeastern United States. Insects 11:1–18

Andersen LH, Kristensen TN, Loeschcke V, Toft S, Mayntz D (2010) Protein and carbohydrate composition of larval food affects tolerance to thermal stress and desiccation in adult *Drosophila melanogaster*. J Insect Physiol 56:336–340

Appel AG, Miller MK, Mack TP (1991) Cutaneous water loss of several stages of the red imported fire ant, *Solenopsis invicta* (Buren). Comp Biochem Physiol Part A Physiol 98:281–283

Bai TT, Pei XJ, Liu TX, Fan YL, Zhang SZ (2022) Melanin synthesis genes BgTH and BgDdc affect body color and cuticle permeability in Blattella germanica. Insect Sci 29:1552–1568

Bates OK, Ollier S, Bertelsmeier C (2020) Smaller climatic niche shifts in invasive than non-invasive alien ant species. Nat Commun 11:1 8

Baumgart L, Wittke M, Morsbach S, Abou B, Menzel F (2022) Why do ants differ in acclimatory ability? Biophysical mechanisms behind cuticular hydrocarbon acclimation across species. J Exp Biol 225

Benoit JB, Patrick KR, Desai K, Hardesty JJ, Krause TB, Denlinger DL (2010) Repeated bouts of dehydration deplete nutrient reserves and reduce egg production in the mosquito *Culex pipiens*. J Exp Biol 213:2763–2769

Bishop TR, Robertson MP, Gibb H, van Rensburg BJ, Braschler B, Chown SL, Foord SH, Munyai TC, Okey I, Tshivhandekano PG, Werenkraut V, Parr CL (2016) Ant assemblages have darker and larger members in cold environments. Glob Ecol Biogeogr 25:1489–1499

Blüthgen N, Verhaagh M, Goitía W, Jaffé K, Morawetz W, Barthlott W (2000) How plants shape the ant community in the Amazonian rainforest canopy: the key role of extrafloral nectaries and homopteran honeydew. Oecologia 125:229–240

Bolton B (1994) Identification guide to the ant genera of the world. Harvard University Press, Cambridge

Braulick LS, Cokendolpher JC, Morrison WP (1988) Effect of acute exposure to relative humidity and temperature on four species of fire ants (Solenopsis: Formicidae: Hymenoptera). Texas J Sci 40:331–340

Brisson JA, De Toni DC, Duncan I, Templeton AR (2005) Abdominal pigmentation variation in *Drosophila polymorpha*: geographic variation in the trait, and underlying phylogeography. Evolution 59:1046–1059

Buellesbach J, Whyte BA, Cash E, Gibson JD, Scheckel KJ, Sandidge R, Tsutsui ND (2018) Desiccation resistance and micro-climate adaptation: cuticular hydrocarbon signatures of different Argentine ant Supercolonies across California. J Chem Ecol 44:1101–1114

Bujan J, Yanoviak SP, Kaspari M (2016) Desiccation resistance in tropical insects: causes and mechanisms underlying variability in a Panama ant community. Ecol Evol 6:6282–6291

Bujan J, Ollier S, Villalta I, Devers S, Cerdá X, Amor F, Dahbi A, Bertelsmeier C, Boulay R (2022) Can thermoregulatory traits and evolutionary history predict climatic niches of thermal specialists? Divers Distrib 28:1081–1092

Burdine JD, McCluney KE (2019) Differential sensitivity of bees to urbanization-driven changes in body temperature and water content. Sci Rep:1–10

Buxton JT, Robert KA, Marshall AT, Dutka TL, Gibb H (2021) A cross-species test of the function of cuticular traits in ants (Hymenoptera: Formicidae). Myrmecol News 31:31–46

Callcott AMA, Collins HL (1996) Invasion and range expansion of imported fire ants (Hymenoptera: Formicidae) in North America from 1918–1995. Fla Entomol 79:240–248

Chen J, Rashid T, Feng G (2014) A comparative study between Solenopsis invicta and Solenopsis richteri on tolerance to heat and desiccation stresses. PLoS One 9:1–6

Church NS (1960) Heat loss and the body temperatures of flying insects II. Heat conduction within the body and its loss by radiation and convection. J Exp Biol 37:186–213

Coenen-Staß D (1986) Investigations on the water balance in the red wood ant, *Formica polyctena* (Hymenoptera, formicidae): workers, their larvae and pupae. Comp Biochem Physiol A Physiol 83:141–147

Colinet H, Lee SF, Hoffmann A (2010) Temporal expression of heat shock genes during cold stress and recovery from chill coma in adult *Drosophila melanogaster*. FEBS J 277:174–185

Currie CR, Poulsen M, Mendenhall J, Boomsma JJ, Billen J (2006) Coevolved crypts and exocrine glands support mutualistic bacteria in fungus-growing ants. Science 311:81–83

Elzen G (1986) Oxygen consumption and water loss in the imported fire ant *Solenopsis invicta* Buren. Comp Biochem Physiol Part A 84:13–17

Feener DH, Lighton JRB (1991) Is foraging in the desert ant, *Messor pergandei* (Hymenoptera: Formiceidae), limited by water? Ecol Entomol 16:183–191

Friedman DA, Greene MJ, Gordon DM (2019) The physiology of forager hydration and variation among harvester ant (*Pogonomyrmex barbatus*) colonies in collective foraging behavior. Sci Rep 9:1–9

Garayoa M, Villaro AC, Lezaun MJ, Sesma P (1999) Light and electron microscopic study of the hindgut of the ant (*Formica nigricans*, Hymenoptera): II. Structure of the rectum. J Morphol 242:205–228

Garcia-Robledo C, Chuquillanqui H, Kuprewicz EK, Escobar-Sarria F (2018) Lower thermal tolerance in nocturnal than in diurnal ants: a challenge for nocturnal ectotherms facing global warming. Ecol Entomol 43:162–167

Gibbs AG (1998) Water-proofing properties of Cuticular lipids. Am Zool 38:471–482

Gibbs AG (2002) Lipid melting and cuticular permeability: new insights into an old problem. J Insect Physiol 48:391–400

Gibbs A, Pomonist JG (1995) Physical properties of insect cuticular hydrocarbons: the effects of chain length, methyl-branching and unsaturation. Comp Biochem Physiol Part B 112:243–249

Gibbs AG, Rajpurohit S (2010) Cuticular lipids and water balance. In: Insect hydrocarbons. Cambridge University Press, pp 100–120

Goulson D (2019) The insect apocalypse, and why it matters. Curr Biol 29:R967–R971

Hadley NF (1994) Water relations of terrestrial arthropods. Elsevier Science

Harvey JA, Tougeron K, Gols R, Heinen R, Abarca M, Abram PK, Basset Y, Berg M, Boggs C, Brodeur J, Cardoso P, de Boer JG, De Snoo GR, Deacon C, Dell JE, Desneux N, Dillon ME, Duffy GA, Dyer LA, Ellers J, Espíndola A, Fordyce J, Forister ML, Fukushima C, Gage MJG, García-Robledo C, Gely C, Gobbi M, Hallmann C, Hance T, Harte J, Hochkirch A, Hof C, Hoffmann AA, Kingsolver JG, Lamarre GPA, Laurance WF, Lavandero B, Leather SR, Lehmann P, Le Lann C, López-Uribe MM, Sen Ma C, Ma G, Moiroux J, Monticelli L, Nice C, Ode PJ, Pincebourde S, Ripple WJ, Rowe M, Samways MJ, Sentis A, Shah AA, Stork N, Terblanche JS, Thakur MP, Thomas MB, Tylianakis JM, Van Baaren J, Van de Pol M, Van der Putten WH, Van Dyck H, Verberk WCEP, Wagner DL, Weisser WW, Wetzel WC, Woods HA, Wyckhuys KAG, Chown SL (2023) Scientists' warning on climate change and insects. Wiley

Heinrich B (1993) The hot-blooded insects. In: The Hot-Blooded Insects. Springer-Verlag, Berlin

Hellenbrand JP, Penick CA (2023) Ant cuticle microsculpturing: diversity, classification, and evolution. Myrmecol News 33:123–138

Heller NE, Sanders NJ, Shors JW, Gordon DM (2008) Rainfall facilitates the spread, and time alters the impact, of the invasive Argentine ant. Oecologia 155:385–395

Hoffmann AA, Hallas RJ, Dean JA, Schiffer M (2003) Low potential for climatic stress adaptation in a rainforest *Drosophila* species. Science (New York, NY) 301:100–102

Hood W, Tschinkel W (1990) Desiccation resistance in arboreal and terrestrial ants. Physiol Entomol 15:23–35

Idec JH, Bishop TR, Fisher BL (2023) Using computer vision to understand the global biogeography of ant color. Ecography 2023:1–14

IPCC (2022) IPCC, 2023: climate change 2023: synthesis report. Contribution of working groups I, II and III to the sixth assessment report of the intergovernmental panel on climate change. IPCC

Itoh T, Yokohari F, Tominaga Y (1984) Two types of antennal Hygro-and Thermoreeeptlve Seosllla of the cricket, *Gryllus bimaculatus* (De Geer). Zoolog Sci 1:533–543

Iwasaki M, Itoh T, Yokohari F, Tominaga Y (1995) Identification of antennal hygroreceptive sensillum and other sensilla of the firefly, *Luciola cruciata*. Zool Sci 12:725–732

Jayatilaka P, Narendra A, Reid SF, Cooper P, Zeil J (2011) Different effects of temperature on foraging activity schedules in sympatric *Myrmecia* ants. J Exp Biol 214:2730–2738

Johnson RA (2000) Water loss in desert ants: caste variation and the effect of cuticle abrasion. Physiol Entomol 25:48–53

Johnson RA, Kaiser A, Quinlan M, Sharp W (2011) Effect of cuticular abrasion and recovery on water loss rates in queens of the desert harvester ant *Messor pergandei*. J Exp Biol 214:3495–3506

Kalmus H (1941) The resistance to desiccation of *Drosophila* mutants affecting body colour. Proc R Soc B 130:185–201

Karan D, Dahiya N, Munjal AK, Gibert P, Moreteau B, Parkash R, David JR (1998) Desiccation and starvation tolerance of adult *Drosophila*: opposite latitudinal clines in natural populations of three different species. Evolution 52:825–831

Kaspari M (1993) Body size and microclimate use in Neotropical granivorous ants. Oecologia 96:500–507

Kay CA, Whitford WG (1975) Influences of temperature and humidity on oxygen consumption of five Chihuahuan desert ants. Comp Biochem Physiol Part A Physiol 52:281–286

King KJ, Sinclair BJ (2015) Water loss in tree weta (Hemideina): adaptation to the montane environment and a test of the melanisation-desiccation resistance hypothesis. J Exp Biol 218:1995–2004

Lahav S, Soroker V, Hefetz A, Vander Meer RK (1999) Direct behavioral evidence for hydrocarbons as ant recognition discriminators. Naturwissenschaften 86:246–249

Lamb SD, McCombe GGI, Lawrence E, Macwan R, Mayer T, Jandt JM (2020) Subterranean nesting behaviour in response to soil moisture conditions in the southern ant, *Monomorium antarcticum* Smith (Hymenoptera: Formicidae). N Z Entomol 43:33–43

Law SJ, Bishop TR, Eggleton P, Griffiths H, Ashton L, Parr C (2020) Darker ants dominate the canopy: testing macroecological hypotheses for patterns in colour along a microclimatic gradient. J Anim Ecol 89:347–359

Leprince O, Buitink J (2010) Desiccation tolerance: from genomics to the field. Plant Sci 179:554–564

Leroi AM, Bennett AF, Lenski RE (1994) Temperature acclimation and competitive fitness: an experimental test of the beneficial acclimation assumption. Proc Natl Acad Sci USA 91:1917–1921

Lighton JRB, Feener DH (1989) Water-loss rate and cuticular permeability in foragers of the desert ant *Pogonomyrmex rugosus*. Physiol Zool 62:1232–1256

Maisov AV, Podlipaeva YI, Kipyatkov VE (2007) Expression of stress proteins of HSP70 family in response to cold in Myrmica ants from various geographic populations. Cell Tissue Biol 1:434–438

Mankowski ME, Morrell JJ (2011) Role of relative humidity in colony founding and queen survivorship in two carpenter ant species. J Econ Entomol 104:740–744

McGill BJ, Enquist BJ, Weiher E, Westoby M (2006) Rebuilding community ecology from functional traits. Trends Ecol Evol 21:178–185

Menke SB, Fisher RN, Jetz W, Holway DA (2007) Biotic and abiotic controls of Argentine ant invasion success at local and landscape scales. Ecology 88:3164–3173

Menzel F, Zumbusch M, Feldmeyer B (2018) How ants acclimate: impact of climatic conditions on the cuticular hydrocarbon profile. Funct Ecol 32:657–666

Menzel F, Morsbach S, Martens JH, Räder P, Hadjaje S, Poizat M, Abou B (2019) Communication versus waterproofing: the physics of insect cuticular hydrocarbons. J Exp Biol 22

Nascimento G, Câmara T, Arnan X (2022) Critical thermal limits in ants and their implications under climate change. Biol Rev 97:1287–1305

Nguyen AD, DeNovellis K, Resendez S, Pustilnik JD, Gotelli NJ, Parker JD, Cahan SH (2017) Effects of desiccation and starvation on thermal tolerance and the heat-shock response in forest ants. J Comp Physiol B Biochem Syst Environ Physiol 187:1107–1116

North RD (1991) Transpiration and humidity preference in a temperate wood ant Formica rufa L. J Insect Physiol 37:279–286

Ostwald MM, Tretter S, Buellesbach J, Calixto JM, Fewell JH, Gadau J, Baudier KM (2023) Body mass and cuticular hydrocarbon profiles, but not queen number, underlie worker desiccation resistance in a facultatively polygynous harvester ant (*Pogonomyrmex californicus*). J Comp Physiol B Biochem Syst Environ Physiol 193:261–269

Parkash R, Singh S, Ramniwas S (2009) Seasonal changes in humidity level in the tropics impact body color polymorphism and desiccation resistance in *Drosophila jambulina*-evidence for melanism-desiccation hypothesis. J Insect Physiol 55:358–368

Parr CL, Bishop TR (2022) The response of ants to climate change. Glob Chang Biol 28:3188–3205

Parr CL, Dunn RR, Sanders NJ, Weiser MD, Photakis M, Bishop TR, Fitzpatrick MC, Arnan X, Baccaro F, Brandão CRF, Chick L, Donoso DA, Fayle TM, Gómez C, Grossman B, Munyai TC, Pacheco R, Retana J, Robinson A, Sagata K, Silva RR, Tista M, Vasconcelos H, Yates M, Gibb H (2017) GlobalAnts: a new database on the geography of ant traits (Hymenoptera: Formicidae). Insect Conserv Diver 10:5–20

Peeters C, Molet M, Lin CC, Billen J (2017) Evolution of cheaper workers in ants: a comparative study of exoskeleton thickness. Biol J Linn Soc 121:556–563

Peters RH (1983) The ecological implications of body size. Cambridge University Press, Cambridge

Peters MK, Peisker J, Steffan-Dewenter I, Hoiss B (2016) Morphological traits are linked to the cold performance and distribution of bees along elevational gradients. J Biogeogr 43:2040–2049

Phillips JE, Hanrahan J, Chamberlin M, Thomson B (1987) Mechanisms and control of reabsorption in insect hindgut. In: Advances in insect physiology. Academic, London, pp 329–422

Phillips SA, Jusino-Atresino R, Thorvilson HG (1996) Desiccation resistance in populations of the red imported fire ant (Hymenoptera: Formicidae). Environ Entomol 25:460–464

Rajpurohit S, Peterson LM, Orr AJ, Marlon AJ, Gibbs AG (2016) An experimental evolution test of the relationship between melanism and desiccation survival in insects. PLoS One 11:1–17

Roeder KA, Roeder DV, Bujan J (2021) Ant thermal tolerance: a review of methods, hypotheses, and sources of variation. Ann Entomol Soc Am 114:459–469
Roeder KA, Benson BR, Weiser MD, Kaspari M (2022) Testing the role of body size and litter depth on invertebrate diversity across six forests in North America. Ecology 103:e03601
Schilman PE, Lighton JRB, Holway DA (2005) Respiratory and cuticular water loss in insects with continuous gas exchange: comparison across five ant species. J Insect Physiol 51:1295–1305
Schilman PE, Lighton JRB, Holway DA (2007) Water balance in the Argentine ant (*Linepithema humile*) compared with five common native ant species from southern California. Physiol Entomol 32:1–7
Sheridan JA, Bickford D (2011) Shrinking body size as an ecological response to climate change. Nat Clim Chang 1:401–406
Shi NN, Tsai C-C, Camino F, Bernard GD, Yu N, Wehner R (2015) Keeping cool: enhanced optical reflection and radiative heat dissipation in Saharan silver ants. Science 349:298–301
Sigal MD, Arlian LG (1982) Water balance of the social insect *Formica Exsectoides* (Hymenoptera: Formicidae) and its ecological implications. Physiol Zool 55:355–366
Simberloff D, Martin JL, Genovesi P, Maris V, Wardle DA, Aronson J, Courchamp F, Galil B, García-Berthou E, Pascal M, Pyšek P, Sousa R, Tabacchi E, Vilà M (2013) Impacts of biological invasions: what's what and the way forward. Trends Ecol Evol 28:58–66
Sinclair BJ, Ferguson LV, Salehipour-Shirazi G, Macmillan HA (2013) Cross-tolerance and cross-talk in the cold: relating low temperatures to desiccation and immune stress in insects. Integr Comp Biol 53:545–556
Slette IJ, Post AK, Awad M, Even T, Punzalan A, Williams S, Smith MD, Knapp AK (2019) How ecologists define drought, and why we should do better. Glob Chang Biol 25:3193–3200
Solis DR, Rossi ML, Fox EGP, Nogueira NDL, Tanaka FAO, Bueno OC (2013) On the morphology of the digestive system of two *Monomorium* ant species. J Insect Sci 13:1–13
Sørensen JG, Kristensen TN, Loeschcke V (2003) The evolutionary and ecological role of heat shock proteins. Ecol Lett 6:1025–1037
Sprenger PP, Burkert LH, Rengere Abou B, Federle W, Menzel F (2018) Coping with the climate: cuticular hydrocarbon acclimation of ants under constant and fluctuating conditions. J Exp Biol 221
Stanley SM, Parsons PA (1981) The response of the cosmopolitan species, *Drosophila melanogaster*, to ecological gradients. In: The proceedings of the ecological society of Australia. Ecological Society of Australia, Canberra, pp 121–130
Talbot M (1934) Distribution of ant species in the Chicago region with reference to ecological factors and physiological toleration. Ecology 15:416–439
Talbot M (1943) Response of the ant Prenolepis imparis say to temperature and humidity changes. Ecology 24:345–352
Tammariello SP, Rinehart JP, Denlinger DL (1999) Desiccation elicits heat shock protein transcription in the flesh fly, *Sarcophaga crassipalpis*, but does not enhance tolerance to high or low temperatures. J Insect Physiol 45:933–938
Torres CW, Brandt M, Tsutsui ND (2007) The role of cuticular hydrocarbons as chemical cues for nestmate recognition in the invasive Argentine ant (*Linepithema humile*). Insect Soc 54:363–373
Tseng M, Kaur KM, Soleimani Pari S, Sarai K, Chan D, Yao CH, Porto P, Toor A, Toor HS, Fograscher K (2018) Decreases in beetle body size linked to climate change and warming temperatures. J Anim Ecol 87:647–659
Välimäki P, Kivelä SM, Raitanen J, Pakanen VM, Vatka E, Mäenpää MI, Keret N, Tammaru T (2015) Larval melanism in a geometrid moth: promoted neither by a thermal nor seasonal adaptation but desiccating environments. J Anim Ecol 84:817–828
Wagner D, Tissot M, Gordon D (2001) Task-related environment alters the cuticular hydrocarbon composition of harvester ants. J Chem Ecol 27:1805–1819
Wagner DL, Grames EM, Forister ML, Berenbaum MR, Stopak D (2021) Insect decline in the Anthropocene: death by a thousand cuts. Proc Natl Acad Sci 118:1–10

Walsh J, Pontieri L, D'Ettorre P, Linksvayer TA (2020) Ant cuticular hydrocarbons are heritable and associated with variation in colony productivity. Proc R Soc B Biol Sci 287:16–18
Watt WB (1968) Adaptive significance of pigment polymorphisms in Colias butterflies. I. Variation of melanin pigment in relation to thermoregulation. Evolution 22:437–458
Whitford WG, Kay CA, Schumacher AM (1975) Water loss in Chihuahuan Desert ants. Physiol Zool 48:390–397
Whyte BA, Sandidge R, Buellesbach J, Cash EI, Scheckel KJ, Gibson JD, Tsutsui ND (2023) The role of body size and cuticular hydrocarbons in the desiccation resistance of invasive Argentine ants (*Linepithema humile*). J Exp Biol 226
Wiescher PT, Pearce-Duvet JMC, Feener DH (2011) Environmental context alters ecological trade-offs controlling ant coexistence in a spatially heterogeneous region. Ecolog Entomol 36:549–559
Wilson RS, Franklin CE (2002) Testing the beneficial acclimation hypothesis. Trends Ecol Evol 17:66–70
Yao CA, Ignell R, Carlson JR (2005) Chemosensory coding by neurons in the coeloconic sensilla of the *Drosophila antenna*. J Neurosci 25:8359–8367
Yokohari F (1978) Hygroreceptor mechanism in the antenna of the cockroach Periplaneta. J Comp Physiol Part A 124:53
Zhao L, Jones WA (2012) Expression of heat shock protein genes in insect stress responses. Invertebr Surviv J 9(1):93–101

Part III
Populations

Chapter 6
Climate Change and Population-Level Responses in Ants

Diego Centeno-Alvarado, Douglas Thierry de Melo Neves, and Inara R. Leal

6.1 Introduction

The rapid intensification of global climate change has triggered a wide range of responses in living organisms. These impacts are seen across many levels of biological organization, from changes in individual traits to shifts in ecosystem functioning (Stange and Ayres 2010; Bellard et al. 2012; Weiskopf et al. 2020). Among these levels, the population holds a central role, as it represents a fundamental unit of biological organization and is the scale at which processes such as birth, death, colonization, local extinction, growth, reproduction, and changes in age structure and adaptive evolution occur (Waples and Gaggiotti 2006; Goudge 2022). Studying climate change impacts at the population level is crucial for identifying the biological processes and mechanisms that drive species' responses, including how climate variability shapes population evolution (Parmesan 2006).

Shifts in climatic conditions (e.g. temperature, precipitation, seasonality), as well as the increased frequency and intensity of extreme events (e.g. heatwaves, droughts, wildfires), can alter microclimatic environments and habitat structure, directly influencing the survival, development, phenology, and reproduction of organisms (Bale 1991; Régnière et al. 2012; Ramniwas et al. 2023). These physiological and behavioural impacts can lead to local fluctuations in population size (Kwon et al. 2024), while acute climatic disturbances may cause sudden

D. Centeno-Alvarado (✉) · D. T. de Melo Neves
Programa de Pós-Graduação em Biologia Vegetal, Universidade Federal de Pernambuco, Recife, PE, Brazil
e-mail: diego.centenoa@ufpe.br; douglas.thierry@ufpe.br

I. R. Leal
Departamento de Botânica, Universidade Federal de Pernambuco, Recife, Pernambuco, Brazil
e-mail: inara.leal@ufpe.br

I. R. Leal et al. (eds.), *Ant Responses to Climate Change*, Entomology in Focus 10,
https://doi.org/10.1007/978-3-032-21127-9_6

demographic collapses or bottlenecks (Ims et al. 2008; Rhee et al. 2025). Phenological shifts may also arise, often leading to mismatches between life cycles and resource availability (Dempster and Pollard 1981). Such mismatches can reduce reproductive success and increase mortality (Singer and Parmesan 2010; Pureswaran et al. 2015). At broader spatial scales, range shifts are driven by metapopulation dynamics, involving the gaining of new populations and/or loss of existing populations (Hanski 1998; Braunisch et al. 2010; Guarenghi et al. 2023). Climate change affects these processes across heterogeneous landscapes, altering the dispersal ability and landscape permeability, factors that are key to maintaining connectivity and viability (e.g. Cao and Shi 2022; Dar et al. 2022). Barriers to dispersal and reduced gene flow between subpopulations may compromise long-term viability and the capacity for adaptive responses (Braunisch et al. 2010). Ultimately, the combined direct and indirect effects of climate change may reshape the genetic structure and adaptive potential, affecting population resilience despite accelerating environmental change (Barbour and Gibert 2021).

The complex organization of social insects like ants (characterized by division of labour, cooperative brood care, and overlapping generations) can influence how their populations respond to climate change, either buffering or amplifying its effects (Menzel and Feldmeyer 2021). Climate change may affect ant populations through direct impacts such as altered colony thermal physiology (e.g. critical temperature thresholds for foraging, Chong and Lee 2009), shifts in reproductive timing and success (e.g. changes in nuptial flight phenology, Bigger and Campbell 2025), or increased stress from extreme weather events (e.g. prolonged drought, Sundaram et al. 2021). Indirect effects may include alterations in resource availability and quality (Nowrouzi et al. 2019), intensified competition (Kwon 2018), and disrupted mutualistic relationships (Romo et al. 2015).

This chapter reviews the literature on the effects of climatic variation on population-level responses in ants, focusing on the reported patterns, proposed mechanisms, and critical methodological gaps. We structure this review around five key dimensions of population-level responses, addressing changes in (1) demography, (2) activity patterns, (3) distribution, (4) reproductive phenology, and (5) genetic structure (Fig. 6.1). By integrating these dimensions, we aim to offer a comprehensive overview of how climate change is reshaping the population biology of ants, which are key players in the functioning of terrestrial ecosystems.

6.2 Reviewing the Literature

We conducted a systematic literature review, searching the Web of Science (Clarivate Analytics) using the following string: TI = ((ant OR ants OR Formicidae) AND (climat* OR temperature OR heat* OR radia* OR precipitation OR rain* OR drought) AND (population* OR structure* OR abundan* OR densit* OR demograph* OR birth* OR death* OR activit* OR distribut* OR phenolog* OR gen*)) OR AB = ((ant OR ants OR Formicidae) AND (climat* OR temperature OR heat*

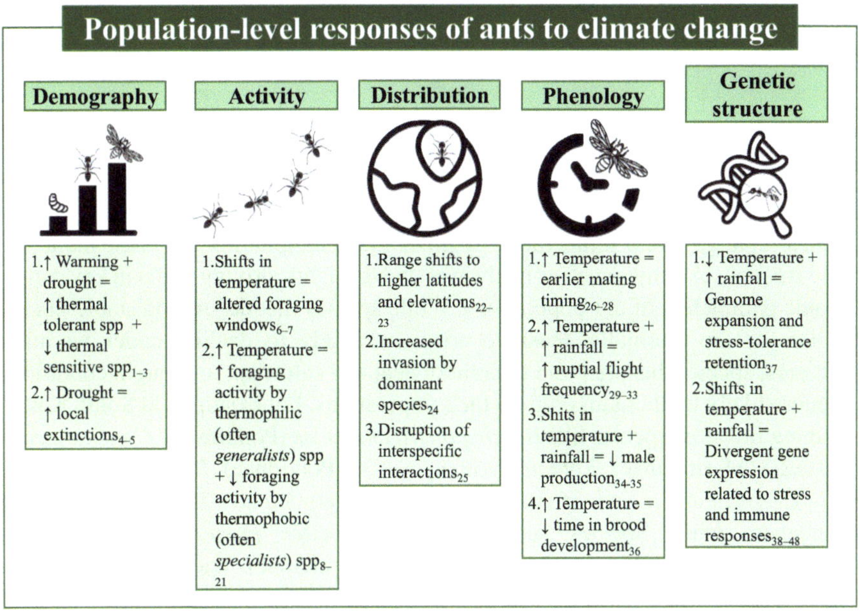

Fig. 6.1 Population-level responses of ants to climate change: Effects on demography ([1] Almeida et al. 2023; [2] Gallé 2017; [3] Kwon et al. 2024; [4] Sundaram et al. 2021; [5] Sanders and Gordon 2004), activity patterns ([6] Gamboa 1976; [7] Jayatilaka et al. 2011; [8] Stuble et al. 2013; [9] Mobbs et al. 1978; [10] Shivashankar et al. 1989; [11] Crist and MacMahon 1991; [12] Fernández-Escudero and Tinaut 1998; [13] Hoffmann 1998; [14] Pol and Casenave 2004; [15] Chong and Lee 2009; [16] Lu et al. 2012; [17] Curbani et al. 2021; [18] Bustamante and Amarillo-Suárez 2019; [19] Bustamante et al. 2020; Tizón et al. 2014; [20] Asfiya et al. 2016; [21] Doblas-Miranda and Reyes-López 2008), distribution ([22] Chap. 8; [23] Chap. 9; [24] Chap. 7; [25] Romo et al. 2015), reproductive phenology ([26] Helms 2023; [27] Hart et al. 2018; [28] Chick et al. 2019; [29] Bigger and Campbell 2025; [30] Rwegasira et al. 2015; [31] Do Nascimento et al. 2011; [32] Hart et al. 2018; [33] Kaspari et al. 2001; [34] Uquillas et al. 2025; [35] Donoso et al. 2022; [36] Boulay et al. 2009), and genetic structure ([37] Lau et al. 2019; [38] Ślipiński et al. 2015; [39] Nguyen et al. 2016; [40] Blatrix et al. 2017; [41] Cahan et al. 2017; [42] Stucki et al. 2017; [43] Willot et al. 2018; [44] Tonione et al. 2020; [45] Vatanparast et al. 2021; [46] Vatanparast and Park 2021; [47] Muñoz-Valencia et al. 2022; [48] Ma et al. 2024)

OR radia* OR precipitation OR rain* OR drought) AND (population* OR structure* OR abundan* OR densit* OR demograph* OR birth* OR death* OR activit* OR distribut* OR phenolog* OR gen*)). This search found 4380 papers, which were then narrowed down to 63 that specifically looked at how ants respond to climate factors at the population level. We included studies examining the following: (1) responses to long-term climate change, (2) responses to short-term climatic variation (e.g. diel and seasonal temperature and rainfall fluctuations), and (3) responses across different contemporary climates (e.g. warmer vs. colder environments). This comprehensive scope allowed us to identify the patterns and mechanisms underlying climate-driven population dynamics in ants. We also examined the references cited within these studies to identify additional relevant literature. Most (46%) considered activity patterns, and others considered reproductive phenology (22.2%), genetic structure (19.1%), demography (9.5%), and distribution (3.2%).

6.3 Demography

Climate change alters temperature regimes and precipitation patterns, including more frequent extremes such as heatwaves and droughts, driving shifts in key demographic parameters of ant populations. While species that depend on stable microhabitats or are thermally sensitive are more likely to decline under warming scenarios, species that are more tolerant of heat and can adapt to drought conditions are more likely to maintain or grow their populations. For example, in South Korea, warming favoured species like *Brachyponera chinensis* (Ponerinae), *Crematogaster osakensis* (Myrmicinae), *Ectomomyrmex javanus* (Ponerinae), *Pristomyrmex punctatus* (Myrmicinae), and *Strumigenys lewisi* (Myrmicinae), which increased in abundance, while species such as *Aphaenogaster japonica* (Myrmicinae), *Camponotus atrox*, *Myrmica transsibirica* (Myrmicinae), and *Pheidole fervida* (Myrmicinae) declined (Kwon et al. 2024). Similarly, experimentally induced drought in the eastern Brazilian Amazon resulted in increased occurrence of thermally and drought-tolerant ants like fungus growers (*Apterostigma* and *Paratrachymyrmex*, Myrmicinae) and certain generalist taxa such as *Azteca* (Dolichoderinae), *Crematogaster*, *Nylanderia* (Formicinae), and *Pheidole* (Almeida et al. 2023). A long-term study in a sandy grassland in central Hungary further supports these patterns: over 37 years, increased drought significantly reduced the nest densities of dune-adapted species (e.g. *Formica cunicularia* and *Lasius niger* [Formicinae]), while the nest density increased for drought-tolerant species such as *Plagiolepis taurica* (Formicinae) (Gallé 2017).

Some species, such as *Pogonomyrmex barbatus* (Myrmicinae), may face local extinctions when they cannot meet metabolic demands under prolonged drought, leading to reduced foraging efficiency and brood production (Sundaram et al. 2021, Fig. 6.1). Long-term monitoring in New Mexico revealed that colony mortality in *P. barbatus* increases following drier summers, especially for newly established colonies (Sanders and Gordon 2004, Fig. 6.1). Interestingly, the survival of *P. barbatus* colonies was influenced by the spatial arrangement of interspecific neighbours: *Novomessor cockerelli* (Myrmicinae) reduced colony survival in some years, whereas *Myrmecocystus mimicus* (Formicinae) had no detectable effect (Sanders and Gordon 2004). Moreover, local landscape heterogeneity can buffer or amplify climate-driven stressors. Variations in vegetation cover, soil moisture, and the availability of thermal refugia create fine-scale mosaics that affect population persistence. For example, in Hawaii, *Linepithema humile* (Dolichoderinae) nests are more abundant under rocks or bare ground than in grassy areas, indicating that certain microhabitats can offer more favourable thermal and moisture regimes for nesting (Hartley et al. 2010).

6.4 Activity Patterns

The amount of energy a colony can obtain through foraging is determined by its activity levels, which are intimately related to its physiological state and long-term population dynamics. By altering foraging behaviour, climate change could alter these dynamics. Due to thermal restrictions or a lack of behavioural flexibility, certain populations may experience altered foraging windows (Gamboa 1976, Jayatilaka et al. 2011, Fig. 6.1). For instance, *Myrmecia croslandi* (Myrmeciinae) usually forages in a unimodal pattern throughout the day in Canberra, Australia. However, when temperatures rise above 38 °C, the activity pattern changes to a bimodal distribution, and foraging stops during peak heat hours (Jayatilaka et al. 2011). On the other hand, *Acromyrmex versicolor* (Myrmicinae) forages mostly at night in Tempe, Arizona, but it can change its activity to include daytime foraging when the temperature drops (Gamboa 1976).

Climatic variation, particularly diel fluctuations in temperature, is a major driver of ant foraging activity, with species-specific responses reflecting physiological tolerances and behavioural strategies. For thermophilic species, sometimes generalists, increased activity tends to be correlated with warming, whereas thermophobic species, which may include some specialists, can become less active (Stuble et al. 2013, Fig. 6.1). Moreover, activity typically increases with temperature up to a thermal optimum, beyond which it declines sharply (see Chap. 4). For example, species like *Anoplolepis gracilipes* (Formicinae), *Dinoponera lucida* (Ponerinae), *Harpegnathos saltator* (Ponerinae), *Iridomyrmex purpureus* (Dolichoderinae), and *Solenopsis invicta* (Myrmicinae) show increased foraging activity in response to warming over narrower temperature ranges (approximately 21–29 °C), while more heat-tolerant species such as *Melophorus aeneovirens* (Formicinae), *Pogonomyrmex occidentalis*, *P. pronotalis*, *P. rastratus*, and *Proformica longiseta* (Formicinae) exhibit intensified foraging at higher temperatures (up to more than 45 °C) (Mobbs et al. 1978; Shivashankar et al. 1989; Crist and MacMahon 1991; Fernández--Escudero and Tinaut 1998; Hoffmann 1998; Pol and Casenave 2004; Chong and Lee 2009; Lu et al. 2012; Curbani et al. 2021). Similarly, leaf-cutting ants such as *Acromyrmex lobicornis* and *Atta cephalotes* (Myrmicinae), thermophilic generalists, show increased foraging and harvesting rates with warming (Bustamante and Amarillo-Suárez 2019; Bustamante et al. 2020; Tizón et al. 2014). *Acromyrmex lobicornis* increases foraging activity when temperatures rise by approximately 4.5 °C under controlled conditions (Tizón et al. 2014), whereas *Atta cephalotes* harvests across a wide temperature spectrum, from as low as 10 °C up to 45 °C (Bustamante and Amarillo-Suárez 2019; Bustamante et al. 2020). This warming--driven increase in activity could amplify herbivory in Neotropical ecosystems (Bustamante and Amarillo-Suárez 2019, Bustamante et al. 2020, Tizón et al. 2014).

In contrast, *Pheidole megacephala* maintains a steady foraging activity with increasing temperature up to approximately 39 °C, beyond which the activity declines sharply (Asfiya et al. 2016). Montane species such as *Pogonomyrmex montanus* exhibit minimal sensitivity to temperature variation, showing little change in foraging behaviour across a range of conditions (MacKay and MacKay 1989). Additionally, the response of the granivorous ant *Messor barbarus* (Myrmicinae) appears more complex: although foraging activity increases with ambient temperatures between 14 and 36 °C (López et al. 1992), higher temperatures have also been associated with reduced foraging trail length, likely as an adaptive response to thermal stress, as observed when the mean ambient temperature rose from 24.4 to 29.2 °C (Doblas-Miranda and Reyes-López 2008).

Contemporary climatic variation across sites or under experimental conditions has also been shown to influence ant foraging activity. In both laboratory and field studies, the foraging activity of *Solenopsis invicta* has been positively correlated with temperature. For instance, in Brazos County, Texas, foraging speed increased with temperature (Drees et al. 2007). Similarly, *Prolasius advena* (Formicinae) colonies demonstrated elevated aggression and foraging activity under warmer conditions, particularly when deprived of carbohydrate-rich food, whereas *Monomorium antarcticum* (Myrmicinae) showed heightened foraging activity at lower temperatures and greater aggression in smaller colonies (Barbieri et al. 2015). However, the responses to thermal environments can be complex. For example, in a recent study, ant occurrence at baits did not indicate higher heat tolerance among urban ants compared to their peri-urban counterparts; instead, cold tolerance was greater among ants from cooler peri-urban habitats (Bujan et al. 2024).

Climate change at the local scale affects various aspects of ant foraging behaviour, nest architecture, and colony development. For instance, *Atta sexdens* significantly increased leaf-cutting activity under low barometric pressure, indicating sensitivity to short-term climatic fluctuations that may precede rainfall (Sujimoto et al. 2020). Laboratory experiments with *Acromyrmex subterraneus* showed that foraging efficiency was suppressed at extreme cold (10 °C) or heat (40 °C) (Ramirez--Olier et al. 2022). Similarly, temperature influences both nest construction and colony growth. In a study in Paris, *Lasius niger* colonies exposed to higher surface temperatures over 100 days excavated deeper and more complex nests (García Ibarra et al. 2024). Field and laboratory studies of *Lasius niger* under urban heat island and artificial light conditions showed that urban ants were generally more active at night, whereas rural ants demonstrated greater locomotor activity under controlled conditions (Trigos-Peral et al. 2024).

Furthermore, foraging success and competitive dynamics can be influenced by species-specific reactions to temperature changes. Colder temperatures in New Zealand significantly slowed some species' ability to find food: under starvation, native *Prolasius advena* exhibited up to an eightfold increase in discovery time when the temperature dropped from 23 to 13 °C, whereas the invasive Argentine ant (*Linepithema humile*) showed no significant impact due to temperature reduction (McGrannachan and Lester 2013).

Ant activity also decreased by extreme rainfall events in addition to temperature (Gan et al. 2021, Fig. 6.1). Following the monsoon, the invasive *Anoplolepis gracilipes* species' foraging activities decreased in Southeast Asia (Gan et al. 2021). This decrease was probably caused by increased mortality and nest flooding, which postponed the construction of new satellite nests (Gan et al. 2021).

These findings suggest that colony activity patterns, which are closely related to the physiological state and long-term population dynamics, can be significantly changed by climate change. Foraging behaviour changes brought on by extreme weather, behavioural flexibility, or thermal limits may alter energy acquisition, which could have an impact on colony survival, growth, and reproduction. Community-level dynamics may change as thermophilic generalists become more active and thermophobic specialists become less active. This could result in changes in species abundances, competitive hierarchies, and ecosystem functions. The stability and resilience of ant populations as well as the ecological services they offer may be impacted by such changes.

6.5 Distribution

Changes in the geographic distributions of ant populations in response to climate change reflect metapopulation dynamics, where shifts are driven by species-specific patterns of local population extinctions and the establishment of new populations. There is strong evidence of range shifts towards higher latitudes and elevations, mainly due to warming temperatures (see Chaps. 8 and 9; Fig. 6.1). Additionally, according to species distribution models, climate change will make it less daunting for aggressive or ecologically dominant ants to invade and establish themselves, especially in disturbed or urbanized environments (see Chap. 7; Fig. 6.1). Even though the majority of recent research on ant distribution modelling under climate change has concentrated on invasive species and agricultural pests (e.g. Bertelsmeier et al. 2015; Song et al. 2021; Chen 2008), there is increasing interest in forecasting the distributions of native species, such as biological control agents and attine ants (e.g. Senula et al. 2019; Ma et al. 2025). One prominent example is the weaver ant genus *Oecophylla* (Formicinae), which includes *O. longinoda* and *O. smaragdina* (Ma et al. 2025). These species are known for their aggressive predatory behaviour and their function in suppressing insect pests (Ma et al. 2025). These ants may be able to spread far beyond their present ranges due to warming, creating new chances for natural pest control (Ma et al. 2025). However, as their introduction into non-native areas may result in competition with local ant species, disturbance of food webs, and unexpected effects on non-target organisms, such expansions also raise ecological concerns (Ma et al. 2025). Furthermore, in contrast to many other ants, attine species have a unique symbiotic relationship with cultivated fungi, which potentially leads to ecologically unique distribution patterns and makes them useful models for determining climate change vulnerability (Senula et al. 2019). Further demonstrating the influence of particular climatic factors in determining the possible distributions of attine ants under climate change, suitability areas for *Mycetomoellerius turrifex* (Myrmicinae) are positively correlated with rising temperatures, while suitability areas for *Trachymyrmex arizonensis*, *T. carinatus*, and *T. smithi* (Myrmicinae) are correlated with greater variation in precipitation seasonality (Ma et al. 2025).

6.6 Reproductive Phenology

Ant reproductive phenology is changing due to climate change, which has an impact on when reproduction takes place as well as how frequently and when colonies produce sexuals or establish new colonies (e.g. Cremer and Heinze 2003, Boulay et al. 2009, Helms and Cahan 2010, Do Nascimento et al. 2011, Kaspari et al. 2001, Rwegasira et al. 2015, Du et al. 2007, Dean and Dean 2018, Hart et al. 2018, Chick et al. 2019, Kuroki et al. 2019, Donoso et al. 2022, Helms 2023, Uquillas et al. 2025, Fig. 6.1). At broad geographic scales, regional temperature shifts are advancing mating seasons. For instance, mating seasons have advanced by about 1 day every 10 years, according to a comprehensive study conducted across 501 species in the contiguous United States (Helms 2023). At smaller scales, this pattern is evident: *Lasius niger* colonies in urban heat islands begin reproductive flights roughly 3 days before those in cooler climates (Hart et al. 2018). Similarly, in the United States, acorn ants (*Temnothorax curvispinosus*, Myrmicinae) have produced reproductive individuals earlier due to warming (Chick et al. 2019).

Climate change also affects the frequency and duration of reproductive events, in addition to their timing. Increased rainfall increased the number of nocturnal nuptial flights among 15 Afrotropical species, according to a multiyear light-trap study conducted in Ghana (Bigger and Campbell 2025). Likewise, in Tanga, Tanzania, the reproductive season of *Oecophylla longinoda* was prolonged by extended rainy periods (Rwegasira et al. 2015). While mating activity among Ecitoninae ants in Brazil's Atlantic Forest increased with temperature (Do Nascimento et al. 2011), mating flights in *Monomorium talbotae* from South-Central Texas peaked during warmer temperatures (Dean and Dean 2018). The frequency of reproductive flights in *Lasius niger* rose linearly with temperature, highlighting the direct impact of climate on reproductive effort (Hart et al. 2018). As seen in Neotropical species, warmer tropical years may also lengthen the reproductive window (Kaspari et al. 2001).

More complex patterns are revealed by long-term observations. Male production on Panama's Barro Colorado Island rose during El Niño years but fell precipitously during periods of intense heat (Uquillas et al. 2025). In contrast, male abundance peaked during warm weeks and brief periods of rain during the dry season in neighbouring seasonally wet rainforests but decreased when rainfall was either too little or too long (Donoso et al. 2022). Changes in precipitation regimes brought on by the climate may also affect reproductive tactics. To mimic the phenology of other desert ants, *Pogonomyrmex rugosus*, for example, has adapted to increasingly arid regions by either producing alates in the summer and overwintering them or timing their mating flights with spring rains (Helms and Cahan 2010).

Caste distribution and reproductive tactics are also impacted by climate change. The proportion of winged males in the ant genus *Cardiocondyla* (Myrmicinae) (which produces both wingless fighter males and winged dispersers) increases with lower temperatures (Cremer and Heinze 2003; Du et al. 2007). There are trade-offs between dispersal and local competition that may be influenced by the climate, as wingless males reproduce within their natal colony, while winged males disperse to mate outside the nest (Du et al. 2007). The timing of development is also influenced by the temperature. While queens consistently developed more slowly, higher temperatures in *Aphaenogaster senilis* accelerated pupation in both workers and queens, indicating that warming maintains caste differentiation while increasing developmental rates (Boulay et al. 2009). On the other hand, *Lasius japonicus* colony development and queen production were suppressed by low temperatures, indicating that cooler conditions may constrain reproductive output (Kuroki et al. 2019).

Population dynamics may be significantly impacted by changes in ant reproductive phenology. On the one hand, reproductive flights that are earlier or more frequent, for instance, may cause temporal mismatches with ideal conditions for mating, dispersal, or colony founding, which would lower recruitment success. On the other hand, depending on the dispersal capacity and resource availability, higher sexual production under ideal circumstances may improve colony spread or encourage gene flow. Changes in morph ratios brought on by climate change may impact reproductive competition and distort mating systems in species with alternative male morphs.

6.7 Genetic Structure

Populations have evolved specific genetic mechanisms to adapt to specific environmental changes (Gerland and Hwa 2009). Present evidence indicates that ant genome size, gene expression, and patterns of genetic differentiation are being impacted by climate change, despite the paucity of direct empirical studies (e.g. Ślipiński et al. 2015, Nguyen et al. 2016, Blatrix et al. 2017, Cahan et al. 2017, Stucki et al. 2017, Willot et al. 2018, Lau et al. 2019, Tonione et al. 2020, Vatanparast and Park 2021, Muñoz-Valencia et al. 2022, Vatanparast et al. 2021, Ma et al. 2024, Fig. 6.1). The relationship between climate and genome size is supported by evidence from contemporary climatic variation. For example, populations of *Aphaenogaster* species in North America that live in colder and wetter climates typically have larger genomes, indicating that climatic extremes may promote genome expansion or the retention of genetic material that helps the species withstand stress (Lau et al. 2019).

Moreover, variations in climate have also been shown to cause population-level differences in cellular stress responses. The differential expression of heat shock proteins (HSPs) is a well-known mechanism. For example, *Aphaenogaster rudis* populations from warmer environments produce significantly less Hsp40 in response to rising temperatures compared with populations from cooler environments (Cahan et al. 2017). This suggests that chronically warmer environments result in less investment in stress proteins (Cahan et al. 2017). Similar to this, populations of *Formica exsecta* show patterns of gene expression that are influenced by both humidity and temperature, indicating local environmental adaptation (Stucki et al. 2017). Under colder, dry, and moist conditions, colonies from arid, sun-exposed island habitats downregulate genes related to stress and immunity, including Hsp75, indicating a suppression of these pathways (Stucki et al. 2017). On the other hand, mainland colonies that live in cooler, shaded woodlands exhibit a stronger stress and immune response when exposed to cold stress, as evidenced by the upregulation of the same genes (Stucki et al. 2017). In contrast, when exposed to rising temperatures, *Formica cinerea*, a xerothermic ant species that is widely distributed, showed no discernible changes in Hsp60, Hsp75, or Hsp90 expression (Ślipiński et al. 2015).

The reactions of other species are more dramatic. High temperatures caused the *lethal(2)essential-for-life-like (l(2)efl)* gene, which codes for a small heat shock protein (sHsp), to be expressed 165 times more in the winter ant *Prenolepis imparis* (Formicinae) (Tonione et al. 2020). Interspecific variations in thermal gene regulation were highlighted by the discovery that this gene is strongly heat-inducible in the Sahara ant *Cataglyphis bombycina* (Formicinae) (Willot et al. 2018) but surprisingly not in the temperate species *Aphaenogaster picea* or the desert ant *Pogonomyrmex barbatus* (Nguyen et al. 2016).

In addition to heat shock proteins, other molecular reactions show flexibility or adaptation to environmental stress. The stress-induced upregulation of some serine protease genes in *Solenopsis japonica* points to a dormant overwintering state known as diapause in response to colder temperatures (Vatanparast and Park 2021). Similar to this, genes involved in the production of glycerol have been shown to express themselves more in *Solenopsis invicta* when temperatures drop, which probably contributes to improved cold tolerance (Vatanparast et al. 2021).

Historical climate change has had a significant impact on the genetic differentiation of ant populations. Because of the dry conditions during the last glacial maximum, which drastically decreased the amount of forest cover, *Tetraponera aethiops* (Pseudomyrmecinae) populations throughout Central Africa, for example, show distinct genetic discontinuities (Blatrix et al. 2017). Despite the lack of visible physical barriers, this fragmentation shows how the historical climate has affected the genetic structure, particularly in species that have obligate symbiotic relationships, such as *T. aethiops* with the tree *Barteria fistulosa* (Passifloraceae) (Blatrix et al. 2017). Pleistocene climatic fluctuations have also shaped the haplotype and nucleotide diversity of *Camponotus japonicus* populations in China (Ma et al. 2024). Although the mountain ranges in this area isolated populations and hindered gene flow, they also preserved suitable habitats by acting as refugia at low elevations during glaciation (Ma et al. 2024). Similar to this, populations of *Atta cephalotes* in the Andes exhibit high levels of genetic variation, with climate having a greater impact on the genetic structure than the actual geographic barriers (Muñoz-Valencia et al. 2022).

Important population dynamics such as survival rates, reproduction, and disease resistance may be impacted by these genetic responses. While populations with suppressed reactions may experience higher mortality and lower reproductive success, which could result in declines or local extinctions under ongoing climate change, populations with enhanced genetic and expression-based stress responses may maintain higher growth rates and stability.

6.8 Conclusion

Although ants are important components of terrestrial ecosystems (Del Toro et al. 2012), our understanding of their population-level responses to climate change is still developing. Much of the existing research focuses on individual traits (see Chaps. 2, 3, 4, and 5), predictive distributions (see Chap. 7), or community-level metrics (Chaps. 8, 9, 10, 11, and 12). However, determining which functional traits are most pertinent for forecasting species' responses is a crucial question in comprehending the effects of climate change at the population level. Thermal tolerance (see Chap. 4), resistance to desiccation (see Chap. 5), and dispersal capacity, including the flight and navigation (e.g. Helms 2017), are likely candidates. Our knowledge of how these characteristics differ and affect climate responses is still growing. For instance, Nascimento et al. (2024) found that tropical ants exhibit significant intra- and inter-colony variation in thermal tolerance ($CT_{max)}$, with within-colony differences that are comparable to those seen at the community level—even in species that are monomorphic and polymorphic. This variability may be crucial in protecting populations from the effects of climate change, highlighting the significance of comprehending trait variation in order to more accurately forecast the persistence of species in the future. In addition, while physiological, behavioural, and ecological responses to climate stress are increasingly studied in the field and lab (e.g. Diamond et al. 2011; Nascimento et al. 2022), methods for scaling this knowledge to inform large-scale predictions remain limited (e.g. Benedetti-Cecchi et al. 2018). Genomic data on climate-related adaptation of ants is still in the early stages. Only a handful of studies have explored genome-environment associations or adaptive variation across climatic gradients in ants (e.g. Stucki et al. 2017; Lau et al. 2019). Moreover, biases in sampling and sequencing methodologies must be addressed to ensure robust inferences.

Although the available literature remains limited, emerging evidence suggests that climate change may shape ant population dynamics through diverse and interdependent mechanisms. To advance our understanding of how ant populations respond to climate change, future research should adopt an integrative approach that spans the key dimensions of population-level change, from the demographic to the genetic structure. Understanding changes in population size will require linking demographic trends to shifts in traits, phenology, range, and genetics while also accounting for interactions with other anthropogenic stressors such as habitat loss and invasive species. Despite their ecological relevance, phenological responses in ants remain underdocumented; more long-term studies are needed to detect climate-driven changes in reproductive timing. Lastly, investigating how climate change affects gene flow, genetic diversity, and population connectivity is essential for assessing the adaptive potential and long-term viability of ant populations. A comprehensive, multidimensional research agenda will be crucial not only for anticipating the responses of ant populations to climate change but also for guiding conservation strategies in an increasingly altered world.

References

Almeida RPS, Silva RR, da Costa ACL, Ferreira LV, Meir P, Ellison AM (2023) Induced drought strongly affects richness and composition of ground-dwelling ants in the eastern Amazon. Oecologia 201:299–309. https://doi.org/10.1007/s00442-023-05316-x

Asfiya W, Yeeles P, Lach L, Majer JD, Heterick B, Didham RK (2016) Abiotic factors affecting the foraging activity and potential displacement of native ants by the invasive African big-headed ant *Pheidole megacephala* (Fabricius, 1793) (Hymenoptera: Formicidae). Myrmecol News 22:43–54

Bale JS (1991) Insects at low temperature: a predictable relationship. Funct Ecol 5:291–298. https://doi.org/10.2307/2389267

Barbieri RF, Granchier J, Lester PJ (2015) Synergistic effects of temperature, diet and colony size on the competitive ability of two ant species. Austral Ecol 40:90–99. https://doi.org/10.1111/aec.12176

Barbour MA, Gibert JP (2021) Genetic and plastic rewiring of food webs under climate change. J Anim Ecol 90(8):1814–1830. https://doi.org/10.1111/1365-2656.13541

Bellard C, Bertelsmeier C, Leadley P, Thuiller W, Courchamp F (2012) Impacts of climate change on the future of biodiversity. Ecol Lett 15(4):365–377. https://doi.org/10.1111/j.1461-0248.2011.01736.x

Benedetti-Cecchi L, Bulleri F, Bello MD, Maggi E, Ravaglioli C, Rindi L (2018) Hybrid datasets: integrating observations with experiments in the era of macroecology and big data. Ecology 99(12):2654–2666. https://doi.org/10.1002/ecy.2504

Bertelsmeier C, Luque GM, Hoffmann BD, Courchamp F (2015) Worldwide ant invasions under climate change. Biodivers Conserv 24:117–128. https://doi.org/10.1007/s10531-014-0794-3

Bigger M, Campbell CAM (2025) Effect of rain and temperature on the nocturnal flight phenology of Afrotropical ants (Hymenoptera: Formicidae). Bull Entomol Res. https://doi.org/10.1017/S0007485325000318

Blatrix R, Peccoud J, Born C, Piatscheck F, Benoit L, Sauve M, Djiéto-Lordon C, Atteke C, Wieringa JJ, Harris D, McKey D (2017) Comparative analysis of spatial genetic structure in an ant–plant symbiosis reveals a tension zone and highlights speciation processes in tropical Africa. J Biogeogr 44(8):1856–1868. https://doi.org/10.1111/jbi.12972

Boulay R, Cerdá X, Fertin A, Ichinose K, Lenoir A (2009) Brood development into sexual females depends on the presence of a queen but not on temperature in an ant dispersing by colony fission, *Aphaenogaster senilis*. Ecol Entomol 34:595–602. https://doi.org/10.1111/j.1365-2311.2009.01108.x

Braunisch V, Segelbacher G, Hirzel AH (2010) Modelling functional landscape connectivity from genetic population structure: a new spatially explicit approach. Mol Ecol 19(17):3664–3678. https://doi.org/10.1111/j.1365-294X.2010.04703.x

Bujan J, Bertelsmeier C, Ješovnik A (2024) Insects in temperate urban parks face stronger selection pressure from the cold than the heat. Ecol Evol 14(8):e11335. https://doi.org/10.1002/ece3.11335

Bustamante S, Amarillo-Suárez A (2019) Foraging plasticity of *Atta cephalotes* (Hymenoptera: Formicidae) in response to temperature differences between forest and pasture. Rev Biol Trop 67(4):963–974

Bustamante S, Amarillo-Suárez A, Wirth R (2020) Effects of pasture and forest microclimatic conditions on the foraging activity of leaf-cutting ants. Biotropica 52:697–708. https://doi.org/10.1111/btp.12783

Cahan SH, Nguyen AD, Staton-Geddes J, Penick CA, Hernáiz-Hernández Y, DeMarco BB, Gotelli NJ (2017) Modulation of the heat shock response is associated with acclimation to novel temperatures but not adaptation to climatic variation in the ants *Aphaenogaster picea* and *A. rudis*. Comp Biochem Physiol A Mol Integr Physiol 204:113–120. https://doi.org/10.1016/j.cbpa.2016.11.017

Cao A, Shi X (2022) The effects of climate change on habitat connectivity: a case study of the brown-eared pheasant in China. Land 11:806. https://doi.org/10.3390/land11060806
Chen Y (2008) Global potential distribution of an invasive species, the yellow crazy ant *(Anoplolepis gracilipes)* under climate change. Integr Zool 3:166–175. https://doi.org/10.1111/j.1749-4877.2008.00095.x
Chick LD, Strickler SA, Perez A, Martin RA, Diamond SE (2019) Urban heat islands advance the timing of reproduction in a social insect. J Therm Biol 80:119–125. https://doi.org/10.1016/j.jtherbio.2019.01.004
Chong K-F, Lee C-Y (2009) Influences of temperature, relative humidity and light intensity on the foraging activity of field populations of the longlegged ant, *Anoplolepis gracilipes* (Hymenoptera: Formicidae). Sociobiology 54(2):531–539
Cremer S, Heinze J (2003) Stress grows wings: environmental induction of winged dispersal males in *Cardiocondyla* ants. Curr Biol 13(3):219–223. https://doi.org/10.1016/S0960-9822(03)00012-5
Crist TO, MacMahon JA (1991) Foraging patterns of *Pogonomyrmex occidentalis* (Hymenoptera: Formicidae) in a shrub-steppe ecosystem: the roles of temperature, trunk trails, and seed resources. Environ Entomol 20(1):265–275. https://doi.org/10.1093/ee/20.1.265
Curbani F, Zocca C, Ferreira RB, Waichert C, Sobrinho TG, Srbek-Araujo AC (2021) Litter surface temperature: a driving factor affecting foraging activity in *Dinoponera lucida* (Hymenoptera: Formicidae). Sociobiology 68(1):e-6030. https://doi.org/10.13102/sociobiology.v68i1.6030
Dar SA, Singh SK, Wan HY, Cushman SA, Bashir T, Sathyakumar S (2022) Future land use and climate change escalate connectivity loss for Himalayan brown bears. Anim Conserv 26(2):199–215. https://doi.org/10.1111/acv.12813
Dean DA, Dean SR (2018) A survey of the ant fauna and seasonal alate nuptial flights at two locations in South-Central Texas. Southwest Entomolog 43(3):639–647. https://doi.org/10.3958/059.043.0309
Del Toro I, Ribbons RR, Pelini SL (2012) The little things that run the world revisited: a review of ant-mediated ecosystem services and disservices (hymenoptera: Formicidae). Myrmecol News 17:133–146
Dempster JP, Pollard E (1981) Fluctuations in resource availability and insect populations. Oecologia 50:412–416. https://doi.org/10.1007/BF00344984
Diamond SE, Sorger M, Hulcr DM, Hulcr J, Pelini SL, Del Toro I, Hirsch C, Oberg E, Dunn RR (2011) Who likes it hot? A global analysis of the climatic, ecological, and evolutionary determinants of warming tolerance in ants. Glob Chang Biol 18(2):448–456. https://doi.org/10.1111/j.1365-2486.2011.02542.x
Do Nascimento IC, Delabie JHC, Lucia TMCD (2011) Phenology of mating flight in Ecitoninae (Hymenoptera: Formicidae) in a Brazilian Atlantic Forest location. Annales de la Société Entomologique de France 47(1–2, 112):–118. https://doi.org/10.1080/00379271.2011.10697702
Doblas-Miranda E, Reyes-López J (2008) Foraging strategy quick response to temperature of *Messor barbarus* (Hymenoptera: Formicidae) in Mediterranean environments. Physiol Ecol 37(4):857–861. https://doi.org/10.1603/0046-225x(2008)37[857:fsqrtt]2.0.co;2
Donoso DA, Basset Y, Shik JZ, Forrister DL, Uquillas A, Salazar-Méndez Y, Arizala S, Polanco P, Beckett S, Dominguez D, Barrios H (2022) Male ant reproductive investment in a seasonal wet tropical forest: consequences of future climate change. PLoS One 17(3):e0266222. https://doi.org/10.1371/journal.pone.0266222
Drees BBM, Summerlin B, Vinson SB (2007) Foraging activity and temperature relationship for the red imported fire ant. Southwest Entomolog 32(3):149–155. https://doi.org/10.3958/0147-1724-32.3.149
Du Y, Schrempf A, Heinze J (2007) Environmental determination of the male morph in the ant *Cardiocondyla obscurior* (Hymenoptera: Formicidae). Eur J Entomol 104:243–246. https://doi.org/10.14411/eje.2007.038

Fernández-Escudero I, Tinaut A (1998) The heat-cold dialectic in the activity of *Proformica longiseta*, a thermophilous ant inhabiting a high mountain (Sierra Nevada, Spain). Int J Biometeorol 41:175–182. https://doi.org/10.1007/s004840050072

Gallé L (2017) Climate change impoverishes and homogenizes ants' community structure: a long term study. Community Ecol 18(2):128–136. https://doi.org/10.1556/168.2017.18.2.2

Gamboa GJ (1976) Effects of temperature on the surface activity of the desert leaf-cutter ant, Acromyrmex versicolor versicolor (Pergande) (Hymenoptera: Formicidae). Am Midl Nat 95(2):485–491

Gan JYS, Pathy TS, Yeo DYC, Lee JM, Yek SH (2021) The population dynamics of the yellow crazy ant *Anoplolepis gracilipes* (Hymenoptera: Formicidae) on a tropical Island in Malaysia. Ecol Entomol 47:339–346. https://doi.org/10.1111/een.13119

García Ibarra F, Jouquet P, Bottinelli N, Bultelle A, Monnin T (2024) Experimental evidence that increased surface temperature affects bioturbation by ants. J Anim Ecol 93(3):247–366. https://doi.org/10.1111/1365-2656.14040

Gerland U, Hwa T (2009) Evolutionary selection between alternative modes of gene regulation. Proc Natl Acad Sci USA 106(22):8841–8846. https://doi.org/10.1073/pnas.0808500106

Goudge TA (2022) What is a population. Philos Sci 22(4):272–279. https://doi.org/10.1086/287442

Guarenghi MM, Walter A, dos Santos RF (2023) Integrating habitat availability, permeability, and configuration in a model of landscape connectivity: the contribution of habitat's site-to-site. Environ Manag 71:998–1010. https://doi.org/10.1007/s00267-022-01783-9

Hanski I (1998) Metapopulation dynamics. Nature 396:41–49. https://doi.org/10.1038/23876

Hart AG, Hesselberg T, Nesbit R, Goodenough AE (2018) The spatial distribution and environmental triggers of ant mating f lights: using citizen-science data to reveal national patterns. Ecography 41:877–888. https://doi.org/10.1111/ecog.03140

Hartley S, Krushelnycky PD, Lester PJ (2010) Integrating physiology, population dynamics, and climate to make multi-scale predictions for the spread of an invasive insect: the Argentine ant at Haleakala National Park. Hawaii Ecogr 33:83–94. https://doi.org/10.1111/j.1600-0587-2009-06037.x

Helms J (2017) The flight ecology of ants (Hymenoptera: Formicidae). Myrmecol News 26:19–30. https://doi.org/10.25849/myrmecol.news_026:019

Helms JA (2023) Climate, geography, and the mating phenology of ants. Insect Soc 70:119–125. https://doi.org/10.1007/s00040-022-00888-y

Helms K, Cahan SH (2010) Divergence in mating-flight patterns of the seed-harvester ant *Pogonomyrmex rugosus* (Hymenoptera: Formicidae) in the western Mojave Desert. Myrmecol News 13:15–17. https://doi.org/10.25849/myrmecol.news_013:015

Hoffmann BD (1998) Thermophilia in a tropical Australian ant of the *Melophorus aeneovirens* (Lowne) species-group (Hymenoptera: Formicidae). Aust J Entomol 37:162–167. https://doi.org/10.1111/j.1440-6055.1998.00037.x

Ims RA, Henden J-A, Killengreen ST (2008) Collapsing population cycles. Trends Ecol Evol 23(2):79–86. https://doi.org/10.1016/j.tree.2007.10.010

Jayatilaka P, Narendra A, Reid SF, Cooper P, Zeil J (2011) Different effects of temperature on foraging activity schedules in sympatric *Myrmecia* ants. J Exp Biol 214:2730–2738. https://doi.org/10.1242/jeb.053710

Kaspari M, Pickering J, Windsor D (2001) The reproductive flight phenology of a neotropical ant assemblage. Ecol Entomol 26(3):245–257. https://doi.org/10.1046/j.1365-2311.2001.00320.x

Kuroki I, Tagawa J, Nakamura K (2019) Geographic variation of temperature effects on initial colony development of *Lasius japonicus* (Hymenoptera: Formicidae). Appl Entomol Zool 54:175–183. https://doi.org/10.1007/s13355-019-00610-8

Kwon T-S (2018) High competition between ant species at intermediate temperatures. J Therm Biol 71:59–66. https://doi.org/10.1016/j.jtherbio.2017.11.015

Kwon T-S, Lee D-S, Choi WI, Kim E-S, Park Y-S (2024) Selection of climate variables in ant species distribution models: case study in South Korea. Int J Biometeorol 63:263–277. https://doi.org/10.1007/s00484-023-02588-z

Lau MK, Ellison AM, Nguyen A, Penick C, DeMarco B, Gotelli NJ, Sanders NJ, Dunn RR, Cahan SH (2019) Draft *Aphaenogaster* genomes expand our view of ant genome size variation across climate gradients. PeerJ 7:e6447. https://doi.org/10.7717/peerj.6447

López F, Serrano JM, Acosta FJ (1992) Temperature-vegetation structure interaction: the effect on the activity of the ant *Messor barbarus* (L.). Vegetatio 99-100:119–128. https://doi.org/10.1007/BF00118217

Lu Y-Y, Wang L, Zeng L, Xu Y-J (2012) The effects of temperature on the foraging activity of red imported fire ant workers (Hymenoptera: Formicidae) in South China. Sociobiology 59(2):573–583. https://doi.org/10.13102/sociobiology.v59i2.619

Ma R, Zhang L, Xu Y, He H (2024) The influence of climate oscillations and geological events on population differentiation of *Camponotus japonicus* in the Chinese mainland. Ecol Evol 14(2):e11077. https://doi.org/10.1002/ece3.11077

Ma R, Liangliang Z, Yang L, He H (2025) Potential distribution of the weaver ant *Oecophylla* under current and future climate. Pest Manag Sci. https://doi.org/10.1002/ps.8887

MacKay WP, MacKay EE (1989) Diurnal foraging patterns of *Pogonomyrmex* harvester ants (Hymenoptera: Formicidae). Southwest Nat 34(2):213–218. https://doi.org/10.2307/3671730

McGrannachan CM, Lester PJ (2013) Temperature and starvation effects on food exploitation by Argentine ants and native ants in New Zealand. J Appl Entomol 137:550–559. https://doi.org/10.1111/jen.12032

Menzel F, Feldmeyer B (2021) How does climate change affect social insects? Curr Opin Insect Sci 46:10–15. https://doi.org/10.1016/j.cois.2021.01.005

Mobbs CK, Tedder G, Wade AM, Williams R (1978) A note on food and foraging in relation to temperature in the meat ant *Iridomyrmex purpureus* form *viridiaeneus*. Aust J Entomol 17:192–197. https://doi.org/10.1111/j.1440-6055.1978.tb00142.x

Muñoz-Valencia V, Montoya-Lerma J, Seppä P, Diaz F (2022) Landscape genetics across the Andes mountains: environmental variation drives genetic divergence in the leaf-cutting ant *Atta cephalotes*. Mol Ecol 32(1):95–109. https://doi.org/10.1111/mec.16742

Nascimento G, Câmara T, Arnan X (2022) Critical thermal limits in ants and their implications under climate change. Biol Rev 97:1287–1305. https://doi.org/10.1111/brv.12843

Nascimento G, Câmara T, Arnan X (2024) Critical thermal maxima in neotropical ants at colony, population, and community levels. Bull Entomol Res:1–10. https://doi.org/10.1017/S0007485324000567

Nguyen AD, Gotelli NJ, Cahan SH (2016) The evolution of heat shock protein sequences, cis-regulatory elements, and expression profiles in the eusocial Hymenoptera. BMC Evol Biol 16:15. https://doi.org/10.1186/s12862-015-0573-0

Nowrouzi S, Bush A, Harwood T, Staunton KM, Robson SKA, Andersen AN (2019) Incorporating habitat suitability into community projections: ant responses to climate change in the Australian wet tropics. Divers Distrib 25(7):1273–1288. https://doi.org/10.1111/ddi.12935

Parmesan C (2006) Ecological and evolutionary responses to recent climate change. Annu Rev Ecol Evol Syst 37:637–669. https://doi.org/10.1146/annurev.ecolsys.37.091305.110100

Pol R, Casenave JL (2004) Activity patterns of harvester ants *Pogonomyrmex pronotalis* and *Pogonomyrmex rastratus* in the Central Monte Desert, Argentina. Journal of Insect Behavior 17(5):647–661. https://doi.org/10.1023/B:JOIR.0000042546.20520.c8

Pureswaran DS, Grandpré L, Paré D, Taylor A, Barrette M, Morin H, Régnière J, Kneeshaw DD (2015) Climate-induced changes in host tree-insect phenology may drive ecological state-shift in boreal forests. Ecology 96(6):1480–1491. https://doi.org/10.1890/13-2366.1

Ramirez-Olier JP, Sanches JJ, Barbosa JVS, Botero LR, Fourcassié V, Zanuncio JC, Zanetti R (2022) Walking and foraging activity of *Acromyrmex subterraneus molestans* (hymenoptera: Formicidae) at different temperatures. Physiol Entomol 47:162–169. https://doi.org/10.1111/phen.12384

Ramniwas S, Tyagi PK, Sharma A, Kumar G (2023) Editorial: abiotic stress and physiological adaptive strategies of insects. Front Physiol 14. https://doi.org/10.3389/fphys.2023.1210052

Régnière J, Powell J, Bentz B, Nealis V (2012) Effects of temperature on development, survival and reproduction of insects: experimental design, data analysis and modeling. J Insect Physiol 58(5):634–647. https://doi.org/10.1016/j.jinsphys.2012.01.010

Rhee H, Naber S, Krehenwinkel H, Hochkirch A (2025) Wildfires induce a reduction in body size and morphological variation of an insular endemic insect. Ecol Entomol 50(3):485–494. https://doi.org/10.1111/een.13418

Romo H, Silvestre M, Munguira ML (2015) Potential distribution models and the effect of climatic change on the distribution of *Phengaris nausithous* considering its food plant and host ants. J Insect Conserv 19:1101–1118. https://doi.org/10.1007/s10841-015-9825-y

Rwegasira RG, Mwatawala M, Rwegasira GM, Offenberg J (2015) Occurrence of sexuals of African weaver ant (*Oecophylla longinoda* Latreille) (hymenoptera: Formicidae) under a bimodal rainfall pattern in eastern Tanzania. Bull Entomol Res 105:182–186. https://doi.org/10.1017/S0007485314000868

Sanders NJ, Gordon DM (2004) The interactive effects of climate, life history, and interspecific neighbours on mortality in a population of seed harvester ants. Ecol Entomol 29:632–637. https://doi.org/10.1111/j.0307-6946.2004.00641.x

Senula SF, Scavetta JT, Banta JA, Mueller UG, Seal JN, Kellner K (2019) Potential distribution of six North American higher-Attine fungus-farming ant (Hymenoptera: Formicidae) species. J Insect Sci 19(6):24. https://doi.org/10.1093/jisesa/iez118

Shivashankar T, Sharathchandra HC, Veeresh GK (1989) Foraging activity and temperature relations in the ponerine ant *Harpegnathos saltator* Jerdon (Formicidae). Proc Anim Sci 98:367–372. https://doi.org/10.1007/BF03179963

Singer MC, Parmesan C (2010) Phenological asynchrony between herbivorous insects and their hosts: signal of climate change or pre-existing adaptive strategy. Philos Trans R Soc B 365:3161–3176. https://doi.org/10.1098/rstb.2010.0144

Ślipiński P, Pomorski JJ, Kowalewska K (2015) Heat shock proteins expression during thermal risk exposure in the temperate xerothermic ant *Formica cinerea*. Sociobiology 62(3):457–459. https://doi.org/10.13102/sociobiology.v62i3.409

Song J, Zhang H, Li M, Han W, Yin Y, Lei J (2021) Prediction of spatiotemporal invasive risk of the red import fire ant, *Solenopsis invicta* (Hymenoptera: Formicidae), in China. Insects 12:874. https://doi.org/10.3390/insects12100874

Stange EE, Ayres MP (2010) Climate change impacts: insects. In: Encyclopedia of life sciences. Wiley, Chichester. https://doi.org/10.1002/9780470015902.a0022555

Stuble KL, Pelini SL, Diamond SE, Fowler DA, Dunn RR, Sanders NJ (2013) Foraging by forest ants under experimental climatic warming: a test at two sites. Ecol Evol 3(3):482–491. https://doi.org/10.1002/ece3.473

Stucki D, Freitak D, Sundström L (2017) Survival and gene expression under different temperature and humidity regimes in ants. PLoS One 12(7):e0181137. https://doi.org/10.1371/journal.pone.0181137

Sujimoto FR, Costa CM, Zitelli CHL, Bento JMS (2020) Foraging activity of leaf-cutter ants is affected by barometric pressure. Ethology 126:290–296. https://doi.org/10.1111/eth.12967

Sundaram M, Steiner E, Gordon DM (2021) Rainfall, neighbors, and foraging: the dynamics of a population of red harvester ant colonies 1988–2019. Ecol Monogr 92:e1503. https://doi.org/10.1002/ecm.1503

Tizón R, Wulff JP, Peláez DV (2014) The effect of increase in the temperature on the foraging of *Acromyrmex lobicornis* (Hymenoptera: Formicidae). Zool Stud 53:40. https://doi.org/10.1186/s40555-014-0040-4

Tonione MA, Bi K, Tsutsui ND (2020) Transcriptomic signatures of cold adaptation and heat stress in the winter ant (*Prenolepis imparis*). PLoS One 15(10):e0239558. https://doi.org/10.1371/journal.pone.0239558

Trigos-Peral G, Maák IE, Schmid S, Chudzik P, Czaczkes TJ, Witek M, Casacci LP, Sánchez-García D, Lőrincz Á, Kochanowski M, Heinze J (2024) Urban abiotic stressors drive changes in the foraging activity and colony growth of the black garden ant *Lasius Niger*. Sci Total Environ 915:170157. https://doi.org/10.1016/j.scitotenv.2024.170157

Uquillas A, Bonilla N, Arizala S, Basset Y, Barrios H, Donoso DA (2025) Climate drives the long-term ant male production in a tropical community. Sci Rep 15:428. https://doi.org/10.1038/s41598-024-84789-z

Vatanparast M, Park Y (2021) Comparative RNA-Seq analyses of *Solenopsis japonica* (Hymenoptera: Formicidae) reveal gene in response to cold stress. Genes 12:1610. https://doi.org/10.3390/genes12101610

Vatanparast M, Sajjadian SM, Park Y (2021) Glycerol biosynthesis plays an essential role in mediating cold tolerance the red imported fire ant, *Solenopsis invicta*. Arch Insect Biochem Physiol 109(2):e21861. https://doi.org/10.1002/arch.21861

Waples RS, Gaggiotti O (2006) What is a population? An empirical evaluation of some genetic methods for identifying the number of gene pools and their degree of connectivity. Mol Ecol 15(6):1419–1439. https://doi.org/10.1111/j.1365-294X.2006.02890.x

Weiskopf SR, Rubenstein MA, Grozier LG, Gaichas S, Griffis R, Halofsky JE, Hyde KJW, Morelli TL, Morisette JT, Muñoz RC, Pershing AJ, Peterson DL, Poudel R, Staudinger MD, Sutton-Grier AE, Thompson L, Vose J, Weltzin JF, Whyte KP (2020) Climate change effects on biodiversity, ecosystems, ecosystem services, and natural resource management in the United States. Sci Total Environ 733:137782. https://doi.org/10.1016/j.scitotenv.2020.137782

Willot Q, Mardulyn P, Defrance M, Gueydan C, Aron S (2018) Molecular chaperoning helps safeguarding mitochondrial integrity and motor functions in the Sahara silver ant *Cataglyphis bombycina*. Sci Rep 8:9220. https://doi.org/10.1038/s41598-018-27628-2

Chapter 7
Ant Invasions Under Climate Change

Cleo Bertelsmeier and Olivia K. Bates

7.1 Introduction

Global trade and travel have transported thousands of insect species around the planet (Simberloff et al. 2013). This human-mediated transport can result from the contamination of commodities such as live plants or wood products, hitchhiking of the species on transport vectors such as trains, ships or containers, or from the deliberate harvesting of the species such as for the pet trade (Gippet et al. 2019; Meurisse et al. 2019; Fenn-Moltu et al. 2023). Among these introduced species, ants are a particularly successful group due to their social lifestyle (Bertelsmeier 2021). Ant species have been transported for a long time by accident, mostly associated with trade in live plants and fruits (Suarez et al. 2005; Ollier and Bertelsmeier 2022). It is likely that the global movement of ant species started as early as the establishment of the first trade routes, such as the silk or spice roads, but evidence of the impact of such ancient globalisation remains scarce (Bertelsmeier 2021). Genetic markers have allowed retracing the invasion history of the tropical fire ant, *Solenopsis geminata*, which is highly consistent with the Spanish trade routes in the sixteenth century (Gotzek et al. 2015). Similarly, the clonal raider ant, *Ooceraea biroi*, has likely spread from Bangladesh via the historically significant Bay of Bengal shipping ports (Trible et al. 2020). Yet, precise dating of these complex introduction routes remains challenging (Garnas et al. 2016). First records of observations of non-native insects at a global scale are, however, abundant from the early nineteenth century on (Seebens et al. 2017), allowing us to analyse the dynamics of non-native ant movement over the last 200 years. During that time, non-native ants have spread following the two waves of globalisation (1850–1914, 1960-present day) with a strong decline in new invasions between the two world wars (Bertelsmeier et al. 2017).

C. Bertelsmeier (✉) · O. K. Bates
Department of Ecology and Evolution, University of Lausanne, Lausanne, Switzerland
e-mail: cleo.bertelsmeier@unil.ch; olivia.bates@ucl.ac.uk

I. R. Leal et al. (eds.), *Ant Responses to Climate Change*, Entomology in Focus 10,
https://doi.org/10.1007/978-3-032-21127-9_7

Overall, more than 520 ant species are known to have been transported outside of their native range thanks to interception records at ports of entry where inspectors check incoming goods and passengers, and a subset of these—more than 300 species—have successfully established self-sustaining outdoor populations with hotspots of ant establishment in the Nearctic and Oceanian regions (Wong et al. 2023). The dispersal of these species has already led to the biotic homogenisation of communities, with particularly strong impacts on the ant faunas of tropical islands (Aulus-Giacosa et al. 2024). In Florida, non-native ants now account for more than 30% of all ant occurrence records (Booher et al. 2023). This is highly concerning for conservation since non-native ants may have devastating impacts on native biodiversity (Holway et al. 2002), in particular through habitat disturbance, hybridisation and direct competition with native species (Siddiqui et al. 2021). Non-native ant species can dramatically reduce native species richness and abundance (Tercel et al. 2023) and alter the phylogenetic (Lessard et al. 2009) and functional (Wong et al. 2020) diversity of native communities. Species with severe ecological impacts also often score high on socio-economic impacts (Gruber et al. 2022). Although the data on global economic impacts are incomplete and difficult to estimate, non--native ants are likely to have caused several billion US dollars in losses, mainly due to the impacts of only two species, *Solenopsis invicta* and *Linepithema humile* (Angulo et al. 2022).

Global invasion rates continue to rise with many new emergent invaders that have never been detected before (Seebens et al. 2018), and invasion "debts" have accumulated due to cryptic establishments and time lags between establishment and discovery (Rouget et al. 2015). Additional drivers of global change might further accelerate biological invasions (Brook et al. 2008). In particular, climate change might alter the climatic suitability of areas for numerous non-native species and accelerate range expansion because it removes thermal barriers to establishment at higher latitudes or altitudes. For example, most non-native ant species are native to tropical, subtropical or Mediterranean regions (Bertelsmeier 2021) and might be able to establish in areas where previously cold winter temperatures have prevented their survival. Furthermore, climate change might also affect other, less studied, stages of the invasion process (Lach 2021), such as increasing the survival of species during transport or open new trade routes due to ice opening in the northern seas (Bertelsmeier et al. 2016). Moreover, climate change may alter the impact of non-native ants on their recipient communities. For example, experimental soil warming has shown that the negative impacts of non-native ants on the foraging activity of seed-dispersing native ants were exacerbated by higher temperatures (Warren et al. 2023). As many native ants will be vulnerable to climate change itself, native communities are likely to become simplified and may therefore provide less biotic resistance to non-native ants (Lach 2021).

In this chapter, we focus on changes in range size in response to climate change. Much research has addressed which ant species are likely to decrease, shift or expand their distributions and try to explore local and global abiotic variables that may be important in determining species' responses to climate change. These forecasts may play an important role in alerting scientists and policymakers to potential future risks and inform discussions about how to mitigate the negative effects of a changing world. Here, we will review the evidence so far for the impacts of climate change on the ranges of invasive ants based on experimental and observational research and modelling forecasts.

7.2 What Do We Know? The Evidence So Far

We conducted a systematic literature review, searching the Web of Science using the search term: TS = (invade* OR invasive OR introdu* OR alien OR non-native) AND TS = (ant OR formicidae) AND TS = (climate change* OR changing climates OR weather change). This resulted in 190 papers, which were further filtered resulting in 25 on the topic of future predictions (see Sect. 7.1), and 21 papers that assessed observations of past range changes (see Sect. 7.2) or biotic interactions and community-scale experiments which might influence the future ranges of non--native ants (see Sect. 7.3).

7.2.1 Future Predictions

Predictions of potential future ranges under climate change have been published for only 16 out of the more than 300 non-native ant species (Fig. 7.1). Although this highlights some clear taxonomic biases, the most studied species include some of the most invasive and widespread species (Fig. 7.1a). The most studied species are *Solenopsis invicta* (13 modelling studies), *Anoplolepis gracilipes* and *Solenopsis geminata* (6 studies each) and *Linepithema humile* and *Wasmannia auropunctata* (5 studies each). However, this also means that we know very little about the potential future impacts and geographical ranges of those species that do not currently have large ecological or socioeconomic impacts, but may potentially do so under climate change. Interest in forecasts has increased over time, peaking over the last years (Fig. 7.1b).

Worryingly, when there are several studies per species, the evidence is often mixed, with different studies coming to opposite conclusions (Fig. 7.2). However, different studies use different modelling approaches (correlational, hybrid or mechanistic), climatic input data, spatial resolutions, geographic extents (global or regional focus) and modelling algorithms, such that model outcomes are not straightforward to compare. Moreover, it is common to predict future distributions for a single time horizon, for example, in the years 2040 (Sung et al. 2018), 2050 (Roura-Pascual et al. 2004; Chen 2008; Lee et al. 2022), 2080 (Bertelsmeier and Courchamp 2014; Bertelsmeier et al. 2015) or 2090 (Wang et al. 2018). However, one should be careful when making management recommendations based on single time horizons. Indeed, nonlinear dynamics are common, and even trend reversals can occur, meaning that an initial range expansion might be followed by a range

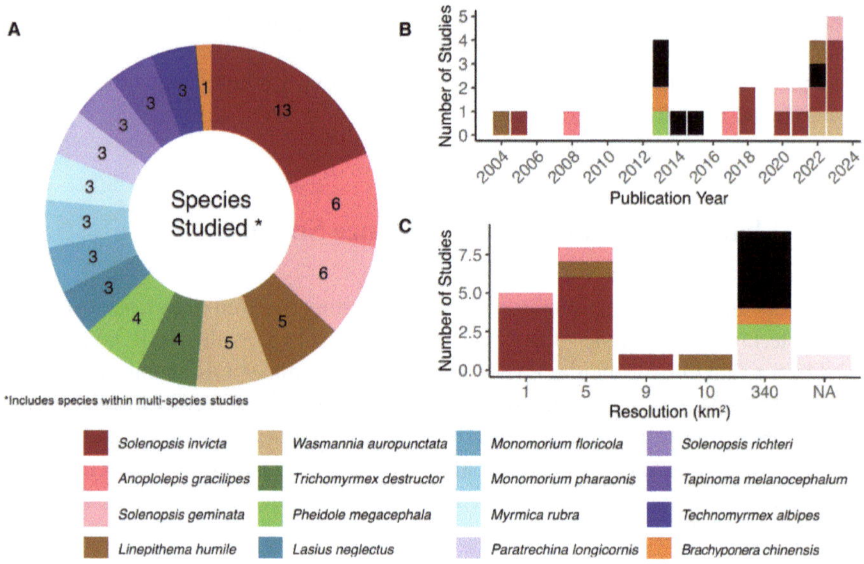

Fig. 7.1 Species studied in climate-change forecasts. Pie chart of the number of studies involving species (**a**), and number of studies present each year (**b**), and at which resolutions (**c**). Black colour represents multi-species studies in (**b**) and (**c**)

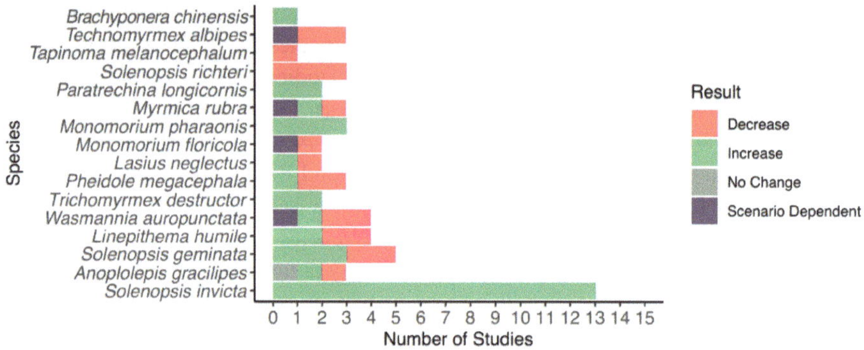

Fig. 7.2 Conclusions of modelling studies forecasting species ranges under climate change

contraction or vice versa. In some cases, these trend reversals can be so extreme that they lead to qualitatively different results. For example, *Technomyrmex albipes* was predicted to increase in suitable range by 36% in 2050 and to lose 64% of its current suitable range in 2080 (Bertelsmeier et al. 2013c), implying that two models differing only by the time horizon chosen for their forecasts would lead to opposite conclusions! One reason for trend reversal may be that changes in different limiting climatic variables are not always linked, for example, some areas may become warmer and drier at first and later increase in rainfall (Walther et al. 2002).

Alternatively, species may first increase because they approach their thermal optimum and exceed it at higher temperatures (Penick et al. 2017).

Although most modelling studies use some sort of model evaluation (such as area under the curve (AUC) or the true skill statistic (TSS)), several models may have "good" performance in predicting the species current distribution but still give quite different future predictions in terms of range size and the location of suitable areas.

7.2.1.1 Correlative Species Distribution Models

The most popular approach to predicting future ranges under climate change are correlative species distribution models (SDMs), which rely on two types of input data: species occurrence points (presence data) and climatic variables that can be easily downloaded in the form of global maps from data providers like Worldclim (Fick and Hijmans 2017) or CHELSA (Karger et al. 2020). Different statistical and machine learning algorithms are available to predict the probability of the species' presence based on these climatic variables and provide as output maps of species' habitat suitability (Guisan and Thuiller 2005; Broennimann and Guisan 2008; Araújo et al. 2019). These models can then be projected onto maps of future climatic conditions. Comparing model projections of future and current climatic conditions allows estimating changes in species potential distributions. Overall, forecasts of future ant invasions do not confirm the widely held belief that non--native species will universally benefit from climate change (Hulme 2017). While some species are indeed predicted to be boosted by climate change (e.g. *Brachyponera chinensis*; Bertelsmeier et al. 2013a) or move to higher latitudes and retract their tropical range margins (e.g. *Anoplolepis gracilipes* (Chen 2008), *Linepithema humile* (Roura-Pascual et al. 2004) and *Solenopsis geminata* (Lee et al. 2021)), other species are predicted to collapse with climate change (e.g. *Pheidole megacephala*; Bertelsmeier et al. 2013b).

One study has predicted the future distribution of 15 of the most invasive ant species using a consistent modelling framework and input climate data: only five species were predicted to increase at a global scale, while the majority was predicted to decrease (by up to −63%) (Bertelsmeier et al. 2015). Even using the same modelling methods, range changes were species specific: for example, *Solenopsis invicta* was predicted to increase in all world regions, while *Pheidole megacephala* was predicted to decrease in all regions. Most species showed a mix of gains and losses, and these patterns were heterogeneous across regions. Instead of simple shifts to higher latitudes or altitudes, most species displayed regionally complex responses, which might be due to the interaction of broad latitudinal climatic gradients with more local features such as distance from water bodies, topography or microclimates (Bertelsmeier et al. 2015). Strikingly, biodiversity hotspots (i.e. the most biodiverse and vulnerable places on Earth) were highly suitable for most invasive ant species and were predicted to remain much more suitable than the global average (Bertelsmeier et al. 2015). A study on climate change impacts on ant invasions in

France also found a large spatial heterogeneity: some regions did not show changes in suitability, while others were predicted to strongly increase in climatic suitability for most species, such as Brittany (Bertelsmeier and Courchamp 2014). Such fine-scale predictions might be useful to inform surveillance strategies. For example, it could be worthwhile to conduct port-of-entry-specific risk assessments since the species are likely to arrive through trade and transport hubs and spread further from there (Bertelsmeier and Courchamp 2014). In a different study, for example, *Solenopsis invicta* has been predicted to increase globally, but the relative risk varied with trade intensity across potential ports of entry (Li et al. 2023). Likewise, studies have assessed local suitability within China to establish early action plans for *Linepithema humile* (Li et al. 2022) or *Wasmannia auropunctata* (Mao et al. 2022).

To decrease the influence of varying methodology on predictions, we advocate using multiple algorithms to predict species distributions. Even when they are popular and give good predictive scores such as for Maxent (e.g. Lee et al. 2021; Song et al. 2021; Li et al. 2022; Mao et al. 2022), single modelling methods can generate spatial predictions that are quite different from other models. For example, a comparison of different machine learning techniques predicting the distributions of *Solenopsis invicta* and *Anoplolepis gracilipes* demonstrated striking differences in predictions. There is no "gold standard" in the field, and no single model is considered superior (Zhao et al. 2023), so an ensemble-based approach is recommended (Lee et al. 2022). But even using ensemble models that aggregate the predictions of multiple models, one has to be careful when deciding on a classification threshold that categorises areas as either suitable or unsuitable. In fact, areas may increase in overall range size while decreasing in relative suitability, or vice versa, as illustrated in a study predicting climate change impacts on *Myrmica rubra* and *Trichomyrmex* (formerly *Monomorium*) *destructor* (Bertelsmeier et al. 2013d).

Finally, the choice of the input climate data is obviously crucial. Unfortunately, studies often use single socio-economic story lines (CO_2 emission scenarios) and global circulation models (GCMs) that model future climate change under different physical hypotheses including atmosphere-ocean coupling. Moreover, these choices are sometimes not reported, making comparisons even more difficult (e.g. Lee et al. 2021; Mao et al. 2022). The input climate data is also often coarse, and grid cells generally cover 1 km^2 to more than 300 km^2. This mismatch with the size of the organisms is problematic since ants experience their environment at very fine spatial scales (Potter et al. 2013). Indeed, temperatures can vary by several degrees in the shade within a few centimetres on a single leaf (Pincebourde and Casas 2019), and ant physiological tolerance is linked to these microclimates (Baudier et al. 2015; see Chap. 4). Indeed, the importance of different variables can change with spatial grain, as suggested by a study on *Linepithema humile* (Menke et al. 2009). Other variables like anthropogenic modifications of the environment are also not taken into account by global scale climate models. For example, *Linepithema humile* thrives in xeric areas but only if there is local irrigation, for example, in vineyards (Schulze-Sylvester et al. 2018). Interestingly, in another study, the presence of this species was also predicted by irrigation at the very fine scale but not by precipitation at the landscape scale (Menke et al. 2007). This in turn contrasts with the findings of another study on the same species where local rainfall accelerated its spread (Heller et al. 2008). Together, these studies illustrate the surprising complexities relating to how abiotic factors can structure invasions at different spatial scales.

7.2.1.2 Hybrid Models

There are different types of semi-mechanistic "hybrid" models that integrate biological knowledge in range size predictions. For example, the popular CLIMEX model (Sutherst and Maywald 2005) estimates an "ecoclimatic index" showing the relative suitability of a grid cell and ranges from 0 to 100, where 0 indicates no growth at all, values below 10 correspond to marginal habitats, 10–20 support substantial populations and values above 20 are highly favourable. This index is estimated using data on species growth rates in relation to temperature and humidity, calculated at a weekly basis for any given location on a map. In addition, the model allows to add stress responses of the species to short and intense periods of adversely cold, hot, dry or wet conditions, reducing population growth and survival. The physiological limits that are used as model input can be measured experimentally (see Chaps. 4 and 5). Alternatively, researchers have estimated them using species occurrence data. For example, a study on *Anoplolepis gracilipes* has estimated physiological limits in an iterative way, measuring at each step the model fit and explanatory power of the species' current distribution (Jung et al. 2017). In a similar study, researchers lacking biological data to parameterise the CLIMEX model for *Solenopsis geminata* instead used a model "template" for a typical species from the wet tropics and then also did an iterative parameter adjustment (Byeon et al. 2020a). However, it is unknown if this approach performs differently from a classic correlational model since it does not add any true physiological measurements and estimates them from the species' distribution (which is conceptually not very different from estimating the probability of occurrence directly from the distributional data).

Some studies have, however, used published experimental data on temperature-dependent developmental rates, metabolic activity, survival and growth, as well as moisture-dependent survival to fit a hybrid model estimating range increases in *Solenopsis invicta* in South Korea due to climate change (Byeon et al. 2020b). Such hybrid models not only have the advantage of making predictions more biologically realistic, but they can also allow better understanding of regional differences in population dynamics. For example, a CLIMEX model for the same species has predicted lower winter mortality but slower growth in Oceania compared to the United States (Sutherst and Maywald 2005).

Experimental data can also be used separately from the modelling approach itself to interpret the predictions of a correlational SDM. For example, experimentally acquired data on thermal tolerance (critical thermal limits and limits for foraging activity) have been used to understand a range prediction for *Wasmannia auropunctata*. The predicted range limits are very coherent with the species' lower thermal tolerance (Coulin et al. 2019).

7.2.1.3 Mechanistic Models

Mechanistic models use first principles to estimate growth and survival based on temperature and to predict which areas fulfil a species' requirements to complete its life cycle. For example, an eco-physiological model of colony growth and the production of sexuals (males and queens) has been built for *Solenopsis invicta* and predicted a possible range expansion in the United States (Morrison et al. 2005). The model assumed that soil temperature is the key determinant of the species' range limits—cold winter temperatures over an extended period of time are the main factor limiting colony survival—and used minimum and maximum daily soil temperature data. Another model has been built to estimate the number of "degree-days", i.e. the number of days where the temperature is above a physiological threshold (15.8 °C) that allows the colony to grow (Hartley et al. 2010). This "bottom-up" approach to modelling has the advantage of incorporating more realistic temperature response curves. This can be particularly useful when a population has not yet reached equilibrium and has not colonised all areas and climates where it could potentially thrive. On the downside, mechanistic models cannot easily account for several potentially limiting variables and interactions among them.

7.2.2 Observations of Past Range Changes

Ideally, one would like to test if past ant invasions have been linked to climate change. Unfortunately, however, field observations akin to the famous study demonstrating the fingerprint of climate change on range changes in butterflies (Parmesan et al. 1999), are lacking in ants. However, the recent spread of *Pseudomyrmex gracilis* has been recorded from its native tropical range that spans from South America to Eastern Texas along the entire Gulf coast from Texas, through Louisiana, Mississippi and Alabama to southern Georgia and Florida. It has been hypothesised that this explosive range expansion might be related with climate and/or habitat changes, but this has not been formally tested (Wetterer et al. 2022). In another time series analysis, a 30-year field study on *Linepithema humile* in California documented the colonisation of some new sites and decreases in abundance at others. The study concluded that site-specific variables can drive these changes over time. However, a future study would be needed to separate the factors that are intrinsic to the process of invasion from long-term environmental changes (such as climate change) that represent extrinsic influence on the invasion dynamics (Menke et al. 2018). Along these lines, another study found a link between multi-year droughts and a slowed spread of *Linepithema humile* (Couper et al. 2021), but it is unclear if these droughts could be attributed to climate change. However, even if a study found a correlation between climate change and the expansion of a species, this is not necessarily proof that climate change is the cause of the expansion. Many species have expanded to novel climates during the invasion process, without a change in the climate itself (Bates and Bertelsmeier 2021).

7.2.3 Community-Scale Experiments

Many studies have explored biotic interactions in the context of climate change. These studies contribute important knowledge about the underlying mechanisms of species responses to climate change, but often do not make an explicit link with species' current or future ranges at larger spatial scales. Nonetheless, some findings have qualitative implications about what we might expect under climate change.

Biotic interactions may sometimes facilitate invasions (Simberloff and Von Holle 1999; Traveset and Richardson 2014). In the case of ants, it is well-known that many non-native species tend hemipteran and benefit from sugary honeydew (Green et al. 2011; Wilder et al. 2011), resulting in higher colony growth rates (Helms and Vinson 2008) and more aggressive behaviours (Grover et al. 2007). Higher temperatures could strengthen the mutualism between *Tapinoma melanocephalum* and mealybugs. The mealybug colony growth and the tending behaviour by the ants are temperature-dependent, leading to greater honeydew production under experimental temperature manipulation in a greenhouse (Zhou et al. 2017).

On the other hand, biotic interactions may contribute to the slowing of biological invasions. Experimental increases in temperature and temperature variability increased the susceptibility of *Lasius neglectus*, a widespread non-native species, to fungal infections. This could be because the higher temperature causes a stress-response decreasing immunity function or because grooming behaviours diminished under elevated temperature (Pamminger et al. 2016). Competitive interactions may also be determinantal to invasions, especially under climate change. Experimental confrontations of *Lasius neglectus* and two dominant Mediterranean species which are present in parts of the invaded range suggested that one native species, *Crematogaster scutellaris*, was a better competitor and particularly so at high temperatures (Frizzi et al. 2017). The other native species, *Tapinoma nigerrimum*, was a weaker competitor but particularly so at low temperatures (Frizzi et al. 2017). This shows that the non-native *Lasius neglectus* is at a disadvantage during interspecific encounters when temperatures rise. However, this may be true for competitive interactions at the warm range limit, but not at the poleward range limit.

In some cases, competitive interactions may also be weakened, thereby reducing biotic resistance and facilitating the spread of non-native ants. For example, a study on an ant community in California found that native species were more vulnerable to drought stress than non-native species, indicating a possible advantage for non-native ants under future conditions (Johnson and Stahlschmidt 2020).

Recently, experimental warming studies have built wired heating devices that can be inserted in the soil in natural environments in order to elevate the temperature of a plot. These devices allow heating the soil following external weather patterns, but elevating soil temperature by a defined number of °C above the external air temperature, thereby reflecting natural weather patterns under climate change. These set-ups allow following community dynamics over different seasons and evaluating impacts of climate change on native and non-native ant species present in those plots. For example, in an ant community in the southern United States, experimental heating of the soil by +5 °C was beneficial to two non-native ant species, *Brachyponera chinensis* and *Solenopsis invicta*, at the expense of the native ant community (Merchlinsky et al. 2023). Another study showed that even within their native range, non-native species may become more dominant following climate change: relative to other ant species of its native community, *Wasmannia auropunctata* became more abundant in warmed plots (Bujan et al. 2022).

7.3 Niche Shifts to Novel Climates

One underlying assumption of correlative SDMs and some hybrid models is that current distributions reflect the entire set of climatic conditions under which species will thrive in the future (Soberón and Peterson 2005; Gallien et al. 2010; Elith et al. 2010) and that they conserve this niche once they are introduced elsewhere (Fig. 7.3). However, in many cases, species do not occupy their entire fundamental niche because of dispersal limitations, species interactions or resource availability (Gaston 2009; Sexton et al. 2009). Once species are transported out of their native range, they are free from many of the constraints in their native range and can fill their fundamental niche, resulting in a "realised niche shift" (Fig. 7.3). It is also possible that species evolve higher thermal tolerances through selection in the introduced range, resulting in a "fundamental niche shift" (Fig. 7.3), but evidence for this process is scarce and difficult to provide (Bates and Bertelsmeier 2021). To demonstrate adaptive evolution, one would need to show that (a) a trait (like thermal tolerance) differs between the introduced and the native source population, (b) that this trait has a genetic basis and (c) that it confers a fitness advantage to the introduced population. In the absence of evidence for adaptive evolution, it is more parsimonious to assume a realised niche shift if the species has colonised novel climates.

Whatever the underlying mechanism, niche shifts can make range predictions difficult. For example, recent surveillance data (Malone et al. 2023) has shown that *Solenopsis invicta* has already spread much further than predicted 15 years ago (Morrison et al. 2005), suggesting niche expansion. The colonisation of novel climates may be a slow process though. In New Zealand, it has taken many non-native ant species several decades to colonise much of the climatically suitable areas, indicating a considerable time lag in their spread (Ward 2007). If niche shifts occur during invasion, the predictions of future ranges under climate change are not reliable. A study investigated the frequency and amplitude of niche shifts across all 82 non-native ants for which sufficient occurrence data was available to do this analysis (Bates et al. 2020). The surprising conclusion was that, contrary to expectations, the most invasive non-native ant species showed the smallest niche shifts. Invasiveness was measured using both impact-based and spatial spread-based indicators. The results can be explained by the fact that the most invasive species had the largest niche hypervolume in their native range. This suggests that these species might have already filled their fundamental niche to a greater extent in their native

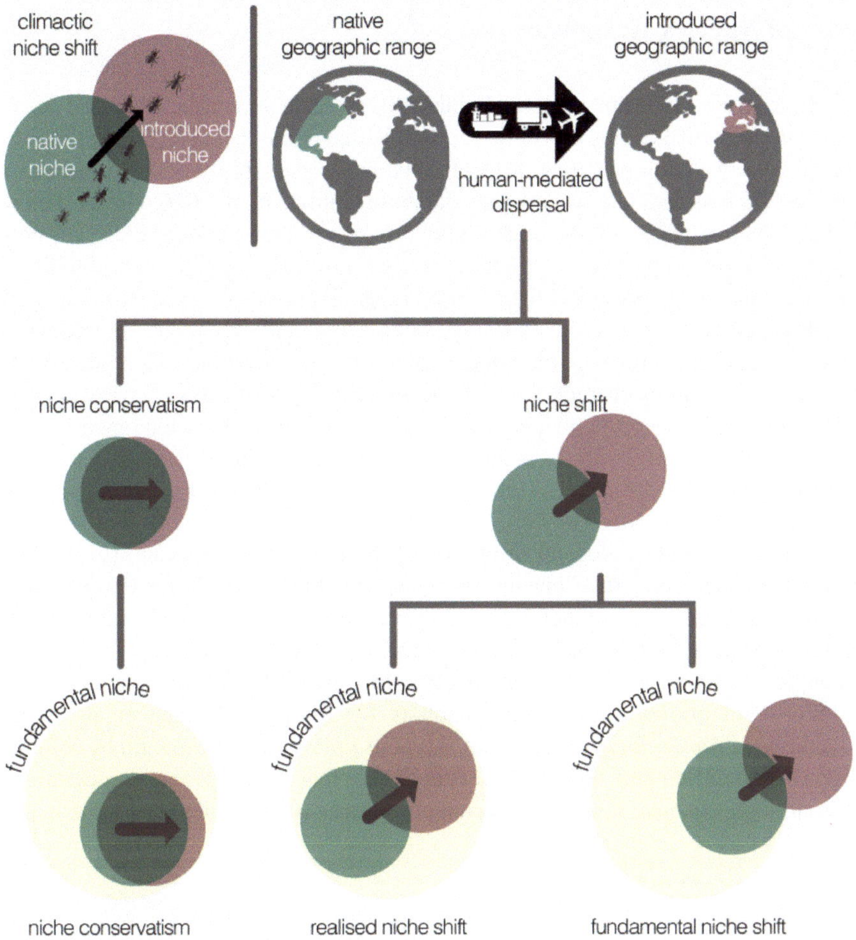

Fig. 7.3 Niche shift dynamics. When species are transported to new areas outside of their native range, they may either conserve or shift their niche. If they shift their niche, this may be a realised niche shift (filling parts of the fundamental niche) or a fundamental niche shift (extending the boundaries of the fundamental niche)

range, leaving less room for a realised niche shift, perhaps because these species are also dominant members of their native communities and not much limited by competition (Bates et al. 2020). But experimental approaches would be needed to explore the mechanisms leading to differences in the extent of niche shifts. In any case, these findings are encouraging for range predictions of the most invasive species—that are also the species which pose the greatest threats to biodiversity and for which we urgently need these forecasts.

7.4 More Complexities: We Need a Better Understanding of Species Responses

Much progress has been made in building large-scale forecasts. These advancements in predictive ecology have been possible due to the increasing complexity of statistical tools (Mouquet et al. 2015) and the ready availability of occurrence data for ants, such as through the Antmaps database (Janicki et al. 2016; Guénard et al. 2017). However, these predictions might surpass our actual understanding of ecological systems. In particular, it remains unclear how species will respond to novel climates that are not present in their current range. For example, species might show a certain degree of phenotypic plasticity under different environments. Indeed, non-native species might differ in their degree of plasticity, and perhaps high phenotypic plasticity could be a characteristic of the most invasive species (Davidson et al. 2011). Indeed, *Solenopsis invicta* does not have the highest heat tolerance (CT_{max}) compared to the native species in an invaded community. However, it is among the most thermally plastic species—able to adjust CT_{max} by several degrees over seasons (Bujan et al. 2020).

In addition to physiological responses of the individual, ants can also thermoregulate through collective behaviours (Jones and Oldroyd 2007; see Chap. 2). For example, ants may choose their nest location according to their preferred microclimates in order to "buffer" against adverse macroclimatic conditions (Fig. 7.4). For example, *Myrmica rubra* selects relatively cool and moist microhabitats for its nest locations but prefers warmer microhabitats for foraging (Warren et al. 2019). Likewise, *Tetramorium immigrans* subsists at its poleward range margin by colonising urban environments that are heat islands, i.e. that their average temperature is several degrees above that of the surrounding areas (Cordonnier et al. 2020). The

Fig. 7.4 Reasons why the macroclimate may be different from the conditions experienced by species: selection of suitable microclimates (diverging from macroclimates within heterogeneous landscapes), behavioural thermoregulation in the nest or foraging activity, the synergistic impacts of different land use types (e.g. urbanised areas) or modified local conditions due to land-use changes

choice of the nest site location can also vary over seasons. In California, *Linepithema humile* colonises coastal sage scrub from the urban environment during periods of high rainfall, but staying within 250 m of the urban edge (Bolger 2007).

Even within their nests, ants can choose the location of their sexuals and brood in order to optimise the temperature for their development. For example, *Solenopsis invicta* workers frequently move brood items up and down in their nest galleries at different seasons or times of the day (Penick and Tschinkel 2008). *Linepithema humile* also moves brood even between different nests of their spatially connected supercolony (Heller and Gordon 2006). Therefore, the microclimate experienced by the ants might substantially differ from that of the macroclimate represented on a grid cell of a map (Fig. 7.4). Ants may also thermoregulate their nest climate: nests are often constructed in ways to assure optimal air flows, and ants can close and open nest openings according to the external temperature (Menzel and Feldmeyer 2021). Moreover, the architecture of nests themselves is complex (Tschinkel 2021), and ants may adjust the depth and architecture of nest galleries when exposed to different climates (Sankovitz and Purcell 2021; García Ibarra et al. 2024).

7.5 Research Priorities

Overall, the increasing availability of global climate data sets, statistical models and ant distribution databases have allowed the prediction of non-native ant distributions under different scenarios of climate change and using different modelling methods. These models suggest species-specific and regionally heterogeneous responses. To make further progress in understanding the complexities of species responses to climate change and to predict future species distributions, we suggest three priorities for future research. First, it is urgent to find ways to combine experimental data and macroecological modelling. In particular, knowledge about the species' biology under different climatic conditions—including physiology, behaviour and species interactions—is accumulating in the literature. However, it is still poorly explored how these lab-based experiments and field observations could scale up to global changes in species ranges. Second, great progress could be made if we could identify the characteristics of species that are able to expand to novel climates. Indeed, it is unrealistic to have precise experimental data on many of the more than 300 non-native ant species that have established self-sustaining populations. Instead, it would be useful to analyse the ecological and morphological traits of species in relation to range dynamics to make predictions how functionally similar species might respond to changing conditions. Finally, our third priority is to study synergistic effects of global changes. Climate change is not acting in isolation, and several drivers of change, including urbanisation, land use change or pollution, may interact with climate change and weaken native communities or promote the spread of non-native species (Fig. 7.4).

References

Angulo E, Hoffmann BD, Ballesteros-Mejia L, Taheri A, Balzani P, Bang A, Renault D, Cordonnier M, Bellard C, Diagne C, Ahmed DA, Watari Y, Courchamp F (2022) Economic costs of invasive alien ants worldwide. Biol Invasions 24:2041–2060

Araújo MB, Anderson RP, Barbosa AM, Beale CM, Dormann CF, Early R, Garcia RA, Guisan A, Maiorano L, Naimi B, O'Hara RB, Zimmermann NE, Rahbek C (2019) Standards for distribution models in biodiversity assessments. Sci Adv 5:1–12

Aulus-Giacosa L, Ollier S, Bertelsmeier C (2024) Non-native ants are breaking down biogeographic boundaries and homogenizing community assemblages. Nat Commun 15:2266

Bates OK, Bertelsmeier C (2021) Climatic niche shifts in introduced species. Curr Biol 31:R1252–R1266

Bates OK, Ollier S, Bertelsmeier C (2020) Smaller climatic niche shifts in invasive than non-invasive alien ant species. Nat Commun 11:5213

Baudier KM, Mudd AE, Erickson SC, O'Donnell S (2015) Microhabitat and body size effects on heat tolerance: implications for responses to climate change (army ants: Formicidae, Ecitoninae). J Anim Ecol 84:1322–1330

Bertelsmeier C (2021) Globalization and the anthropogenic spread of invasive social insects. Curr Opin Insect Sci 46:16–23

Bertelsmeier C, Courchamp F (2014) Future ant invasions in France. Environ Conserv 41:217–228

Bertelsmeier C, Guénard B, Courchamp F (2013a) Climate change may boost the invasion of Asian needle ants. PLoS One 8:e75438

Bertelsmeier C, Luque GM, Courchamp F (2013b) Global warming may freeze the invasion of big-headed ants. Biol Invasions 15:1561–1582

Bertelsmeier C, Luque GM, Courchamp F (2013c) The impact of climate change changes over time. Biol Conserv 16:107–115

Bertelsmeier C, Luque GM, Courchamp F (2013d) Increase in quantity and quality of suitable areas for invasive species as climate changes. Conserv Biol 27:1458–1467

Bertelsmeier C, Luque GM, Hoffmann BD, Courchamp F (2015) Worldwide ant invasions under climate change. Biodivers Conserv 24:117–128

Bertelsmeier C, Blight O, Courchamp F (2016) Invasions of ants (Hymenoptera: Formicidae) in the light of global climate change. Myrmecol News 22:25–42

Bertelsmeier C, Ollier S, Liebhold A, Keller L (2017) Recent human history governs global ant invasion dynamics. Nat Ecol Evol 1:0184

Bolger DT (2007) Spatial and temporal variation in the Argentine ant edge effect: implications for the mechanism of edge limitation. Biol Conserv 136:295–305

Booher DB, Gotelli NJ, Nelsen MP, Ohyama L, Deyrup M, Moreau CS, Suarez AV (2023) Six decades of museum collections reveal disruption of native ant assemblages by introduced species. Curr Biol 33:2088–2094

Broennimann O, Guisan A (2008) Predicting current and future biological invasions: both native and invaded ranges matter. Biol Lett 4:585–589

Brook BW, Sodhi NS, Bradshaw CJA (2008) Synergies among extinction drivers under global change. Trends Ecol Evol 23:453–460

Bujan J, Roeder KA, Yanoviak SP, Kaspari M (2020) Seasonal plasticity of thermal tolerance in ants. Ecology 101:1–6

Bujan J, Nottingham AT, Velasquez E, Meir P, Kaspari M, Yanoviak SP (2022) Tropical ant community responses to experimental soil warming. Biol Lett 18:3–7

Byeon DH, Jung JM, Park Y, Lee HS, Lee JH, Jung S, Lee WH (2020a) Model-based assessment of changes in the potential distribution of *Solenopsis geminata* (Hymenoptera: Formicidae) according to climate change scenarios. J Asia-Pacific Biodivers 13:331–338

Byeon DH, Lee JH, Lee HS, Park Y, Jung S, Lee WH (2020b) Prediction of spatiotemporal invasive risk by the red imported fire ant (Hymenoptera: Formicidae) in South Korea. Agronomy 10:1–15

Chen Y (2008) Global potential distribution of an invasive species, the yellow crazy ant (*Anoplolepis gracilipes*) under climate change. Integr Zool 3:166–175
Cordonnier M, Bellec A, Escarguel G, Kaufmann B (2020) Effects of urbanization–climate interactions on range expansion in the invasive European pavement ant. Basic Appl Ecol 44:46–54
Coulin C, de la Vega GJ, Chifflet L, Calcaterra LA, Schilman PE (2019) Linking thermo-tolerances of the highly invasive ant, *Wasmannia auropunctata*, to its current and potential distribution. Biol Invasions 21:3491–3504
Couper LI, Sanders NJ, Heller NE, Gordon DM (2021) Multiyear drought exacerbates long-term effects of climate on an invasive ant species. Ecology 102:1–7
Davidson AM, Jennions M, Nicotra AB (2011) Do invasive species show higher phenotypic plasticity than native species and, if so, is it adaptive? A meta-analysis. Ecol Lett 14:419–431
Elith J, Kearney M, Phillips S (2010) The art of modelling range-shifting species. Methods Ecol Evol 1:330–342
Fenn-Moltu G, Ollier S, Caton B, Liebhold AM, Nahrung H, Pureswaran DS, Turner RM, Yamanaka T, Bertelsmeier C (2023) Alien insect dispersal mediated by the global movement of commodities. Ecol Appl 33:e2721
Fick SE, Hijmans RJ (2017) WorldClim 2: new 1-km spatial resolution climate surfaces for global land areas. Int J Climatol 37:4302–4315
Frizzi F, Bartalesi V, Santini G (2017) Combined effects of temperature and interspecific competition on the mortality of the invasive garden ant, Lasius neglectus: a laboratory study. J Therm Biol 65:76–81
Gallien L, Münkemüller T, Albert CH, Boulangeat I, Thuiller W (2010) Predicting potential distributions of invasive species: where to go from here? Divers Distrib 16:331–342
García Ibarra F, Jouquet P, Bottinelli N, Bultelle A, Monnin T (2024) Experimental evidence that increased surface temperature affects bioturbation by ants. J Anim Ecol 93:319–332
Garnas JR, Auger-Rozenberg M-A, Roques A, Bertelsmeier C, Wingfield MJ, Saccaggi DL, Roy HE, Slippers B (2016) Complex patterns of global spread in invasive insects: eco-evolutionary and management consequences. Biol Invasions 18:935–952
Gaston K (2009) Geographic range limits of species. Proc R Soc B Biol Sci 276:1391–1393
Gippet JM, Liebhold AM, Fenn-Moltu G, Bertelsmeier C (2019) Human-mediated dispersal in insects. Curr Opin Insect Sci 35:96–102
Gotzek D, Axen HJ, Suarez AV, Helms Cahan S, Shoemaker D (2015) Global invasion history of the tropical fire ant: a stowaway on the first global trade routes. Mol Ecol 24:374–388
Green PT, O'Dowd DJ, Abbott KL, Jeffery M, Retallick K, Mac Nally R (2011) Invasional meltdown: invader-invader mutualism facilitates a secondary invasion. Ecology 92:1758–1768
Grover CD, Kay AD, Monson JA, Marsh TC, Holway DA (2007) Linking nutrition and behavioural dominance: carbohydrate scarcity limits aggression and activity in Argentine ants. Proc R Soc B Biol Sci 274:2951–2957
Gruber MAM, Santoro D, Cooling M, Lester PJ, Hoffmann BD, Boser C, Lach L (2022) A global review of socioeconomic and environmental impacts of ants reveals new insights for risk assessment. Ecol Appl 32:e2577
Guénard B, Weiser MD, Gómez K, Narula N, Economo EP (2017) The global ant biodiversity informatics (GABI) database: synthesizing data on the geographic distribution of ant species (Hymenoptera: Formicidae). Myrmecol News 24:83–89
Guisan A, Thuiller W (2005) Predicting species distribution: offering more than simple habitat models. Ecol Lett 8:993–1009
Hartley S, Krushelnycky PD, Lester PJ (2010) Integrating physiology, population dynamics and climate to make multi-scale predictions for the spread of an invasive insect: the Argentine ant at Haleakala National Park, Hawaii. Ecography (Cop) 33:83–94
Heller NE, Gordon DM (2006) Seasonal spatial dynamics and causes of nest movement in colonies of the invasive Argentine ant (*Linepithema humile*). Ecol Entomol 31:499–510
Heller NE, Sanders NJ, Shors JW, Gordon DM (2008) Rainfall facilitates the spread, and time alters the impact, of the invasive Argentine ant. Oecologia 155:385–395

Helms KR, Vinson SB (2008) Plant resources and colony growth in an invasive ant: the importance of honeydew-producing Hemiptera in carbohydrate transfer across trophic levels. Environ Entomol 37:487–493
Holway DA, Lach L, Suarez AV, Tsutsui ND, Case TJ, Suarez AV, Holway DA, Lach L, Tsutsui ND, Casel TJ (2002) The causes and consequences of ant invasions. Annu Rev Ecol Syst 33:181–233
Hulme PE (2017) Climate change and biological invasions: evidence, expectations, and response options. Biol Rev 92:1297–1313
Janicki J, Narula N, Ziegler M, Guénard B, Economo EP (2016) Visualizing and interacting with large-volume biodiversity data using client-server web-mapping applications: the design and implementation of antmaps.org. Ecol Inform 32:185–193
Johnson DJ, Stahlschmidt ZR (2020) City limits: heat tolerance is influenced by body size and hydration state in an urban ant community. Ecol Evol 10:4944–4955
Jones JC, Oldroyd BP (2007) Nest thermoregulation in social insects. Adv Insect Phys 33:153–191
Jung JM, Byeon DH, Jung SH, Yu YM, Yasunaga-Aoki C, Lee WH (2017) Global prediction of geographical change of yellow crazy ant (*Anoplolepis gracilipes*) distribution in response to climate change scenario. J Fac Agric Kyushu Univ 62:403–410
Karger DN, Schmatz DR, Dettling G, Zimmermann NE (2020) High-resolution monthly precipitation and temperature time series from 2006 to 2100. Sci Data 7:248
Lach L (2021) Invasive ant establishment, spread, and management with changing climate. Curr Opin Insect Sci 47:119–124
Lee CM, Lee DS, Kwon TS, Athar M, Park YS (2021) Predicting the global distribution of *Solenopsis geminata* (Hymenoptera: Formicidae) under climate change using the maxent model. Insects 12:229
Lee WH, Song JW, Yoon SH, Jung JM (2022) Spatial evaluation of machine learning-based species distribution models for prediction of invasive ant species distribution. Appl Sci 12:10260
Lessard J, Fordyce JA, Gotelli NJ, Sanders NJ (2009) Invasive ants alter the phylogenetic structure of ant communities. Ecology 90:2664–2669
Li M, Xian X, Zhao H, Xue L, Chen B, Huang H, Wan F, Liu W (2022) Predicting the potential suitable area of the invasive ant *Linepithema humile* in China under future climatic scenarios based on optimized MaxEnt. Diversity 14:921
Li D, Li Z, Wang X, Wang L, Khoso AG, Liu D (2023) Climate change and international trade can exacerbate the invasion risk of the red imported fire ant Solenopsis invicta around the globe. Entomol Gen 43:315–323
Malone M, Ivanov K, Taylor SV, Schürch R (2023) Fast range expansion of the red imported fire ant in Virginia and prediction of future spread in the United States. Ecosphere 14:e4652
Mao M, Chen S, Ke Z, Qian Z, Xu Y (2022) Using MaxEnt to predict the potential distribution of the little fire ant (*Wasmannia auropunctata*) in China. Insects 13:1008
Menke SB, Fisher RN, Jetz W, Holway DA (2007) Biotic and abiotic controls of Argentine ant invasion success at local and landscape scales. Ecology 88:3164–3173
Menke SB, Holway DA, Fisher RN, Jetz W (2009) Characterizing and predicting species distributions across environments and scales: Argentine ant occurrences in the eye of the beholder. Glob Ecol Biogeogr 18:50–63
Menke SB, Ward PS, Holway DA (2018) Long-term record of Argentine ant invasions reveals enduring ecological impacts. Ecology 99:1194–1202
Menzel F, Feldmeyer B (2021) How does climate change affect social insects? Curr Opin Insect Sci 46:10–15
Merchlinsky A, Frankson PT, Gitzen R, Lepczyk CA, Mohan JE, Warren RJ (2023) Warming promotes non-native invasive ants while inhibiting native ant communities. Ecol Entomol 48:588–596
Meurisse N, Rassati D, Hurley BP, Brockerhoff EG, Haack RA (2019) Common pathways by which non-native forest insects move internationally and domestically. J Pest Sci 92:13–27

Morrison LW, Korzukhin MD, Porter SD (2005) Predicted range expansion of the invasive fire ant, Solenopsis invicta, in the eastern United States based on the VEMAP global warming scenario. Divers Distrib 11:199–204

Mouquet N, Lagadeuc Y, Devictor V, Doyen L, Duputié A, Eveillard D, Faure D, Garnier E, Gimenez O, Huneman P, Jabot F, Jarne P, Joly D, Julliard R, Kéfi S, Kergoat GJ, Lavorel S, Le Gall L, Meslin L, Morand S, Morin X, Morlon H, Pinay G, Pradel R, Schurr FM, Thuiller W, Loreau M (2015) Predictive ecology in a changing world. J Appl Ecol 52:1293–1210

Ollier S, Bertelsmeier C (2022) Precise knowledge of commodity trade is needed to understand invasion flows. Front Ecol Environ 20:467–473

Pamminger T, Steier T, Tragust S (2016) High temperature and temperature variation undermine future disease susceptibility in a population of the invasive garden ant Lasius neglectus. Sci Nat 103:46

Parmesan C, Ryrholm N, Stefanescu C, Hill JK, Thomas CD, Descimon H, Huntley B, Kaila L, Kullberg J, Tammaru T, Tennent WJ, Thomas JA, Warren M (1999) Poleward shifts in geographical ranges of butterfly species associated with regional warming. Nature 399:579–583

Penick CA, Tschinkel WR (2008) Thermoregulatory brood transport in the fire ant, Solenopsis invicta. Insect Soc 55:176–182

Penick CA, Diamond SE, Sanders NJ, Dunn RR (2017) Beyond thermal limits: comprehensive metrics of performance identify key axes of thermal adaptation in ants. Funct Ecol 31:1091–1100

Pincebourde S, Casas J (2019) Narrow safety margin in the phyllosphere during thermal extremes. Proc Natl Acad Sci USA 116:5588–5596

Potter KA, Arthur Woods H, Pincebourde S (2013) Microclimatic challenges in global change biology. Glob Chang Biol 19:2932–2939

Rouget M, Robertson MP, Wilson JRU, Hui C, Essl F, Renteria JL, Richardson DM (2015) Invasion debt – quantifying future biological invasions. Divers Distrib 22:445–456

Roura-Pascual N, Suarez AV, Gómez C, Pons P, Touyama Y, Wild AL, Peterson AT, Gomez C, Pons P, Touyama Y, Wild AL, Gómez C (2004) Geographical potential of Argentine ants (*Linepithema humile* Mayr) in the face of global climate change. Proc Biol Sci 271:2527–2535

Sankovitz M, Purcell J (2021) Ant nest architecture is shaped by local adaptation and plastic response to temperature. Sci Rep 11:23053

Schulze-Sylvester M, Corronca JA, Paris CI (2018) Growing industries, growing invasions? The case of the Argentine ant in vineyards of northern Argentina. Insects 9:11

Seebens H, Blackburn TM, Dyer E, Genovesi P, Hulme PE, Jeschke JM, Pagad S, Pyšek P, Winter M, Arianoutsou M, Bacher S, Blasius B, Brundu G, Capinha C, Celesti-Grapow L, Dawson W, Dullinger S, Fuentes N, Jäger H, Kartesz J, Kenis M, Kreft H, Kühn I, Lenzner B, Liebhold A, Mosena A, Moser D, Nishino M, Pearman D, Pergl J, Rabitsch W, Rojas-Sandoval J, Roques A, Rorke S, Rossinelli S, Roy HE, Scalera R, Schindler S, Štajerová K, Tokarska-Guzik B, van Kleunen M, Walker K, Weigelt P, Yamanaka T, Essl F (2017) No saturation in the accumulation of alien species worldwide. Nat Commun 8:14435

Seebens H, Blackburn TM, Dyer EE, Genovesi P, Hulme PE, Jeschke JM, Pagad S, Pyšek P, van Kleunen M, Winter M, Ansong M, Arianoutsou M, Bacher S, Blasius B, Brockerhoff EG, Brundu G, Capinha C, Causton CE, Celesti-Grapow L, Dawson W, Dullinger S, Economo EP, Fuentes N, Guénard B, Jäger H, Kartesz J, Kenis M, Kühn I, Lenzner B, Liebhold AM, Mosena A, Moser D, Nentwig W, Nishino M, Pearman D, Pergl J, Rabitsch W, Rojas-Sandoval J, Roques A, Rorke S, Rossinelli S, Roy HE, Scalera R, Schindler S, Štajerová K, Tokarska-Guzik B, Walker K, Ward DF, Yamanaka T, Essl F (2018) Global rise in emerging alien species results from increased accessibility of new source pools. Proc Natl Acad Sci USA 115:201719429

Sexton JP, McIntyre PJ, Angert AL, Rice KJ (2009) Evolution and ecology of species range limits. Annu Rev Ecol Evol Syst 40:415–436

Siddiqui JA, Bamisile BS, Khan MM, Islam W, Hafeez M, Bodlah I, Xu Y (2021) Impact of invasive ant species on native fauna across similar habitats under global environmental changes. Environ Sci Pollut Res 28:54362–54382

Simberloff D, Von Holle B (1999) Positive interactions of nonindigenous species. Biol Invasions 1:21–32

Simberloff D, Martin J-L, Genovesi P, Maris V, Wardle DA, Aronson J, Courchamp F, Galil B, García-Berthou E, Pascal M, Pyšek P, Sousa R, Tabacchi E, Vilà M (2013) Impacts of biological invasions: what's what and the way forward. Trends Ecol Evol 28:58–66

Soberón J, Peterson AT (2005) Interpretation of models of fundamental ecological niches and species' distributional areas. Biodivers Inform 2:1–10

Song J, Zhang H, Li M, Han W, Yin Y, Lei J (2021) Prediction of spatiotemporal invasive risk of the red import. Insects 12:874

Suarez AV, Holway DA, Ward PS (2005) The role of opportunity in the unintentional introduction of nonnative ants. Proc Natl Acad Sci USA 102:17032–17035

Sung S, Kwon YS, Lee DK, Cho Y (2018) Predicting the potential distribution of an invasive species, *Solenopsis invicta* Buren (Hymenoptera: Formicidae), under climate change using species distribution models. Entomol Res 48:505–513

Sutherst RW, Maywald G (2005) A climate model of the red imported fire ant, *Solenopsis invicta* Buren (Hymenoptera: Formicidae): implications for invasion of new regions, particularly Oceania. Environ Entomol 34:317–335

Tercel MPTG, Cuff JP, Symondson WOC, Vaughan IP (2023) Non-native ants drive dramatic declines in animal community diversity: a meta-analysis. Insect Conserv Divers 16:733–744

Traveset A, Richardson DM (2014) Mutualistic interactions and biological invasions. Annu Rev Ecol Evol Syst 45:89–113

Trible W, McKenzie SK, Kronauer DJC (2020) Globally invasive populations of the clonal raider ant are derived from Bangladesh: source of the clonal raider ant invasion. Biol Lett 16:1–5

Tschinkel WR (2021) Ant architecture: the wonder, beauty and science of underground nests. Princeton University Press, Princeton

Walther GR, Post E, Convey P, Menzel A, Parmesan C, Beebee TJC, Fromentin JM, Hoegh-Guldberg O, Bairlein F (2002) Ecological responses to recent climate change. Nature 416:389–395

Wang H, Wang H, Tao Z, Ge Q (2018) Potential range expansion of the red imported fire ant (*Solenopsis invicta*) in China under climate change. J Geogr Sci 28:1965–1975

Ward DF (2007) Modelling the potential geographic distribution of invasive ant species in New Zealand. Biol Invasions 9:723–735

Warren RJ, Mathew A, Reed K, Bayba S, Krupp K, Spiering DJ (2019) Myrmica rubra microhabitat selection and putative ecological impact. Ecol Entomol 44:239–248

Warren RJ, Frankson PT, Mohan JE (2023) Global change drivers synergize with the negative impacts of non-native invasive ants on native seed-dispersing ants. Biol Invasions 25:773–786

Wetterer JK, Hill JG, MacGown JA, Booher DB (2022) Populations of the graceful twig ant, *Pseudomyrmex gracilis*, now bridge the distribution gap between their native range in Texas and non-native range in Florida, USA. Trans Am Entomol Soc 148:51–58

Wilder SM, Holway DA, Suarez AV, LeBrun EG, Eubanks MD (2011) Intercontinental differences in resource use reveal the importance of mutualisms in fire ant invasions. Proc Natl Acad Sci 108:20639–20644

Wong MKL, Guénard B, Lewis OT (2020) The cryptic impacts of invasion: functional homogenization of tropical ant communities by invasive fire ants. Oikos 129:585–597

Wong MKL, Economo EP, Guénard B (2023) The global spread and invasion capacities of alien ants. Curr Biol 33:566–571

Zhao H, Xian X, Guo J, Yang, Zhang Y, Chen B, Huang H, Liu W (2023) Monitoring the little fire ant, *Wasmannia auropunctata* (Roger 1863), in the early stage of its invasion in China: predicting its geographical distribution pattern under climate change. J Integr Agric 22:2783–2795

Zhou A, Qu X, Shan L, Wang X (2017) Temperature warming strengthens the mutualism between ghost ants and invasive mealybugs. Sci Rep 7:1–10

Part IV
Communities

Chapter 8
Using Latitudinal Climatic Gradients to Inform Ant Responses to Climate Change

Alan N. Andersen, Stefan Caddy-Retalic, and Heraldo L. Vasconcelos

8.1 Introduction

Climatic gradients are extensively used for understanding the importance of climatic drivers of biodiversity. This makes them powerful tools for informing potential impacts of climate change, using space-for-time substitution as a window into future biodiversity change. If we want to know what biodiversity will be like under a new climate, then it is informative to look at biodiversity occurring *now* under that climate elsewhere in the broader region.

Elevation gradients are commonly used for investigating climatic drivers of biodiversity, including for ants (see Chap. 9), because the associated climatic gradients are so highly compressed. This is especially the case for temperature, a globally dominant factor influencing ant distribution. However, insights from elevational gradients cannot necessarily be generalised beyond montane faunas. A broader understanding of the role of climate as a driver of geographic patterns of biodiversity can be revealed by latitudinal climatic gradients (de Frenne et al. 2013; Caddy-Retalic et al. 2017). The importance of temperature as a primary factor influencing ant distribution is reflected by strong latitudinal gradients of ant diversity at global (Dunn et al. 2009; Dantas and Fonseca 2023) and continental (Cushman et al. 1993)

A. N. Andersen (✉)
Research Institute for the Environment and Livelihoods, Charles Darwin University, Brinkin, Northern Territory, NT, Australia
e-mail: alan.andersen@cdu.edu.au

S. Caddy-Retalic
University of Adelaide, North Terrace, Adelaide, SA, Australia
e-mail: stefan.caddy-retalic@adelaide.edu.au

H. L. Vasconcelos
Universidade Federal de Uberlândia, Uberlândia, MG, Brazil
e-mail: heraldo@ufu.br

I. R. Leal et al. (eds.), *Ant Responses to Climate Change*, Entomology in Focus 10,
https://doi.org/10.1007/978-3-032-21127-9_8

scales. However, climatic gradients at a regional scale are especially useful for informing potential biodiversity responses to climate change because they provide a climatic and biotic context that is directly relevant to a particular locality, including information on patterns of species turnover in addition to diversity (Andersen et al. 2015; Caddy-Retalic et al. 2019; del Toro et al. 2019).

Latitudinal gradients at regional scales can be used to address the following five key questions relating to responses of ants to climate change. First, how does ant diversity vary in relation to changes in temperature and rainfall that fall within the range forecast under climate change? Second, do responses to climatic variation vary among different components of ant faunas? Third, to what extent do responses to climatic variation vary among different attributes of diversity? Fourth, how do ant responses to climatic variation vary within and among biomes? Finally, can climatic zones with particularly high rates of species turnover, making them especially sensitive to climate change, be identified?

Despite the attractiveness of regional-scale latitudinal gradients for informing ant responses to climate change, there have been surprisingly few such studies. In this chapter, we illustrate how the limited available studies inform the five key questions.

8.2 Ant-Climate Relationships: An Exemplar Latitudinal Gradient

The 550 km TREND transect in south-eastern Australia (Caddy-Retalic et al. 2019) illustrates how a latitudinal gradient can be used to document ant-climate associations and thus inform potential ant responses to climate change. The transect spans 800 mm in mean annual rainfall and 5 °C in mean annual temperature, from the warm, arid north (−31° latitude) to the cool, mesic south (−36°, Fig. 8.1). Ants were surveyed at multiple sites at each of 15 locations along the transect, recording 227 species from 39 genera (Caddy-Retalic et al. 2019). Ant species richness showed no significant variation with rainfall, reflecting the unusually high ant diversity of arid Australia (Andersen 2016), but rainfall was the strongest correlate of variation in species composition and therefore a key driver of species turnover.

Ant species distributions along TREND were used to make predictions of compositional change under potential future climates. Baseline compositional variation explained by environmental variables was delimited through canonical correspondence analysis (CCA), and this was compared with predicted CCAs produced for the years 2050, 2070, and 2090 based on different climate scenarios. Ant communities were shown to have a strong and consistent response to a changing climate, with the polygon representing the entire baseline gradient in ant composition space migrating extensively along an axis relating to temperature and rainfall (Fig. 8.2).

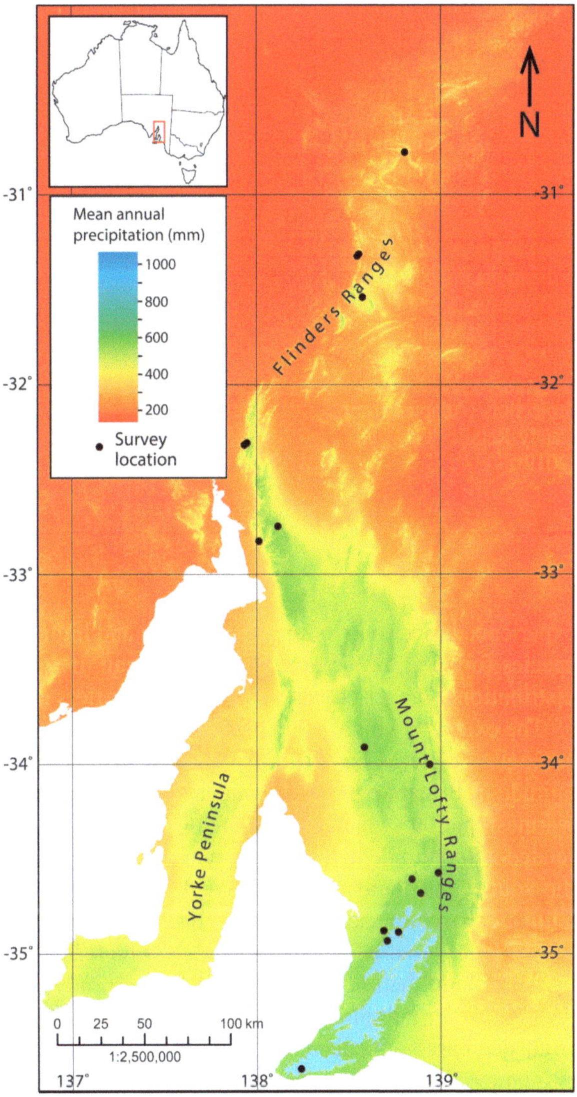

Fig. 8.1 The TREND latitudinal rainfall gradient, from 200 mm in the north to 1000 mm in the south, and its location in south-eastern Australia (inset). (Adapted from Caddy-Retalic et al. (2019))

8.3 Climatic Responses of Different Components of Ant Faunas

Ant communities can be strongly stratified vertically (see Chap. 12), comprising specialist cryptobiotic (living in soil and litter) and arboreal components in addition to the epigaeic taxa that dominate in most biomes. These different components experience very different microclimates and therefore can be expected to show different relationships to macroclimatic variation. For example, arboreal species

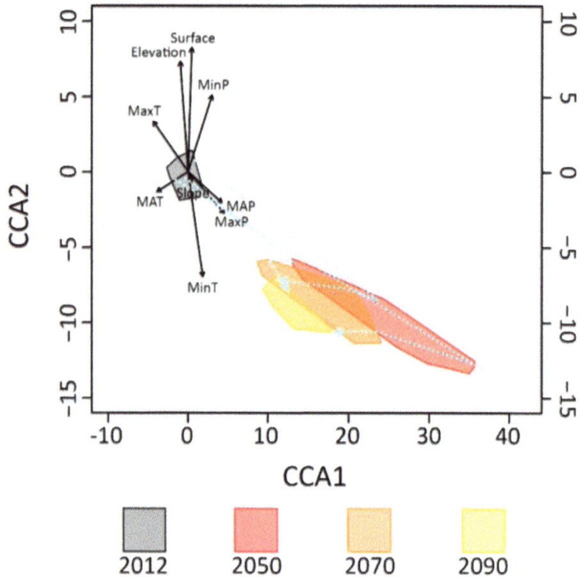

Fig. 8.2 CCA ordinations of environmental and ant species matrices along the TREND latitudinal gradient in South Australia at the time of sampling in 2012 (grey polygons, delimiting ordination spaces occupied by all sites) and predicted CCA ordinations in 2050 (red), 2070 (orange), and 2090 (yellow) based on the RCP8.5 (2.6–4.8 °C warming by 2100) climate projection. (Adapted from Caddy-Retalic et al. (2019))

experience higher maximum temperatures and higher temperature variation than do species foraging on the ground, and they can therefore be expected to have larger distributional ranges (Leahy et al. 2021; Scheffers et al. 2013).

The extent to which ground and arboreal ant assemblages show different macroecological patterns has been examined using a climatic gradient spanning 22° of latitude in Brazilian tropical savanna (Vasconcelos et al. 2023). Rainfall during the warmest quarter (which is correlated with net primary productivity) is strongly correlated with latitude in the region, ranging from <100 mm in the north to 700 mm in the south, and there is a north-south gradient of decreasing mean temperature. Climatic variation across latitude had different effects on ground and arboreal assemblages. Overall ant species richness increases with latitude, but more sharply on the ground than in trees (Fig. 8.3). The turnover of species between strata was lower at lower latitudes, such that the compositional dissimilarity between vertical strata decreased with decreasing latitude. These findings suggest that under the forecast hotter and drier climate, ant communities of Brazilian savanna will become less diverse, especially on the ground, and arboreal assemblages will become less differentiated from those on the ground.

8.4 Climatic Responses of Different Attributes of Diversity

There is increasing recognition of the conservation value of biodiversity attributes other than those relating to species, such as functional and phylogenetic diversity (Faith 1992; Swenson et al. 2012). Functional diversity is relevant to the ecological

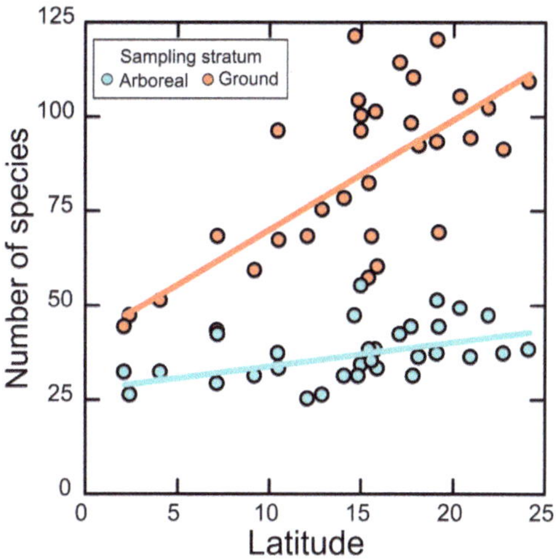

Fig. 8.3 Latitudinal variation in ant species richness of ground and arboreal assemblages in tropical savannas of Brazil. (From Vasconcelos et al. 2023)

roles of species (Swenson and Weiser 2014), and phylogenetic diversity incorporates evolutionary relationships, addressing whether species are drawn from similar or from distantly related lineages (Cadotte et al. 2013; Webb et al. 2002; Cavender-Bares et al. 2009). Macroecological relationships between the different attributes of diversity are often not congruent with each other. For example, functional and phylogenetic diversity are only weakly correlated with species diversity in European ant faunas (Arnan et al. 2017).

The latitudinal gradient in Brazilian savannas has revealed complex climatic relationships among different attributes of diversity and their interactions with vertical strata (Neves et al. 2024). As for species richness, functional and phylogenetic richness are both also positively correlated with latitude. However, the broad macroecological patterns of functional and phylogenetic diversity are substantially different from those of species richness, and the relationships vary between ground and canopy assemblages. Climate change will therefore likely have complex interactive effects on different attributes of diversity and different components of ant faunas.

8.5 Variation in Climatic Predictors of Ant Diversity Within and Among Biomes

There are too few available studies to analyse how climatic predictors of ant diversity vary among biomes. However, ant-climate relationships can vary markedly *within* a single biome in different continents due to differences in the functional composition of faunas that reflect different biogeographical histories. This has been well-documented for tropical savannas (Andersen and Vasconcelos 2022). Australian savannas evolved in association with the continent's desert biome, and its fauna is part of an exceptionally diverse and compositionally similar, arid-adapted fauna that occurs throughout most of the continent's inland (Andersen 2016). The fauna defies the typical macroecological pattern of declining diversity with increasing aridity by maintaining exceptionally high diversity throughout its latitudinal rainfall gradient from >1500 mm near the northern coast to 500 mm at its southern boundary with the central arid zone (Fig. 8.4; Andersen et al. 2015; del Toro et al. 2019). In contrast, Brazilian savannas have a forest-derived ant fauna due to its evolutionary history in association with tropical forest, arid-adapted taxa are largely absent, and diversity declines with increasing aridity (Andersen and Vasconcelos 2022). The ant fauna of African savannas has a stronger representation of arid-adapted taxa than in Brazil, but these are relatively minor elements, and diversity also declines with increasing aridity (Andersen and Vasconcelos 2022). These patterns suggest that under future climates of lower rainfall, ant diversity is likely to decline in savannas of Brazil and Africa, but not in Australia.

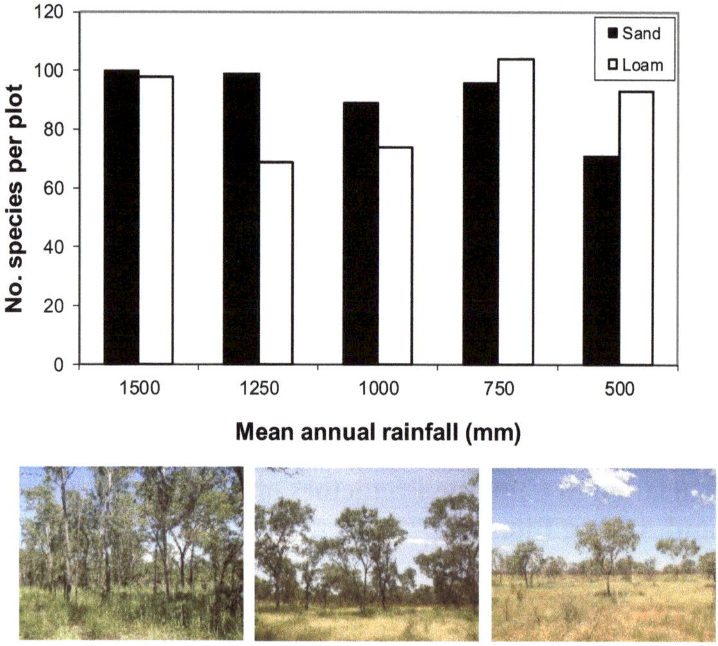

Fig. 8.4 Exceptionally high ant diversity is maintained with increasing aridity in Australian tropical savannas. (Data from Andersen et al. (2015))

8.6 Identifying Sensitive Zones

Caution is required when using macroecological relationships to inform future changes in particular places because of the different scale of climatic change. Macroecological relationships hold broadly over extensive climatic gradients whereas projected change in climatic variables represent tiny fractions of such gradients. For example, ant species richness often declines with increasing aridity (see Chap. 11), but this occurs over gradients of hundreds of millimetres in mean annual rainfall and so can't necessarily predict impacts of far smaller future increases in different locations within the gradient. This is even more so given that macroecological relationships are typically non-linear (Andersen et al. 2015; García-Navas et al. 2025). In this context, latitudinal gradients are particularly informative when they identify climate-sensitive zones where rates of species turnover are particularly high and therefore biodiversity change is likely to be most pronounced under a changing climate (Williams et al. 1995; Kreyling et al. 2014). For example, particularly rapid species and phylogenetic turnover of plants occurs along the TREND transect in the range of 15–16 °C in mean annual temperature and 400–600 mm in mean annual rainfall, corresponding to the arid-mesic transition (Guerin et al. 2013). An analysis of ant compositional change along the rainfall gradient of Australian tropical savannas reveals that the primary discontinuity occurs at the arid/monsoonal transition (500–600 mm) in the south and that a secondary discontinuity occurs at about 1000 mm, corresponding to the transition between the semi-arid and mesic monsoonal zones in the north (Andersen et al. 2015; del Toro et al. 2019). These transitions are consistent with previously documented biogeographic boundaries for plants (Bowman et al. 1988) and birds (Whitehead et al. 1992), indicating that they are likely to be particularly sensitive to future changes in rainfall for biodiversity broadly. These climate-sensitive zones should be priority locations for ongoing monitoring.

8.7 Conclusion

We have illustrated how latitudinal gradients can reveal potentially important insights into ant responses to climate change through an improved understanding of the role of climate as a driver of ant distribution. This includes an understanding of how ant-climate associations can vary according to different components of ant faunas, to different attributes of diversity, and in different biogeographic regions. It also includes the identification of climate-sensitive zones where rates of species turnover are particularly high and therefore biodiversity change is likely to be most pronounced under a changing climate.

We note, however, that caution must be applied when using climatic gradients for informing biodiversity responses to climate change (Caddy-Retalic et al. 2017). Rainfall and temperature typically co-vary, and so it is often not possible to identify their relative importance (Meirmans 2015). In addition, the assumption that current distributional patterns provide a robust basis for predicting future change ignores the role of historical factors (Warren et al. 2014), and it does not consider the role of biotic interactions and how these might change due to differential climate sensitivity of interacting species. The importance of changed biotic environments under future climates has been demonstrated in the TREND transect through a predicted decoupling of responses of ants and plants. Compared with ants, plants show significantly different climate relationships, and the predicted responses to climate change of the two taxa are therefore not congruent (Caddy-Retalic et al. 2019). Such decoupling of species associations implies the creation of novel communities under future climates. This has potentially important implications for the many mutualisms between ants and plants and for ecological services provided by ants more generally.

References

Andersen AN (2016) Ant megadiversity and its origins in arid Australia. Austral Entomol 55:132–147

Andersen AN, Vasconcelos HL (2022) Historical biogeography shapes functional ecology: inter-continental contrasts in responses of savanna ant communities to stress and disturbance. J Biogeogr 49:590–599

Andersen AN, Del Toro I, Parr CL (2015) Savanna ant species richness is maintained along a bioclimatic gradient of decreasing rainfall and increasing latitude in northern Australia. J Biogeogr 42:2313–2322

Arnan X, Cerdá X, Retana J (2017) Relationships among taxonomic, functional, and phylogenetic ant diversity across the biogeographic regions of Europe. Ecography 40:448–457

Bowman DMJS, Wilson BA, Dunlop CR (1988) Preliminary biogeographic analysis of the Northern Territory vascular flora. Aust J Bot 36:503–517

Caddy-Retalic S, Andersen A, Aspinwall M, Breed M, Byrne M, Christmas M, Dong N, Evans B, Fordham D, Guerin G, Hoffmann A, Hughes A, van Leeuwen S, McInerney F, Prober S, Rossetto M, Rymer P, Steane D, Wardle G, Lowe A (2017) Bioclimatic transect networks: powerful observatories of ecological change. Ecol Evol 7:4607–4619

Caddy-Retalic S, Hoffmann B, Guerin G, Andersen A, Wardle G, McInerney C, Lowe A (2019) Plant and ant assemblages predicted to decouple under climate change. Divers Distrib 25:551–567

Cadotte M, Albert CH, Walker SC (2013) The ecology of differences: assessing community assembly with trait and evolutionary distances. Ecol Lett 16:1234–1244

Cavender-Bares J, Kozak KH, Fine PVA, Kembel SW (2009) The merging of community ecology and phylogenetic biology. Ecol Lett 12:693–715

Cushman JH, Lawton JH, Manly BFJ (1993) Latitudinal patterns in European ant assemblages: variation in species richness and body size. Oecologia 95:30–37

Dantas A, Fonseca CR (2023) Global biogeographical patterns of ants and their abiotic determinants. Perspect Ecol Conserv 21:237–246

de Frenne P, Graae BJ, Rodríguez-Sánchez F, Kolb A, Chabrerie O, Decocq G et al (2013) Latitudinal gradients as natural laboratories to infer species' responses to temperature. J Ecol 101:784–795 del Toro I, Ribbons RR, Hayward J, Andersen AN (2019) Are stacked species distribution models accurate at predicting multiple levels of diversity along a rainfall gradient? Austral Ecol 44:105–113

Dunn RR, Agosti D, Andersen AN, Arnan X, Bruhl CA, Cerada X, Ellison AM, Fisher BL, Fitzpatrick MC, Gibb H, Gotelli NJ, Gove AD, Guenard B, Janda M, Kaspari M, Laurent EJ, Lessard J-P, Longino JT, Majer JD, Menke SB, McGlynn TP, Parr CL, Philpott SM, Pfeiffer M, Retana J, Suarez AV, Vasconcelos HL, Weiser MD, Sanders NJ (2009) Climatic drivers of hemispheric asymmetry in global patterns of ant species richness. Ecol Lett 12:324–333

Faith DP (1992) Conservation evaluation and phylogenetic diversity. Biol Conserv 61:1–10

García-Navas V, Martínez-Núñez C, Christidis L, Ozgul A (2025) A stroll along Australian ecosystems: using bioclimatic transects to examine environmental drivers of community assembly in birds. J Biogeogr 52:136–147

Guerin GR, Biffin E, Lowe AJ (2013) Spatial modelling of species turnover identifies climate ecotones, climate change tipping points and vulnerable taxonomic groups. Ecography 36:1086–1096

Kreyling J, Jentsch A, Beier C (2014) Beyond realism in climate change experiments: gradient approaches identify thresholds and tipping points. Ecol Lett 17:125

Leahy L, Scheffers BR, Andersen AN, Hirsch BT, Williams SE (2021) Vertical niche and elevation range size in tropical ants: implications for climate resilience. Divers Distrib 27:485–496

Meirmans PG (2015) Seven common mistakes in population genetics and how to avoid them. Mol Ecol 24:3223–3231

Neves KC, Andersen AN, Schultz TR, Vasconcelos HL (2024) Macroecological patterns of functional and phylogenetic diversity vary between ground and arboreal assemblages in Neotropical savanna ants. Ecography 2024:e07122

Scheffers BR, Phillips BL, Laurance WF, Sodhi NS, Diesmos A, Williams SE (2013) Increasing arboreality with altitude: a novel biogeographic dimension. Proc Roy Soc B 280:20131581

Swenson NG, Weiser MD (2014) On the packing and filling of functional space in eastern North American tree assemblages. Ecography 37:1056–1062

Swenson NG, Erickson DL, Mi X, Bourg NA, Forero-Montana J, Ge X et al (2012) Phylogenetic and functional alpha and beta diversity in temperate and tropical tree communities. Ecology 93:112–125

Vasconcelos HL, Neves KC, Andersen AN (2023) Vertical stratification of ant assemblages varies along a latitudinal gradient in Brazilian savanna. J Biogeogr 50:1331–1340

Warren DL, Cardillo M, Rosauer DF, Bolnick DI (2014) Mistaking geography for biology: inferring processes from species distributions. Trends Ecol Evol 29:572–580

Webb CO, Ackerly DD, McPeek MA, Donoghue MJ (2002) Phylogenies and community ecology. Ann Rev Ecol Syst 33:475–505

Whitehead PJ, Bowman DMJS, Tidemann SC (1992) Biogeographic patterns, environmental correlates and conservation of avifauna in the Northern Territory, Australia. J Biogeogr 19:151–161

Williams RJ, Cook GD, Braithwaite RW, Andersen AN, Corbett LK (1995) Australia's wet-dry tropics: identifying the sensitive zones. In: Pernatta J, Leemans R, Elder D, Humphrey S (eds) The impact of climate change on ecosystems and species: terrestrial Bomes. IUCN, Gland, Switzerland, pp 39–65

Chapter 9
Elevation Gradients in Montane Ant Species

Robert J. Warren II

9.1 Ant Thermophilia and Elevational Gradients

Conditions can shift rapidly and somewhat predictably, along elevational gradients (Greenland 1987; Fridley 2009), making montane slopes well-established laboratories for studying how environmental drivers determine the composition and distribution of species (von Humboldt and Bonpland 1807; Whittaker 1956). Indeed, from the beginning, early naturalists and ecologists recognized dramatic shifts across elevational distances (von Humboldt and Bonpland 1807; Whittaker 1956). From these early studies of plant dynamics along mountain slopes, elevational gradients have become common natural experiments for studying the putative drivers of biodiversity in general (Whittaker 1956; Körner 2007; Sundqvist et al. 2013)—particularly given that many mountain ranges are hotspots of biodiversity (Lomolino 2001; Jankowski et al. 2009; Suissa et al. 2021). Researchers assumed relatively consistent and systematic decreases in temperature and increases in moisture with elevation to explain corresponding shifts in montane species (von Humboldt and Bonpland 1807; Whittaker 1956), and the potential link between montane environmental conditions and biodiversity patterns remains an active area of study (Rahbek 1995; Sanders et al. 2007; Bishop et al. 2014). Moreover, given that taxa often vary by elevation in much the same way as they vary across other macroscales, such as latitude and area (McCoy 1990; Brown 2001; McCain 2010), elevational gradients often are considered microcosms of larger biodiversity patterns (Rahbek 1995;

Supplementary Information The online version contains supplementary material available at https://doi.org/10.1007/978-3-032-21127-9_9.

R. J. Warren II (✉)
Buffalo State University, Buffalo, NY, USA
e-mail: warrenrj@buffalostate.edu

I. R. Leal et al. (eds.), *Ant Responses to Climate Change*, Entomology in Focus 10,
https://doi.org/10.1007/978-3-032-21127-9_9

McCain and Grytnes 2010; Peters et al. 2016). Yet, environmental conditions along montane elevational gradients are not as predictable or consistent as assumed (Goulden et al. 2006; Daly et al. 2010; Elsen and Tingley 2015), and there is inconsistent variation in biodiversity shifts along elevation gradients depending on taxa, community structure and region (Rahbek 2004; Guo et al. 2013; Peters et al. 2016).

Most insects are poikilothermic ectotherms—their body temperature largely dependent on the temperature of their surroundings (Wilson 1971; Heinrich 1996). Ants are no exception (Kaspari et al. 2000; Roeder et al. 2021b; Parr and Bishop 2022; see Chap. 4). Ant life history, physiology, behaviour and distributions are greatly influenced by micro- and macroscale temperature. Indeed, temperature impacts all key ant life history stages, underlying crucial functions in ant fitness and evolution. Ants exhibit a wide range of morphological and physiological adaptations to various thermal conditions (Sommer and Wehner 2012; Shi et al. 2015; Shik et al. 2019; see Chaps. 3 and 4) as temperature, and its effect on ant metabolism controls their activity levels, such as foraging and brood care (Challet et al. 2005; Kaspari et al. 2016; Roeder et al. 2022). Ant colony growth depends on brood production which, in turn, depends on nest temperature (Porter and Tschinkel 1993; Abril et al. 2008; Penick et al. 2017). As such, ants often choose locations, such as by aspect, soil column and sunlight, that maintain an optimal thermal environment for reproduction (Jones and Oldroyd 2006; King et al. 2018). Moreover, ants will relocate nests during the reproductive season to adjust to changing thermal conditions (Banschbach et al. 1997; Kaspari and Valone 2002; Heller and Gordon 2006), and they will translocate brood within the nest to thermoregulate development (Penick and Tschinkel 2008; Bollazzi et al. 2008; Baudier and O'Donnell 2016; see Chap. 2).

Ants generally are described as thermophilic (heat loving); concomitantly, they also could be described as cryophobic [cold fearing] (Hölldobler and Wilson 1990; Parr and Bishop 2022). As a result, ant abundance and richness typically are higher in warm environments at both microclimatic and global scales (Kaspari et al. 2000; Sanders et al. 2007; Parr and Bishop 2022). At global scales, ant diversity decreases with increased latitude (Dunn et al. 2009; Jenkins et al. 2011; Economo et al. 2018), a taxonomic pattern observed in most organisms (Huey 1978; Rohde 1997; Mittelbach et al. 2007). Given that temperature decreases with higher elevation as it does with higher latitude, elevational gradients often have been used as smaller-scale equivalents of latitudinal gradients: "elevation-for-latitude substitution" (Nogués-Bravo et al. 2008; Jump et al. 2009; Loewen et al. 2023). Generally, air temperature decreases about 0.6 °C with each 100 m increase in altitude (adiabatic lapse rate)—though the lapse rate can be complicated somewhat along montane gradients due to high elevation cloud formation, tree loss (timberline) and seasonal snow cover (Whiteman 2000; Barry 2008). Numerous studies have linked montane ant species and communities with elevational temperature gradients (e.g. Bharti et al. 2013; Bishop et al. 2014; Joseph et al. 2019), though many assumed temperature shifts as a proxy of elevation shifts (e.g. Nowrouzi et al. 2016; Tiede et al. 2017; Plowman et al. 2020) or estimated it from large-scale climate models (e.g. Del Toro 2013; Longino and Branstetter 2018; Marathe et al. 2020).

As temperature decreases with montane elevation, moisture generally increases due to precipitation intensification with cooling (condensation) and the effect of cloud interception [which increases humidity and precipitation] (Whiteman 2000; Barry 2008). Whereas precipitation does not get the same attention as temperature in explaining ant species richness and abundance, many ant species are associated with higher moisture (Kaspari and Weiser 2000; Warren II et al. 2019; Johnson 2021), and some research suggests that ants segregate by moisture along elevational gradients as well (Nowrouzi et al. 2016; Longino and Branstetter 2018; Pérez-Toledo et al. 2021). The elevational climate model suggests that the combination of temperature and water availability gradients together influence species diversity, including ants (McCain and Grytnes 2010; Szewczyk and McCain 2016; Uhey et al. 2020).

Whereas ant diversity consistently declines with higher montane elevation (Rahbek 2004; Smith 2015; Szewczyk and McCain 2016), the loss of species diversity moving up the mountains is not random (Sanders et al. 2003; Bishop et al. 2015; Nowrouzi et al. 2016). That is, specific ant species often occur within specific altitudinal zones rather than ranging across elevation bands along elevation gradients (Robertson 2002; Bishop et al. 2015; Nowrouzi et al. 2016). Indeed, ant species dissimilarity can range from 50% to 90% across 500 m changes in elevation (Olson 1994; Fisher 1996; Nowrouzi et al. 2016). The shifts in ant species composition coincide with temperature (Bharti et al. 2013; Bishop et al. 2014; Joseph et al. 2019), moisture (Olson 1994; Burwell and Nakamura 2016; Nowrouzi et al. 2016) and vegetation structure (Sanders et al. 2003; Munyai and Foord 2012; Plowman et al. 2020). Given that these montane habitat bands decrease in size and increase in isolation with elevation, it is not surprising that endemism is common in montane habitats worldwide (Malhi et al. 2010; Steinbauer et al. 2012; Noroozi et al. 2018). For ants, high elevation zones contain most of the unique species in the montane gradients (Robertson 2002; Longino and Branstetter 2018; Joseph et al. 2019). That is, the high elevation ant species are not subsets of low elevation ant communities but unique assemblages all their own adapted to the colder, isolated habitat (Machac et al. 2011; Chick et al. 2020; Pérez-Toledo et al. 2021).

Classic ecological theory suggests trade-offs exist along environmental gradients—in both aquatic and terrestrial habitats—between species that can tolerate extreme climates and species that are good competitors (Wallace 1878; Connell 1961; Grubb 1977a). As a result, harsh climates filter out competitive species that cannot tolerate the environmental extremes and negative biotic interactions eliminate the stress-tolerant but poor competitors—a pattern often observed in ants (Fellers 1987; Bestelmeyer 2000; Adler et al. 2007). In general, stress-tolerant, subordinate ants occupy the harsh extremes of environmental gradients whereas stress-intolerant, dominant ants occur in benign environments (Cerdà et al. 1997; Bestelmeyer 2000; Kwon 2018). Along montane elevation gradients, cold-adapted ants generally occupy the higher elevations whereas more widespread, warm-adapted ants occupy the lower elevations (Machac et al. 2011; Warren II and Chick 2013; Chick et al. 2020). Whether these patterns also correspond with dominant/subordinate trade-offs has not been investigated.

9.2 Patterns of Diversity Along Elevational Gradients

Given that species richness generally decreases with latitude regardless of taxa, equator to poles (Forster 1778; Wallace 1878; Hillebrand 2004), including ants (Cushman et al. 1993; Dunn et al. 2009; Economo et al. 2018), and temperature similarly decreases, a similar pattern might be expected at a smaller scale when moving from high to low elevations (Rahbek 2004; Smith 2015; Szewczyk and McCain 2016). However, whereas ant diversity and abundance generally decrease with increasing montane elevation worldwide, temperate to tropical, the decreases are not always linear (Rahbek 2004; Smith 2015; Szewczyk and McCain 2016).

Current research reports three general patterns for the decrease in ant richness with elevation: (1) a steady (monotonic) decline with increased elevation (hereafter, "monotonic decline"), (2) a mid-elevation peak (hump-shaped) [hereafter, "mid-elevation peak"] or (3) no discernible pattern (Rahbek 1995; Guo et al. 2013; Peters et al. 2016). The mid-elevation peak in species diversity is most often found along montane elevation gradients, and it often is proposed as the prototypical biodiversity pattern across elevation, yet, a substantial number of studies show monotonic declines (Rahbek 2004; McCain and Grytnes 2010; Peters et al. 2016). In ant studies, Smith (2015) reported that 28% of montane elevation studies show a mid-elevation peak and 50% a monotonic decline whereas Szewczyk and McCain (2016) reported that 72% reported a mid-elevation peak and 14% monotonic decline. Approximately 14–22% of the studies showed no ant diversity patterning with elevation (Smith 2015; Szewczyk and McCain 2016). By comparison, of the empirical studies of ant richness along elevation gradients reviewed for this chapter ($n = 47$), including many of those previously reviewed (Smith 2015; Szewczyk and McCain 2016), 34% reported a monotonic decrease in ant richness with increased elevation, 58% reported a mid-elevation peak, and 8% reported no pattern (Sect. 9.1; Electronic Supplementary Material). The inconsistency in biodiversity patterning is at least partly an artefact of incomplete elevational sampling. Sampling at the highest elevations can be difficult, and, depending on latitude, the mountain itself might not span the scale of intent, resulting in "gradient truncation" whereby only a portion of the potential gradient is sampled (Nogués-Bravo et al. 2008; McCain and Grytnes 2010; Pérez-Toledo et al. 2021)—truncated at either the high (Nogués-Bravo et al. 2008) or low (McCoy 2002) end of the gradient. The results reviewed here suggest that gradient truncation occurs in ant elevational studies as the (mean) topographical relief, which represents the spatial range of the elevation gradient, for sampling in studies reporting mid-elevation peaks was almost twice as high (2466 m) as those reporting monotonic declines in ant species richness (1269 m) (Fig. 9.1a). Moreover, the truncation appears to occur at higher elevations as the mean minimum sampling elevation did not differ greatly between studies that found monotonic declines (438 m) and mid-elevation peaks (505 m) in ant species richness (Fig. 9.1b), whereas the mean maximum elevation was higher for those that found mid-elevation peaks (2905) than those that found monotonic declines (1775) (Fig 9.1c). These results suggest that studies that find a monotonic decline might not capture the full potential of ant biodiversity patterning. An alternate possibility is that the underly climate

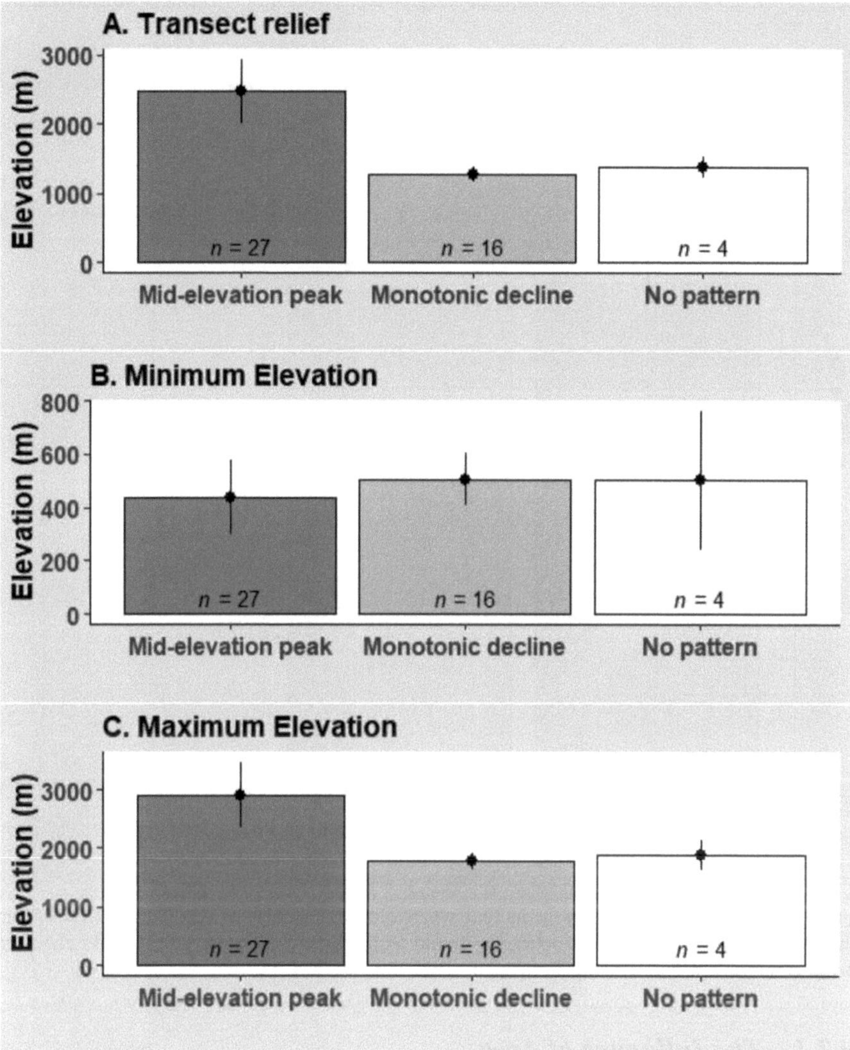

Fig. 9.1 Barplots (mean ± SE) of (**a**) transect relief (difference between highest and lowest elevations in elevation sampling), (**b**) minimum elevation of transect sampling and (**c**) maximum elevation of transect sampling as functions of the type of ant species richness patterning found along the elevational gradients: mid elevation peak ("hump-shaped" pattern), monotonic decline (linear decrease) or no pattern at all. These data were compiled from peer-reviewed journal articles that reported ant-elevation results ($n = 47$)

is truncated rather than the gradient truncation; in which case, the patterning in ant species richness would be expected to shift between temperate elevational gradients (which have greater temperature ranges) and tropical regions. The studies reviewed here do not support that possibility as the mean latitude for studies both reporting mid-elevation peaks and those reporting monotonic declines was 24° (Fig. 9.2).

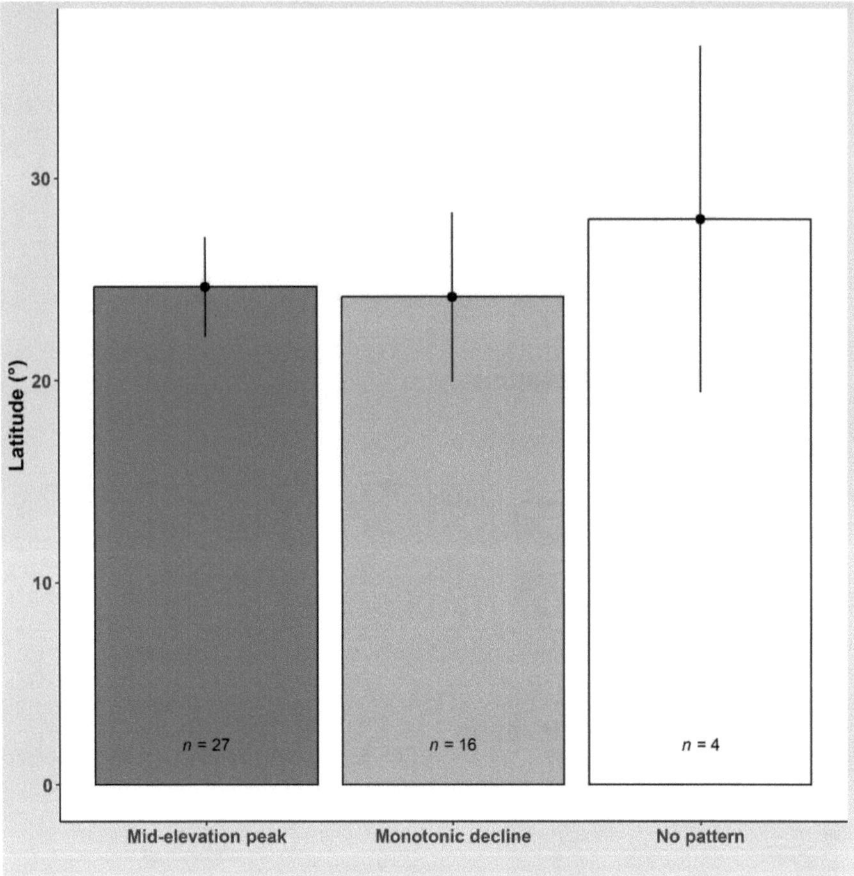

Fig. 9.2 Barplot (mean ± SE) of the latitude where elevation studies were conducted (n = 47) per the type of ant species richness patterning found along the elevational gradients: mid elevation peak, monotonic decline or no pattern at all

9.2.1 The Influence of Area

Given that, in almost all taxa, species richness declines with habitat area (MacArthur and Wilson 1967), the most parsimonious explanation for the monotonic decline patterning in ants might be that habitat area also declines monotonically with elevation. That is, between conical uplifting and erosion, mountains form roughly pyramidal shapes. As such, it is presumed that with each isothermal increase in mountain height comes an associated decrease in surface area. As species richness almost universally decreases with decreased area (Lawton 1999; Lomolino 2000; Dodds 2009), the same may be expected of ant species (Sanders 2002; Bishop et al. 2014;

Szewczyk and McCain 2016). For example, within the larger, lower elevational areas there is greater variation in microhabitats that support disparate ant species (Brühl et al. 1999). The montane area hypothesis rarely has been tested, and, when it has, the results have been weak and inconsistent (Sanders 2002; Bishop et al. 2014; Szewczyk and McCain 2016). Given that the actual montane area by elevational rarely is measured, these results may reflect sampling error as, unexpectedly, a decrease in area with increasing montane elevation is not necessarily monotonic (Elsen and Tingley 2015). Indeed, montane area peaks at mid-elevation about 40% of the time, suggesting that (a) assumptions of a monotonic decline in area with increased montane elevation are erroneous and (b) area *could* explain monotonic declines and mid-elevation peaks in ant species richness depending on the mid-elevational area of the mountain. Moreover, the inconsistency in findings between studies showing monotonic declines and mid-elevational peaks may represent montane area variation as much as temperature.

Regardless of mid-slope area, mountains, by nature, contain relatively less space at the highest-elevation mountaintops than mountain bottoms. This landscape heterogeneity results in isolated, high elevation communities within a matrix of unsuitable warmer habitat (Bishop et al. 2017; Joseph et al. 2019; Fontcuberta et al. 2022)—the mountaintops essentially forming an archipelago of sky islands. Given that high elevation zones harbour unique, cold-adapted ant species (Robertson 2002; Longino and Branstetter 2018; Joseph et al. 2019), a metapopulation perspective may be needed for understanding high-elevation ant communities. For example, both the biotic (competition) and abiotic (temperature, moisture) conditions at low elevations appear to eliminate or severely reduce terrestrial connectivity for high-elevation ant species (Brühl et al. 1999; Longino and Colwell 2011; Bishop et al. 2015), making arboreal dispersal across mountaintops critical for high-elevation metapopulations. The majority of ant species engage in aerial mating flights, and the maximum flight distance for these flights, which also result in colony dispersion, ranges from 30 to 30,000 m (Helms 2017), though little-to-no data are available on high-elevation dispersal events. The dispersal ability of high elevations ants remains largely unknown, but it would be expected to be difficult within the extreme climates and short breeding seasons. Limited genetic work on high-elevation ants does suggest that they are more isolated than low elevation ants (Sanllorente et al. 2015; Fontcuberta et al. 2022). The local populations may be formed by one or few dispersal events so that the offspring of a single queen monopolize the habitat patches (Sanllorente et al. 2015). Furthermore, high elevation ants might be sink populations of lower elevation ants with dispersal restricted to upward movement (Fontcuberta et al. 2022). If mountaintops indeed form isolated habitats for high-elevation ants, the mountain ranges are sky island archipelagos across which metapopulation dispersal dynamics and island biogeography (e.g. richness determined by the distance between and areal size of sky islands) might be as important as elevational gradients in understanding ant diversity in montane habitats.

9.2.2 The Influence of Temperature

In studies where temperature and area both were tested, temperature better explained ant diversity patterning than area (Sanders 2002; Bishop et al. 2014; Szewczyk and McCain 2016). As with area, temperature also generally is assumed to decrease monotonically (as precipitation/moisture inversely increases) with montane elevation. And, as with area, the assumption of a monotonic shift in environmental conditions is more assumed than measured. Whereas temperature pretty reliably decreases with altitude (adiabatic lapse rate), this pattern is complicated by montane geolithic influences on local climate such as cloud formation from air uplift and, at some latitudes, timberline shifts from canopy cover to open habitat (Whiteman 2000; Barry 2008). Indeed, mid-elevation shifts in vegetation physiognomy can be quite abrupt across elevation gradients in mountains worldwide (Grubb 1977b; Bruijnzeel and Veneklaas 1998; Longino and Branstetter 2018), and, in the few studies that measured vegetation structure, ant diversity does appear to change with physiographic shifts (Sanders et al. 2003; Munyai and Foord 2012; Plowman et al. 2020), though both ant diversity and vegetation structure may be the effect of the same environmental shifts. For example, the interaction of temperature and precipitation might remain monotonic across elevation gradients in wet environments where moisture is relatively constant, but in drier, warmer environments, moisture might be limiting at lower elevations as is temperature at higher elevations (Dvorský et al. 2015; Uhey et al. 2020; Pérez-Toledo et al. 2022). Moreover, in some mountain ranges, precipitation peaks mid-slope rather than at the highest elevations (Rosenzweig 1992; Camberlin et al. 2014; Yu et al. 2021). Finally, an underlying, implicit if not explicit, assumption in linking species richness patterns with elevational gradients along mountain slopes is that the species richness patterns are invariant. However, the few studies reviewed here that considered temporal and spatial variation in ant species richness patterns indicated that they are not (Munyai and Foord 2012; Bishop et al. 2014). For example, Bishop et al. (2014) found that the patterning of ant species richness along montane slopes differed depending on the season and year. Similar variation in species richness patterning have been documented in moths and mammals as well (McCain 2004; Beck et al. 2010). However, of the papers reviewed here, 66% were based on a single season of climate sampling.

Cold air drainage further complicates montane temperatures by creating daily shifts in temperature conditions along mountain slopes from the temperate regions to the tropics (Goulden et al. 2006; Daly et al. 2010; Novick et al. 2016). Nighttime cooling at higher elevations in mountain ranges creates a stable layer of cold air that drains downslope and pools at the lower elevations and valleys. Botanists and horticulturists have long noted a mid-slope frost refugia in mountains (hereafter, "thermal belt") whereby the cold air of high elevations drains through the middle elevations and pools at the lower elevations (Chickering Jr. 1884; Cox 1920; Bergen 1969). Fruit tree growers often take advantage of montane thermal belts by planting at mid-elevations where warm-season temperatures are moderate and cool-season temperatures are moderated by cold air drainage (Cox 1920; Gu et al. 2008; Sahu et al. 2020). Hence, mid-elevations may contain the mildest climate conditions, along a mountain slope, which would allow the persistence of ants adapted to both high- and low-elevation conditions—much the same as what is predicted with the mid-domain effect, which is based on geometric theory (Colwell and Lees 2000). The mid-domain theory proposes that species richness peaks at mid-elevations because these domains are the intersection where high-elevation and low-elevation species distributions overlap independent of environmental conditions. The mid-domain effect has been employed to explain mid-elevation peaks in richness for plants and animals (Grytnes 2003; McCain 2004; Chaladze et al. 2014), including ants (Sanders 2002; Bharti et al. 2013; Pérez-Toledo et al. 2021), but its verification is based on observing a mid-elevation peak in species richness, which supports the pattern but not a process (Zapata et al. 2005; Brehm et al. 2007; Currie and Kerr 2008). None of the elevation gradient papers reviewed here considered cold air drainage in explaining ant species patterns.

Elevation studies utilize vertical shifts in montane climate, but dramatic shifts in environmental conditions also occur horizontally within elevation bands with changes in slope aspect (direction of slope face). Equatorial-facing slopes (south-facing in the Northern Hemisphere, north-facing in the Southern Hemisphere) receive considerably more solar irradiation than non-equatorial slopes, making them relatively warmer and drier than non-equatorial slopes. Indeed, the mean annual difference in temperature across slope aspects from Equatorial Africa to arctic permafrost is approximately 1.5–5.5 °C (Rorison et al. 1986; Burnett et al. 2008; Pepin et al. 2017)—roughly equivalent to 200–800 m shifts in altitude (Whiteman 2000; Barry 2008). It is not surprising, then, that across slope aspects, there are dramatic shifts in plant and animal communities (Warren 2010; Dobrowski 2011; Nahidan et al. 2015), including insects (Lantschner and Corley 2015; Jones et al. 2019; Huang et al. 2022). Few ant studies include aspect when assessing ant elevational patterns, and, interestingly enough, in the few that did, the pattern of ant species richness and ant communities differed across slope aspects (Munyai and Foord 2012; Muluvhahothe et al. 2021). For example, Munyai and Foord (2012) found that ant species richness peaked at mid-elevation on the equatorial-facing slope and exhibited a monotonic decline on the non-equatorial-facing slope. Given the ubiquity of environmental shifts impacts on organisms and ecosystems across slope aspects, aspect may be an undervalued variable in determining the drivers of ant richness along elevation gradients.

A particular problem with connecting environmental variables with ant diversity on montane slopes is that direct measurement of montane environmental variables is uncommon. In the research reviewed for this manuscript, only 13% directly measured temperature where they collected ants. In 21% of the studies, researchers used macroclimate models (e.g. Worldclim). McCain and Garfinkel (2021) noted that predictive modelling outweighs empirical data in assessing insect responses to climate and that the current trend favours more predictive models and fewer empirical studies. This trend may be particularly problematic for montane elevation studies as the models interpolate climate incorporating elevation but do not reflect important montane parameters such as aspect and cold air drainage (Daly 2006; Lembrechts et al. 2019), do not account for the spatial microtopography and temporal climate extremes that drive ecological dynamics (Parmesan et al. 2000; Easterling et al. 2000; Timmermans et al. 2019) and do not account for spatial autocorrelation and measurement error (Ploton et al. 2020; Wyborn and Evans 2021; Meyer and Pebesma 2022). Using predicted values without consideration of inherent error and underlying mechanism may lack scientific integrity (Meyer and Pebesma 2022) or, even more provocatively, be considered "pseudoscience" (Chomsky et al. 2023). In the end, this is part of a larger problem with elevational studies and biogeographic patterns in general, whereby data from single transects might not generalize to larger spatial scales (Sanders 2002; Machac et al. 2011), and fitting data patterns to species richness patterns generates innumerous, untestable hypotheses without mechanisms (Warren II et al. 2022; Diniz-Filho et al. 2023).

9.3 Climate Change and Montane Ants

Climate change generally prompts upward and poleward shifts in species distributions (Parmesan and Yohe 2003; La Sorte and Jetz 2010; Chen et al. 2011), and ants are no exception (Parr and Bishop 2022), though species-level ant upward and poleward responses are more often modelled (Bertelsmeier et al. 2013; Del Toro et al. 2015; Kwon et al. 2016) than measured (Warren II and Chick 2013; McCain and Garfinkel 2021). The lack of empirical data on upward ant movement can be somewhat attributed to the difficulty of gathering field data and a lack of historical distribution data to compare against. Clearly, continued research efforts in this area are needed, and they should include on-the-ground climate measurements (versus modelled) paired with ant abundance and richness. Moreover, consideration of montane area, aspect and cold air drainage would be quite useful.

Measuring ant climate response is not without great challenge, however. Broad-scale ant climate responses might be muddied by local-scale ant behaviour (Parr and Bishop 2022; see Chap. 2). That is, many ant species can capitalize on local habitat heterogeneity, as well as vertical movement through the soil and vegetation column, to ameliorate broad-scale cold and heat extremes (Porter and Tschinkel 1993; McGlynn et al. 2010; King et al. 2018; see Chap. 12). Finally, some ant species disperse via aerial mating flights at distances that can be difficult to track (King and Tschinkel 2016; Helms 2017).

Challenges aside, ants clearly segregate by montane elevation, notably with unique, sometimes endemic, cold-adapted species clustered at higher elevations (Robertson 2002; Longino and Branstetter 2018; Joseph et al. 2019). An underlying mechanism for ant elevational segregation appears to be the physiological match between ant thermal tolerance and montane temperature gradients (Warren II and Chick 2013; Bishop et al. 2017; Chick et al. 2020). Ant physiological tolerance of temperature extremes is described by their physiological thermal limits [i.e. retain muscular coordination] (Roeder et al. 2021b; see Chap. 4). These tolerances appear to shape ant distributions, but the tolerances also are shaped by them. That is, ant critical thermal limits generally decrease with elevation, both across species (suggesting environmental filtering of warm- and cold-tolerant ants) and within species [suggesting adaptation or acclimation] (Chick et al. 2020; Roeder et al. 2021a; Parr and Bishop 2022). Along elevation gradients, ant species and ant physiology appear to sort more by minimum rather than maximum temperature tolerances as indicated by closer links between ant minimum temperature tolerance and elevation than maximum tolerance and elevation (Warren II and Chick 2013; Bishop et al. 2017; Chick et al. 2020; see Chap. 4).

One implication of ant thermal optima is that individual ant species will move upwards with climate change (Parr and Bishop 2022), but the species richness patterns should shift as well. Given that ants specifically adapted to cold temperatures occur at the highest elevations (Machac et al. 2011; Warren II and Chick 2013; Chick et al. 2020), and those ants generally are less competitive than warm-adapted low-elevation ants (Cerdà et al. 1997; Bestelmeyer 2000; Kwon 2018), the cold temperatures may create a barrier against competitive, low-elevation ants (Dunn et al. 2007; Urban et al. 2012; Warren II and Chick 2013). Moreover, the potential negative effects of upward ant movement may be exacerbated by the greater presence of highly aggressive non-native species (see Chap. 7), generally from tropical origins, at lower elevations (Lowe et al. 2004; Bharti et al. 2013; Lytle et al. 2020). Whereas theoretical models project upward movement of most non-native ant species (e.g. Bertelsmeier et al. 2015), empirical research indicates that some non-native invasive ant species appear limited in their ability to acclimate to colder temperatures (Coulin et al. 2019; Warren II et al. 2020; Nelson et al. 2023), which might make mountains a refuge for native ants—unless continued warming obliterates the refugia. Moreover, some non-native ant species have demonstrated considerable plasticity in cold tolerance, allowing them to invade the highest montane elevations in their range (Lytle et al. 2020).

9.4 Conclusions

Elevation gradients provide fertile ground for studying species distributions and diversity, links that naturalists and researchers have connected for centuries. The scale of measurement, and multiple sampling errors, weakens inference from these links without rigorous methodology and conservative interpretation. Some suggestions for moving forward include (1) boots-on-the-ground empirical data (e.g. temperature, moisture, area), (2) replication across slope aspects instead of just across elevation bands and (3) long-term sampling to capture non-linear temporal variation in temperature and other environmental drivers. Even if inference is weak, climate warming portends a threatening future for high-elevation ants. First, the cold-adapted species at the highest elevations will have finite options for upward movement in maintaining their thermal optima, and, second, for those that can acclimate, the warmer temperatures may allow upward movement by more competitive native species and highly aggressive non-native species that are currently limited to lower elevations.

References

Abril S, Oliveras J, Gomez C (2008) Effect of temperature on the oviposition rate of Argentine ant queens (*Linepithema humile* Mayr) under monogynous and polygynous experimental conditions. J Insect Physiol 54:265–272. https://doi.org/10.1016/j.jinsphys.2007.09.009
Adler FR, LeBrun EG, Feener DH Jr (2007) Maintaining diversity in an ant community: modeling, extending, and testing the dominance-discovery trade-off. Am Nat 169:323–333. https://doi.org/10.1086/510759
Banschbach VS, Levit N, Herbers JM (1997) Nest temperatures and thermal preferences of a forest ant species: is seasonal polydomy a thermoregulatory mechanism? Insect Soc 44:109–122. https://doi.org/10.1007/s000400050034
Barry RG (2008) Mountain weather and climate. Cambridge University Press, Cambridge, MA
Baudier KM, O'Donnell S (2016) Structure and thermal biology of subterranean army ant bivouacs in tropical montane forests. Insect Soc 63:467–476. https://doi.org/10.1007/s00040-016-0490-2
Beck J, Altermatt F, Hagmann R, Lang S (2010) Seasonality in the altitude–diversity pattern of alpine moths. BAAE 11:714–722. https://doi.org/10.1016/j.baae.2010.08.009
Bergen JD (1969) Cold air drainage on a forested mountain slope. JAMC 8:884–895. http://www.jstor.org/stable/26174696
Bertelsmeier C, Luque GM, Courchamp F (2013) Global warming may freeze the invasion of big-headed ants. Biol Invasions 15:1561–1572. https://doi.org/10.1007/s10530-012-0390-y
Bertelsmeier C, Luque GM, Hoffmann BD, Courchamp F (2015) Worldwide ant invasions under climate change. Biodivers Conserv 24:117–128. https://doi.org/10.1007/s10531-014-0794-3
Bestelmeyer BT (2000) The trade-off between thermal tolerance and behavioural dominance in a subtropical South American ant community. J Anim Ecol 69:998–1009. https://www.jstor.org/stable/2647160
Bharti H, Sharma YP, Bharti M, Pfeiffer M (2013) Ant species richness, endemicity and functional groups, along an elevational gradient in the Himalayas. Asian Myrmecol 5:79–101. https://doi.org/10.20362/am.005010
Bishop TR, Robertson MP, van Rensburg BJ, Parr CL (2014) Elevation–diversity patterns through space and time: ant communities of the Maloti-Drakensberg Mountains of southern Africa. J Biogeogr 41:2256–2268. https://doi.org/10.1111/jbi.12368
Bishop TR, Robertson MP, van Rensburg BJ, Parr CL (2015) Contrasting species and functional beta diversity in montane ant assemblages. J Biogeogr 42:1776–1786. https://doi.org/10.1111/jbi.12537
Bishop TR, Robertson MP, Van Rensburg BJ, Parr CL (2017) Coping with the cold: minimum temperatures and thermal tolerances dominate the ecology of mountain ants. Ecol Entomol 42:105–114. https://doi.org/10.1111/een.12364
Bollazzi M, Kronenbitter J, Roces F (2008) Soil temperature, digging behaviour, and the adaptive value of nest depth in South American species of Acromyrmex leaf-cutting ants. Oecologia 158:165–175. https://doi.org/10.1007/s00442-008-1113-z
Brehm G, Colwell RK, Kluge J (2007) The role of environment and mid-domain effect on moth species richness along a tropical elevational gradient. Glob Ecol Biogeogr 16:205–219. https://doi.org/10.1111/j.1466-8238.2006.00281.x
Brown JH (2001) Mammals on mountainsides: elevational patterns of diversity: mammals on mountainsides. Glob Ecol Biogeogr 10:101–109. https://doi.org/10.1046/j.1466-822x.2001.00228.x
Brühl CA, Mohamed M, Linsenmair KE (1999) Altitudinal distribution of leaf litter ants along a transect in primary forests on Mount Kinabalu, Sabah, Malaysia. J Trop Ecol 15:265–277. https://doi.org/10.1017/S0266467499000802
Bruijnzeel LA, Veneklaas EJ (1998) Climatic conditions and tropical montane forest productivity: the fog has not lifted yet. Ecology 79:3–9. https://doi.org/10.1890/0012-9658(1998)079[0003:CCATMF]2.0.CO;2

Burnett BN, Meyer GA, McFadden LD (2008) Aspect-related microclimatic influences on slope forms and processes, northeastern Arizona. J Geophys Res 113. https://doi.org/10.1029/2007JF000789
Burwell CJ, Nakamura A (2016) Can changes in ant diversity along elevational gradients in tropical and subtropical Australian rainforests be used to detect a signal of past lowland biotic attrition? Austral Ecol 41:209–218. https://doi.org/10.1111/aec.12322
Camberlin P, Boyard-Micheau J, Philippon N et al (2014) Climatic gradients along the windward slopes of Mount Kenya and their implication for crop risks. Part 1: climate variability. Int J Climatol 34:2136–2152. https://doi.org/10.1002/joc.3427
Cerdà X, Retana J, Cros S (1997) Thermal disruption of transitive hierarchies in Mediterranean ant communities. J Anim Ecol 66:363–374. https://doi.org/10.2307/5982
Chaladze G, Otto S, Tramp S (2014) A spider diversity model for the Caucasus ecoregion. J Insect Conserv 18:407–416. https://doi.org/10.1007/s10841-014-9649-1
Challet M, Jost C, Grimal A et al (2005) How temperature influences displacements and corpse aggregation behaviors in the ant *Messor sancta*. Insect Soc 52:309–315. https://doi.org/10.1007/s00040-005-0821-1
Chen I-C, Hill JK, Ohlemüller R et al (2011) Rapid range shifts of species associated with high levels of climate warming. Science 333:1024–1026. https://doi.org/10.1126/science.1206432
Chick LD, Lessard J-P, Dunn RR, Sanders NJ (2020) The coupled influence of thermal physiology and biotic interactions on the distribution and density of ant species along an elevational gradient. Diversity 12:456. https://doi.org/10.3390/d12120456
Chickering JW Jr (1884) Thermal belts. Am Meteorol J 1:213–218
Chomsky N, Roberts I, Watumull J (2023) Opinion | Noam Chomsky: the false promise of ChatGPT. The New York Times March 08
Colwell RK, Lees DC (2000) The mid-domain effect: geometric constraints on the geography of species richness. Trends Ecol Evol 15:70–76. https://doi.org/10.1016/S0169-5347(99)01767-X
Connell JH (1961) The influence of interspecific competition and other factors on the distribution of the barnacle Chthamalus stallatus. Ecology 42:710–723
Coulin C, De La Vega GJ, Chifflet L et al (2019) Linking thermo-tolerances of the highly invasive ant, *Wasmannia auropunctata*, to its current and potential distribution. Biol Invasions 21:3491–3504. https://doi.org/10.1007/s10530-019-02063-0
Cox HJ (1920) Weather conditions and thermal belts in the North Carolina Mountain region and their relation to fruit growing. Ann Assoc Am Geogr 10:57–68. https://doi.org/10.1080/00045602009357013
Currie DJ, Kerr JT (2008) Tests of the mid-domain hypothesis: a review of the evidence. Ecol Monogr 78:3–18. https://doi.org/10.1890/06-1302.1
Cushman JH, Lawton JH, Manly BFJ (1993) Latitudinal patterns in European ant assemblages: variation in species richness and body size. Oecologia 95:30–37. https://doi.org/10.1007/BF00649503
Daly C (2006) Guidelines for assessing the suitability of spatial climate data sets. Int J Climatol 26:707–721. https://doi.org/10.1002/joc.1322
Daly C, Conklin DR, Unsworth MH (2010) Local atmospheric decoupling in complex topography alters climate change impacts. Int J Climatol 30:1857–1864. https://doi.org/10.1002/joc.2007
Del Toro I (2013) Diversity of Eastern North American ant communities along environmental gradients. PLoS One 8:e67973. https://doi.org/10.1371/journal.pone.0067973
Del Toro I, Silva RR, Ellison AM (2015) Predicted impacts of climatic change on ant functional diversity and distributions in eastern North American forests. Divers Distrib 21:781–791. https://doi.org/10.1111/ddi.12331
Diniz-Filho JAF, Coelho MTP, Terribile LC (2023) Unpacking underdetermination: theoretical challenges in understanding macroecological and biogeographic patterns. J Biogeogr 50:1890–1898. https://doi.org/10.1111/jbi.14699
Dobrowski SZ (2011) A climatic basis for microrefugia: the influence of terrain on climate. Glob Chang Biol 17:1022–1035. https://doi.org/10.1111/j.1365-2486.2010.02263.x

Dodds WK (2009) Laws, theories and patterns in ecology. University of California Press, Berkeley, CA

Dunn RR, Parker CR, Sanders NJ (2007) Temporal patterns of diversity: assessing the biotic and abiotic controls on ant assemblages. Biol J Linn Soc 91:191–201. https://doi.org/10.1111/j.1095-8312.2007.00783.x

Dunn RR, Agosti D, Andersen AN et al (2009) Climatic drivers of hemispheric asymmetry in global patterns of ant species richness. Ecol Lett 12:324–333. https://doi.org/10.1111/j.1461-0248.2009.01291.x

Dvorský M, Altman J, Kopecký M et al (2015) Vascular plants at extreme elevations in eastern Ladakh, Northwest Himalayas. Plant Ecol Divers 8:571–584. https://doi.org/10.1080/17550874.2015.1018980

Easterling DR, Meehl GA, Parmesan C et al (2000) Climate extremes: observations, modeling, and impacts. Science 289:2068–2074. https://doi.org/10.1126/science.289.5487.206

Economo EP, Narula N, Friedman NR et al (2018) Macroecology and macroevolution of the latitudinal diversity gradient in ants. Nat Commun 9:1–8. https://doi.org/10.1038/s41467-018-04218-4

Elsen PR, Tingley MW (2015) Global mountain topography and the fate of montane species under climate change. Nature Clim Change 5:772–776. https://doi.org/10.1038/nclimate2656

Fellers JH (1987) Interference and exploitations in a guild of woodland ants. Ecology 68:1466–1478. https://doi.org/10.2307/1939230

Fisher BL (1996) Ant diversity patterns along an elevational gradient in the Réserve Naturelle Integrale Andohahela, Madagascar. Fieldiana Zool 94:129–147

Fontcuberta A, Kapun M, Tran Van P et al (2022) Effects of social organization and elevation on spatial genetic structure in a montane ant. Ecol Evol 12:e8813. https://doi.org/10.1002/ece3.8813

Forster JR (1778) Observations made during a voyage round the world, on physical geography, natural history and ethic philosophy. G. Robinson, London

Fridley JD (2009) Downscaling climate over complex terrain: high finescale (<1000 m) spatial variation of near-ground temperatures in a montane forested landscape (Great Smoky Mountains). JAMC 48:1033–1049. https://doi.org/10.1175/2008JAMC2084.1

Goulden ML, Miller SD, da Rocha HR (2006) Nocturnal cold air drainage and pooling in a tropical forest. J Geophys Res Atmos 111. https://doi.org/10.1029/2005JD006037

Greenland D (1987) Mountain climates, upland. In: Oliver JE, Fairbridge RW (eds) The encyclopedia of climatology. Springer US, Boston, MA, pp 594–601

Grubb PJ (1977a) Maintenance of species-richness in plant communities - importance of regeneration niche. Biol Rev Camb Philos Soc 52:107–145. https://doi.org/10.1111/j.1469-185X.1977.tb01347.x

Grubb PJ (1977b) Control of forest growth and distribution on wet tropical mountains: with special reference to mineral nutrition. Annu Rev Ecol Syst 8:83–107. https://www.jstor.org/stable/2096722

Grytnes JA (2003) Species-richness patterns of vascular plants along seven altitudinal transects in Norway. Ecography 26:291–300. https://doi.org/10.1034/j.1600-0587.2003.03358.x

Gu L, Hanson PJ, Post WM et al (2008) The 2007 Eastern US spring freeze: increased cold damage in a warming world? Bioscience 58:253–262. https://doi.org/10.1641/B580311

Guo Q, Kelt DA, Sun Z et al (2013) Global variation in elevational diversity patterns. Sci Rep 3:1–7. https://doi.org/10.1038/srep03007

Heinrich B (1996) The thermal warriors: strategies of insect survival. Harvard University Press, Cambridge, MA

Heller N, Gordon D (2006) Seasonal spatial dynamics and causes of nest movement in colonies of the invasive Argentine ant (*Linepithema humile*). Ecol Entom. https://doi.org/10.1111/j.1365-2311.2006.00806.x

Helms J (2017) The flight ecology of ants (Hymenoptera: Formicidae). Myrmecol News 26. https://doi.org/10.25849/myrmecol.news_026:019

Hillebrand H (2004) On the generality of the latitudinal diversity gradient. Am Nat 163:192–211. https://doi.org/10.1086/381004
Hölldobler B, Wilson EO (1990) The ants. Belknap, Cambridge, Mass
Huang J, Miao X, Wang Q et al (2022) Metabarcoding reveals massive species diversity of Diptera in a subtropical ecosystem. Ecol Evol 12:e8535. https://doi.org/10.1002/ece3.8535
Huey RB (1978) Latitudinal patterns of between-altitude faunal similarity: mountains might be "higher" in the tropics. Am Nat 112:225–229. https://doi.org/10.1086/283262
Jankowski JE, Ciecka AL, Meyer NY, Rabenold KN (2009) Beta diversity along environmental gradients: implications of habitat specialization in tropical montane landscapes. J Anim Ecol 78:315–327. https://doi.org/10.1111/j.1365-2656.2008.01487.x
Jenkins CN, Sanders NJ, Andersen AN et al (2011) Global diversity in light of climate change: the case of ants. Divers Distrib 17:652–662. https://doi.org/10.1111/j.1472-4642.2011.00770.x
Johnson RA (2021) Desiccation limits recruitment in the pleometrotic desert seed-harvester ant *Veromessor pergandei*. Ecol Evol 11:294–308. https://doi.org/10.1002/ece3.7039
Jones JC, Oldroyd BP (2006) Nest thermoregulation in social insects. Adv Insect Physiol 33:153–191. https://doi.org/10.1016/S0065-2806(06)33003-2
Jones JA, Hutchinson R, Moldenke A et al (2019) Landscape patterns and diversity of meadow plants and flower-visitors in a mountain landscape. Landsc Ecol 34:997–1014. https://doi.org/10.1007/s10980-018-0740-y
Joseph GS, Muluvhahothe MM, Seymour CL et al (2019) Stability of Afromontane ant diversity decreases across an elevation gradient. GECCO 17:e00596. https://doi.org/10.1016/j.gecco.2019.e00596
Jump AS, Mátyás C, Peñuelas J (2009) The altitude-for-latitude disparity in the range retractions of woody species. Trends Ecol Evol 24:694–701. https://doi.org/10.1016/j.tree.2009.06.007
Kaspari M, Valone T (2002) On ectotherm abundance in a seasonal environment - studies of a desert ant assemblage. Ecology 83:2991–2996. https://doi.org/10.1890/0012-9658(2002)083[2991:OEAIAS]2.0.CO;2
Kaspari M, Weiser MD (2000) Ant activity along moisture gradients in a Neotropical forest. Biotropica 32:703–711. https://doi.org/10.1111/j.1744-7429.2000.tb00518.x
Kaspari M, Alonso L, O'Donnell S (2000) Three energy variables predict ant abundance at a geographical scale. Proc R Soc B 267:485. https://doi.org/10.1098/rspb.2000.1026
Kaspari M, Clay NA, Lucas J et al (2016) Thermal adaptation and phosphorus shape thermal performance in an assemblage of rainforest ants. Ecology 97:1038–1047. https://doi.org/10.1890/15-1225.1
King JR, Tschinkel WR (2016) Experimental evidence that dispersal drives ant community assembly in human-altered ecosystems. Ecology 97:236–249. https://doi.org/10.1890/15-1105.1
King JR, Warren IIRJ, Maynard DS, Bradford MA (2018) Ants: ecology and impacts in dead wood | SpringerLink. In: Ulyshen M (ed) Saproxylic insects. SprinkerLink, pp 237–262
Körner C (2007) The use of "altitude" in ecological research. Trends Ecol Evol 22:569–574. https://doi.org/10.1016/j.tree.2007.09.006
Kwon T-S (2018) High competition between ant species at intermediate temperatures. J Therm Biol 72:59–66. https://doi.org/10.1016/j.jtherbio.2017.11.015
Kwon T-S, Li F, Kim S-S et al (2016) Modelling vulnerability and range shifts in ant communities responding to future global warming in temperate forests. PLoS One 11:e0159795. https://doi.org/10.1371/journal.pone.0159795
La Sorte FA, Jetz W (2010) Projected range contractions of montane biodiversity under global warming. Proc R Soc B 277:3401–3410. https://doi.org/10.1098/rspb.2010.0612
Lantschner MV, Corley JC (2015) Spatial pattern of attacks of the invasive woodwasp Sirex noctilio, at landscape and standscales. PLoS One 10:e0127099. https://doi.org/10.1371/journal.pone.0127099
Lawton JH (1999) Are there general laws in ecology? Oikos 84:177. https://www.jstor.org/stable/i283220

Lembrechts JJ, Nijs I, Lenoir J (2019) Incorporating microclimate into species distribution models. Ecography 42:1267–1279. https://doi.org/10.1111/ecog.03947
Loewen CJG, Jackson DA, Gilbert B (2023) Biodiversity patterns diverge along geographic temperature gradients. Glob Chang Biol 29:603. https://doi.org/10.1111/gcb.16457
Lomolino MV (2000) Ecology's most general, yet protean pattern: the species-area relationship. J Biogeogr 27:17–26. https://www.jstor.org/stable/2655979
Lomolino MV (2001) Elevation gradients of species-density: historical and prospective views. Glob Ecol Biogeogr 10:3–13. https://doi.org/10.1046/j.1466-822x.2001.00229.x
Longino JT, Branstetter MG (2018) The truncated bell: an enigmatic but pervasive elevational diversity pattern in Middle American ants. Ecography 42:272–283. https://doi.org/10.1111/ecog.03871
Longino JT, Colwell RK (2011) Density compensation, species composition, and richness of ants on a neotropical elevational gradient. Ecosphere 2:1–20. https://doi.org/10.1890/ES10-00200.1
Lowe S, Browne M, Boudjelas S, De Poorter M (2004) 100 of the world's worst invasive alien species. The Invasive Species Specialist Group of the Species Survival Commission of the World Conservation Union, Auckland, New Zealand
Lytle AJ, Costa JT, Warren RJ II (2020) Invasion and high-elevation acclimation of the red imported fire ant, *Solenopsis invicta*, in the southern Blue Ridge Escarpment region of North America. PLoS One 15:e0232264. https://doi.org/10.1371/journal.pone.0232264
MacArthur RH, Wilson EO (1967) The theory of Island biogeography. Princeton University Press, Princeton
Machac A, Janda M, Dunn RR, Sanders NJ (2011) Elevational gradients in phylogenetic structure of ant communities reveal the interplay of biotic and abiotic constraints on diversity. Ecography 34:364–371. https://doi.org/10.1111/j.1600-0587.2010.06629.x
Malhi Y, Silman M, Salinas N et al (2010) Introduction: elevation gradients in the tropics: laboratories for ecosystem ecology and global change research. Glob Chang Biol 16:3171–3175. https://doi.org/10.1111/j.1365-2486.2010.02323.x
Marathe A, Priyadarsanan DR, Krishnaswamy J, Shanker K (2020) Spatial and climatic variables independently drive elevational gradients in ant species richness in the Eastern Himalaya. PLoS One 15:e0227628. https://doi.org/10.1371/journal.pone.0227628
McCain CM (2004) The mid-domain effect applied to elevational gradients: species richness of small mammals in Costa Rica. J Biogeogr 31:19–31. https://doi.org/10.1046/j.0305-0270.2003.00992.x
McCain CM (2010) Global analysis of reptile elevational diversity. Glob Ecol Biogeogr 19:541–553. https://doi.org/10.1111/j.1466-8238.2010.00528.x
McCain CM, Garfinkel CF (2021) Climate change and elevational range shifts in insects. Curr Opin Insect Sci 47:111–118. https://doi.org/10.1016/j.cois.2021.06.003
McCain CM, Grytnes J (2010) Elevational gradients in species richness. ELS. https://doi.org/10.1002/9780470015902.a0022548
McCoy ED (1990) The distribution of insects along elevational gradients. Oikos 58:313. https://doi.org/10.2307/3545222
McCoy ED (2002) The "veiled gradients" problem in ecology. Oikos 99:189–192. https://www.jstor.org/stable/3547765
McGlynn TP, Dunn T, Wayman E, Romero A (2010) A thermophile in the shade: light-directed nest relocation in the Costa Rican ant *Ectatomma ruidum*. J Trop Ecol 26:559–562
Meyer H, Pebesma E (2022) Machine learning-based global maps of ecological variables and the challenge of assessing them. Nat Commun 13:2208. https://doi.org/10.1038/s41467-022-29838-9
Mittelbach GG, Schemske DW, Cornell HV et al (2007) Evolution and the latitudinal diversity gradient: speciation, extinction and biogeography. Ecol Lett 10:315–331. https://doi.org/10.1111/j.1461-0248.2007.01020.x

Muluvhahothe MM, Joseph GS, Seymour CL et al (2021) Repeated surveying over 6 years reveals that fine-scale habitat variables are key to tropical mountain ant assemblage composition and functional diversity. Sci Rep 11:56. https://doi.org/10.1038/s41598-020-80077-8

Munyai TC, Foord SH (2012) Ants on a mountain: spatial, environmental and habitat associations along an altitudinal transect in a centre of endemism. J Insect Conserv 16:677–695. https://doi.org/10.1007/s10841-011-9449-9

Nahidan S, Nourbakhsh F, Mosaddeghi MR (2015) Variation of soil microbial biomass C and hydrolytic enzyme activities in a rangeland ecosystem: are slope aspect and position effective? Arch Agron Soil Sci 61:797–811. https://doi.org/10.1080/03650340.2014.958819

Nelson RA, MacArthur-Waltz DJ, Gordon DM (2023) Critical thermal limits and temperature-dependent walking speed may mediate coexistence between the native winter ant (*Prenolepis imparis*) and the invasive Argentine ant (*Linepithema humile*). J Therm Biol 111:103392. https://doi.org/10.1016/j.jtherbio.2022.103392

Nogués-Bravo D, Araújo MB, Romdal T, Rahbek C (2008) Scale effects and human impact on the elevational species richness gradients. Nature 453:216–219. https://doi.org/10.1038/nature06812

Noroozi J, Talebi A, Doostmohammadi M et al (2018) Hotspots within a global biodiversity hotspot - areas of endemism are associated with high mountain ranges. Sci Rep 8:10345. https://doi.org/10.1038/s41598-018-28504-9

Novick KA, Oishi AC, Miniat CF (2016) Cold air drainage flows subsidize montane valley ecosystem productivity. Glob Chang Biol 22:4014–4027. https://doi.org/10.1111/gcb.13320

Nowrouzi S, Andersen AN, Macfadyen S et al (2016) Ant diversity and distribution along elevation gradients in the Australian wet tropics: the importance of seasonal moisture stability. PLoS One 11:e0153420. https://doi.org/10.1371/journal.pone.0153420

Olson DM (1994) The distribution of leaf litter invertebrates along a neotropical altitudinal gradient. J Trop Ecol 10:129–150. https://doi.org/10.1017/S0266467400007793

Parmesan C, Yohe G (2003) A globally coherent fingerprint of climate change impacts across natural systems. Nature 421:37–42. https://doi.org/10.1038/nature01286

Parmesan C, Root TL, Willig MR (2000) Impacts of extreme weather and climate on terrestrial biota. BAMS 81:443–450. https://doi.org/10.1175/1520-0477

Parr CL, Bishop TR (2022) The response of ants to climate change. Glob Chang Biol 28:3161–3479. https://doi.org/10.1111/gcb.16140

Penick CA, Tschinkel WR (2008) Thermoregulatory brood transport in the fire ant, Solenopsis invicta. Insect Soc 55:176–182. https://doi.org/10.1007/s00040-008-0987-4

Penick CA, Diamond SE, Sanders NJ, Dunn RR (2017) Beyond thermal limits: comprehensive metrics of performance identify key axes of thermal adaptation in ants. Funct Ecol 31:1091–1100. https://doi.org/10.1111/1365-2435.12818

Pepin NC, Pike G, Schaefer M et al (2017) A comparison of simultaneous temperature and humidity observations from the SW and NE slopes of Kilimanjaro: the role of slope aspect and differential land-cover in controlling mountain climate. Glob Planet Change 157:244–258. https://doi.org/10.1016/j.gloplacha.2017.08.006

Pérez-Toledo GR, Valenzuela-González JE, Moreno CE et al (2021) Patterns and drivers of leaf-litter ant diversity along a tropical elevational gradient in Mexico. J Biogeogr 48:2512–2523. https://doi.org/10.1111/jbi.14217

Pérez-Toledo GR, Villalobos F, Silva RR et al (2022) Alpha and beta phylogenetic diversities jointly reveal ant community assembly mechanisms along a tropical elevational gradient. Sci Rep 12:1–13. https://doi.org/10.1038/s41598-022-11739-y

Peters MK, Hemp A, Appelhans T et al (2016) Predictors of elevational biodiversity gradients change from single taxa to the multi-taxa community level. Nat Commun 7:1–11. https://doi.org/10.1038/ncomms13736

Ploton P, Mortier F, Réjou-Méchain M et al (2020) Spatial validation reveals poor predictive performance of large-scale ecological mapping models. Nat Commun 11:4540. https://doi.org/10.1038/s41467-020-18321-y

Plowman NS, Mottl O, Novotny V et al (2020) Nest microhabitats and tree size mediate shifts in ant community structure across elevation in tropical rainforest canopies. Ecography 43:431–442. https://doi.org/10.1111/ecog.04730

Porter SD, Tschinkel WR (1993) Fire ant thermal preferences: behavioral control of growth and metabolism. Behav Ecol Sociobiol 32:321–329. https://doi.org/10.1007/BF00183787

Rahbek C (1995) The elevational gradient of species richness: a uniform pattern? Ecography:97–205. https://doi.org/10.1111/j.1600-0587.1995.tb00341.x

Rahbek C (2004) The role of spatial scale and the perception of large-scale species-richness patterns: scale and species-richness patterns. Ecol Lett 8:224–239. https://doi.org/10.1111/j.1461-0248.2004.00701.x

Robertson HG (2002) Comparison of leaf litter ant communities in woodlands, lowland forests and montane forests of North-Eastern Tanzania. Biodivers Conserv 11:1637–1652. https://doi.org/10.1023/A:1016883901065

Roeder KA, Bujan J, de Beurs KM et al (2021a) Thermal traits predict the winners and losers under climate change: an example from North American ant communities. Ecosphere 12:e03645. https://doi.org/10.1002/ecs2.3645

Roeder KA, Roeder DV, Bujan J (2021b) Ant thermal tolerance: a review of methods, hypotheses, and sources of variation. Ann Entomol Soc Am 114:459–469. https://doi.org/10.1093/aesa/saab018

Roeder DV, Paraskevopoulos AW, Roeder KA (2022) Thermal tolerance regulates foraging behaviour of ants. Ecol Entom 47:331–338. https://doi.org/10.1111/een.13118

Rohde K (1997) The larger area of the tropics does not explain latitudinal gradients in species diversity. Oikos 79:169–172. https://doi.org/10.2307/3546102

Rorison IH, Sutton F, Hunt R (1986) Local climate, topography and plant-growth in Lathkill-Dale nnr. I. A 12-year summary of solar-radiation and temperature. PC&E 9:49–56. https://doi.org/10.1111/1365-3040.ep11612961

Rosenzweig ML (1992) Species diversity gradients: we know more and less than we thought. J Mammal 73:715–730. https://doi.org/10.2307/1382191

Sahu N, Saini A, Behera SK et al (2020) Why apple orchards are shifting to the higher altitudes of the Himalayas? PLoS One 15:e0235041. https://doi.org/10.1371/journal.pone.0235041

Sanders NJ (2002) Elevational gradients in ant species richness: area, geometry, and Rapoport's rule. Ecography 25:25–32. https://doi.org/10.1034/j.1600-0587.2002.250104.x

Sanders NJ, Moss J, Wagner D (2003) Patterns of ant species richness along elevational gradients in an arid ecosystem. Glob Ecol Biogeogr 13:93–102. https://www.jstor.org/stable/3697559

Sanders NJ, Lessard J-P, Fitzpatrick MC, Dunn RR (2007) Temperature, but not productivity or geometry, predicts elevational diversity gradients in ants across spatial grains. Glob Ecol Biogeogr 16:640–649. https://doi.org/10.1111/j.1466-8238.2007.00316.x

Sanllorente O, Ruano F, Tinaut A (2015) Large-scale population genetics of the mountain ant *Proformica longiseta* (Hymenoptera: Formicidae). Popul Ecol 57:637–648. https://doi.org/10.1007/s10144-015-0505-2

Shi NN, Tsai C-C, Camino F et al (2015) Keeping cool: enhanced optical reflection and radiative heat dissipation in Saharan silver ants. Science. https://doi.org/10.1126/science.aab3564

Shik JZ, Arnan X, Oms CS et al (2019) Evidence for locally adaptive metabolic rates among ant populations along an elevational gradient. J Anim Ecol 88:1240–1249. https://doi.org/10.1111/1365-2656.13007

Smith MA (2015) Ants, elevation, phylogenetic diversity and community structure. Ecosphere 6:1–17. https://doi.org/10.1890/ES14-00473.1

Sommer S, Wehner R (2012) Leg allometry in ants: extreme long-leggedness in thermophilic species. Arthropod Struct Dev 41:71–77. https://doi.org/10.1016/j.asd.2011.08.002

Steinbauer MJ, Otto R, Naranjo-Cigala A et al (2012) Increase of Island endemism with altitude - speciation processes on oceanic islands. Ecography 35:23–32. https://doi.org/10.1111/j.1600-0587.2011.07064.x

Suissa JS, Sundue MA, Testo WL (2021) Mountains, climate and niche heterogeneity explain global patterns of fern diversity. J Biogeogr 48:1296–1308. https://doi.org/10.1111/jbi.14076
Sundqvist MK, Sanders NJ, Wardle DA (2013) Community and ecosystem responses to elevational gradients: processes, mechanisms, and insights for global change. Annu Rev Ecol Evol Syst 44:261–280. https://doi.org/10.1146/annurev-ecolsys-110512-135750
Szewczyk T, McCain CM (2016) A systematic review of global drivers of ant elevational diversity. PLoS One 11:e0155404. https://doi.org/10.1371/journal.pone.0155404
Tiede Y, Schlautmann J, Donoso DA et al (2017) Ants as indicators of environmental change and ecosystem processes. Ecol Indic 83:527–537. https://doi.org/10.1016/j.ecolind.2017.01.029
Timmermans B, Wehner M, Cooley D et al (2019) An evaluation of the consistency of extremes in gridded precipitation data sets. Clim Dyn 52:6651–6670. https://doi.org/10.1007/s00382-018-4537-0
Uhey DA, Hofstetter RW, Remke M et al (2020) Climate and vegetation structure shape ant communities along elevational gradients on the Colorado Plateau. Ecol Evol 10:8313–8322. https://doi.org/10.1002/ece3.6538
Urban MC, Tewksbury JJ, Sheldon KS (2012) On a collision course: competition and dispersal differences create no-analogue communities and cause extinctions during climate change. Proc Biol Sci 279:2072–2080. https://doi.org/10.1098/rspb.2011.2367
von Humboldt A, Bonpland A (1807) Essay on the geography of plants. University of Chicago Press, Chicago
Wallace AR (1878) Tropical nature, and other essays. MacMillan & Co., London & New York
Warren RJ (2010) An experimental test of well-described vegetation patterns across slope aspects using woodland herb transplants and manipulated abiotic drivers. New Phytol 185:1038–1049. https://doi.org/10.1111/j.1469-8137.2009.03147.x
Warren RJ II, Chick L (2013) Upward ant distribution shift corresponds with minimum, not maximum, temperature tolerance. Glob Change Biol 19:2082–2088. https://doi.org/10.1111/gcb.12169
Warren RJ II, Candeias M, Lafferty A, Chick LD (2020) Regional-scale environmental resistance to non-native ant invasion. Biol Invasions 22:813–825. https://doi.org/10.1007/s10530-019-02133-3
Warren RJ II, Costa JT, Bradford MA (2022) Seeing shapes in clouds: the fallacy of deriving ecological hypotheses from statistical distributions. Oikos 2022:e09315. https://doi.org/10.1111/oik.09315
Warren IIRJ, Mathew A, Reed K et al (2019) *Myrmica rubra* microhabitat selection and putative ecological impact. Ecol Entom 44:239–248. https://doi.org/10.1111/een.12700
Whiteman CD (2000) Mountain meteorology - fundamentals and applications, ebook. Oxford University Press, Cambridge, UK
Whittaker RH (1956) Vegetation of the Great Smoky Mountains. Ecol Monogr 26:1–69. https://doi.org/10.2307/1943577
Wilson EO (1971) The insect societies. Harvard University Press, Cambridge, MA
Wyborn C, Evans MC (2021) Conservation needs to break free from global priority mapping. Nat Ecol Evol 5:1322–1324. https://doi.org/10.1038/s41559-021-01540-x
Yu Z, Zou S, Li N et al (2021) Elevation-related climatic factors dominate soil free-living nematode communities and their co-occurrence patterns on Mt. Halla, South Korea. Ecol Evol 11:18540–18551. https://doi.org/10.1002/ece3.8454
Zapata FA, Gaston KJ, Chown SL (2005) The mid-domain effect revisited. Am Nat 166:E144–E148. https://doi.org/10.1086/491685

Chapter 10
Functional and Phylogenetic Change in Ant Communities Across Elevational Gradients: Informing Impacts of Climate Change

Tom R. Bishop and Alana Zammit

10.1 Introduction

Mountains hold immense cultural, spiritual, and practical value for human society (Grêt-Regamey et al. 2012). Mountain regions encompass huge environmental variation. Warm foothills and lowlands give way to increasingly colder and wetter conditions in the highlands. Importantly, this environmental change occurs over a relatively small geographic distance. Temperatures decline by 0.6 °C per 100 m of elevational change, and so within just a few kilometres, organisms can experience a range of climatic conditions—from tropical lowland rainforest to near-alpine. To encounter this range of environmental regimes while remaining at sea level would require movement across hundreds of kilometres of latitude, north or south. In addition, the partial pressures of O_2 and CO_2 decline with elevation, and UV-B irradiance tends to increase, while anthropogenic land-use pressure tends to decrease (Wymann von Dach et al. 2018). This environmental diversity underpins the public's fascination with mountains, as well as the variety of ecosystem services that they provide us.

Mountain regions are also disproportionately important for biodiversity. Despite making up ~25% of the Earth's land surface area, mountain regions are home to 87% of terrestrial vertebrate species (Rahbek et al. 2019) and are likely to form important refugia in the future (Morelli et al. 2020). This rich biodiversity is intrinsically linked to the environmental variation that mountains exhibit. In

T. R. Bishop (✉)
School of Biosciences, Cardiff University, Cardiff, UK

Department of Zoology and Entomology, University of Pretoria, Pretoria, South Africa
e-mail: bishoptr@cardiff.ac.uk

A. Zammit
Keble College, University of Oxford, Oxford, UK

I. R. Leal et al. (eds.), *Ant Responses to Climate Change*, Entomology in Focus 10,
https://doi.org/10.1007/978-3-032-21127-9_10

general, relatively few species can tolerate the colder, generally wetter, and "harsher" conditions that prevail at high elevations. As a result, the number of species comprising a local ecological community tends to decline with increasing elevation (McCain and Grytnes 2010), although there are nuances to this general picture (Grytnes et al. 2006; McCain and Grytnes 2010; Szewczyk and McCain 2016). It is not just species richness that changes with elevation, however. The identities of the species living at different elevations, and therefore the overall community composition, also change (Rahbek et al. 2019; Steinwandter and Seeber 2023). Typically, species are adapted to a particular range of environmental conditions and live only at a limited number of elevations. In this way, there is a turnover of species as low elevation specialists disappear and high elevation specialists take their place—often linked to parallel, abrupt vegetation changes (McCain and Grytnes 2010). These changes in species number and identity over relatively small geographic areas, as well as the low levels of disturbance at high elevation, have made elevational gradients a model system for ecologists to understand the general rules underpinning patterns of species diversity.

In this context, ants (Hymenoptera: Formicidae) have proven a useful taxon in the study of elevation-diversity gradients (see Chap. 9). Ants are important soil organisms distributed across every continent except Antarctica and are central to many terrestrial food webs and ecological networks. They are also abundant and easy to sample. In general, ants conform to the elevational diversity patterns of other taxa: different compositions of communities tend to be found across elevations, and at higher elevations, fewer species are typically found. In this chapter, we seek to move beyond the taxonomic and species-centric view of ant diversity, focusing on the functional and phylogenetic changes that ant communities undergo across elevational gradients. We synthesise this literature and relate it to the challenge of climate change: what do functional and phylogenetic perspectives provide us in the context of a rapidly changing twenty-first century?

10.2 Functional and Phylogenetic Diversity

Traditionally, biodiversity science at the whole organism level has adopted a taxonomic perspective (Violle et al. 2014) that focuses on the number and identity of species. For instance, alpha diversity can be estimated as the species richness of a local community or as some combination of species richness and the relative abundances of species such as Shannon's diversity or evenness metrics (Magurran et al. 2010). Beta diversity can be estimated in various ways by analysing the change in species identities from one community to the next (Magurran et al. 2010; Anderson et al. 2011). While this approach is clearly important for understanding variation in biodiversity across space and time, it provides limited information on *why* changes occur. In response to this limited mechanistic understanding, biodiversity science has turned to functional trait and phylogenetic perspectives over the last three decades (McGill et al. 2006; Violle et al. 2014; Chown 2022).

Functional and phylogenetic perspectives on diversity seek to uncover the processes that drive patterns in taxonomic diversity. A functional trait perspective aims to quantify the physiological, morphological, and behavioural characteristics (i.e. traits) that strongly influence an individual's performance and survival (McGill et al. 2006; Violle et al. 2007). By studying traits, we can gain a clearer mechanistic understanding into why biodiversity varies. For instance, physiological traits such as thermal tolerance vary predictably across temperature gradients (see Chap. 4), explaining why community composition also changes (Addo-Bediako et al. 2000; Boyle et al. 2021). Similarly, many morphological traits such as body size or relative leg length influence how efficiently an organism can move through an environment (see Chap. 3), helping us to understand why community composition changes in parallel with changes in habitat complexity (Gibb and Parr 2013; Gibb et al. 2018). These examples are of "response" traits that describe how organisms respond to the environments in which they find themselves. "Effect" traits can also be quantified which estimate the impact that organisms have on the environment, such as the amount or prey consumed, dung buried, or water filtered (Streit and Bellwood 2022). For the papers we review here, only response traits have been measured. The phylogenetic perspective, on the other hand, uses the degree of relatedness between species within a community and the relatedness between entire communities, to infer ecological processes. Phylogenetic information can be used as a proxy for trait similarity—more closely related species are likely to share very similar traits, assuming that traits are relatively strongly conserved across the phylogeny—or it can be used to understand the role of chance and historical contingency in driving the assembly of local communities. For instance, communities that are strongly clustered phylogenetically indicate that only lineages with particular traits can survive under the prevailing environmental conditions or that historical chance allowed a single lineage to colonise and subsequently radiate in a given location (Cavender-Bares et al. 2009). Conversely, very high phylogenetic diversity indicates that competition amongst closely related and ecologically similar species is an important driver of the assembly process (Cavender-Bares et al. 2009). Ultimately, information on species niches and interactions with each other (via their traits) or historical patterns of dispersal and radiation (via their phylogenies) are crucial for understanding *why* biodiversity patterns look the way that they do. The biology of most ant species is poorly known, and so in most cases, taxonomic identity does not provide this insight.

Broadly, there are three ways in which functional trait and phylogenetic diversity can be conceptualised and quantified. The first is a multidimensional approach and applies only to functional traits. In this approach, often referred to as "functional diversity", individual traits make up different axes of a multidimensional space. Individual species or organisms are then plotted into this space based on their trait values. The multidimensional volume (or area with just two traits or range with a single trait) can then be estimated and used as a measure of functional alpha diversity (i.e. functional richness) for a single community (Swenson 2014; Mammola and Cardoso 2020). In addition, various measures of the spread and positioning of species in the space, potentially weighted by relative abundance, can be used as measures of functional evenness or divergence for single communities (Mason et al. 2005). Beta diversity can be quantified by calculating the overlap in volume (or area or range) of two separate communities. While this approach is popular, it is not clear how measures of multivariate richness and spread relate in a general sense to changing environmental conditions. The greater mechanistic understanding alluded to above refers to the explicit analysis of traits across elevation (or any other gradient). Further, there is often a focus on response traits at the expense of effect traits. The second approach is tree-based and can incorporate either functional trait or phylogenetic information. Genetic relatedness is commonly represented as a dendrogram (i.e. a phylogeny), but trait differences between species can also be represented in this way. Typically, Euclidean distances between species, based on one or more traits, are used as the basis for a trait-dendrogram. Measuring alpha and beta diversities using phylogenetic trees or trait trees is the same. Alpha diversity can be quantified as the sum of the branch lengths that exist with a given community, potentially weighted by relative abundance. Beta diversity can be quantified by comparing the number of common branches between two communities. A variety of metrics exist to measure the functional or phylogenetic richness, divergence, or evenness of individual communities in a tree-based framework (Cardoso et al. 2014). Finally, the trait composition of individual communities can be summarised as a community weighted mean: the average value of a trait for a given community. Community weighted means can be occurrence- or abundance-weighted and typically use data from just a single trait. These values can therefore be interpreted as the average value of a given trait for a community, representing the most common or dominant trait value or state. These can then be modelled with respect to space, time, or environmental variables in the same way as any other alpha diversity metric. A large part of the literature has been dedicated to developing ever more complex metrics based on these three broad approaches.

10.3 Climate Change Through Functional and Phylogenetic Perspectives

In the context of climate change, these functional trait and phylogenetic perspectives on biodiversity promise to help us better predict and manage our changing planet (Violle et al. 2014). Will species migrate away from rising temperatures and into previously cooler conditions? Will they remain in situ and adapt? Or will they go extinct as a result of new, intolerable climatic and biotic conditions? Answering these questions is critical if we are to prioritise the right areas for nature conservation (Jones et al. 2016) and to best manage ecosystems to adapt to climate change or recover appropriately from human-induced disturbance. It is simply not practical, however, to collect data on how each species on Earth responds to environmental change. Instead, the functional and phylogenetic approach is to model species traits (potentially constrained by phylogeny) as continuous variables. Under this approach, knowing the cross-species relationships between a given trait and a given environmental variable (i.e. between thermal tolerance and mean annual temperature) allows us to make predictions as to which lineages will move, adapt, or go extinct in response to changing climates (Violle et al. 2014; Chown 2022). Given that mountain regions and elevational gradients are likely to provide important climate refugia in the future (Morelli et al. 2020), understanding existing patterns of functional and phylogenetic diversity in these locations is an important first step in characterising how they may be impacted by climate change.

10.4 The Ant-Elevation Functional and Phylogenetic Literature

The use of functional and phylogenetic perspectives has rapidly increased over the last three decades. This has been due to the development of new practical methods to gather trait and genetic data from organisms, but also due to the new conceptual frameworks that have been proposed to analyse and interpret these new forms of data. Ant ecologists have also taken up these new approaches with fervour. For instance, a paper attempting to standardise the morphological traits collected from ant species published in 2017 (Parr et al. 2017) has been cited 158 times by September 2024. In general, functional trait approaches in ants have been focussed on the worker caste. For instance, data on worker physiology, worker morphology, or worker behaviour is often collected, while the traits of the reproductive caste or those of the whole colony largely ignored. This is a known conceptual issue (Gibb et al. 2022) and is largely a practical one—it is exceedingly difficult to measure traits from queen ants or from entire colonies.

In this chapter, we review the ant functional and phylogenetic literature as it relates to elevation gradients, in the context of informing likely responses to climate change. For our review, we searched *Web of Science* in September 2024 using the following string: (TI = ((ant OR ants OR formicidae) AND (mountain OR elevation* OR altitud* or alpine) AND (functional diversity OR phylogenetic diversity OR function* OR phylogenetic OR trait* OR trait diversity))) OR AB = ((ant OR ants OR formicidae) AND (mountain* OR elevation* OR altitude* OR alpine) AND (trait* OR function* OR phylogenetic OR body size OR size OR morpholog* OR colour OR color OR physiolog* OR behaviour*)).

This initial search yielded 458 papers. We screened these for studies that examined changes in functional or phylogenetic metrics across elevational gradients of more the 300 m that focussed on individual ant species or entire ant communities. We did not consider papers that used traits derived from community data such as habitat generalists/specialists or commonness/rarity. We added a single paper which we knew about that did not appear in our original search. This process left us with 59 papers in total. These studies come from across the globe (Fig. 10.1) and sampled gradients from 0 to 4750 m a.s.l. with an average gradient size of 1730 m

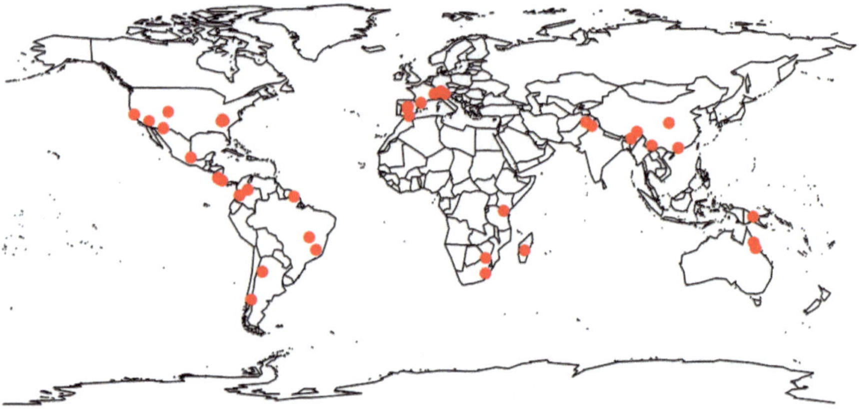

Fig. 10.1 World map of locations where ant-elevational studies investigating functional and phylogenetic changes have occurred

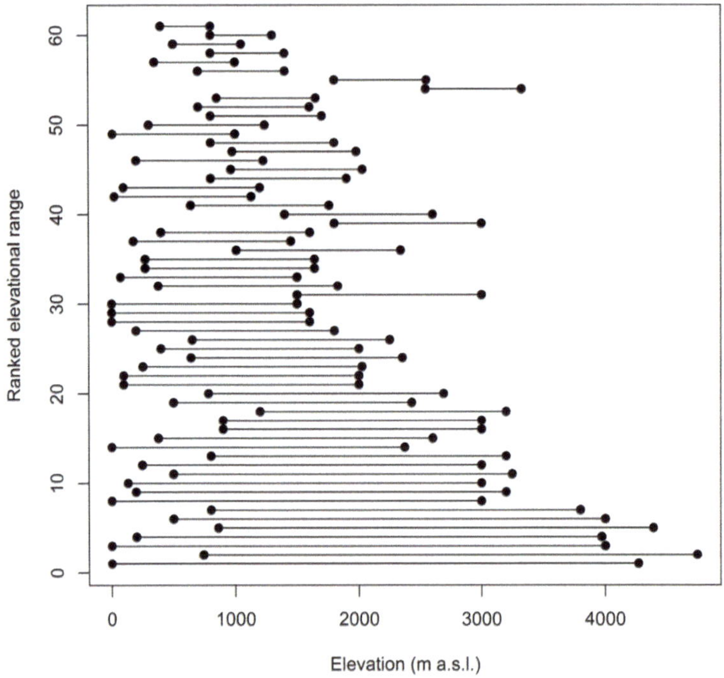

Fig. 10.2 Distribution of elevational range across all selected studies

(Fig. 10.2). Most (79%) of the papers investigated only functional trait changes across elevation, 13% investigated phylogenetic changes, and 8% looked at both (Fig. 10.3). Of the 43 papers looking at functional trait changes, 11 took an explicit "functional diversity" approach.

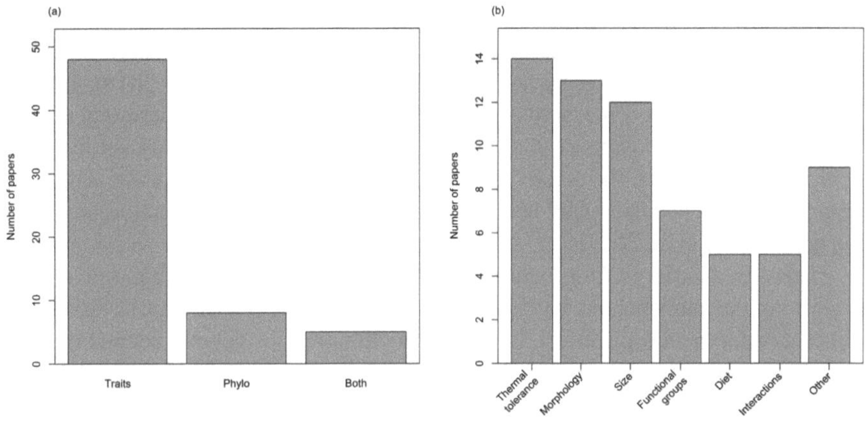

Fig. 10.3 Plots showing (**a**) the distribution of studies investigating functional traits, phylogenetic diversity, or both and (**b**) the distribution of trait categories investigated across studies

10.5 Functional Trait Variation Across Elevation

We assigned each paper investigating ant traits across elevation into one or more of the following groups: thermal tolerance, morphology, size, functional groups, diet, interactions, and other (Fig. 10.3). We distinguished between papers that investigated morphology in general from those that explicitly looked at size (even if this was a morphological measure of size) as there are typically very different hypotheses linking them to elevational change.

10.5.1 Thermal Tolerance

Thermal tolerance was the most studied ant trait across elevation (Fig. 10.3). The majority of studies quantified the thermal tolerance of worker ants as critical temperatures, CT_{min} and CT_{max}, the minimum and maximum temperature that an organism experiences before it loses its ability to right itself and maintain proper motor functioning (see Chap. 4). Unsurprisingly, ants are more tolerant to the cold and less tolerant to the heat in upland areas (Warren and Chick 2013; Chick et al. 2020; Khaliq et al. 2023; O'Donnell and Baudier 2023). These headline patterns are not universal, however. A number of studies find evidence of only lower thermal tolerance limits declining with elevation (Bishop et al. 2017; Baudier and O'Donnell 2018; Lytle et al. 2020; Tonione et al. 2020; Leahy et al. 2022) or of no detectable changes in upper thermal tolerance limits (Nowrouzi et al. 2018). In a minority of cases, the highest levels of thermal tolerance are found at mid-elevations (Villalta et al. 2020) or non-significant elevational changes in either upper or lower thermal tolerance limits are detected (Yela et al. 2020). Changes in thermal tolerance can be microhabitat dependent: in Costa Rica upper and lower thermal tolerance limits decline with elevation for subterranean species, but only lower tolerance limits decreases for epigaeic (i.e. surface active) species (Baudier et al. 2018). Finally, there is also evidence that ant colonies as a whole alter their thermoregulatory behaviours at high elevations in an attempt to conserve energy. Army ant bivouacs at high elevation allow the colony to cool down when rearing their larvae, which are less thermally sensitive, and heat up again when rearing temperature-sensitive pupae (Baudier et al. 2019).

There are a variety of mechanisms driving these patterns at different biological scales. At the intraspecific level, changes in thermal tolerance can be driven by either developmental plasticity, long-term acclimation, or genetic adaptation over multiple generations (Warren and Chick 2013; Baudier et al. 2018; Lytle et al. 2020). The extent of intraspecific variation in these thermal tolerance traits, and whether all species are capable of exhibiting similar levels of variation, remains unclear (Nascimento et al. 2024). At the community level, changes in thermal tolerance with elevation are driven by species loss and gain. For instance, high-elevation specialists with low thermal tolerances replace low-elevation specialists with higher thermal tolerances as one climbs a mountain (Bishop et al. 2017; Chick et al. 2020; Leahy et al. 2022; Khaliq et al. 2023). These replacements lead to a decrease in the thermal tolerance of the entire ant community on average. These community patterns could also be driven by intraspecific changes.

Overall, thermal tolerance patterns of mountain ants conform with theories from the wider literature. For instance, the thermal adaptation hypothesis (Janzen 1967; Kaspari et al. 2015) posits that organisms should be physiologically adapted to the environments in which they live—cold, high elevation environments should select for ants with low thermal tolerance and vice versa. The reason for this is that is costly to maintain tolerances to extreme conditions if these conditions are never, or rarely, encountered (Angilletta 2009).

10.5.2 Morphology

Morphology (see Chap. 3) was the second most studied category of ant trait in our literature search (Fig. 10.3). Research on morphological changes across elevation fall into two categories: those that investigate how specific morphological metrics vary with elevation and those that summarise many morphological variables as functional diversity and investigate how this changes with elevation.

Only five studies investigate how individual morphologies change with elevation and they vary in their findings. For instance, ants at higher elevations have been reported as having longer limbs and an overall more "gracile" appearance over extensive (>1600 m) elevational gradients in Australia and Europe (Villalta et al. 2020; Gibb et al. 2024), but not across smaller gradients in South America (Fichaux et al. 2021). Gracile morphology is often associated with other traits such as dorsally positioned eyes and relatively few hairs. The reasons why this suite of traits may be selected for at high elevations are unclear but we urge caution in interpreting them—these signals come from two elevational transects out of three from a single study in Australia. Many typical high-elevation genera are not gracile at all (see "Phylogenetic Variation Across Elevation" below), and so these ideas should be tested for in other contexts. A further study investigating a range of morphological traits found limited evidence for intraspecific changes across elevation, with only a single species showing significant increases in mandible length, eye position (more dorsal), and leg length (Gaudard et al. 2019). Finally, ants appear to have darker cuticle colours at high elevations (Bishop et al. 2016; Gibb et al. 2024). Darker colours absorb a greater fraction of incoming solar radiation, allowing high elevation ants to heat up more quickly in an otherwise cool environment (Bishop et al. 2016). In addition, dark colours can act as sunscreen and protect organisms from harmful UV-B rays at high elevation or elsewhere (Bishop et al. 2016).

Eight studies combine suites of morphological measures into functional diversity. Typically, these use dimension reduction techniques such as principal component analysis to reduce many morphological traits into a few key "axes" of variation and then calculate functional diversity measures form the resulting multidimensional space. These studies tend to find decreases in functional diversity (while controlling for species richness) with elevation (Fontanilla et al. 2019; Silvestre et al. 2021; Neves et al. 2024; Zhao et al. 2024). In some cases, these relationships are dependent on the season as a consequence of changes in ant foraging activity (de Castro et al. 2020), and in other cases, they are not found at all (Muluvhahothe et al. 2021). These results mirror those found for species richness: we see a smaller range of traits (and lower functional diversity values) than would be expected given the reduced species richness in high elevation ant communities. One study has investigated whether functional beta diversity (i.e. the trait composition of different communities) within elevational bands changes across mountains. They found no evidence for this phenomenon (Nunes et al. 2020), suggesting that within-elevation differences are consistent in both high and low elevation environments. Functional beta diversity, however, can appear to change across elevational bands. In southern Africa, for example, high elevation communities appeared to represent a nested subset of the morphologies found at low elevations (Bishop et al. 2015). In this case, high elevation communities are represented by average trait values rather than extreme ones—highlighting that outlier morphologies such as long mandibles, large eyes, and long legs tend to be found at lower elevations. This finding in particular contradicts that of the Australian study of Gibb et al. (2024).

In sum, while it is clear that morphological diversity is greatly reduced at high elevations, it is unclear exactly which morphological traits are selected for or against in high elevation environments. The datasets exist to answer this question but have largely been used to assess popular functional diversity metrics, rather than applying existing trait-environment theory to test how individual morphological metrics may vary with elevation. In our view, it is crucial that the ant-elevational literature move beyond "functional diversity" and instead return to the individual traits of interest.

10.5.3 Size

Twelve studies investigated changes in worker, queen, or total colony size across elevation. Many of these draw on Bergmann's ecogeographic rule that organisms are larger in colder environments because of smaller surface area to volume ratios and therefore lower heat loss (Bergmann 1847). There is wide support for this rule in homeotherms (He et al. 2023) but inconsistent support in ectotherms (Vinarski 2014).

The five relevant ant studies at the community level find no support for Bergmann's rule in average worker body size, with size either decreasing with elevation, peaking at mid-elevation, or not changing at all (Nowrouzi et al. 2018; Silvestre et al. 2019; Alcantara et al. 2024; Gibb et al. 2024; Schifani et al. 2024). Two additional studies investigating individual species also found no evidence for Bergmann's rule in ants (Ruppell et al. 1998; Sandoval-Arango et al. 2020). Four additional intraspecific studies, however, have found support for Bergmann's rule in worker ants (Bernadou et al. 2016; Purcell et al. 2016; Shik et al. 2019; He et al. 2024) along with a single community-level study (Bishop et al. 2016, 2019). Notably, in the one study that measured colony size (the total size of the superorganism) across elevation, there was no relationship (Bernadou et al. 2016). Finally, a single study investigated how the degree of polymorphism within a species changes with elevation. On average (i.e. at the community level), they found that polymorphism decreased with elevation on average, at higher elevations (Reymond et al. 2013).

These findings are inconclusive and arguably more work needs to be done to fully understand how ant body (or colony) size varies with elevation. This is in contrast to latitudinal patterns across the globe where it seems that average ant body size decreases with increasing temperature, conforming to Bergmann's rule (Gibb et al. 2018). The difference between two kinds of gradients in terms of the ant body size is confusing given that other traits (i.e. colour and thermal tolerance) appear to respond to elevation. It may be that body size changes are harder to evolve relative to other traits and that continued gene flow and migratory pressure along elevational gradients (which is not present across much larger latitudinal gradients) prevents body size changes from taking place. This argument is speculative, however. Ultimately, more dedicated studies testing Bergmann's rule in ants across elevation are needed, and the data is likely already available.

10.5.4 Functional Groups

A qualitative approach to understanding the ecological strategies of ants has been to use categorical functional groups. Within the ants the most popular classification is based on the Australian ant fauna (Andersen 1990), and this classification has been deployed widely across the globe in a range of contexts (Andersen 1995, 1997;

Lach et al. 2010; Hoffmann and Andersen 2023). Andersen's functional groups are in relation to environmental stress and disturbance, combining higher-level taxonomic identity with broad ecological traits such as species' response to environmental variation, their feeding guild, and their behavioural dominance relative to other ants.

Relatively few studies have investigated ant functional groups across elevational gradients. Our literature search found seven studies that used a functional group classification of ants across elevation, but two of these used functional groups as part of a wider assessment of functional beta diversity alongside morphological measures and have been discussed above (de Castro et al. 2020; Nunes et al. 2020). Of the remaining five, there is largely a consensus. Firstly, all five studies demonstrate that functional group composition changes with elevation (Andersen 1997; Bernadou et al. 2013; Bharti et al. 2013; Werenkraut et al. 2015; Marathe et al. 2021). Second, four of these studies find that high elevation communities tend to be dominated by the opportunist and cold climate specialists groups, with some representation of the generalised Myrmicinae. Further, they tend to find that hot climate specialists and specialist predators tend to occupy low elevations, but drop out as elevation increases (Andersen 1997; Bernadou et al. 2013; Bharti et al. 2013; Marathe et al. 2021). These studies come from North America, Europe, and Asia, illustrating that functional group composition of ant communities is largely consistent across these different biogeographic regions. The one exception is a study from Argentina which found that there was a reduction in cold climate specialists at the highest elevations and only limited changes in opportunist species (Werenkraut et al. 2015).

These functional group changes with elevation make ecological sense. High elevation environments are typically low energy and low productivity systems meaning that specialist predators are unlikely to survive while opportunist species are more likely to sustain themselves. Further, the thermal changes that elevational gradients represent parallel the shift from tropical climate specialists in the lowlands to cold climate specialists in the highlands.

10.5.5 Diet

While functional group classifications often infer or assume particular diets of species, they do not quantify diet directly. Given the changes in environmental conditions that occur across elevational gradients, we may also expect the dietary preferences of ants to change. A handful of studies have investigated how ant diet varies across elevation, with mixed findings.

Four studies used "cafeteria"-style experiments to assess the overall dietary preferences of ant communities across elevational gradients. In these experiments, baits with artificial or natural food sources are laid out and checked periodically for foraging ants. In natural habitats in Tanzania, the use of water, salt, and carbohydrate baits decreased with elevation, but lipid bait usage increased (Peters et al. 2014).

Despite this, carbohydrate baits remained the most frequently used food source at high elevation. The increased use of fats at high elevation is hypothesised to be linked to their use as long-term food stores and as a thermal insulator—properties that may serve ants well in cold, food-scare upland environments. A study in Papua New Guinea used a range of naturally foraged items, rather than artificial baits, and found that preferences for chicken faeces and live termite baits declined with elevation, but that there was no change in foraging for sugars or crushed insect items (Orivel et al. 2018). The authors speculate that there is no limitation of sugar resources in this environment, which could explain the low and invariant preference for sugar. In Ecuador, relative carbohydrate use increased with elevation in the wet season but not in the dry, and lipid use declined with elevation in both seasons (Moses et al. 2023). Finally, a cafeteria study in Spain examined the niche width and overlap of ant species by measuring their attendance at different baits. They found that niche width and overlap both decreased with increasing elevation, suggesting that ant species were foraging for a smaller and shared set of resources in upland sites (Seoane et al. 2021). Consequently, these cafeteria experiments provide few general conclusions as to how ant diets change across elevational gradients—the only consistent finding being that carbohydrate, salt, and protein baits were typically the most frequently visited across studies and elevations.

One study used stable isotopes of carbon and nitrogen to study the relationships between *Azteca* ants and their *Cecropia* plant mutualists in Costa Rica (Trimble and Sagers 2004). Ant $\delta^{15}N$ declined with increasing elevation. $\delta^{15}N$ is used to indicate the trophic position of organisms, and so a decline in this metric with elevation suggests that ants occupy lower trophic positions at higher elevations. This conclusion should not be applied widely, however, given its limited taxonomic and geographic dataset. However, the use of stable isotopes may be a more useful way forward in terms of understanding ant diet-elevation patterns rather than cafeteria experiments, which may be highly context dependent and offer contrasting results depending on the natural resources available in a given environment, season, and colony developmental state (Moses et al. 2023).

10.5.6 Ecosystem Functions

A key set of ecosystem functions and services provided by ants hinge on their relationships with other species (Del Toro et al. 2012). Ants provide seed dispersal and protective functions to a range of plant and insect species, and these appear to change consistently across elevation. For instance, seed removal rates decline with elevation in both North America (Del Toro and Ribbons 2019) and Europe (Silvestre et al. 2019). Further, the effectiveness of ants at protecting aphid mutualists (Nelson et al. 2019), the number of ant-treehopper aggregations (Camacho et al. 2022), and the number of ant-plant interactions all decline with increasing elevation (Plowman et al. 2017). These changes may be explained in part by lower ant activity at high elevations, which itself is probably driven by temperature (Bishop et al. 2014;

Nelson et al. 2019; Moses et al. 2023). However, they are also likely to be driven by the reduced need for these interactions at high elevations—predator abundances are lower at high elevations, reducing the need for ant protection, and ant-dispersed plants may simply not be present which reduces the need for specialised ant dispersal partners (Del Toro and Ribbons 2019). These findings, while few, feed into a broader narrative that abiotic conditions shape ecosystems and communities in high elevation and high latitude areas, while biotic factors shape them in low elevation and low latitude areas (Roslin et al. 2017).

10.5.7 Other Traits

In our review, nine studies investigated a suite of other ant traits that did not fall into the major categories discussed above. These include studies on foraging, architecture, non-thermal physiology, and phenology.

Two studies investigated changes in ant foraging across elevation. The first found that the size of colonies' foraging area, averaged across the entire community, decreased with increasing elevation (Bernstein and Gobbel 1979). This result, however, was explained by an increase in the vegetation density with elevation. The logic being that denser vegetation leads to more resources being available, meaning that colonies need to forage shorter distances to gather enough food. It is unlikely that this result generalises across the majority of elevational transects across the globe, where resource availability tends to decrease at the highest elevations. Low temperatures are likely to constrain ant foraging more broadly (Roeder et al. 2022). A second study, investigating a range of ant traits, found that ant foraging schedules shifted to match the prevailing temperature profiles at different elevations (Villalta et al. 2020). While these data come from a single species—*Aphaenogaster iberica*—similar shifts in foraging schedules have been reported in a variety of contexts, highlighting the plasticity that ants have in this trait (Parr and Bishop 2022).

Three studies investigated how ant nest architecture changes across elevation. Studies in Papua New Guinea (Plowman et al. 2020) and Switzerland (Reymond et al. 2013) measured the diversity of nesting substrates and nest structures across elevation. They found decreases (Switzerland) and mid-elevational peaks (Papua New Guinea) in nest structure diversity. Both studies argue that harsher environmental conditions and a need to insulate nests select for lower architectural diversity at high elevation. A third study tested whether the fungal gardens of leaf cutter ants changed position with elevation—predicting that they should be shallower at higher elevations in order to optimally thermoregulate (Yela et al. 2020). This prediction held for only one of the two species studied.

Non-thermal physiology papers investigated a range of traits. Fat content and development time were studied along a Swiss elevational gradient, but neither trait varied systematically with elevation (Purcell et al. 2016). Similarly, lipid content did not change across elevation in the south eastern USA (Lytle et al. 2020). In Spain, metabolic rates increased with elevation, thought to be due to the metabolic cold adaptation hypothesis which suggests that metabolic rates are elevated in cold conditions in order to allow organisms to quickly take advantage of brief warm periods (Shik et al. 2019). Finally, low elevation ants tend to engage in their mating flights earlier in the year than those at high elevations, implying that high elevation ants tended to overwinter as solitary queens following a late-summer nuptial flight (Dunn et al. 2007). These changes in flight phenology were not plastic within species, but rather reflected changes in which species lived at different elevations—early fliers did not occupy high elevations, probably because temperatures were too low for colony founding or worker survival.

10.6 Phylogenetic Variation Across Elevation

There are relatively few studies investigating phylogenetic diversity across elevation (Fig. 10.3), but their findings are largely consistent. The majority of studies using species-level phylogenies find a decrease in phylogenetic diversity with increasing elevation (Machac et al. 2011; Smith et al. 2014; Liu et al. 2018; Camacho et al. 2021; Pérez-Toledo et al. 2022; Neves et al. 2024; Zhao et al. 2024). The finding of fewer phylogenetic lineages of ants at higher elevations does not simply reflect the decline in species richness with elevation, but rather identifies a high-elevation "clustered" fauna: *Formica*, *Myrmica*, *Temnothorax*, *Lasius*, *Stenamma*, and *Adelomyrmex* (Machac et al. 2011; Smith et al. 2014; Liu et al. 2018; Pérez-Toledo et al. 2022; Zhao et al. 2024). These generic identities, unsurprisingly, align with Andersen's (1997) classification of cold-climate specialists. This raises an important question as to the usefulness of phylogenetic studies: what information are they providing beyond purely "taxonomic" approaches, at least with respect to climate change? A purely taxonomic approach would often identify clustering patterns in community data. In terms of understanding climate change, phylogenetic data may best be able to shed light on the "evolvability" of various traits that allow different taxa to respond to changing conditions.

Phylogenetic clustering in the highlands can be replaced by phylogenetic overdispersion in the lowlands (Liu et al. 2018). Overdispersion describes a situation where a much more diverse set of lineages is represented in a local community relative to what would be expected by chance. Again, this hint of stronger biotic controls on community structure in the lowlands, and stronger abiotic controls in the highlands, mirrors findings across latitude (Roslin et al. 2017). This pattern may provide a unique climate change impact monitoring tool. For instance, as climates warm species will be pressured to move into higher latitudes and elevations. These higher elevations will be warmer than they were, matching the thermal envelope of previously lowland species. Consequently, we may expect higher elevations to become increasingly phylogenetically overdispersed as temperatures rise. Simultaneously, taxa that were previously clustered may be targeted as being ripe for translocation or more directed conservation efforts in upland areas (Bellis et al. 2019).

There are exceptions to these broad patterns. For instance, one global-scale study found no strong evidence for clustering effects and a mix of negative, positive, and neutral relationships between phylogenetic diversity and elevation (Smith 2015). The reason for this may be that a genus-level phylogeny was used throughout, rather than a species-level phylogeny, and that this lack of resolution may obscure fine-scale patterns that emerge between different species. Three studies found no relationship (Reymond et al. 2013; Fichaux et al. 2021) or a weak positive relationship (Silvestre et al. 2021) between phylogenetic diversity and elevation. These studies were in the minority, however.

10.7 Conclusion

Functional and phylogenetic perspectives have revealed the constrained set of traits that allow ants to exist in high elevation areas—namely, tolerance to low temperatures, a reduction in morphological diversity, dark colouration, and a decline in the reliance on mutualistic partners. These overall findings highlight the strong environmental filter that elevational gradients impose on ants and other organisms. Despite this, however, the functional and phylogenetic elevation literature is small and contains results that are at odds with one another. Inconclusive findings likely reflect complex interactions between elevation, latitude, and prevailing climate that we currently do not have enough data on to assess. Consequently, we argue for a more targeted approach using functional and phylogenetic data to help us pinpoint when, where, and why communities will respond to global change. Given this, we suggest four core areas that future research should focus on:

1. Combined taxonomic, functional, and phylogenetic approaches. The majority of studies that we review here look at taxonomic and functional diversity or taxonomic and phylogenetic diversity. Rarely do studies investigate all three perspectives on biodiversity. There is value, however, in understanding each of these components. For instance, why do high elevation ant communities tend to look phylogenetically clustered? What is the environment (presumably) selecting for? Parallel functional trait data, to complement phylogenies, would reveal the precise traits that confer survival at high elevation. Currently, we can only surmise which traits drive such phylogenetic clustering from other studies (i.e. thermal tolerance, colour, body size). Further, combining trait data with phylogenetic data more often would allow us to better understand how particular trait values and combinations have evolved and how they may evolve in the future. This would add value to phylogenetic data that otherwise finds similar patterns and conclusions to purely taxonomic approaches. In the context of climate change, pinning down the exact mechanisms driving shifts in communities is crucial to predicting future changes and generalising our understanding across different geographic regions. This point is particularly important given the assumption that lowland ant communities will shift into the highlands as temperatures rise—

this shift may not happen uniformly across species depending on exactly how increased temperatures interact with existing trait values and combinations.

2. Time-series data. All studies reviewed here are snapshots of what ant-elevational diversity patterns look like. These systems, however, can be dynamic and change through time via consistent season effects or seemingly random inter-annual differences (Donoso 2017). It is unknown whether these taxonomic fluctuations through time will also drive functional and phylogenetic changes, but capturing natural intra- and interannual fluctuations is crucial if we are to estimate the additional "climate change" effect (Bishop et al. 2019). Estimates of natural community fluctuations are an important part of contextualising future climate-related shifts and determining whether changes are simply "natural" or part of a longer term trend.
3. Further unpacking morphological change. Many studies investigate functional diversity or functional richness and observe changes across elevation. Rarely do they explore which trait values are more or less common in high elevation environments. This is not a data availability issue but an analysis issue. We argue that it is time to move away from functional diversity measures and to investigate how individual traits change and to link these changes to ecological theory. As we argue above, pinning down the exact mechanisms driving community change is critical if we are to predict future community states following climate change. Broad measures of functional diversity are not that helpful in this context.
4. Measuring the right traits. Which traits are meaningful? Traits such as thermal tolerance and body size have clear conceptual and empirical links to variables such as temperature that systematically change across elevation (and time, with climate change). It is less clear how other sets of traits relate to elevation or other climatically important variables, and studying them without clear a priori hypotheses may not be the most productive way forward. We urge ant functional ecologists to consider what their studied traits mean (Gibb et al. 2022) and how they relate to the environmental changes captured by elevational gradients. For instance, might queen dispersal and founding traits be important in dictating which species survive at different elevations? What about nest architecture? Or desiccation resistance? These kinds of traits are understudied yet we predict would reveal a lot about how mountain ant biodiversity is structured. Crucially, many of them will have important roles to play in our understanding of climate change risk—temperatures are rising, but droughts are also becoming more frequent suggesting that desiccation risk (see Chap. 5) and thermal tolerance (see Chap. 4) traits will need to be understood to characterise the response of ants to climate change (Bujan et al. 2016).

References

Addo-Bediako A, Chown SL, Gaston KJ (2000) Thermal tolerance, climatic variability and latitude. Proc R Soc Lond Ser B Biol Sci 267(1445):739–745. https://doi.org/10.1098/rspb.2000.1065

Alcantara M et al (2024) Bugs and Bergmann's rule: a cross-taxon large-scale study reveals idiosyncratic altitudinal and latitudinal body size patterns for different insect taxa. Entomologia Generalis 44(3):715–725. https://doi.org/10.1127/entomologia/2024/2246

Andersen AN (1990) The use of ant communities to evaluate change in Australian terrestrial ecosystems: a review and a recipe. Proc Ecol Soc Aust 16:347–357

Andersen AN (1995) A classification of Australian ant communities, based on functional groups which parallel plant life-forms in relation to stress and disturbance. J Biogeogr 22(1):15–29. https://doi.org/10.2307/2846070

Andersen A (1997) Functional groups and patterns of organization in North American ant communities: a comparison with Australia. J Biogeogr 24(4):433–460. https://doi.org/10.1111/j.1365-2699.1997.00137.x

Anderson MJ et al (2011) Navigating the multiple meanings of β diversity: a roadmap for the practicing ecologist. Ecol Lett 14(1):19–28. https://doi.org/10.1111/j.1461-0248.2010.01552.x

Angilletta MJ (2009) Thermal adaptation: a theoretical and empirical synthesis. Oxford University Press, Oxford, UK

Baudier K, O'Donnell S (2018) Complex body size differences in thermal tolerance among army ant workers (*Eciton burchellii* parvispinum). J Therm Biol 78:277–280. https://doi.org/10.1016/j.jtherbio.2018.10.011

Baudier KM et al (2018) Extreme insolation: climatic variation shapes the evolution of thermal tolerance at multiple scales. Am Nat 192(3):347–359. https://doi.org/10.1086/698656

Baudier K et al (2019) Plastic collective endothermy in a complex animal society (army ant bivouacs: *Eciton burchellii* parvispinum). Ecography 42(4):730–739. https://doi.org/10.1111/ecog.04064

Bellis J et al (2019) Identifying factors associated with the success and failure of terrestrial insect translocations. Biol Conserv 236:29–36. https://doi.org/10.1016/j.biocon.2019.05.008

Bergmann C (1847) Über die Verhältnisse der Wärmeökonomie der Thiere zu ihrer Grösse. Vandenhoeck und Ruprecht

Bernadou A et al (2013) Physical and land-cover variables influence ant functional groups and species diversity along elevational gradients. Landsc Ecol 28(7):1387–1400. https://doi.org/10.1007/s10980-013-9892-y

Bernadou A et al (2016) Body size but not colony size increases with altitude in the holarctic ant, *Leptothorax acervorum*. Ecol Entomol 41(6):733–736. https://doi.org/10.1111/een.12338

Bernstein R, Gobbel M (1979) Partitioning of space in communities of ants. J Anim Ecol 48(3):931–942. https://doi.org/10.2307/4205

Bharti H et al (2013) Ant species richness, endemicity and functional groups, along an elevational gradient in the Himalayas. Asian Myrmecol 5:79–101

Bishop TR et al (2014) Elevation-diversity patterns through space and time: ant communities of the Maloti-Drakensberg Mountains of southern Africa. J Biogeogr. Edited by L. Gillman 41(12):2256–2268. https://doi.org/10.1111/jbi.12368

Bishop TR et al (2015) Contrasting species and functional beta diversity in montane ant assemblages. J Biogeogr 42(9):1776–1786. https://doi.org/10.1111/jbi.12537

Bishop TR et al (2016) Ant assemblages have darker and larger members in cold environments. Glob Ecol Biogeogr 25(12):1489–1499. https://doi.org/10.1111/geb.12516

Bishop TR et al (2017) Coping with the cold: minimum temperatures and thermal tolerances dominate the ecology of mountain ants: thermal tolerances of mountain ants. Ecol Entomol 42(2):105–114. https://doi.org/10.1111/een.12364

Bishop TR et al (2019) Thermoregulatory traits combine with range shifts to alter the future of montane ant assemblages. Glob Chang Biol 25(6):2162–2173. https://doi.org/10.1111/gcb.14622

Boyle MJW et al (2021) Localised climate change defines ant communities in human-modified tropical landscapes, Edited by D. Allen. Funct Ecol 35(5):1094–1108. https://doi.org/10.1111/1365-2435.13737

Bujan J, Yanoviak SP, Kaspari M (2016) Desiccation resistance in tropical insects: causes and mechanisms underlying variability in a Panama ant community. Ecol Evol 6(17):6282–6291. https://doi.org/10.1002/ece3.2355

Camacho G et al (2021) Spatial phylogenomics of acrobat ants in Madagascar-Mountains function as cradles for recent diversity and endemism. J Biogeogr 48(7):1706–1719. https://doi.org/10.1111/jbi.14107

Camacho L, Lessard J, Avilés L (2022) Partner supply mediates elevational gradients in the outcome of ant-hemipteran mutualistic associations. J Biogeogr 49(11):1967–1977. https://doi.org/10.1111/jbi.14469
Cardoso P et al (2014) Partitioning taxon, phylogenetic and functional beta diversity into replacement and richness difference components. J Biogeogr 41(4):749–761. https://doi.org/10.1111/jbi.12239
Cavender-Bares J et al (2009) The merging of community ecology and phylogenetic biology. Ecol Lett 12(7):693–715. https://doi.org/10.1111/j.1461-0248.2009.01314.x
Chick L et al (2020) The coupled influence of thermal physiology and biotic interactions on the distribution and density of ant species along an elevational gradient. Diversity 12(12):456. https://doi.org/10.3390/d12120456
Chown SL (2022) Macrophysiology for decision-making. J Zool 319:1–22. https://doi.org/10.1111/jzo.13029
de Castro F et al (2020) Environmental drivers of taxonomic and functional diversity of ant communities in a tropical mountain. Insect Conserv Divers 13(4):393–403. https://doi.org/10.1111/icad.12415
Del Toro I, Ribbons R (2019) Variation in ant-mediated seed dispersal along elevation gradients. PeerJ 7:e6686. https://doi.org/10.7717/peerj.6686
Del Toro I, Ribbons RR, Pelini SL (2012) The little things that run the world revisited: a review of ant-mediated ecosystem services and disservices (Hymenoptera: Formicidae). Myrmecol News 17(0):133–146
Donoso DA (2017) Tropical ant communities are in long-term equilibrium. Ecol Indic 83:515–523. https://doi.org/10.1016/j.ecolind.2017.03.022
Dunn R et al (2007) Reproductive phenologies in a diverse temperate ant fauna. Ecol Entomol 32(2):135–142. https://doi.org/10.1111/j.1365-2311.2006.00839.x
Fichaux M et al (2021) Environmental determinants of leaf litter ant community composition along an elevational gradient. Biotropica 53(1):97–109. https://doi.org/10.1111/btp.12849
Fontanilla A et al (2019) Taxonomic and functional ant diversity along tropical, subtropical, and subalpine elevational transects in Southwest China. Insects 10(5). https://doi.org/10.3390/insects10050128
Gaudard CA, Robertson MP, Bishop TR (2019) Low levels of intraspecific trait variation in a keystone invertebrate group. Oecologia 190(4):725–735. https://doi.org/10.1007/s00442-019-04426-9
Gibb H, Parr CL (2013) Does structural complexity determine the morphology of assemblages? An experimental test on three continents. PLoS One 8(5):e64005. https://doi.org/10.1371/journal.pone.0064005
Gibb H et al (2018) Habitat disturbance selects against both small and large species across varying climates. Ecography 41(7):1184–1193. https://doi.org/10.1111/ecog.03244
Gibb H et al (2022) Ecological strategies of (pl)ants: towards a world-wide worker economic spectrum for ants. Funct Ecol 37(1):13–25. https://doi.org/10.1111/1365-2435.14135
Gibb H et al (2024) Morphological strategies in ant communities along elevational gradients in three mountain ranges. Diversity 16(1):48. https://doi.org/10.3390/d16010048
Grêt-Regamey A, Brunner SH, Kienast F (2012) Mountain ecosystem services: who cares? Mt Res Dev 32(S1). https://doi.org/10.1659/MRD-JOURNAL-D-10-00115.S1
Grytnes JA, Heegaard E, Ihlen PG (2006) Species richness of vascular plants, bryophytes, and lichens along an altitudinal gradient in western Norway. Acta Oecol 29(3):241–246. https://doi.org/10.1016/j.actao.2005.10.007
He J et al (2023) A global assessment of Bergmann's rule in mammals and birds. Glob Chang Biol 29(18):5199–5210. https://doi.org/10.1111/gcb.16860
He X et al (2024) Intraspecific functional traits and stable isotope signatures of ground-dwelling ants across an elevational gradient. Soil Ecol Lett 6(4). https://doi.org/10.1007/s42832-024-0230-x
Hoffmann B, Andersen A (2023) Patterns of European ant communities reveal a functionally coherent Holarctic Fauna. Diversity-Basel 15(3). https://doi.org/10.3390/d15030341

Janzen DH (1967) Why mountain passes are higher in the tropics. Am Nat 101(919):233–249. https://doi.org/10.1086/282487

Jones KR et al (2016) Incorporating climate change into spatial conservation prioritisation: a review. Biol Conserv 194:121–130. https://doi.org/10.1016/j.biocon.2015.12.008

Kaspari M et al (2015) Thermal adaptation generates a diversity of thermal limits in a rainforest ant community. Glob Chang Biol 21(3):1092–1102. https://doi.org/10.1111/gcb.12750

Khaliq I et al (2023) The role of thermal tolerance in determining elevational distributions of four arthropod taxa in mountain ranges of southern Asia. J Anim Ecol 92(10):2052–2066. https://doi.org/10.1111/1365-2656.13996

Lach L, Parr C, Abbott K (2010) Ant ecology. Oxford University Press

Leahy L et al (2022) Arboreality drives heat tolerance while elevation drives cold tolerance in tropical rainforest ants. Ecology 103(1). https://doi.org/10.1002/ecy.3549

Liu C et al (2018) Mountain metacommunities: climate and spatial connectivity shape ant diversity in a complex landscape. Ecography 41(1):101–112. https://doi.org/10.1111/ecog.03067

Lytle A, Costa J, Warren R (2020) Invasion and high-elevation acclimation of the red imported fire ant, *Solenopsis invicta*, in the southern Blue Ridge Escarpment region of North America. PLoS One 15(5). https://doi.org/10.1371/journal.pone.0232264

Machac A et al (2011) Elevational gradients in phylogenetic structure of ant communities reveal the interplay of biotic and abiotic constraints on diversity. Ecography 34(3):364–371. https://doi.org/10.1111/j.1600-0587.2010.06629.x

Magurran AE et al (2010) Biological diversity: frontiers in measurement and assessment. Oxford University Press, Oxford, New York

Mammola S, Cardoso P (2020) Functional diversity metrics using kernel density n-dimensional hypervolumes. Methods Ecol Evol 11(8):986–995. https://doi.org/10.1111/2041-210X.13424

Marathe A et al (2021) Species and functional group composition of ant communities across an elevational gradient in the Eastern Himalaya. J Asia Pac Entomol 24(4):1244–1250. https://doi.org/10.1016/j.aspen.2021.08.009

Mason NWH et al (2005) Functional richness, functional evenness and functional divergence: the primary components of functional diversity. Oikos 111(1):112–118. https://doi.org/10.1111/j.0030-1299.2005.13886.x

McCain CM, Grytnes J-A (2010) Elevational gradients in species richness, in eLS. John Wiley & Sons, Ltd. https://doi.org/10.1002/9780470015902.a0022548

McGill BJ et al (2006) Rebuilding community ecology from functional traits. Trends Ecol Evol 21(4):178–185. https://doi.org/10.1016/j.tree.2006.02.002

Morelli TL et al (2020) Climate-change refugia: biodiversity in the slow lane. Front Ecol Environ 18(5):228–234. https://doi.org/10.1002/fee.2189

Moses J et al (2023) Nutrient use by tropical ant communities varies among three extensive elevational gradients: a cross-continental comparison. Glob Ecol Biogeogr 32(12):2212–2229. https://doi.org/10.1111/geb.13757

Muluvhahothe M et al (2021) Repeated surveying over 6 years reveals that fine-scale habitat variables are key to tropical mountain ant assemblage composition and functional diversity. Sci Rep 11(1). https://doi.org/10.1038/s41598-020-80077-8

Nascimento G, Câmara T, Arnan X (2024) Critical thermal maxima in neotropical ants at colony, population, and community levels. Bull Entomol Res:1–10. https://doi.org/10.1017/S0007485324000567

Nelson A et al (2019) Progressive sensitivity of trophic levels to warming underlies an elevational gradient in ant-aphid mutualism strength. Oikos 128(4):540–550. https://doi.org/10.1111/oik.05650

Neves K et al (2024) Macroecological patterns of functional and phylogenetic diversity vary between ground and arboreal assemblages in Neotropical savanna ants. Ecography 2024(8). https://doi.org/10.1111/ecog.07122

Nowrouzi S et al (2018) Is thermal limitation the primary driver of elevational distributions? Not for montane rainforest ants in the Australian Wet Tropics. Oecologia 188(2):333–342. https://doi.org/10.1007/s00442-018-4154-y

Nunes CA et al (2020) High temporal beta diversity in an ant metacommunity, with increasing temporal functional replacement along the elevational gradient. Front Ecol Evol 8:571439. Available at: https://www.frontiersin.org/articles/10.3389/fevo.2020.571439 (Accessed: 21 September 2023)

O'Donnell S, Baudier K (2023) Significant colony differences in thermal tolerances and decoupling of high and low critical temperatures in the army ant *Eciton burchellii* parvispinum. Ecol Entomol 48(5):622–626. https://doi.org/10.1111/een.13258

Orivel J et al (2018) Resource use and food preferences in understory ant communities along a complete elevational gradient in Papua New Guinea. Biotropica 50(4):641–648. https://doi.org/10.1111/btp.12539

Parr CL, Bishop TR (2022) The response of ants to climate change. Glob Chang Biol 28(10):3188–3205. https://doi.org/10.1111/gcb.16140

Parr CL et al (2017) GlobalAnts: a new database on the geography of ant traits (Hymenoptera: Formicidae), insect conservation and diversity. Edited by R. Didham and J. Orivel. Insect Conserv Divers 10(1):5–20. https://doi.org/10.1111/icad.12211

Pérez-Toledo GR et al (2022) Alpha and beta phylogenetic diversities jointly reveal ant community assembly mechanisms along a tropical elevational gradient. Sci Rep 12(1):7728. https://doi.org/10.1038/s41598-022-11739-y

Peters MK et al (2014) Variation in nutrient use in ant assemblages along an extensive elevational gradient on Mt Kilimanjaro. J Biogeogr 41(12):2245–2255. https://doi.org/10.1111/jbi.12384

Plowman N et al (2017) Network reorganization and breakdown of an ant - plant protection mutualism with elevation. Proc R Soc B Biol Sci 284(1850). https://doi.org/10.1098/rspb.2016.2564

Plowman N et al (2020) Nest microhabitats and tree size mediate shifts in ant community structure across elevation in tropical rainforest canopies. Ecography 43(3):431–442. https://doi.org/10.1111/ecog.04730

Purcell J et al (2016) Environmental influence on the phenotype of ant workers revealed by common garden experiment. Behav Ecol Sociobiol 70(3):357–367. https://doi.org/10.1007/s00265-015-2055-1

Rahbek C et al (2019) Humboldt's enigma: what causes global patterns of mountain biodiversity? Science 365(6458):1108–1113. https://doi.org/10.1126/science.aax0149

Reymond A et al (2013) Functional diversity decreases with temperature in high elevation ant fauna. Ecol Entomol 38(4):364–373. https://doi.org/10.1111/een.12027

Roeder DV, Paraskevopoulos AW, Roeder KA (2022) Thermal tolerance regulates foraging behaviour of ants. Ecol Entomol 47(3):331–338. https://doi.org/10.1111/een.13118

Roslin T et al (2017) Higher predation risk for insect prey at low latitudes and elevations. Science 356(6339):742–744. https://doi.org/10.1126/science.aaj1631

Ruppell O, Heinze J, Holldobler B (1998) Size-dimorphism in the queens of the North American ant *Leptothorax rugatulus* (Emery). Insect Soc 45(1):67–77. https://doi.org/10.1007/s000400050069

Sandoval-Arango S, Henao H, Montoya-Lerma J (2020) Divergence in Bergmann's clines: elevational variation and heritability of body size in a leaf-cutting ant. Insect Soc 67(3):355–366. https://doi.org/10.1007/s00040-020-00771-8

Schifani E et al (2024) Ant diversity along elevational gradients in the European Alps: insights for conservation under a changing climate. J Insect Conserv 28(3):401–413. https://doi.org/10.1007/s10841-023-00546-z

Seoane J et al (2021) Abiotic controls, but not species richness, shape niche overlap and breadth of ant assemblages along an elevational gradient in central Spain. Acta Oecologia 110. https://doi.org/10.1016/j.actao.2020.103695

Shik JZ et al (2019) Evidence for locally adaptive metabolic rates among ant populations along an elevational gradient. J Anim Ecol 88(8):1240–1249. https://doi.org/10.1111/1365-2656.13007

Silvestre M et al (2019) Responses of seed size, ant worker size, and seed removal rate to elevation in Mediterranean grasslands. Oecologia 189(3):781–793. https://doi.org/10.1007/s00442-019-04356-6

Silvestre M et al (2021) Diverging facets of grassland ant diversity along a Mediterranean elevational gradient. Ecol Entomol 46(6):1301–1314. https://doi.org/10.1111/een.13077

Smith M (2015) Ants, elevation, phylogenetic diversity and community structure. Ecosphere 6(11). https://doi.org/10.1890/ES14-00473.1

Smith M, Hallwachs W, Janzen D (2014) Diversity and phylogenetic community structure of ants along a Costa Rican elevational gradient. Ecography 37(8):720–731. https://doi.org/10.1111/j.1600-0587.2013.00631.x

Steinwandter M, Seeber J (2023) Ground-dwelling invertebrates of the high alpine: changes in diversity and community composition along elevation (1500–3000 m). Appl Soil Ecol 190:104988. https://doi.org/10.1016/j.apsoil.2023.104988

Streit RP, Bellwood DR (2022) To harness traits for ecology, let's abandon "functionality". Trends Ecol Evol [preprint]. https://doi.org/10.1016/j.tree.2022.11.009

Swenson NG (2014) Functional and phylogenetic ecology in R. Springer (Use R!), New York. https://doi.org/10.1007/978-1-4614-9542-0

Szewczyk T, McCain CM (2016) A systematic review of global drivers of ant elevational diversity. PLoS One 11(5):e0155404. https://doi.org/10.1371/journal.pone.0155404

Tonione MA et al (2020) Intraspecific variation in thermal acclimation and tolerance between populations of the winter ant, *Prenolepis imparis*. Ecol Evol 10(11):4749–4761. https://doi.org/10.1002/ece3.6229

Trimble S, Sagers C (2004) Differential host use in two highly specialized ant-plant associations: evidence from stable isotopes. Oecologia 138(1):74–82. https://doi.org/10.1007/s00442-003-1406-1

Villalta I et al (2020) Does social thermal regulation constrain individual thermal tolerance in an ant species? J Anim Ecol. Edited by J.Z. Shik 89(9):2063–2076. https://doi.org/10.1111/1365-2656.13268

Vinarski MV (2014) On the applicability of Bergmann's rule to ectotherms: the state of the art. Biol Bull Rev 4(3):232–242. https://doi.org/10.1134/S2079086414030098

Violle C et al (2007) Let the concept of trait be functional! Oikos 116(5):882–892. https://doi.org/10.1111/j.0030-1299.2007.15559.x

Violle C et al (2014) The emergence and promise of functional biogeography. Proc Natl Acad Sci 111(38):13690–13696. https://doi.org/10.1073/pnas.1415442111

Warren R, Chick L (2013) Upward ant distribution shift corresponds with minimum, not maximum, temperature tolerance. Glob Chang Biol 19(7):2082–2088. https://doi.org/10.1111/gcb.12169

Werenkraut V, Fergnani P, Ruggiero A (2015) Ants at the edge: a sharp forest-steppe boundary influences the taxonomic and functional organization of ant species assemblages along elevational gradients in northwestern Patagonia (Argentina). Biodivers Conserv 24(2):287–308. https://doi.org/10.1007/s10531-014-0808-1

Wymann von Dach S, Bracher CTP, Peralvo M, Perez K, Adler C (2018) Leaving no one in mountains behind: Localizing the SDGs for resilience of mountain people and ecosystems (Issue Brief on Sustainable Mountain Development). Centre for Development and Environment and Mountain Research Initiative, with Bern Open Publishing (BOP), Bern, Switzerland. info:eu-repo/semantics/report. Available at: https://boris.unibe.ch/120130/ (Accessed: 31 May 2022)

Yela N, Calcaterra L, Aranda-Rickert A (2020) Coping with temperature extremes: thermal tolerance and behavioral plasticity in desert leaf-cutting ants (Hymenoptera: Formicidae) across an altitudinal gradient. Myrmecology 30:139–150. https://doi.org/10.25849/myrmecol.news_030:139

Zhao M et al (2024) Taxonomic, functional and phylogenetic diversity of ant communities in response to elevational gradients in the Qinling Mountains. Biodivers Conserv 33:2833–2851. https://doi.org/10.1007/s10531-024-02887-w

Chapter 11
Effects of Increased Aridity on Ant Diversity in Arid Lands

Xavier Arnan, Alan N. Andersen, and Inara R. Leal

11.1 Introduction

Arid lands in the broad sense (hyper-arid, arid, semi-arid, and seasonally arid lands) are regions that suffer water scarcity, either persistently or over long periods of the year (Wickens 1998; Lioubimtseva 2004). This creates stressful conditions for the biota, limiting its diversity but at the same time being an important evolutionary force that can lead to species diversification and the emergence of unique adaptive strategies (Wickens 1998). Arid lands encompass more than 30% of the terrestrial ecosystems of the world (Wickens 1998; UNDEMG 2011; Huang et al. 2016) and support an important part of the world's diversity, with high rates of endemism (Wickens 1998; Silva et al. 2017).

Global climate change stands out as one of the most pressing environmental challenges confronting the Earth in the twenty-first century (IPCC 2022). Increasing aridity is one major consequence of ongoing climate change and is expected to cause widespread changes in key ecosystem attributes, functions and dynamics (Huang et al. 2016). This is especially the case in arid lands (Whitford 2002; IPCC 2022; Abel et al. 2023), where water is already limited. Many arid lands are situated

X. Arnan (✉)
Universidade de Pernambuco – Campus Garanhuns, Garanhuns, Pernambuco, Brazil
e-mail: xavier.arnan@upe.br

A. N. Andersen
Research Institute for the Environment and Livelihoods, Charles Darwin University, Brinkin, Northern Territory, NT, Australia
e-mail: alan.andersen@cdu.edu.au

I. R. Leal
Departamento de Botânica, Universidade Federal de Pernambuco, Recife, Pernambuco, Brazil
e-mail: inara.leal@ufpe.br

I. R. Leal et al. (eds.), *Ant Responses to Climate Change*, Entomology in Focus 10,
https://doi.org/10.1007/978-3-032-21127-9_11

in developing countries, where they are home to more than two billion people (35% of the global population) and encompass approximately 44% of the world's cultivated systems. The detrimental impacts of aridification on human welfare and global biodiversity are therefore potentially profound. There is therefore a pressing need to understand ecosystem responses in different arid regions of the world.

Ants are an ecologically dominant faunal group globally (Hölldobler and Wilson 1990). They are highly diverse (>20,000 species), are tremendously abundant, and occupy virtually every terrestrial habitat in the world (Schultheiss et al. 2022). The ecological dominance of ants is manifested through multiple interactions with animals, plants, fungi, microbes and the abiotic environment, making them key modulators of biodiversity (Parker and Kronauer 2021) and the provision of ecosystem services (Del Toro et al. 2012; Elizalde et al. 2020; see Chap. 15). Given the generally thermophilic nature of ants, their ecological importance is especially high in hot arid ecosystems (Bestelmeyer and Schooley 1999; Boulay et al. 2017). In addition to the ecological roles they play in other ecosystems, ants are dominant granivores in most arid lands (Brown et al. 1979; Morton and Davidson 1988).

In this chapter, we review macroecological patterns of ant diversity and composition in relation to rainfall in arid ecosystems, as a basis for predicting their responses to an increasingly arid future. We focus on the question: What increases in aridity are likely to result in taxonomically and functionally significant changes in ant communities? We are particularly interested in identifying any thresholds leading to stepwise change from one state to another. Based on regional climate-change predictions, we then forecast rates of ant diversity change in different arid regions of the world.

11.2 Arid Lands and Increasing Aridity

The level of aridity of a region is driven primarily by climate, but also depends on edaphic factors (that affect water infiltration and water retention capacity) and to a lesser extent vegetation (which also affects the water cycle). Here we distinguish between hyper-arid, arid, semi-arid, and seasonally arid lands (Fig. 11.1). *Hyper-arid lands* typically receive less than 100 mm mean annual precipitation (MAP) and are very sparsely vegetated. Such lands usually experience hot climates and include sandy deserts (e.g. the Sahara Desert, Arabian Desert), rocky deserts (e.g. Atacama Desert, Mojave Desert) and salt deserts (e.g. Dasht-e Kavir, Great Salt Desert). However, some hyper-arid lands experience cold climates, with long, harsh winters and short, dry summers. This includes the Gobi Desert, where MAP is only 40 mm. *Arid regions* typically receive 100–300 mm MAP. They have higher (although still relatively sparse) vegetation cover, sometimes including small trees. 'Hot' examples include the Sonoran Desert in North America, the Kalahari Desert in Africa and the Great Victoria Desert in Australia, and 'cold' examples include parts of Central Asia, the high-altitude regions of the Andes Mountains and portions of the Great Basin in the United States. *Semi-arid regions* typically receive between 300 and

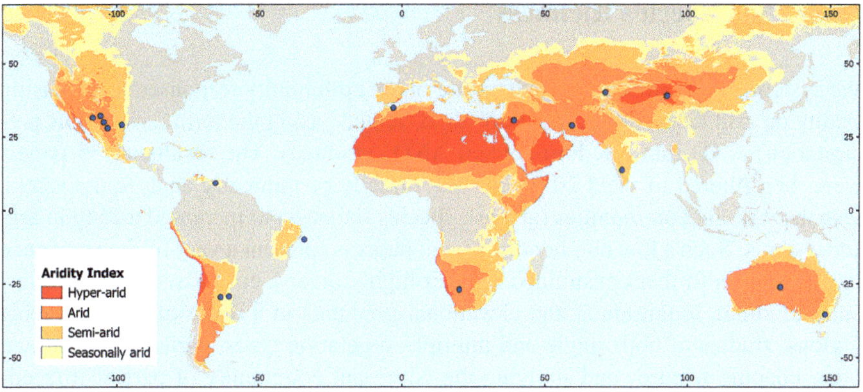

Fig. 11.1 Global distribution of the four broad classes (according to Zomer et al. 2022) of arid lands, showing the locations (blue dots) from the studies included in our review. Note that the point in the centre of Australia corresponds to a study that virtually sampled locations from the entire arid region of Australia

600 mm each year. For any given level of rainfall, the extent of ecosystem water stress is strongly influenced by rainfall seasonality. For example, regions receiving 600 mm MAP in the tropics are considered semi-arid, because almost all of the rain falls in a summer wet season. Semi-arid vegetation types include grasslands, shrublands and open woodlands. In addition to tropical savannas, examples include the Eurasian Steppe, the Great Plains of North America, the Argentine Pampa, the Mediterranean Basin and parts of California, Chile, South Africa and southern Australia. Other *seasonally arid lands* can receive up to 2000 mm MAP. These include many tropical savannas (including all of those in Brazil and far northern Australia), as well as scrublands and dry forests, such as Caatinga and Chaco in South America, and dry deciduous forests in southern Asia.

The United Nations Aridity Index (AI, Middleton and Thomas 1992; Zomer et al. 2022) most accurately measures the aridity levels of a site and is computed as the ratio of the mean annual precipitation over the mean annual potential evapotranspiration. Index values vary between 0 and > 0.65 and allow to the identification of five aridity classes: hyper-arid (AI < 0.03), arid (0.03 ≤ AI < 0.2), semi-arid (0.2 ≤ AI < 0.5), seasonally arid (0.5 ≤ AI < 0.65) and humid (AI ≥ 0.65) (Huang et al. 2016). However, few studies of arid ant communities use this index (Arnan et al. 2018; Eldridge et al. 2020; Wendt et al. 2020, 2021), and most use mean annual precipitation (MAP). However, MAP is strongly correlated with the AI (Arnan et al. 2018), and predictions of future climates are typically based on MAP rather than AI. We also consider relevant studies of ant communities along gradients of productivity (Bernstein 1971; Seely 1982; Paknia and Pfeiffer 2012) or habitat types (Bestelmeyer and Schooley 1999; Staubus et al. 2019; Begum et al. 2021), as these tend to be correlated with MAP (Wickens 1998).

11.3 Ant Species Richness

We systematically reviewed the literature of ant community responses to increasing aridity on arid lands, searching the Web of Science using the term (aridity OR precipitation) AND (ant OR Formicidae) AND (gradient). The resultant 135 papers were then filtered to yield 20 papers that directly or indirectly analyse the spatial responses of ant communities (at least, species richness) to increased aridity in arid ecosystems. Such a low number of relevant papers represent a very limited evidence base, which is further constrained by their highly diverse contexts, covering different latitudinal, longitudinal and elevational gradients in a range of biogeographic regions, studies of both single and multiple vegetation types, variation in soil and other edaphic features and studying the entire ant community or part of it (seed-harvester species).

Most macroecological studies find a negative correlation between ant species richness and the measure of aridity (mean annual precipitation, aridity index, productivity or habitat type) (see Table 11.1 and Fig. 11.2 for studies with MAP). This was the case in the Namib desert of southern Africa (15–95 mm MAP; Marsh 1986), savanna in South Africa's Kruger National Park (450–700 mm MAP; Parr et al. 2004), Venezuela's Araya Peninsula (244–800 mm MAP; Pérez-Sánchez et al. 2013), arid Mongolia (50–600 mm MAP; Bayartogtokh et al. 2014), the USA's Mojave Desert (100–700 mm MAP; Sanders et al. 2003) and central Iran (100–800 mm MAP; Hajian et al. 2024). Similarly, Davidson (1977) reported a positive association between the richness of harvester ants and MAP (ranging from 76 to 276 mm MAP) in North American deserts (Fig. 11.2). A negative relationship between ant species richness and other measures of aridity have been found in the Namib Desert (Seely 1982), North American deserts (Bernstein 1971), Iran (Paknia and Pfeiffer 2012), semi-arid southeastern Australia (Eldridge et al. 2020), the North American Sonoran Desert (Bestelmeyer and Schooley 1999) and the semi-arid Karnataka region of India (Begum et al. 2021). The results from these spatial gradients are supported by longitudinal studies of temporal change at individual sites, where periods of low rainfall are associated with lower ant diversity (e.g. Whitford 1978; Silva et al. 2014; Radnan 2018; Gibb et al. 2019). The negative responses can be attributed to the effects of decreasing primary productivity on habitat complexity and resource availability (e.g. Delsinne et al. 2010; Segev 2010; Pérez-Sánchez et al. 2013; Parr and Bishop 2022).

Other macroecological relationships between ant species richness and aridity have been reported. For example, in Israel the relationship is unimodal, with a peak at 300 mm MAP (Fig. 11.2; Segev 2010). This peak was attributed to the intersection of two biogeographic regions (the Mediterranean and Saharo-Arabian) and thus co-occurrence of species from different faunas. A number of studies have found no marcoecological relationship between ant species richness and aridity, such as in arid Argentina and Chile (43–332 mm MAP, Medel 1995), Paraguayan dry Chaco (460–887 mm MAP, Delsinne et al. 2010) and Brazilian Caatinga (658–1086 mm MAP, Arnan et al. 2018), and for harvester ants in arid Australia (119–568 mm MAP; Morton and Davison 1988) and Mongolia's Gobi Desert (84–197 mm MAP;

Table 11.1 Descriptive summary of the available studies that analyse the effects of spatial increasing aridity on ant species richness in arid lands

Study	Region	Habitat	Proxy of aridity gradient	Gradient range	Spatial gradient	Ant community	Species richness range	Response	Observations	Data for figure
Marsh (1986)	Namib (Africa)	Desert	MAP	15–95 mm	~100 km (coast to inland)	Seed harvesters	6–23	Positive		Yes
Segev (2010)	Israel (North Africa)	Desert	MAP	30–600 mm	~180 km (latitudinal)	All	5–15	Unimodal	Decrease from 0 to 300 mm, increase from 300 to 600 mm	Yes
Davidson (1977)	California, Arizona, New Mexico (North America)	Deserts	MAP	76–276 mm	~1500 km (longitudinal)	Seed harvesters	2–8	Positive		Yes
Morton and Davidson (1938)	Australia	Shrublands and grasslands	MAP	119–569 mm	~3606 km (latitudinal and longitudinal)	Seed harvesters	6–13	Neutral	This study also uses the North America data from Davidson (1977)	Yes

(continued)

Table 11.1 (continued)

Study	Region	Habitat	Proxy of aridity gradient	Gradient range	Spatial gradient	Ant community	Species richness range	Response	Observations	Data for figure
Medel (1995)	Argentina and Chile (South America)	Deserts and arid zones	MAP	43–332 mm	~950 km (latitudinal and longitudinal)	Seed harvesters	7.62 ± 1.76*	Neutral	This study also uses the North America data from Davidson (1977) and the Australian data from Morton and Davidson (1988)	No
Hajian et al. (2024)	Central Iran (Asia)	Deserts and steppes	MAP	100–800 mm	~600 km (altitudinal)	All	2–11	Positive		No
Delsinne et al. (2010)	Paraguay (South America)	Dry Chaco	MAP	460–887 mm	~400 km (latitudinal and longitudinal)	All	36–74	Neutral		Yes
Pfeiffer et al. (2003)	Mongolia (Asia)	Steppe and desert	MAP	84–197 mm	~800 km (latitudinal)	Seed harvester	3–7	Neutral		Yes
Sanders et al. (2003)	Spring Mountains, Nevada, USA (North America)	Desert and mountain shrublands and woodlands	MAP	100–700 mm	NA (altitudinal)	All	6–14	Positive		No

Pérez-Sánchez et al. (2013)	Arab Peninsula, Venezuela (South America)	Xerophilous shrublands, mangroves, grasslands	MAP	244–800 mm	~35 km (longitudinal)	All	7–30	Positive		No
Bayartogtokh et al. (2014)	Mongolia (Asia)	Taiga, steppe, semi-desert, desert	MAP	50–600 mm	~ 2370 km (latitudinal and longitudinal)	All	3–19	Positive		No
Parr et al. (2004)	South Africa (Africa)	Savanna	MAP	450–700 mm	NA (latitudinal)	All	10–40	Positive		Yes
Arnan et al. (2018)	Catimbau NP, Brazil (South America)	Caatinga	MAP	515–994 mm	~ 30 km (NA)	All	15–26	Neutral		Yes
Eldridge et al. (2020)	Eastern Australia	Broad vegetation communities	AI	0.26–0.39	~ 900 km (longitudinal)	All	NA	Negative		No
Wendt et al. (2020)	Portugal (Western Europe)	Mediterranean drylands (man-shaped agroforestry ecosystem)	AI	0.42–0.54	~150 km (NA)	All	NA	Neutral		No
Bernstein (1971)	North America	Desert	Productivity	NA	NA	Seed harvester	NA	Positive		No
Seely (1982)	Namib (Africa)	Desert	Productivity	NA	NA	All	NA	Positive		No

(continued)

Table 11.1 (continued)

Study	Region	Habitat	Proxy of aridity gradient	Gradient range	Spatial gradient	Ant community	Species richness range	Response	Observations	Data for figure
Paknia and Pfeiffer (2012)	Iran (Asia)	Desert	Productivity (NDVI)	0.05–019	~550 km (latitudinal and longitudinal)	All	0–16	Positive (with asymptote)		No
Bestelmeyer and Schooley (1999)	Sonoran Desert (North America)	Desert	Habitat types	(tree/open), 346 mm	9.7 ha	All	NA	Positive (Higher S than close more arid deserts)		No
Begum et al. (2021)	Karnataka, India (Asia)	Semi-arid	Habitat types	Human, vegetative, agricultural, and dry land	NA	All	5–33	Positive (less species in the most arid habitats)		No

Abbreviations: *AI* aridity index, *MAP* mean annual precipitation, *NA* not available

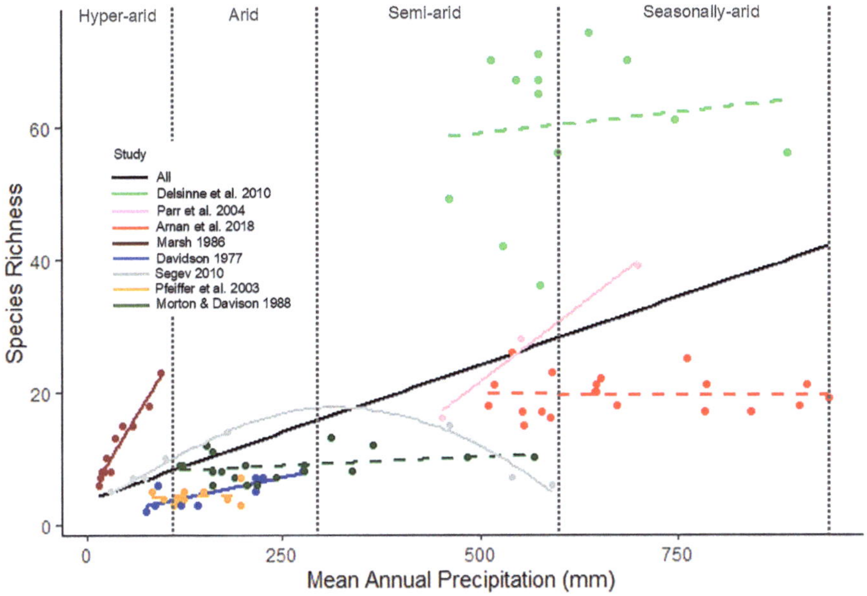

Fig. 11.2 Relationship between ant species richness and mean annual precipitation (MAP) for those communities with available data. The relationship is shown for the overall data from all studies combined (black line) and for each study separately (coloured lines). For detailed characteristics of each study, please see Table 11.1. Dashed lines indicate non-significant relationship. Dotted lines highlight the boundaries between arid land categories

Pfeiffer et al. 2003) (Fig. 11.2). Despite these exceptions, when results are combined from all studies along precipitation gradients, the relationship between species richness and decreasing rainfall is negative ($S = 3.76 + 0.04 \times \text{MAP}$; $t = 6.8$; $p < 0.0001$) (and a quadratic function was not better than the linear one: AIC linear function = 791.8 vs AIC quadratic function = 791.5) (Fig. 11.2).

Our findings do not support previous conclusions that arid ant community richness decreases with increasing aridity only when considering long (>400 km) spatial gradients within the same biome and whenever extreme arid environments (<200–300 mm) are considered (Delsinne et al. 2010; Pérez-Sánchez et al. 2013) (Table 11.1). However, there appears to be consistent variation among continents: there is usually a negative relationship between increasing aridity and ant species richness in North America (Bernstein 1971; Davidson 1977; Bestelmeyer and Schooley 1999; Sanders et al. 2003), Africa (Seely 1982; Marsh 1986; Parr et al. 2004) and Asia (Segev 2010; Bayartogtokh et al. 2014; Begum et al. 2021; Hajian et al. 2024, but see Pfeiffer et al. 2003; Paknia and Pfeiffer 2012), but typically not in South America (Medel 1995; Delsinne et al. 2010; Arnan et al. 2018, but see Pérez-Sánchez et al. 2013), Australia (Morton and Davidson 1988, but see Eldridge et al. 2020) and Europe (Wendt et al. 2020) (Table 11.1). One factor contributing to this is intercontinental differences in evolutionary history, as is the case for the maintenance of diversity with increasing aridity in the arid-adapted ant fauna of Australia's tropical savannas (Andersen and Vasconcelos 2022).

11.4 Ant Community Composition

All papers that analyse variation in ant species composition with increasing aridity find marked changes, regardless of its effect on species richness. Taxonomic changes are often at the genus level. For example, *Formica* is dominant in Mongolian steppe but absent from semi-desert, and *Messor* and *Cataglyphis* dominate the desert zone (Pfeiffer et al. 2003). Similarly, along a MAP gradient from 244 to 800 mm in Venezuela, *Pheidole* and *Solenopsis* dominated the more mesic habitat and were absent from the most xeric regions, where *Dorymyrmex*, *Camponotus* and *Crematogaster* were predominant (Pérez-Sánchez et al. 2013). Each biogeographic realm has a distinctive assemblage of specialist arid-adapted genera, comprising two main functional types. The first consists of highly thermophilic, fast-moving taxa that are often the only ants foraging under extremely high temperatures. These include the formicines *Melophorus* in Australia (Christian and Morton 1992); *Cataglyphis* in northern Africa, the Palearctic and Asia (Agosti 1990); and *Myrmecocystus* in North America (Snelling 1976), along with the myrmicine *Ocymyrmex* in southern Africa (Marsh 1985). The second functional group comprises specialist granivores, such as *Pogonomyrmex*, *Messor* and *Veromessor* in the New World (Brown et al. 1979). Seeds are a particularly valuable resource in desert ecosystems, and granivorous ants can store seeds to survive long dry periods (Davidson 1977; Marsh 1986). An additional functional type is represented by honeypot ants, such as species of *Myrmecocystus* in North America and *Camponotus inflatus* in Australia, where a specialist replete caste stores nectar that can be accessed during droughts.

Species replacement along aridity gradients can also lead to substantial morphological change. In particular, there is often increased body size, leg length and polymorphism in ants of drier areas (Davidson 1977; Marsh 1986; Pérez Sánchez et al. 2013). Larger ants dominate the most arid areas due to better heat tolerance and foraging efficiency (Gibb et al. 2018; Wendt et al. 2020)—they can forage longer in hot conditions and gather resources more efficiently in resource-limited arid habitats (Davidson 1977; Retana et al. 2015). Polymorphism, such as that occurs in *Messor*, *Cataglyphis*, *Melophorus*, and *Veromessor*, enhances adaptability in arid environments (La Richelière et al. 2022). Longer legs help ants navigate high soil temperatures and improve foraging efficiency, as seen in *Cataglyphis* (Boulay et al. 2017), *Ocymyrmex* (Marsh 1985), *Melophorus* (Christian and Morton 1992) and *Dorymyrmex* (Silva et al. 2019).

More generally, species turnover along aridity gradients can result in shifts in functional and phylogenetic diversity. For example, although increased aridity did not affect ant species richness in Brazilian Caatinga, it significantly reduced both functional and phylogenetic diversity (Arnan et al. 2018). Conversely, despite shifts in ant functional traits, Wendt et al. (2020) observed no relationship between functional diversity and aridity in southeast Portugal. Further studies by Wendt et al. (2021) and Ribeiro-Neto et al. (2023) indicated a strong homogenization of ant communities due to increased aridity, leading to high similarity in species, ecological roles and evolutionary lineages. This homogenization could reduce response capacity to future disturbances, challenging ecosystem resilience and sustainability. However, studies by Arnan et al. (2006) and Parr et al. (2004) noted increased resilience of ant communities after forest fires in arid Mediterranean and savanna regions, respectively. These findings may suggest that while ant communities in the most arid areas may exhibit reduced richness, they could also demonstrate heightened resistance and resilience to disturbances, maintaining ecosystem stability and contributing to overall resilience in arid regions.

11.5 Ant Diversity Change in an Increasingly Arid Future

There is substantial biogeographical variation in forecast changes in aridity under future climates. Recent studies (UNEP 2019; Wang et al. 2021) indicate an overall expansion of arid landscapes with global temperature increases of 2 and 4 °C, but this expansion concentrated in tropical and mid-latitudinal zones (Huang et al. 2016). Hyper-arid expansions are predicted mainly for northern Africa and the Middle East, while significant arid expansions are forecast for Australia, southern Africa and Argentina. Semi-arid and dry-subhumid expansions are expected in southwestern North America, parts of South America, southern Africa, and Central Asia. The observed expansions of hyper-arid and arid lands in northern Africa, alongside contractions in the southern Sahara, suggest a poleward migration of northern African drylands.

Given the biogeographic variation in predicted changes in aridity, along with the different macroecological relationships between aridity and ant diversity, the impacts of an increasingly arid global climate on ant diversity can be expected to vary markedly among biogeographic realms. In Australia, there are no hyper-arid regions, and this is predicted to remain the case under climate change, despite significant projected expansions of arid regions (Wang et al. 2021). Australia has by far the highest diversity of arid-adapted elements (Andersen 2016), and there is likely to be substantial geographic expansion of these. In contrast, Africa and the Middle East are expected to experience substantial expansions of hyper-arid regions (Wang et al. 2021), where there are clear negative macroecological relationships between aridity and ant diversity. The arid regions of Argentina also appear to be especially vulnerable, as a significant expansion of arid areas is predicted and South America has very few specialist arid-adapted ant taxa (Andersen and Vasconcelos 2022). Expansions of hyper-arid, semi-arid and seasonal arid regions are predicted for North America, where, despite its high diversity of arid-adapted elements, there is a negative macroecological relationship between aridity and ant diversity. The arid regions of Europe are not expected to undergo significant changes in aridity (Wang et al. 2021). The overlap of regions where ant communities are most negatively affected by increased aridity and regions experiencing the most significant aridity increases suggests that Africa and central Asia are where ant communities may be most affected in an increasingly arid future.

11.6 Final Considerations

Macroecological relationships between biodiversity and increasing aridity are not necessarily reliable predictors of climate-change impacts because the former operate over broad gradients rather than necessarily at the scale of predicted change. A critical issue in understanding climate change impacts is identifying if there are thresholds along aridity gradients where abrupt shifts in ant diversity and composition occur, as such climate-sensitive zones can be expected to be the most at risk (see Chap. 8). In relation to this, a key question is whether the four aridity classes (hyper-arid, arid, semi-arid, and seasonally arid lands) support discrete ant faunas or variation among them is continuous. In all the studies we have reviewed, declines in species richness with increased aridity appear to be continuous rather than involving thresholds of rapid change. However, there is evidence of discontinuous change in species and functional composition.

In northern Australia, a primary discontinuity in species composition occurs at the transition between the central arid zone and the seasonal tropics, at about 500 mm MAP (see Chap. 8). There is also a clear discontinuity within the seasonal tropics at about 1000 mm MAP, marking a transition from semi-arid to mesic. In the Davidson (1977) study of seed-harvesting ants in deserts of the southwestern USA, there was a transition threshold at 100–150 mm MAP, from small-bodied (*Pheidole* and *Solenopsis* spp.) to large body sized or highly polymorphic species (*Pogonomyrmex*, *Novomessor* and *Veromessor* spp.) in more xeric regions. This leads to communities increasingly dominated by larger ants with longer legs and a higher degree of polymorphism. There is a priority need for more studies on compositional responses to increasing aridity in arid regions worldwide, particularly in North Africa and Central Asia, where studies are scarce and climate change predictions are the most pessimistic.

As is the case in other taxonomic groups (Sánchez-Bermejo et al. 2022), different regional pools respond to aridity in different ways. Hence, comprehending the future responses of ant communities to the gradual increase in aridity spurred by climate change necessitates unravelling the interplay between the magnitude of the change in aridity, the traits of species in the regional pool and the assembly processes influenced by aridity.

Acknowledgements We are very grateful to Lluís Comas for his valuable assistance in preparing Fig. 11.1. XA thanks the Brazilian National Council for Scientific and Technological Development (CNPq) for his productivity grant (PQ-1D, Process 313589/2023-2).

References

Abel C, Abdi AM, Tagesson T, Horion S, Fensholt R (2023) Contrasting ecosystem vegetation response in global drylands under drying and wetting conditions. Glob Change Biol 29:3954–3969. https://doi.org/10.1111/gcb.16745

Agosti D (1990) Review and reclassification of Cataglyphis (Hymenoptera, Formicidae). J Nat Hist 24:1457–1505

Andersen AN (2016) Ant megadiversity and its origins in arid Australia. Austral Entomol 55:132–137. https://doi.org/10.1111/aen.12203

Andersen AN, Vasconcelos HL (2022) Historical biogeography shapes functional ecology: inter-continental contrasts in responses of savanna ant communities to stress and disturbance. J Biogeogr 49:590–599. https://doi.org/10.1111/jbi.14343

Arnan X, Rodrigo A, Retana J (2006) Post-fire recovery of Mediterranean ground ant communities follows vegetation and dryness gradients. J Biogeogr 33:1246–1258

Arnan X, Arcoverde GB, Pie MR, Ribeiro-Neto JD, Leal IR (2018) Increased anthropogenic disturbance and aridity reduce phylogenetic and functional diversity of ant communities in Caatinga dry forest. Sci Total Environ 631-632:429–438. https://doi.org/10.1016/j.scitotenv.2018.03.037

Bayartogtokh B, Aibek U, Yamane S, Pfeiffer M (2014) Diversity and biogeography of ants in Mongolia (Hymenoptera: Formicidae). Asian Myrmecol 6:63–82. https://doi.org/10.20362/am.006005

Begum R, Majagi SH, Vijaykumar K (2021) Ants species richness and diversity in relation to different ecological habitat in selected localities of the semi-arid region of Karnataka, India. Environ Monit Assess 193:145. https://doi.org/10.1007/s10661-021-08941-2

Bernstein RA (1971) The ecology of ants in the Mojave Desert: their interspecific relationships, resource utilization and diversity. PhD Thesis, University of California, Los Angeles, 129 pp

Bestelmeyer BT, Schooley RL (1999) The ants of the southern Sonoran Desert: community structure and the roles of trees. Biodivers Conserv 8:643–657

Boulay R, Aron S, Cerdá X, Doums C, Graham P, Hefetz A, Monnin T (2017) Social life in arid environments: the case study of Cataglyphis ants. Annu Rev Entomol 62:305–321. https://doi.org/10.1146/annurev-ento-031616-034941

Brown JH, Reichman O, Davidson DW (1979) Granivory in desert ecosystems. Annu Rev Ecol Syst 10:201–227

Christian KA, Morton SR (1992) Extreme thermophilia in a central Australian ant, Melophorus bagoti. Physiol Zool 65:885–905

Davidson DW (1977) Species diversity and community organization in desert seed-eating ants. Ecology 58:711–724. https://doi.org/10.2307/1936208

Del Toro I, Ribbons RR, Pelini SL (2012) The little things that run the world revisited: a review of ant-mediated ecosystem services and disservices (Hymenoptera: Formicidae). Myrmecol News 17:133–146

Delsinne T, Roisin Y, Herbauts J, Leponce M (2010) Ant diversity along a wide rainfall gradient in the Paraguayan dry Chaco. J Arid Environ 74:1149–1155. https://doi.org/10.1016/j.jaridenv.2010.03.014

Eldridge DJ, Oliver I, Val J, Travers SK, Delgado-Baquerizo M (2020) Grazing and aridity have contrasting effects on the functional and taxonomic diversity of ants. Basic Appl Ecol 48:73–82. https://doi.org/10.1016/j.baae.2020.07.003

Elizalde L, Arbetman M, Arnan X, Eggleton P, Leal IR, Lescano MN, Saez A, Werenkraut V, Pirk GI (2020) The ecosystem services provided by social insects: traits, management tools and knowledge gaps. Biol Rev 95:1418–1441. https://doi.org/10.1111/brv.12616

Gibb H, Sanders NJ, Dunn RR, Arnan X, Vasconcelos HL, Donoso DA, Andersen AN, Silva RR, Bishop TR, Gómez C, Grossman BF, Myusah K, Luke SH, Pacheco R, Pearce-Duvet J, Retana J, Tista M, Parr K (2018) Habitat disturbance selects against both small and large body size across varying climates. Ecography 41:1184–1193. https://doi.org/10.1111/ecog.03244

Gibb H, Grossman BF, Dickman CR, Decker O, Wardle GM (2019) Long-term responses of desert ant assemblages to climate. J Anim Ecol 88:1549–1563. https://doi.org/10.1111/1365-2656.13052

Hajian M, Sadeghi S, Eslami Barzoki Z, Moradmand M, Gholamhosseini A, Ebrahimi M (2024) Ant diversity and species assemblages along an elevational gradient in the arid area of Central Iran. J Insect Biodivers Syst 10:143–159

Hölldobler B, Wilson EO (1990) The ants. Springer-Verlag, Berlin-Heidelberg, Germany

Huang J, Yu H, Guan X, Wang G, Guo R (2016) Accelerated dryland expansion under climate change. Nat Clim Chang 6:166–171. https://doi.org/10.1038/nclimate2837

Intergovernmental Panel on Climate Change (2022) Technical summary. In: Climate change and land: IPCC special report on climate change, desertification, land degradation, sustainable land management, food security, and greenhouse gas fluxes in terrestrial ecosystems. Cambridge University Press, pp 37–74. https://doi.org/10.1017/9781009157988.002

La Richelière F, Muñoz G, Guénard B, Dunn RR, Economo EP, Powell S, Sanders NJ, Weiser MD, Abouheif E, Lessard JP (2022) Warm and arid regions of the world are hotspots of superorganism complexity. Proc R Soc B 289:20211899. https://doi.org/10.1098/rspb.2021.1899

Lioubimtseva E (2004) Climate change in arid environments: revisiting the past to understand the future. Peer Reviewed Articles Paper 1. http://scholarworks.gvsu.edu/gpy_articles/1

Marsh AC (1985) Thermal responses and temperature tolerance in a diurnal desert ant, Ocymyrmex barbiger. Physiol Zool 58:629–636

Marsh AC (1986) Ant species richness along a climatic gradient in the Namib Desert. J Arid Environ 11:235–241

Medel RG (1995) Convergence and historical effects in harvester ant assemblages of Australia, North America, and South America. Biol J Linn Soc 55:29–44

Middleton N, Thomas DSG (1992) World atlas of desertification: United Nations environmental programme. Arnold

Morton SR, Davidson DW (1988) Comparative structure of harvester ant communities in arid Australia and North America. Ecol Monogr 58:19–38

Paknia O, Pfeiffer M (2012) Productivity alone does not explain species richness of ants—an example from Central Persian deserts. J Arid Environ 85:86–92. https://doi.org/10.1016/j.jaridenv.2012.06.008

Parker J, Kronauer DJC (2021) How ants shape biodiversity. Curr Biol 31:R1141–R1224

Parr CL, Bishop TR (2022) The response of ants to climate change. Glob Chang Biol 28:3188–3205. https://doi.org/10.1111/gcb.16140

Parr CL, Robertson HG, Biggs HC, Chown SL (2004) Response of African savanna ants to long-term fire regimes. J Appl Ecol 41:630–642. https://doi.org/10.1111/j.0021-8901.2004.00920.x

Pérez-Sánchez AJ, Lattke JE, Viloria AL (2013) Patterns of ant (Hymenoptera: Formicidae) richness and relative abundance along an aridity gradient in Western Venezuela. Neotrop Entomol 42:128–136. https://doi.org/10.1007/s13744-012-0096-y

Pfeiffer M, Chimedregzen L, Ulykpan K (2003) Community organization and species richness of ants (Hymenoptera/Formicidae) in Mongolia along an ecological gradient from steppe to Gobi desert. J Biogeogr 30:1921–1935. https://doi.org/10.1046/j.0305-0270.2003.00977.x

Radnan G (2018) The effect of changes in landscape and habitat complexity resulting from shrub encroachment and climatic factors (aridity and ENSO) on semi-arid ant communities. PhD. University of New South Wales, Sydney. https://doi.org/10.26190/unsworks/20704

Retana J, Arnan X, Cerdá X (2015) A multidimensional functional trait analysis of resource exploitation in European ants. Ecology 96:2781–2793

Ribeiro-Neto JD, Oliveira FMP, Arcoverde G, Tabarelli M, Leal IR, Arnan X (2023) Aridity and chronic anthropogenic disturbances cause a taxonomic, functional and phylogenetic homogenization of ant communities in a Caatinga dry forest. Biol Conserv 284:110151. https://doi.org/10.1016/j.biocon.2023.110151

Sánchez-Bermejo PC, de Castro-Arrazola I, Cuesta E, Davis ALV, Moreno CE, Sánchez-Piñero F, Hortal J (2022) Aridity drives the loss of dung beetle taxonomic and functional diversity in three contrasting deserts. J Biogeogr 49:2243–2255. https://doi.org/10.1111/jbi.14506

Sanders NJ, Moss J, Wagner D (2003) Patterns of ant species richness along elevational gradients in an arid ecosystem. Glob Ecol Biogeogr 12:93–102. https://doi.org/10.1046/j.1466-822X.2003.00324.x

Schultheiss P, Nooten SS, Wang R, Wong MKL, Brassard F, Guénard B (2022) The abundance, biomass, and distribution of ants on earth. PNAS 119:e2201550119. https://doi.org/10.1073/pnas.2201550119

Seely MK (1982) Species diversity and Namib dune environmental patterns. The Zoological Society of Southern Africa, Occasional Bulletin 2:37

Segev U (2010) Regional patterns of ant-species richness in an arid region: the importance of climate and biogeography. J Arid Environ 74:646–652. https://doi.org/10.1016/j.jaridenv.2009.11.001

Silva EM, Medina AM, Nascimento IC, Lopes PP, Carvalho KS, Santos GMM (2014) Does ant richness and composition respond to phytophysiognomical complexity and seasonality in xeric environments? Sociobiology 61:155–163. https://doi.org/10.13102/sociobiology.v61i2.155-163

Silva JMC, Leal IR, Tabarelli M (2017) Caatinga: the largest tropical dry forest region in South America. Springer International Publishing, Switzerland

Silva ILH, Leal IR, Ribeiro-Neto JD, Arnan X (2019) Spatiotemporal responses of ant communities across a disturbance gradient: the role of behavioral traits. Insect Soc 66:623–635. https://doi.org/10.1007/s00040-019-00717-9

Snelling RR (1976) A revision of the honey ants, genus Myrmecocystus (Hymenoptera: Formicidae). Nat Hist Mus Los Angel Cty Sci Bull 24:1–163

Staubus WJ, Bird S, Meadors S, Meyer WM III (2019) Distributions of invasive arthropods across heterogeneous urban landscapes in Southern California: aridity as a key component of ecological resistance. Insects 10:29. https://doi.org/10.3390/insects10010029

United Nations Department of Environment Management Group (2011) Global Drylands: a UN system-wide response

United Nations Environment Programme (2019) World Atlas of Desertification. Edward Arnold

Wang X, Jiang D, Lang X (2021) Future changes in Aridity Index at two and four degrees of global warming above preindustrial levels. Int J Climatol 41:278–294. https://doi.org/10.1002/joc.6620

Wendt CF, Nunes A, Verble R, Santini G, Boieiro M, Branquinho C (2020) Using a space-for-time approach to select the best biodiversity-based indicators to assess the effects of aridity on Mediterranean drylands. Ecol Indic 113:106250

Wendt CF, Ceia-Hasse A, Nunes A, Verble R, Santiniz G, Boieiros M, Branquinho C (2021) Local environmental variables are key drivers of ant taxonomic and functional beta-diversity in a Mediterranean dryland. Sci Rep 11:2292. https://doi.org/10.1038/s41598-021-82059-w

Whitford WG (1978) Structure and seasonal activity of Chihuahuan Desert and communities. Insect Soc 25:79–88

Whitford W (2002) Ecology of desert systems. Academic Press, London. 343 pp

Wickens GE (1998) Ecophysiology of economic plants in arid and semi-arid lands. Springer-Verlag Berlin Heidelberg GmbH., 343 pp, New York. https://doi.org/10.1007/978-3-662-03700-3

Zomer RJ, Xu J, Trabucco A (2022) Version 3 of the global aridity index and potential evapotranspiration database. Sci Data 9:409. https://doi.org/10.1038/s41597-022-01493-1

Chapter 12
Ant Responses to Climate Change: The Importance of the Vertical Dimension

Lily Leahy

12.1 Introduction

From deep underground to the towering heights of treetops, ants inhabit every vertical stratum in terrestrial ecosystems. The vertical dimension represents a strong climatic gradient moving from the cool, moist, and stable environment of subterranean and leaf-litter habitats to hot, dry, and variable conditions in the canopy. Across this vertical dimension ants occupy distinct niches with unique food and nesting opportunities. The vertical stratification of ants has long fascinated community ecologists who seek to understand how these specific habitat templets influence the assemblage and structure of ant communities. Yet there is a strong possibility that ants did not always live in trees. One hypothesis suggests that ants came "Out of the Ground", originally evolving to inhabit the soil and subsequently moving upwards in vertical space towards the canopy, occupying new niches along the way (Moreau et al. 2006; Lucky et al. 2013).

How ants have adapted to the opportunities and challenges of the vertical realm throughout their evolutionary history is an important story in our current time. While ants have faced numerous large-scale changes in global climate over their ~150 Mya history, these climate change events do not compare to the speed of anthropogenic caused climate change of today (Moreau et al. 2006; Loarie et al. 2009). What clues lie hidden in the physiological and behavioural adaptations of ants to their distinct vertical strata that could reveal which species will be resilient, and which most vulnerable? How does the vertical dimension interact with other temporal and spatial gradients of climate such as daily cycles, seasons, elevation,

L. Leahy (✉)
Department of Ecological, Plant, and Animal Sciences, La Trobe University, Melbourne, Victoria, Australia
e-mail: lilyleahy4@gmail.com

I. R. Leal et al. (eds.), *Ant Responses to Climate Change*, Entomology in Focus 10,
https://doi.org/10.1007/978-3-032-21127-9_12

and latitude to shape ant responses to climate change, and how will future communities differ from what we see today?

In this chapter, I address these questions by drawing on a framework of climate change vulnerability that focuses on exposure and sensitivity to climate and ability to adapt to change (Williams et al. 2008; Huey et al. 2012). I begin by characterising the vertical environment and patterns of vertical stratification from subterranean to canopy habitats. I then delve into the climate sensitivity and exposure that will shape how ant species respond to climate change across the vertical dimension. I place the vertical dimension into a broader climatic context, looking at how verticality is nested within both the local temporal scale of daily climatic fluctuations and the larger-scale climatic gradients of elevation and latitude. Finally, I explore potential climate change trajectories for ant populations, species, and communities contingent on their vertical ecology.

12.2 Ants in the Vertical Dimension

12.2.1 The Vertical Environment

Above-ground vegetation structure is highly diverse, ranging from grass blades to towering 100 m tall *Sequoia* forests. The layering of these structures creates stratification of niche space resulting in heightened taxonomic and functional diversity through the process of niche packing (MacArthur and Wilson 1967; Feng et al. 2020). Along the vertical gradient, novel food and nesting resources become available to ants and climate conditions change dramatically (Davidson et al. 2003; Pincebourde et al. 2007; Scheffers et al. 2017a; Leahy et al. 2021).

In forests, tree canopies are hotter, drier, and more variable than on the forest floor below (Fig. 12.1). The vertical climate gradient can be orders of magnitude steeper than climate gradients of elevation or latitude but over mere metres of distance (Scheffers et al. 2013). From subterranean to canopy habitats, there is also an increase in sunlight interception and wind speed, making canopies more variable and drier than lower strata (Scheffers et al. 2014; Bujan et al. 2016; Wong and Guénard 2017; De Frenne et al. 2019; Law et al. 2020). Air temperature variance can average four times higher in the canopy than the forest floor (Leahy et al. 2021), while vapour pressure deficits (drying power of the air) can be nine times higher in the canopy than in the understorey (Bujan et al. 2016). In contrast, subterranean environments are characterised by low light, limited spatial availability, and extremely stable temperatures (Wong and Guénard 2017). Soil temperatures tend to be warmer than the surface in cold and dry environments (+3.6 ± 2.3 °C, global average) and cooler in warm and humid environments (−0.7 ± 2.3 °C, global average) (Lembrechts et al. 2022).

At night, the vertical climate gradient becomes more homogenised and less climatically variable (Fig. 12.1, (Leahy 2022)). The shift to nighttime results in a

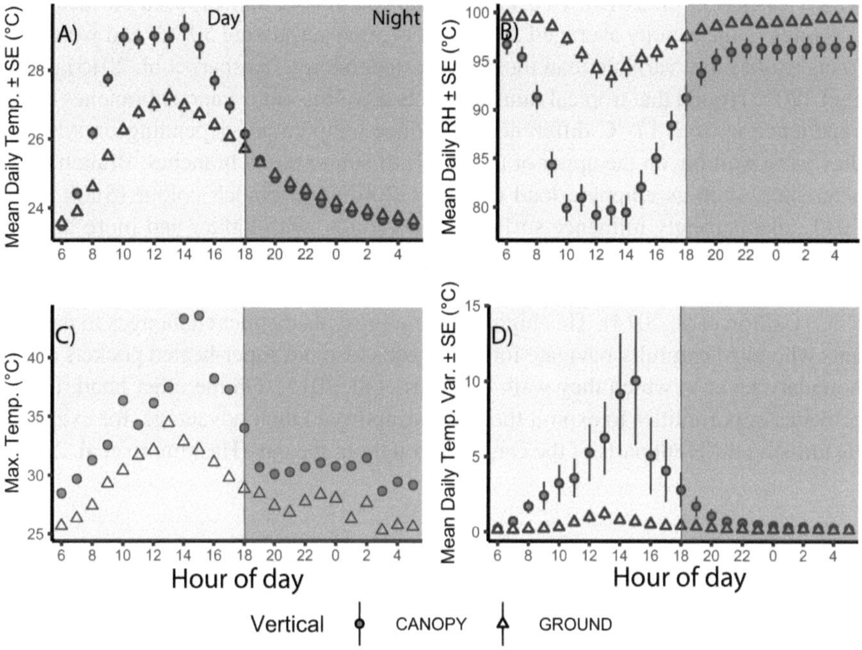

Fig. 12.1 Temperature and relative humidity (RH) for the summer wet season month of January from 2 years (2020–2021) of data recorded by two HOBO Pro v2 (U23–001) climate data loggers set at ground (~0.5 m) and canopy (~20 m) positions in tropical rainforest at 100 m asl at Mossman Gorge (Daintree National Park), Queensland, Australia. Daytime hours are shaded in light grey, and nighttime hours are shaded in dark grey. (**a**, **b**, and **d**) show means and standard errors for daily values during January; (**c**) shows maximum values for January. The canopy is hotter (**a** and **c**), drier (**b**), and more variable (**d**) than the ground. At night the vertical gradient becomes more homogenised: ground and canopy habitats are more similar in temperature and humidity compared to the daytime. (Modified from Leahy et al. 2022)

subtle inversion of the daytime gradient, with canopy temperatures often slightly cooler than the understorey or leaf litter layers (Freiberg 1997). This is demonstrated for mean daily temperatures (Fig. 12.1a), but nighttime maximums can still be warmer in the canopy on the hottest nights (Fig. 12.1c). At night, the canopy becomes wetter, resulting in a homogenisation of the vapour pressure deficit gradient (Fig. 12.1b; see also Bujan et al. 2016). Vegetation layers also buffer the surface of the soil from the cooler night air and limit the upward emission of long-wave thermal radiation from the soil to the atmosphere (Paul et al. 2004). This buffering effect results in warmer soil temperatures in vegetated compared with open habitat during the night (Lembrechts et al. 2022).

The gradient in surface temperatures along the vertical dimension is even more extreme than ambient air temperatures and can be highly variable at microscales. This is particularly important for ants as their operative temperature—the temperature that an organism perceives in the environment independent of thermoregulation (Angilletta 2009)—is strongly influenced by the surface temperatures on which they walk (Kaspari et al. 2015). In a Panamanian rainforest, sun-exposed surface temperatures in the canopy averaged 8 °C higher (often surpassing 50 °C) and were four to eightfold more variable than those in the understorey (Kaspari et al. 2015). Stark et al. (2017) found that tropical rainforest ants traveling along canopy branches could experience up to a 17 °C difference in surface temperature depending on whether they were walking on the upper or underside of sun-exposed branches. Branch characteristics, such as epiphyte load (Freiberg 2001) and branch colour (Stark et al. 2017), also strongly influence surface temperatures, with lighter and more heavily covered branches generally having lower surface temperatures. On an even more microscale, different parts of a single leaf can vary in temperature by as much as 7 °C (Caillon et al. 2014). This high variation presents distinct challenges to canopy ants who must carefully navigate foraging paths to avoid super-heated pockets of the boundary layer in which they walk (Kaspari et al. 2015). On the other hand, it also provides opportunities to exploit thermal variability to their advantage, for example, by utilising different parts of the canopy throughout the day (Hashimoto et al. 2010).

12.2.2 Nesting and Foraging Across the Vertical Dimension

The vertical dimension can be broken into distinct habitat strata with unique ant assemblages inhabiting different zones. Yanoviak and Kaspari (2000) referred to these habitat strata as templets after Southwood (1988) to indicate that a suite of resources, climate, and physical attributes cohesively define each habitat. The way that ant species are classed into vertical strata is complicated by different combinations of nesting and foraging location. Nesting and foraging niche will influence climate change responses in different ways by determining exposure, sensitivity, and adaptation potential.

In the subterranean zone, ants that nest and forage exclusively underground are referred to as hypogaeic and often have distinct morphological adaptations such as small eyes and light colouration (Wong and Guénard 2017). Other subterranean ants can either nest underground or forage in subterranean and surface habitats (Subterranean nesting—ground/subterranean foraging) or nest above-ground but frequent the subterranean zone (ground nesting—ground/subterranean foraging). Surface communities can be defined by their foraging niche, epigaeic (foraging on the surface of the soil or leaf litter), or leaf litter ants (which use the small interstitial spaces between decomposing leaves) and by their nesting niche, ground nesting (in dirt mounds, under rocks, in logs and woody debris), or leaf litter nesting. Generally, ground-nesting ants are epigaeic but leaf litter nesting ants can be epigaeic or leaf-litter foraging or combinations thereof. Another unique surface niche is the column raiding ants such as the infamous army ants (e.g. *Eciton* and *Dorylus*). These species are nomadic nesters, but they could be considered epigaeic foragers as they raid along the surface of the ground (Sosiak and Barden 2021).

Moving upwards, the arboreal zone starts with the shrub layer and moves up to the canopy. Some arboreal ants create nests from plant and ant bioproducts. Examples include the weaver nests of the genus *Oecophylla*, which are made by sticking leaves together with the silk produced by their own larvae, and the carton nests of genera such as *Azteca* and *Crematogaster*. Most arboreal ants are lignicolous, nesting in tree hollows, cracks in tree bark, or hollowed out stems. Other arboreal ants nest in epiphytes or under moss beds (Yanoviak et al. 2011). Arboreal ants can forage either exclusively in the canopy and/or on shrubs (e.g. species of *Pseudomyrmex*, *Tetraponera*, and *Cephalotes*), while some are known to forage along the full vertical gradient of ground to canopy (e.g. species of *Anonychomyrma* and some *Camponotus* spp.) (Leahy et al. 2020; Kirchner et al. 2024).

There has been a strong emphasis in ant research on the distinctiveness of vertical assemblages, but many species also show a large degree of ecological flexibility in their vertical position both in nests and foraging. There is in fact a spectrum of vertical specialists to vertical generalists for both nesting and foraging and where different species fall along this spectrum has important consequences for climate-change vulnerability.

12.3 Climate Change Impacts Across the Vertical Dimension

Climate change is expected to cause increases in mean temperature, the frequency of extreme heat waves and extreme cold events, and the frequency of droughts and will generally produce a more unpredictable and variable climate (Stott 2016). To understand how changes in climatic variability will influence vulnerability to climate change, we need to look both to the past and to the future (Moritz and Agudo 2013; Nadeau et al. 2017; Buckley and Kingsolver 2021). Climate change vulnerability is determined by exposure and sensitivity to climate, resilience to extreme climatic events, and ability to adapt to change (Williams et al. 2008; Huey et al. 2012). Historic climatic variability determines a species' intrinsic sensitivity by influencing genetic diversity, propensity for phenotypic plasticity, physiological tolerances, and current geographic distributions (Kingsolver and Huey 1998; Graham et al. 2006; Davison and Chiba 2008; Moussalli et al. 2009; Rodríguez-Trelles et al. 2013; Huang et al. 2015; Otero et al. 2015). Current and future climatic variability will determine extrinsic exposure and dictate when and where species will feel the effects of rising global temperatures (Nadeau et al. 2017). Degree of climatic exposure under changing conditions will be strongly influenced by geographic distribution, habitat specialisation, behaviour, and ecology (Sunday et al. 2014).

12.3.1 Sensitivity to Climate Change Along the Vertical Dimension

12.3.1.1 Thermal Physiology

Sensitivity to temperature can be indexed by cold tolerance (CT_{min}) and heat tolerance (CT_{max}), which respectively are the lowest and highest temperatures at which an individual can maintain muscular coordination (Roeder et al. 2021; see Chap. 4). For ants, these thermal tolerances have largely been measured on individual workers with limited studies on other castes (Roeder et al. 2021; Nascimento et al. 2022; Parr and Bishop 2022). A key pattern to emerge from these data is that variation in cold tolerance tends to map well onto large-scale geographic gradients of elevation and latitude: as it gets colder, ants have adapted to withstand colder temperatures (Bishop et al. 2017; Diamond and Chick 2018; Nowrouzi et al. 2018). Interestingly, however, heat tolerance is poorly predicted by temperature across large-scale gradients of elevation and latitude but is strongly predicted at the local scale by vertical microhabitat (Baudier et al. 2015; Bujan et al. 2020; Leahy et al. 2022).

In general, hypogaeic species are intolerant of high temperatures (Bujan et al. 2020). Baudier et al. (2018) (compared CT_{max} for 128 colonies of fifteen species of army ants (subfamily: Ecitoninae) that were assigned a surface activity value based

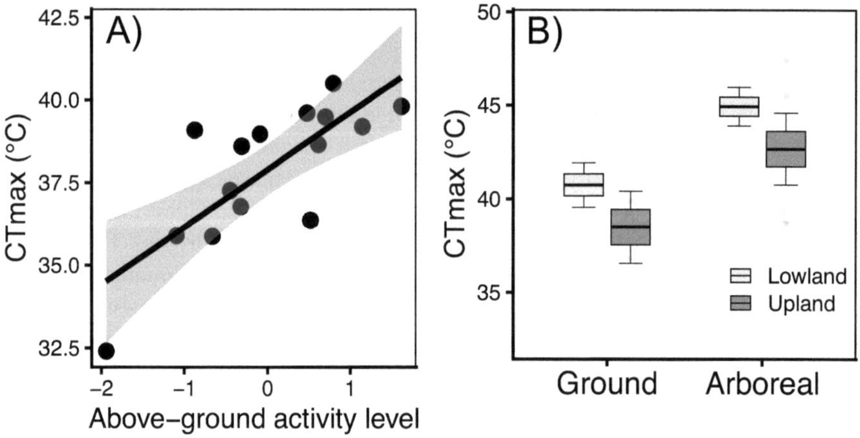

Fig. 12.2 CT_{max} of ants from different vertical habitats from two studies in tropical rainforest. In (**A**) mean species CT_{max} and index of surface activity for 128 colonies of fifteen species of army ants (subfamily: Ecitoninae) from Costa Rican tropical forests. Lower (more negative) values indicate more subterranean than surface activity, this is associated with lower tolerance of hot temperatures (lower CT_{max}). In (**B**), mean CT_{max} of worker ants from 73 colonies of 40 species from ground and arboreal microhabitats and two elevations, lowland (100 m asl) and upland (1200 m asl) in the Australian Wet Tropics. This demonstrates lower tolerance of hot temperatures for ground-dwelling compared to arboreal ants. Modified from (A) Baudier et al. (2018) and (**B**) from Leahy et al. (2022)

on their behaviour and relative eye size, ranging from mostly subterranean to mostly surface activity. They found CT_{max} to be strongly associated with degree of surface activity (Fig. 12.2a). Echoing this finding, a comparative study of North American ants found hypogaeic species to have the lowest thermal tolerance in the ant communities sampled averaging 41 °C compared to >46 °C for ground and arboreal species (Bujan et al. 2020). Likewise, understorey and ground ants tend to have lower CT_{max} than arboreal and canopy specialist species: CT_{max} averaged 3.5–5 °C higher for canopy ants compared to understorey ants in a Panamanian rainforest (Kaspari et al. 2015), while in the Wet Tropics of Australia, arboreal ants averaged a 4.2 °C (95% CI: 2.7–5.6 °C) higher CT_{max} than ground-dwelling ants (Fig. 12.2b; Leahy et al. 2022).

The vertical dimension presents a strong gradient in relative humidity and several studies have shown that arboreal ants have higher desiccation resistance (see Chap. 5) than surface or subterranean ant species (Hood and Tschinkel 1990; Bujan et al. 2016). Bujan et al. (2016) found that the average desiccation resistance of canopy ants was 2.8 times higher than understorey ants. In a temperate pine forest, an environment with lower humidity than in tropical forests, Hood and Tschinkel (1990) found canopy ants were eight times more desiccation-resistant than understorey ants. Some, but not all, of the variation in desiccation resistance is attributable to body size, as canopy ants tend to be larger than understorey ants (Kaspari et al. 2015; Bujan et al. 2016; Law et al. 2020). The difference appears to lie in the protective cuticle of ants, which is more effective in preventing water loss for arboreal than understorey ants (Hood and Tschinkel 1990). If the cuticular hydrocarbon layer was removed, arboreal and understorey ants had similar water loss rates (Hood and Tschinkel 1990). Further, darker insect cuticles have reduced cuticular permeability (Parkash et al. 2008), and canopy and understorey ants tend to be darker in colour compared with ground and subterranean ants (Law et al. 2020). These studies indicate that cuticle protection is a key mechanism in resisting desiccation for ants across the vertical dimension (Duarte et al. 2019).

Intriguingly, Menzel et al. (2018) discovered plasticity in the chemical profile and waterproofing capacity of cuticular hydrocarbons within a generation of ants exposed to contrasting temperature and humidity environments. They found higher survival under desiccation stress for warm and dry acclimated ants (Menzel et al. 2018). We could predict that if cuticular hydrocarbons are a plastic trait with high acclimation potential, this could map well onto a range of ant niche and behavioural traits, such as foraging habitat (e.g. subterranean, surface, arboreal) and temporal activity (nocturnal versus diurnal and wet season versus dry season). This has not been systematically investigated across the vertical dimension but would be a valuable avenue for further research, particularly given the increase in drought frequency predicted for this century.

Vertical niche (subterranean, surface/ground, arboreal) tends to be an evolutionary conserved trait when compared across genera (Lucky et al. 2013). It is very likely that the transition to a canopy lifestyle was accompanied by adaptation of thermal physiological traits such as heat tolerance and desiccation resistance, and this plays some part in why heat tolerance, which is also highly conserved (Araújo et al. 2013; Bujan et al. 2020; Leahy et al. 2022), correlates strongly with vertical microhabitat. Diamond and Chick (2018) invoked Janus, a Roman god, with one face looking forward (cold tolerance) and another back (heat tolerance), to describe how cold tolerance is shaped by current species ranges at macro-scales, but heat tolerance is constrained by evolutionary history. This analogy sheds light on the implications of thermal trait conservation: heat tolerance in ants is unlikely to be malleable to rapid adaptation in the context of climate change (Araújo et al. 2013). This is important because it will determine whether arboreal ants will be able to broaden their heat tolerance to reduce sensitivity within the timeframe of major climatic changes. Current phylogenetic tools used to explore whether vertical niche and physiological traits covary have been somewhat limited, but new multi-response models which model the phylogenetic and non-phylogenetic covariation between traits may provide new options for exploring these questions (see Westoby et al. 2023).

12.3.1.2 Morphological Traits

Sensitivity to climate change will be mediated through a set of traits that dictate how an organism exchanges heat with the thermal environment in which it is situated (Huey et al. 2012). The operative temperature of an organism is the steady-state temperature of an organism in the absence of metabolic heating and evaporative cooling (Bakken et al. 1985). To illustrate this, imagine an ant resting in a subterranean nest without solar radiation and no wind; its operative temperature would be equal or close to the ambient air temperature (Kearney and Porter 2009). Environmental conditions (solar radiation, temperature, and wind speed) change dramatically across the vertical gradient (Xing et al. 2023). We could therefore predict that subterranean, surface, and arboreal ants have faced differential selection on traits that mediate radiation to ensure that ants can maintain a safe operative temperature in these varying environments. Some key morphological traits that influence temperature mediation include body size and body colour (see Chap. 3), and there is evidence that these traits do vary systematically along the vertical dimension.

A typical ant community can have a range of ant sizes and body masses spanning several orders of magnitude. A large study of 222 tropical ant species found that canopy and understorey ants have greater body length than do ground and subterranean ants (Fig. 12.3). Body length is allometrically scaled with leg length, placing large ants above the boundary layer that covers surfaces on which they run, while smaller ants are subsumed within it (Kaspari et al. 2015). The boundary layer can become super-heated when in direct sunlight and low wind speed, and it would be highly advantageous to be a larger ant cruising above this layer in the hot and variable canopy.

By mass, canopy ants tend to be larger as well: a study of 88 species showed canopy ants were four times as heavy as understorey ants (Kaspari et al. 2015). Kaspari et al. (2015) tested how body mass influences the time it takes for an ant to heat up to ambient air temperature (i.e. reach thermal equilibrium) if suddenly

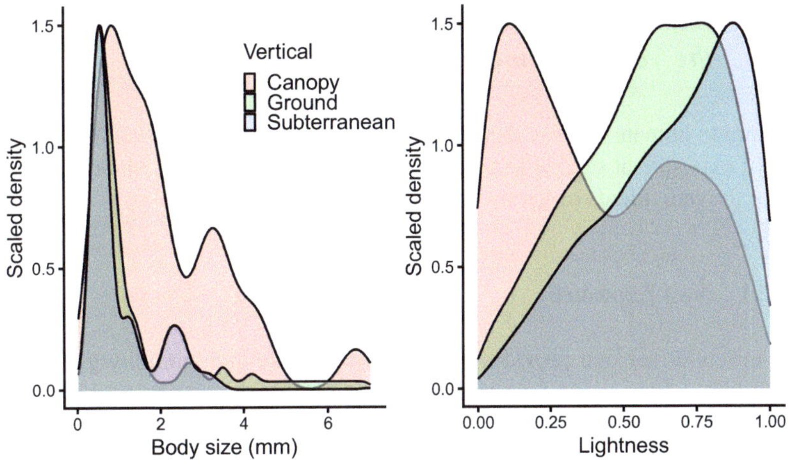

Fig. 12.3 Density plots showing the distribution of body size (Weber's length) and cuticle lightness (higher values indicate lighter colouration) for 222 ant species from soil, ground, and canopy vertical habitats in the tropical rainforests of Sabah, Malaysia. Showing that with increasing verticality, ant communities are made up of larger, darker coloured species. Modified from Law et al. (2020)

exposed to the sun from shade. They found that the largest ant tested (42.5 mg), a canopy species, took five times longer to heat up than the smallest ant tested (1 mg). This could allow a larger ant more time to escape sudden thermal exposure after wandering into a super-heated sun fleck in the canopy. But the story isn't so simple, as larger ants also absorb greater amounts of solar radiation due to their greater surface area and would need to balance these factors to maintain safe operative temperatures.

Body colouration strongly influences absorption of solar radiation: black colours will absorb heat faster than light colours. We might predict that the hot canopy would select for ants of lighter colour to avoid rapid heat absorption, but Law et al. (2020) found the opposite: canopy and understorey ant assemblages were twice as dark as assemblages from ground or subterranean habitats indicating that darker ants dominate the canopy. Globally, ant assemblages tend to become lighter in colour with increasing temperature and then darker again at the most extreme temperatures (Bishop et al. 2016). These contradictory findings are due to the high levels of UV-B radiation found in sunny places, which are also hot, such as forest canopies and desert landscapes. Ants need to trade off protection from UV-B radiation exposure with the risks of absorbing a greater load of thermal radiation that could seem them overheat in the hot canopy. Overall, the vertical dimension provides different environmental challenges to subterranean, ground, and arboreal ants. Mapping the covariation or trade-offs in traits such as body size/mass and colouration that strongly influence heat exchange will be a critical tool for indexing climate change sensitivity in our rapidly changing world.

12.3.2 Exposure to Climate Change Along the Vertical Dimension

Exposure to climate is a key factor mediating climate-change vulnerability. We can think of exposure at several levels including inside the nest and while out foraging and for the individual worker as well as the entire colony.

12.3.2.1 Nest Exposure

The vertical dimension provides a strong temperature and humidity gradient with many options for nesting location. Within each of the main vertical habitats, nest site selection is strongly tied to temperature regulation (Jones and Oldroyd 2006). Canopy nests will be more exposed to temperature extremes under a warming climate than nests in lower strata (Bujan and Yanoviak 2022). Meanwhile, leaf litter nesting species of forests, which rely heavily on the high humidity of the forest floor, are likely to be impacted by drying out of forest systems (Grimbacher et al. 2018; Almeida et al. 2023). Within each strata, the nest type will strongly influence exposure. For example, canopy ants nesting in moist tree holes or deep within tree crevices are likely to be more protected from temperature extremes compared to ants nesting in twigs (Mottl et al. 2020; Plowman et al. 2020).

In one experimental study involving experimentally heating canopy nests, the volume of twig nest cavities was an important predictor of heating rate (Bujan and Yanoviak 2022). The canopy specialists, *Cephalotes* spp., nested in smaller twig cavities which had more surrounding wood protection and therefore took longer to heat up compared to larger cavities. These species were less likely to abscond from their nests following experimental heating (Bujan and Yanoviak 2022). The effect of extreme canopy temperatures on nest abandonment by carton- and silk-building ants has not been investigated. Species such as the weaver ants (*Oecophylla*

Fig. 12.4 Two types of arboreal nests. On the left, workers of *Oecophylla smaragdina* construct a nest from leaves bound together with silk from their larva. On the right, workers of *Gesomyrmex howardi* live in a hollowed out live stem in a mangrove forest. Thermal protective properties of these different arboreal nests will strongly influence exposure to risky temperatures under a climate change future. (Photos: François Brassard)

smaragdina) and carton-nesting ants (*Azteca* spp.) have polydomous (multiple) nests distributed vertically and horizontally in trees (Tanaka et al. 2010; Sangma and Prasad 2021). This nesting behaviour spreads the risk both from nest disturbance but also likely from thermal risks. From twig nests to ant-made structures (Fig. 12.4), the thermal protective ability of nesting structures is likely to play an important role in regulating exposure to climate in the canopy.

The ability to move nest location throughout the life of the colony is a key feature of ant ecology and is associated with changes in temperature and humidity as well as disturbance (McGlynn 2006, Seifert et al. 2017; see Chap. 2). For example, subterranean and surface nesting ants were shown to shift deeper into the soil matrix during the dry season to avoid brood desiccation (Levings 1983; Jacquemin et al. 2015). Many ant species take advantage of hyper-local vertical thermal gradients within their nests to move brood to warmer position where development will occur at a faster rate (Penick and Tschinkel 2008). Across elevation, intra-specific variation in vertical nest location also indicates that temperature exposure is a strong factor mediating nest site choice. For example, with increasing elevation more army ant nests (*Eciton burchellii*) were located underground with fewer positioned in exposed trees (Soare et al. 2011), while in a Papua New Guinea ant assemblage, canopy nests became smaller and were more frequently in protected twigs and epiphytes (Plowman et al. 2020). Where species nest along the vertical gradient will strongly influence exposure and vulnerability. The vertical gradient, however, also provides many opportunities to shift nest site location to buffer against climate extremes which is likely to play a key role in mitigating climate change impacts.

12.3.2.2 Foraging Exposure

Ant foraging is strongly influenced by temperature and humidity gradients across all major ecosystems (Carroll and Janzen 1973; Kaspari and Weiser 2000; Hashimoto et al. 2010; Andrew et al. 2013; Farji-Brener et al. 2018). In biomes with shrubs or trees, we can think of two climate gradients at a localised scale along which species can forage: vertical space from subterranean to canopy (Leahy et al. 2021) and the time of day (Kronfeld-Schor and Dayan 2003; Yusah et al. 2018). These local thermal gradients represent spatial and temporal niche axes that colonies can exploit for maintaining optimal thermal exposure (Kronfeld-Schor and Dayan 2003; Holt 2009). As humans are a diurnally active species, we often think of exposure and climate risk in terms of daytime patterns of activity and habitat use and ignore what happens at night (Gaston 2019). Many ant species are active throughout the day and night (Fellers 1989; Bestelmeyer 2000; Albrecht and Gotelli 2001; Houadria et al. 2015), and some shift their ground to canopy foraging patterns depending on time of day (Pfeiffer and Linsenmair 2000; Tanaka et al. 2010).

To fully capture this variability in foraging locations and schedules, we can consider local vertical and temporal climate gradients as a thermal landscape in which ants are active and therefore exposed to ambient climate conditions. An important question then is how much thermal conditions differ across the four-dimensional

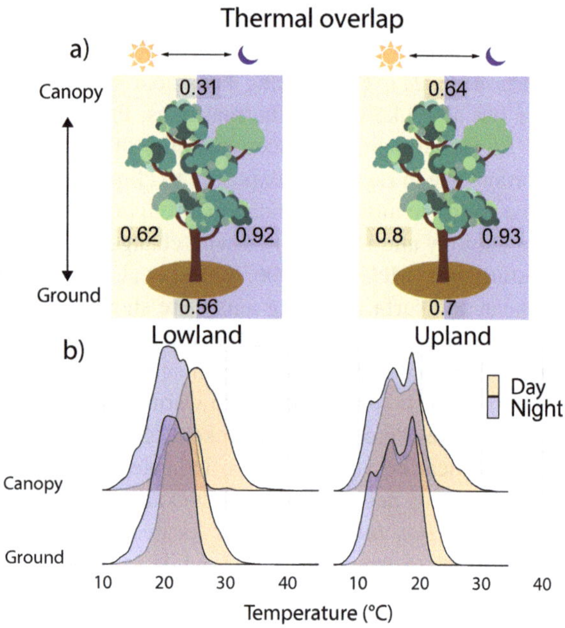

Fig. 12.5 Thermal overlap along thermal niche dimensions of vertical (ground and canopy), time-period (day and night), and their different combinations at a lowland (100 m a.s.l.) and upland (1200 m a.s.l.) rainforest site in the Australian Wet Tropics Bioregion. Showing (**a**) thermal overlap values from (**b**) kernel density distributions (y-axis) of microclimate air temperature recorded every half hour in ground and canopy in each elevation site for 2 years (March 2019–March 2021). (Reproduced with permission from Lily Leahy 2022)

thermal landscape. One way to approach this is to consider the thermal profile of vertical and temporal gradients as a density plot of temperature over a 24-h period and then consider the degree of thermal overlap in profiles across habitats and times of day. Low thermal overlap would indicate that ant species are exposed to very different thermal conditions if they moved to a different part of the thermal landscape, i.e. moving upwards to the canopy or shifting activity to nighttime. High overlap would indicate that there is a lot of shared climate space between spatio-temporal niches. We can then extend this concept to compare spatio-temporal thermal landscapes for different sites across biomes, elevations, and latitudes.

One such study from the Australian Wet Tropics Bioregion compared thermal profiles across vertical and temporal dimensions in lowland and upland elevation sites. At both elevations, ground and canopy thermal profiles were nearly identical at night but diverged substantially during the day (Fig. 12.5). There was also more overlap in thermal profiles between ground and arboreal habitats and between day and night in the cooler uplands (1200 m asl) compared to the hot lowlands (100 m asl). These findings were similar to another study which compared open and closed forested areas: while temperatures declined with elevation, temperature profiles of open and forested areas became more similar with increasing elevation (Kumar and O'Donnell 2009). Lowland canopies have greater breadth of climatic space due to pushing up into higher daytime temperatures and cooler nighttime temperatures than the ground (Fig. 12.5b). This indicates that in hotter lowland environments, the transition from day to night and ground to canopy could be more dramatic than in the uplands.

These thermal profiles show that canopy environments in hot locations, such as lowland rainforest or grassland savannahs, have the most distinct microclimates across the four-dimensional thermal landscape. If species are diurnal canopy specialists without nocturnal activity or the option to shift downwards, they will have to deal with this thermal exposure directly. Luckily, the intense thermal variability of the canopy, created by the sun, passing clouds and movement of branches, presents a complex tapestry of surface temperatures (Stark et al. 2017). Ants draw on this extensive thermal variability to avoid extreme temperatures in the canopy and thereby reduce risky temperature exposure.

This phenomenon was elegantly demonstrated by Spicer et al. (2017) who used artificial super-heated sunspots on arboreal foraging trails to observe patterns of behavioural avoidance. Ants moved out of the super-heated patch when the surface temperature reached between 36 and 44 °C, but their response time varied based on body size; larger ants responded more slowly than small ants, probably for reasons to do with time taken to reach thermal equilibrium. The ants escaped the super-heated patch at temperatures at 5–10 °C below their CT_{max} indicating a preference for safe foraging temperatures with a wide margin to overheating.

The combination of the vertical gradient and the diel cycle from day to night sets up a four-dimensional thermal landscape at the local level with a total temperature range that far surpasses the change in average temperature across many degrees of latitude (Scheffers et al. 2013). Under climate change, vertical-temporal dimensions could provide options to buffer rising temperatures in situ without the need for population-wide range shifts (Bonebrake and Deutsch 2012; Moritz and Agudo 2013; Levy et al. 2019; Montejo-Kovacevich et al. 2020). If canopy species have flexibility in foraging and can shift into nighttime activity or downwards to the ground in the hotter parts of the day, there is ample cooler climatic space which they can move into (Fig. 12.5b). Buffering temperatures in situ along vertical and temporal gradients depends on the degree of specialisation to generalisation in behavioural, morphological, and physiological traits (Levy et al. 2019; Román-Palacios and Wiens 2020). In addition, both inter- and intraspecific competition will dictate the available opportunities for vertical and temporal niche plasticity (Law and Parr 2020; Mottl et al. 2020).

12.4 Dynamic Vertical Communities in a Changing World

What happens along the vertical dimension of terrestrial ecosystems is a fascinating topic that can give us much insight into ecological and evolutionary adaptation to climatic gradients at small spatial scales. Vertical gradients, however, do not exist in isolation, but are nested within the context of the more traditional studied biogeographic gradients of elevation and latitude (Fig. 12.6). The thermal exposure of a population is therefore contingent on the combination of climatic variation produced at micro-, meso-, and macrogeographic scales. These spatial scales also interact. Soil to canopy patterns at the microgeographic scale can scale up to influence

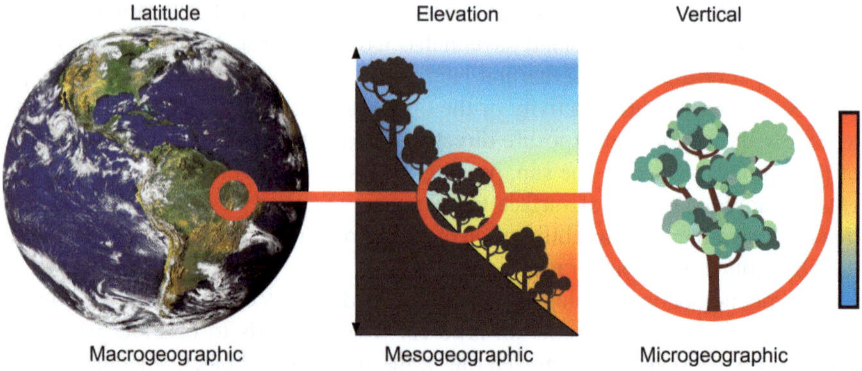

Fig. 12.6 Trees create a microgeographic climate gradient from soil to canopy that is then nested in its position at the mesogeographic climate gradient of elevation. This in turn is nested in the macrogeographic latitudinal gradient. Colour gradient represents cool (blue) to warm (red). (Reproduced with permission from Lily Leahy 2022)

biogeographic patterns at mesoscales of elevation and macroscales of region and latitude (Scheffers et al. 2013; Oliveira and Scheffers 2019; Klinges and Scheffers 2021). As I will outline in this section, contextualising climate change vulnerability in a multi-scale framework can strengthen our predictions of climate futures given variable species responses.

12.4.1 Scaling Up: Vertical Responses Embedded in Meso- and Macroscale Contexts

12.4.1.1 Vertical Niches Across Mountains

Tropical species tend to have narrow elevational and geographic ranges compared to temperate species, as was initially observed by Janzen (1967). It was proposed that temperate species are adapted to highly seasonal environments, are therefore more tolerant of variation, and thereby achieve broader spatial distributions compared to tropical species that experience low seasonality. The notion that species from climatically variable environments have broader geographic ranges has been generalised as the 'climate variability hypothesis' and has received considerable theoretical and empirical support. This hypothesis forms a conceptual framework for why tropical species are predicted to be more vulnerable to climate change: climatic stability selects for narrow thermal tolerance range and small range size, both of which will increase sensitivity to climate change (Deutsch et al. 2008; Huey et al. 2009, 2012; Wright et al. 2009; Corlett 2011; Sunday et al. 2011; Diamond et al. 2012).

Recent research has focused on applying the same principles of the climate variability hypothesis to the vertical dimension. In a tropical Australian rainforest,

Leahy et al. (2021) reported that arboreal ant species and ant species with a wide vertical niche breadth (i.e. they forage from ground to canopy) tend to have broader elevational ranges. This pattern was not attributable to local abundance or body size, two factors that can influence geographic range size (Leahy et al. 2021), but they found a strong positive correlation between exposure to temperature variability and elevation range size. This provides support for the hypothesis that broad tolerance of climatic variation at the local population scale could facilitate broader range sizes across geographic space (Addo-Bediako et al. 2000; Scheffers and Williams 2018).

On the other side of the world, in a Meso-American forest, Longino et al. (2019) also investigated elevation diversity patterns of ants based on their vertical niches: ground/leaf litter versus arboreal. Specifically, they were interested in how local site diversity changed with elevation and predicted a mid-elevation peak in diversity (Sanders et al. 2007; Szewczyk and McCain 2016; Plowman et al. 2020). Their sampling revealed that arboreal ants had a less pronounced mid-elevation peak than leaf-litter and ground ants, tending to remain relatively diverse at both low and mid-elevations. The authors attributed these different patterns of elevational diversity to niche conservatism harking back to the Pleistocene when temperature profiles were approximately 500 m lower in elevation than they are today in Meso-America (Grauel et al. 2016). As the climate shifted into our current interglacial period, the lowlands would have become too warm for many species, who subsequently shifted upwards to mid-elevations. Arboreal species, on the other hand, are generally tolerant of higher temperatures and of climatic variance and were able to persist or to successfully recolonise the lowlands in greater numbers (see also Burwell and Nakamura (2015)). This theory is supported from other locations and for other taxa (Graham et al. 2006; Davison and Chiba 2008). For example, there is a strong positive correlation between verticality and resilience to climatic instability of vertebrate taxa in the Australian Wet Tropics, whereby arboreal species tend to dominate in areas that were historically unstable in their climates since the last glacial maximum (c. 20,000 years ago) (Scheffers et al. 2017b).

How does this relate to climate change today? Species with the ability to persist across broader climatic space, e.g. across elevation, or across latitude, are likely to be more resilient to climate change compared to those with restricted niches with small range size (Calosi et al. 2010; Moritz and Agudo 2013; Saupe et al. 2015; Pintanel et al. 2019). An arboreal lifestyle may have provided ant species with the tools to overcome climatic instability in the past and could therefore provide resilience to future climate change (Thuiller et al. 2005; Davison and Chiba 2008; Nadeau et al. 2017; Scheffers et al. 2017b). Arboreal or vertical generalist species may have the capacity to shift their vertical position along their elevational ranges to compensate for broader scale shifts in the climate-environment (Scheffers et al. 2013) or shift their geographic distribution further up the mountain as the lowland climate warms. On the other hand, ant species with a narrow vertical niche breadth and distribution could be vulnerable to climate change due to the synergistic effects of restricted elevation range size, sensitivity to climate variation, and limited ability to compensate behaviourally through vertical movement (Thuiller et al. 2005; Williams et al. 2008; Huey et al. 2009).

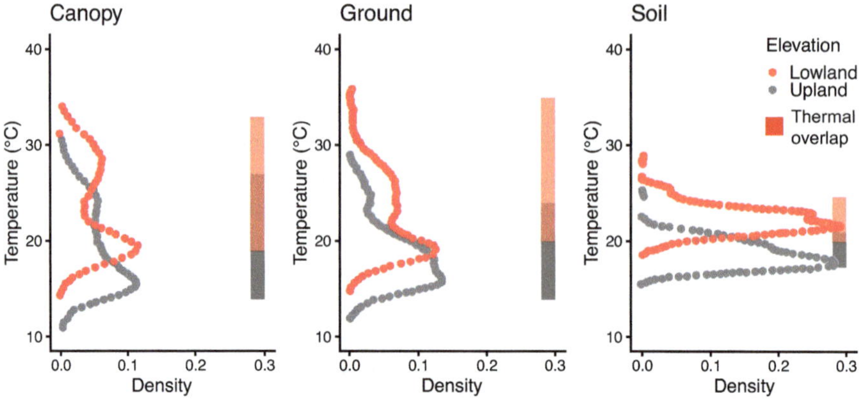

Fig. 12.7 The temperature distribution and overlap amongst lowlands and uplands for canopy, ground, and soil microenvironments combining temperature profiles from three global sites (Madagascar, Australia, and the Philippines). The bars represent the interquartile range (Q1 at 25% and Q3 at 75%) and the extent of thermal overlap between lowlands and uplands. Canopies exhibit a higher degree of overlap (wider dark bar) in thermal conditions compared to ground and soil. Generated using kernel-density estimation. Soil temperature data were available only for Madagascar and Australia. (Modified from Scheffers and Williams 2018)

12.4.1.2 Vertical Community Turnover Across Climate Gradients

When we conceptualise temperature change across climate gradients such as mountains, we often think of how temperature changes along the ground as distinct non-overlapping bands of climate like a layered cake. However, vertical climate gradients make climatic change across landscapes far more complicated. Due to the high climatic variance in the canopy, lowland and upland canopies have a higher degree of thermal overlap than do lowland and upland ground or soil environments (Fig. 12.7; Scheffers and Williams (2018)). All else being equal, climates should be more similar between lowlands and uplands for canopy species than they are for ground species (Scheffers et al. 2017b; Klinges and Scheffers 2021). This has implications for species distributions and community patterns across elevation and latitude (see Chaps. 8, 9, and 10). We would expect that arboreal communities who occupy a climatically variable environment with high thermal overlap can persist and thrive at multiple climatic-elevational bands; this should lead to lower rates of species turnover across elevation for that assemblage compared with more elevationally restricted ground species (Fagan et al. 2006).

There have been limited tests of this hypothesis in ants, but arboreal rainforest communities in the Australian Wet Tropics show lower rates of species turnover compared with ground-based communities (Leahy et al. 2024). Similarly, Plowman et al. (2020) found relatively low dissimilarity in arboreal communities from low to mid-elevations (200 versus 900 m asl) but very high dissimilarity between mid and high (900 versus 1800 m asl) elevations in a Papua New Guinean rainforest (ground communities were not compared within that study). There is a difference in the maximum elevation compared within these studies which could explain the more distinct arboreal montane community in the Plowman et al. (2020) study.

The opposite pattern was found in the savannas of Brazil, a very different biome to the Australian rainforests. Vasconcelos et al. (2023) sampled ants in ground and arboreal habitats at 32 sites spanning 22° of latitude. With increasing geographic and climatic difference between sites, arboreal species showed a trend of increasing rates of species turnover compared to ground assemblages. In addition, with increasing latitude (i.e. decreasing mean temperatures), species richness of ground ants increased at a much faster rate than for arboreal species. Close to the equator, ground species richness was half as diverse compared to mid-latitudes (e.g. 20–25 °C) and significantly more ground species were vertical generalists, foraging up into the trees. These vertical generalist ground species at the tropical sites were a nested subset of the species diversity at mid-latitudes. In comparison, the higher turnover of arboreal species indicates that species composition changed in step with climatic changes due to latitude.

These savanna ants are not exhibiting the usual trends of latitudinal diversity gradients that we would expect from other biomes, nor the elevational trends in species turnover shown for ground and arboreal species in tropical rainforests (Dunn et al. 2009; Economo et al. 2018; Lasmar et al. 2021; Leahy et al. 2024). Further, we would expect that ground species closer to the equator should avoid foraging in the exposed canopy of savanna trees (Kaspari et al. 2015; Leahy et al. 2022). One explanation is that these ground ants may be accessing important food resources in the canopy making the climatic risk worthwhile (Vasconcelos et al. 2023). The diverging community patterns between tropical rainforest and savanna habitats presented by these few examples could be due to changes in net primary productivity: with decreasing latitude, net primary productivity tends to increase in forested habitats, while it may well decrease for drier more extreme savanna environments (Kaspari et al. 2000; Murphy et al. 2016; Lasmar et al. 2021; Vasconcelos et al. 2023).

Currently, studies testing the interactions between vertical, elevation, and latitudinal gradients on community turnover are limited. With the accumulation of data from more bioregions of the world, the precise climatic mechanisms underlying community turnover patterns will hopefully become clearer leading to more robust predictions of community shifts under climate change.

12.5 Synthesis and Future Directions

The vertical dimension presents a strong climatic gradient at the local (microgeographic) scale that is situated within the context of broader climatic gradients such as elevation and latitude. Most studies of ants across the vertical dimension have been focused on above-ground communities of tropical forests, and there is a significant gap in research in other terrestrial biomes and for subterranean environments (Wong and Guénard 2017; Xing et al. 2023). Putting this geographical limitation aside, we can summarise the following factors that are likely to influence

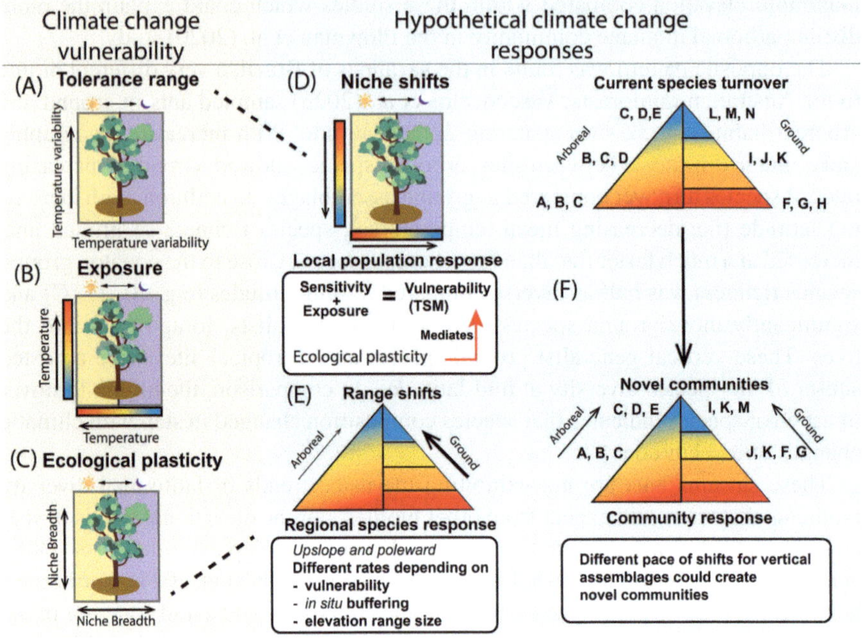

Fig. 12.8 Schematic of the links between microscale climate variability, vulnerability, and potential population, species, and community responses to climate change. Colour gradient: red, warm temperatures; blue, cool temperatures. At the microscale of a tree, vertical (subterranean to canopy) and temporal (day to night) climatic gradients set up a four-dimensional climatic landscape, along which populations have different (**a**) thermal sensitivity, (**b**) exposure, and (**c**) degree of niche breadth or plasticity. We could hypothesise the following climate change responses (**d**) species which can move into cooler parts of the local thermal landscape will do so; (**e**) the climatic variability along the vertical gradient sets up different levels of thermal overlap across elevation, in addition, contrasting levels of vulnerability, potential for in situ buffering; and current range size will exert different pressures to execute range shifts, strength of shifts indicated by arrows. Finally, in (**f**), species turnover across climatic gradients is likely to be contingent of vertical niches, e.g. arboreal species could have different turnover compared to ground or subterranean species with hypothetical ant species indicated by letters (**h**, **i**). The different paces of range shifts as well as the underlying differences in community composition will create different possibilities for novel communities. Competition, resource distribution, and dispersal are other key elements important to ant ecology which will interact with the climate change trajectories outlined here. (Modified from Lily Leahy 2022)

climate change vulnerability and capacity for response of ants across vertical gradients (Fig. 12.8).

Firstly, vertical niche influences thermal sensitivity through a correlative relationship with thermal limits (Baudier et al. 2015; Garcia-Robledo et al. 2018; Bujan et al. 2020; Leahy et al. 2022). In general, subterranean and ground-dwelling ant species experience stable climates and have lower heat tolerance and a narrower thermal tolerance range than arboreal species that experience high climatic variability (Fig. 12.8a). This appears to be the case for ant species in hot lowlands and cool upland rainforests (Leahy et al. 2022).

Arboreal species, and particularly lowland diurnal canopy species, are exposed to surface and ambient air temperatures that are far higher than those experienced on or under the forest floor (Fig. 12.8b). Although, on average, canopy species have smaller thermal safety margins (warming tolerance) than ground species, they may have more capacity for behavioural thermoregulation and other mechanisms to resist desiccation resistance (Diamond et al. 2012; Leahy et al. 2022; Parr and Bishop 2022). Trait syndromes associated with vertical habitats such as body size/mass and body colouration may mediate heat exchange with the environment but are subject to trade-offs as ants balance multiple environmental risks such as desiccation and UV-B exposure making it complicated to predict how these traits will determine sensitivity (Kaspari 1993; Law et al. 2020).

Climate exposure across the vertical dimension could be mitigated through behaviour in both space and time (Fig. 12.8c). Ant species with a wide niche breadth along vertical and temporal gradients occupy more of the available climatic space are likely to have a greater response capacity. Meanwhile, nesting ecology has an important role to play with more thermally buffered or vertical flexible nesting options a safer bet in a warming world (Bujan and Yanoviak 2022). At the local scale, species with the ability to shift activity into cooler parts of their niche space may do so (Fig. 12.8d). This may manifest through changes in foraging location during the hottest parts of the day, shifting to more nocturnal activity, or seasonal shifts in vertical foraging (Fellers 1989; Wetterer 1990; Harada 2005; Hashimoto et al. 2010). The species that are being exposed to novel temperatures, without these options (e.g. diurnal, canopy-specialists), will be impacted first, with potential reductions in resource acquisition and therefore colony fitness.

Impacted groups of species (which do not have in situ response options) may lose the warmer end (lower elevation, lower latitude distribution) of their ranges and therefore initiate range shifts. These shifts will be upslope and towards higher latitudes with cooler climates (Fig. 12.8e). Upland specialists may initially benefit from temperature increases but in the long-term could have their ranges increasingly contract to smaller areas of the mountain tops, placing these species under pressure (Urban 2018). Finally, we can predict that different rates of climate tracking and range shifts based on vertical niches linked to thermal variability exposure and niche breadth at the local scale which will likely result in novel communities in the future (Fig. 12.8f). Note that other factors such as competition, often cited as a hallmark of ant ecology, resource specialisation, reproductive strategy, and dispersal ability, are additional elements which will interact with the climate change trajectories outlined here but are beyond the scope of this chapter (Heinze and Tsuji 1995; Camarota et al. 2016; Plowman et al. 2017; Hakala et al. 2019).

In a study comparing the incidence of North American ant genera across a 20-year period, Roeder et al. (2021) reported a general trend that species with lower CT_{max} had decreased, but, at the same time, species with low and high CT_{max} had increased during this time-period. In other words, some species that are intolerant of high temperatures had also done well in the past 20 years. This was a pivotal finding, as it clearly indicates that thermal limits are informative, but not entirely sufficient to predict population changes under climate change (Guo et al. 2020). Moving forward, climate change science will need to move beyond simply testing for thermal limits and embrace the complexities of multi-scale climatic gradients which interact with species physiology, behaviour, and morphological traits. Ant responses to climate change are likely to be contingent on niche position and breadth along the vertical gradient from soil to canopy. Finally, the vertical dimension must be included amongst the more traditionally studied biogeographic gradients of elevation and latitude and will be important arena for forecasting climate change outcomes.

References

Addo-Bediako A, Chown SL, Gaston KJ (2000) Thermal tolerance, climatic variability and latitude. Proc R Soc B Biol Sci 267:739–745

Albrecht M, Gotelli NJ (2001) Spatial and temporal niche partitioning in grassland ants. Oecologia 126:134–141

Almeida RPS, Silva RR, da Costa ACL, Ferreira LV, Meir P, Ellison AM (2023) Induced drought strongly affects richness and composition of ground-dwelling ants in the eastern Amazon. Oecologia 201:299–309

Andrew NR, Hart RA, Jung M-P, Hemmings Z, Terblanche JS (2013) Can temperate insects take the heat? A case study of the physiological and behavioural responses in a common ant, *Iridomyrmex purpureus* (Formicidae), with potential climate change. J Insect Physiol 59:870–880

Angilletta MJ (2009) Thermal adaptation: a theoretical and empirical synthesis. Oxford University Press

Araújo MB, Ferri-Yáñez F, Bozinovic F, Marquet PA, Valladares F, Chown SL (2013) Heat freezes niche evolution. Ecol Lett 16:1206–1219

Bakken GS, Santee WR, Erskine DJ (1985) Operative and standard operative temperature: tools for thermal energetics studies. Am Zool 25:933–943

Baudier KM, Mudd AE, Erickson SC, S. O'donnell. (2015) Microhabitat and body size effects on heat tolerance: implications for responses to climate change (army ants: Formicidae, Ecitoninae). J Anim Ecol 84:1322–1330

Baudier KM, D'Amelio CL, Malhotra R, O'Connor MP, O'Donnell S (2018) Extreme insolation: climatic variation shapes the evolution of thermal tolerance at multiple scales. Am Nat 192:347–359

Bestelmeyer BT (2000) The trade-off between thermal tolerance and behavioural dominance in a subtropical South American ant community. J Anim Ecol 69:998–1009

Bishop TR, Robertson MP, Gibb H, Van Rensburg BJ, Braschler B, Chown SL, Foord SH, Munyai TC, Okey I, Tshivhandekano PG (2016) Ant assemblages have darker and larger members in cold environments. Glob Ecol Biogeogr 25:1489–1499

Bishop TR, Robertson MP, Van Rensburg BJ, Parr CL (2017) Coping with the cold: minimum temperatures and thermal tolerances dominate the ecology of mountain ants. Ecol Entomol 42:105–114

Bonebrake TC, Deutsch CA (2012) Climate heterogeneity modulates impact of warming on tropical insects. Ecology 93:449–455

Buckley LB, Kingsolver JG (2021) Evolution of thermal sensitivity in changing and variable climates. Annu Rev Ecol Evol Syst 52:563

Bujan J, Yanoviak SP (2022) Behavioral response to heat stress of twig-nesting canopy ants. Oecologia 198:947–955

Bujan J, Yanoviak SP, Kaspari M (2016) Desiccation resistance in tropical insects: causes and mechanisms underlying variability in a Panama ant community. Ecol Evol 6:6282–6291

Bujan J, Roeder KA, de Beurs K, Weiser MD, Kaspari M (2020) Thermal diversity of North American ant communities: cold tolerance but not heat tolerance tracks ecosystem temperature. Glob Ecol Biogeogr 29:1486–1494

Burwell CJ, Nakamura A (2015) Can changes in ant diversity along elevational gradients in tropical and subtropical Australian rainforests be used to detect a signal of past lowland biotic attrition? Austral Ecol 41:209–218

Caillon R, Suppo C, Casas JO, Arthur Woods H, Pincebourde S (2014) Warming decreases thermal heterogeneity of leaf surfaces: implications for behavioural thermoregulation by arthropods. Funct Ecol 28:1449–1458

Calosi P, Bilton DT, Spicer JI, Votier SC, Atfield A (2010) What determines a species' geographical range? Thermal biology and latitudinal range size relationships in European diving beetles (Coleoptera: Dytiscidae). J Animal Ecol 79:194–204

Camarota F, Powell S, Melo AS, Priest G, Marquis RJ, Vasconcelos HL (2016) Co-occurrence patterns in a diverse arboreal ant community are explained more by competition than habitat requirements. Ecol Evol 6:8907–8918

Carroll CR, Janzen DH (1973) Ecology of foraging by ants. Annu Rev Ecol Syst 4:231–257

Corlett RT (2011) Impacts of warming on tropical lowland rainforests. Trends Ecol Evol 26:606–613

Davidson DW, Cook SC, Snelling RR, Chua TH (2003) Explaining the abundance of ants in lowland tropical rainforest canopies. Science 300:969–972

Davison A, Chiba S (2008) Contrasting response to Pleistocene climate change by ground-living and arboreal Mandarina snails from the oceanic Hahajima archipelago. Philos Trans R Soc London B Biol Sci 363:3391–3400

De Frenne P, Zellweger F, Rodriguez-Sanchez F, Scheffers BR, Hylander K, Luoto M, Vellend M, Verheyen K, Lenoir J (2019) Global buffering of temperatures under forest canopies. Nat Ecol Evol 3:744–749

Deutsch CA, Tewksbury JJ, Huey RB, Sheldon KS, Ghalambor CK, Haak DC, Martin PR (2008) Impacts of climate warming on terrestrial ectotherms across latitude. Proc Natl Acad Sci 105:6668–6672

Diamond SE, Chick LD (2018) The Janus of macrophysiology: stronger effects of evolutionary history, but weaker effects of climate on upper thermal limits are reversed for lower thermal limits in ants. Curr Zool 64:223–230

Diamond SE, Sorger DM, Hulcr J, Pelini SL, Toro ID, Hirsch C, Oberg E, Dunn RR (2012) Who likes it hot? A global analysis of the climatic, ecological, and evolutionary determinants of warming tolerance in ants. Glob Chang Biol 18:448–456

Duarte BF, Michelutti KB, Antonialli-Junior WF, Cardoso CAL (2019) Effect of temperature on survival and cuticular composition of three different ant species. J Therm Biol 80:178–189

Dunn RR, Agosti D, Andersen AN, Arnan X, Bruhl CA, Cerdá X, Ellison AM, Fisher BL, Fitzpatrick MC, Gibb H, Gotelli NJ, Gove AD, Guenard B, Janda M, Kaspari M, Laurent EJ, Lessard J-P, Longino JT, Majer JD, Menke SB, McGlynn TP, Parr CL, Philpott SM, Pfeiffer M, Retana J, Suarez AV, Vasconcelos HL, Weiser MD, Sanders NJ (2009) Climatic drivers of hemispheric asymmetry in global patterns of ant species richness. Ecol Lett 12:324–333

Economo EP, Narula N, Friedman NR, Weiser MD, Guénard B (2018) Macroecology and macroevolution of the latitudinal diversity gradient in ants. Nat Commun 9:1–8

Fagan LL, Didham RK, Winchester NN, Behan-Pelletier V, Clayton M, Lindquist E, Ring RA (2006) An experimental assessment of biodiversity and species turnover in terrestrial vs canopy leaf litter. Oecologia 147:335–347

Farji-Brener AG, Dalton MC, Balza U, Courtis A, Lemus-Domínguez I, Fernández-Hilario R, Cáceres-Levi D (2018) Working in the rain? Why leaf-cutting ants stop foraging when it's raining. Insect Soc 65:233–239

Fellers JH (1989) Daily and seasonal activity in woodland ants. Oecologia 78:69–76

Feng G, Zhang J, Girardello M, Pellissier V, Svenning J-C (2020) Forest canopy height co-determines taxonomic and functional richness, but not functional dispersion of mammals and birds globally. Glob Ecol Biogeogr 29:1350–1359

Freiberg M (1997) Spatial and temporal pattern of temperature and humidity of a tropical premontane rain forest tree in Costa Rica. Selbyana 18:77–84

Freiberg M (2001) The influence of epiphyte cover on branch temperature in a tropical tree. Plant Ecol 153:241–250

Garcia-Robledo C, Chuquillanqui H, Kuprewicz EK, Escobar-Sarria F (2018) Lower thermal tolerance in nocturnal than in diurnal ants: a challenge for nocturnal ectotherms facing global warming. Ecol Entomol 43:162–167

Gaston KJ (2019) Nighttime ecology: the "nocturnal problem" revisited. Am Nat 193:481–502

Graham CH, Moritz C, Williams SE (2006) Habitat history improves prediction of biodiversity in rainforest fauna. Proc Natl Acad Sci 103:632–636

Grauel A-L, Hodell DA, Bernasconi SM (2016) Quantitative estimates of tropical temperature change in lowland Central America during the last 42 ka. Earth Planet Sci Lett 438:37–46

Grimbacher PS, Edwards W, Liddell MJ, Nelson PN, Nichols C, Wardhaugh CW, Stork NE (2018) Temporal variation in abundance of leaf litter beetles and ants in an Australian lowland tropical rainforest is driven by climate and litter fall. Biodivers Conserv 27:2625–2640

Guo F, Guénard B, Economo EP, Deutsch CA, Bonebrake TC (2020) Activity niches outperform thermal physiological limits in predicting global ant distributions. J Biogeogr 47:829–842

Hakala SM, Seppä P, Helanterä H (2019) Evolution of dispersal in ants (Hymenoptera: Formicidae): a review on the dispersal strategies of sessile superorganisms. Myrmecol News 29:35–55

Harada Y (2005) Diel and seasonal patterns of foraging activity in the arboreal ant Crematogaster matsumurai Forel. Entomol Sci 8:167–172

Hashimoto Y, Morimoto Y, Widodo ES, Mohamed M, Fellowes JR (2010) Vertical habitat use and foraging activities of arboreal and ground ants (Hymenoptera: Formicidae) in a Bornean tropical rainforest. Sociobiology 56:435–448

Heinze J, Tsuji K (1995) Ant reproductive strategies. Popul Ecol 37:135–149

Holt RD (2009) Bringing the Hutchinsonian niche into the 21st century: ecological and evolutionary perspectives. Proc Natl Acad Sci 106:19659–19665

Hood GW, Tschinkel WR (1990) Desiccation resistance in arboreal and terrestrial ants. Physiol Entomol 15:23–35

Houadria M, Salas-Lopez A, Orivel J, Blüthgen N, Menzel F (2015) Dietary and temporal niche differentiation in tropical ants—can they explain local ant coexistence? Biotropica 47:208–217

Huang Y, Stinchcombe JR, Agrawal AF (2015) Quantitative genetic variance in experimental fly populations evolving with or without environmental heterogeneity. Evolution 69:2735–2746

Huey RB, Deutsch CA, Tewksbury JJ, Vitt LJ, Hertz PE, Pérez HJÁ, Garland T (2009) Why tropical forest lizards are vulnerable to climate warming. Proc R Soc B Biol Sci 276:1939–1948

Huey RB, Kearney MR, Krockenberger A, Holtum JAM, Jess M, Williams SE (2012) Predicting organismal vulnerability to climate warming: roles of behaviour, physiology and adaptation. Philos Trans R Soc B Biol Sci 367:1665–1679

Jacquemin J, Roisin Y, Leponce M (2015) Spatio-temporal variation in ant (Hymenoptera: Formicidae) communities in leaf-litter and soil layers in a premontane tropical forest. Myrmecol News 22:129–139

Jones JC, Oldroyd BP (2006) Nest thermoregulation in social insects. Adv Insect Physiol 33:153–191

Kaspari M (1993) Body size and microclimate use in Neotropical granivorous ants. Oecologia 96:500–507

Kaspari M, Weiser MD (2000) Ant activity along moisture gradients in a neotropical forest. Biotropica 32:703–711

Kaspari M, Alonso L, O'Donnell S (2000) Three energy variables predict ant abundance at a geographical scale. Proc R Soc Lond Ser B Biol Sci 267:485–489

Kaspari M, Clay NA, Lucas J, Yanoviak SP, Kay A (2015) Thermal adaptation generates a diversity of thermal limits in a rainforest ant community. Glob Chang Biol 21:1092–1102
Kearney M, Porter W (2009) Mechanistic niche modelling: combining physiological and spatial data to predict species' ranges. Ecol Lett 12:334–350
Kingsolver JG, Huey RB (1998) Evolutionary analyses of morphological and physiological plasticity in thermally variable environments. Am Zool 38:545–560
Kirchner M, Sorenson C, Blaimer BB, Youngsteadt E (2024) Reaching new heights: arboreal ant diversity in a north American temperate forest ecosystem. Insect Conserv Diversity. https://doi.org/10.1111/icad.12788
Klinges DH, Scheffers BR (2021) Microgeography, not just latitude, drives climate overlap on mountains from tropical to polar ecosystems. Am Nat 197:75–92
Kronfeld-Schor N, Dayan T (2003) Partitioning of time as an ecological resource. Annu Rev Ecol Evol Syst 34:153–181
Kumar A, O'Donnell S (2009) Elevation and forest clearing effects on foraging differ between surface, and subterranean, foraging army ants (Formicidae: Ecitoninae). J Anim Ecol 78:91–97
Lasmar CJ, Bishop TR, Parr CL, Queiroz AC, Schmidt FA, Ribas CR (2021) Geographical variation in ant foraging activity and resource use is driven by climate and net primary productivity. J Biogeogr 48:1448–1459
Law SJ, Parr C (2020) Numerically dominant species drive patterns in resource use along a vertical gradient in tropical ant assemblages. Biotropica 52:101–112
Law SJ, Bishop TR, Eggleton P, Griffiths H, Ashton L, Parr C (2020) Darker ants dominate the canopy: testing macroecological hypotheses for patterns in colour along a microclimatic gradient. J Anim Ecol 89:347–359
Leahy L (2022) Climatic variability and the vulnerability of rainforest biota to a warming world. PhD thesis. James Cook University
Leahy L, Scheffers BR, Williams SE, Andersen AN (2020) Diversity and distribution of the dominant ant genus Anonychomyrma (Hymenoptera: Formicidae) in the Australian wet tropics. Diversity 12:474
Leahy L, Scheffers BR, Andersen AN, Hirsch BT, Williams SE (2021) Vertical niche and elevation range size in tropical ants: implications for climate resilience. Divers Distrib 27:485–496
Leahy L, Scheffers BR, Williams SE, Andersen AN (2022) Arboreality drives heat tolerance while elevation drives cold tolerance in rainforest ants. Ecology 103:e03549
Leahy L, Scheffers BR, Andersen AN, Williams SE (2024) Rates of species turnover across elevation vary with vertical stratum in rainforest ant assemblages. Ecography 2024:e06972
Lembrechts JJ, van den Hoogen J, Aalto J, Ashcroft MB, De Frenne P, Kemppinen J, Kopecký M, Luoto M, Maclean IMD, Crowther TW, Bailey JJ, Haesen S, Klinges DH, Niittynen P, Scheffers BR, Van Meerbeek K, Aartsma P, Abdalaze O, Abedi M, Aerts R, Ahmadian N, Ahrends A, Alatalo JM, Alexander JM, Allonsius CN, Altman J, Ammann C, Andres C, Andrews C, Ardö J, Arriga N, Arzac A, Aschero V, Assis RL, Assmann JJ, Bader MY, Bahalkeh K, Barančok P, Barrio IC, Barros A, Barthel M, Basham EW, Bauters M, Bazzichetto M, Marchesini LB, Bell MC, Benavides JC, Benito Alonso JL, Berauer BJ, Bjerke JW, Björk RG, Björkman MP, Björnsdóttir K, Blonder B, Boeckx P, Boike J, Bokhorst S, Brum BNS, Brůna J, Buchmann N, Buysse P, Camargo JL, Campoe OC, Candan O, Canessa R, Cannone N, Carbognani M, Carnicer J, Casanova-Katny A, Cesarz S, Chojnicki B, Choler P, Chown SL, Cifuentes EF, Čiliak M, Contador T, Convey P, Cooper EJ, Cremonese E, Curasi SR, Curtis R, Cutini M, Dahlberg CJ, Daskalova GN, de Pablo MA, Della Chiesa S, Dengler J, Deronde B, Descombes P, Di Cecco V, Di Musciano M, Dick J, Dimarco RD, Dolezal J, Dorrepaal E, Dušek J, Eisenhauer N, Eklundh L, Erickson TE, Erschbamer B, Eugster W, Ewers RM, Exton DA, Fanin N, Fazlioglu F, Feigenwinter I, Fenu G, Ferlian O, Fernández Calzado MR, Fernández-Pascual E, Finckh M, Higgens RF, Forte TAGW, Freeman EC, Frei ER, Fuentes-Lillo E, García RA, García MB, Géron C, Gharun M, Ghosn D, Gigauri K, Gobin A, Goded I, Goeckede M, Gottschall F, Goulding K, Govaert S, Graae BJ, Greenwood S, Greiser C, Grelle A, Guénard B, Guglielmin

M, Guillemot J, Haase P, Haider S, Halbritter AH, Hamid M, Hammerle A, Hampe A, Haugum SV, Hederová L, Heinesch B, Helfter C, Hepenstrick D, Herberich M, Herbst M, Hermanutz L, Hik DS, Hoffrén R, Homeier J, Hörtnagl L, Høye TT, Hrbacek F, Hylander K, Iwata H, Jackowicz-Korczynski MA, Jactel H, Järveoja J, Jastrzębowski S, Jentsch A, Jiménez JJ, Jónsdóttir IS, Jucker T, Jump AS, Juszczak R, Kanka R, Kašpar V, Kazakis G, Kelly J, Khuroo AA, Klemedtsson L, Klisz M, Kljun N, Knohl A, Kobler J, Kollár J, Kotowska MM, Kovács B, Kreyling J, Lamprecht A, Lang SI, Larson C, Larson K, Laska K, le Maire G, Leihy RI, Lens L, Liljebladh B, Lohila A, Lorite J, Loubet B, Lynn J, Macek M, Mackenzie R, Magliulo E, Maier R, Malfasi F, Máliš F, Man M, Manca G, Manco A, Manise T, Manolaki P, Marciniak F, Matula R, Mazzolari AC, Medinets S, Medinets V, Meeussen C, Merinero S, Mesquita R d CG, Meusburger K, Meysman FJR, Michaletz ST, Milbau A, Moiseev D, Moiseev P, Mondoni A, Monfries R, Montagnani L, Moriana-Armendariz M, Morra di Cella U, Mörsdorf M, Mosedale JR, Muffler L, Muñoz-Rojas M, Myers JA, Myers-Smith IH, Nagy L, Nardino M, Naujokaitis-Lewis I, Newling E, Nicklas L, Niedrist G, Niessner A, Nilsson MB, Normand S, Nosetto MD, Nouvellon Y, Nuñez MA, Ogaya R, Ogée J, Okello J, Olejnik J, Olesen JE, Opedal ØH, Orsenigo S, Palaj A, Pampuch T, Panov AV, Pärtel M, Pastor A, Pauchard A, Pauli H, Pavelka M, Pearse WD, Peichl M, Pellissier L, Penczykowski RM, Penuelas J, Petit Bon M, Petraglia A, Phartyal SS, Phoenix GK, Pio C, Pitacco A, Pitteloud C, Plichta R, Porro F, Portillo-Estrada M, Poulenard J, Poyatos R, Prokushkin AS, Puchalka R, Pușcaș M, Radujković D, Randall K, Ratier Backes A, Remmele S, Remmers W, Renault D, Risch AC, Rixen C, Robinson SA, Robroek BJM, Rocha AV, Rossi C, Rossi G, Roupsard O, Rubtsov AV, Saccone P, Sagot C, Sallo Bravo J, Santos CC, Sarneel JM, Scharnweber T, Schmeddes J, Schmidt M, Scholten T, Schuchardt M, Schwartz N, Scott T, Seeber J, Segalin de Andrade AC, Seipel T, Semenchuk P, Senior RA, Serra-Diaz JM, Sewerniak P, Shekhar A, Sidenko NV, Siebicke L, Collier LS, Simpson E, Siqueira DP, Sitková Z, Six J, Smiljanic M, Smith SW, Smith-Tripp S, Somers B, Sørensen MV, Souza JJLL, Souza BI, Dias AS, Spasojevic MJ, Speed JDM, Spicher F, Stanisci A, Steinbauer K, Steinbrecher R, Steinwandter M, Stemkovski M, Stephan JG, Stiegler C, Stoll S, Svátek M, Svoboda M, Tagesson T, Tanentzap AJ, Tanneberger F, Theurillat J-P, Thomas HJD, Thomas AD, Tielbörger K, Tomaselli M, Treier UA, Trouillier M, Turtureanu PD, Tutton R, Tyystjärvi VA, Ueyama M, Ujházy K, Ujházyová M, Uogintas D, Urban AV, Urban J, Urbaniak M, Ursu T-M, Vaccari FP, Van de Vondel S, van den Brink L, Van Geel M, Vandvik V, Vangansbeke P, Varlagin A, Veen GF, Veenendaal E, Venn SE, Verbeeck H, Verbrugggen E, Verheijen FGA, Villar L, Vitale L, Vittoz P, Vives-Ingla M, von Oppen J, Walz J, Wang R, Wang Y, Way RG, Wedegärtner REM, Weigel R, Wild J, Wilkinson M, Wilmking M, Wingate L, Winkler M, Wipf S, Wohlfahrt G, Xenakis G, Yang Y, Yu Z, Yu K, Zellweger F, Zhang J, Zhang Z, Zhao P, Ziemblińska K, Zimmermann R, Zong S, Zyryanov VI, Nijs I, Lenoir J (2022) Global maps of soil temperature. Glob Chang Biol 28:3110–3144

Levings SC (1983) Seasonal, annual, and among-site variation in the ground ant community of a deciduous tropical forest: some causes of patchy species distributions. Ecol Monogr 53:435–455

Levy O, Dayan T, Porter WP, Kronfeld-Schor N (2019) Time and ecological resilience: can diurnal animals compensate for climate change by shifting to nocturnal activity? Ecol Monogr 89:e01334

Loarie SR, Duffy PB, Hamilton H, Asner GP, Field CB, Ackerly DD (2009) The velocity of climate change. Nature 462:1052

Longino JT, Branstetter MG, Ward PS (2019) Ant diversity patterns across tropical elevation gradients: effects of sampling method and subcommunity. Ecosphere 10:e02798

Lucky A, Trautwein MD, Guénard BS, Weiser MD, Dunn RR (2013) Tracing the rise of ants – out of the ground. PLoS One 8:e84012

MacArthur RH, Wilson EO (1967) The theory of Island biogeography. Princeton University Press, Princeton

McGlynn TP (2006) Ants on the move: resource limitation of a litter-nesting ant community in Costa Rica1. Biotropica 38:419–427

Menzel F, Zumbusch M, Feldmeyer B (2018) How ants acclimate: impact of climatic conditions on the cuticular hydrocarbon profile. Funct Ecol 32:657–666
Montejo-Kovacevich G, Martin SH, Meier JI, Bacquet CN, Monllor M, Jiggins CD, Nadeau NJ (2020) Microclimate buffering and thermal tolerance across elevations in a tropical butterfly. J Exp Biol 223:jeb220426
Moreau CS, Bell CD, Vila R, Archibald SB, Pierce NE (2006) Phylogeny of the ants: diversification in the age of angiosperms. Science 312:101–104
Moritz C, Agudo R (2013) The future of species under climate change: resilience or decline? Science 341:504–508
Mottl O, Yombai J, Fayle TM, Novotný V, Klimeš P (2020) Experiments with artificial nests provide evidence for ant community stratification and nest site limitation in a tropical forest. Biotropica 52:277–287
Moussalli A, Moritz C, Williams SE, Carnaval AC (2009) Variable responses of skinks to a common history of rainforest fluctuation: concordance between phylogeography and paleo-distribution models. Mol Ecol 18:483–499
Murphy BP, Andersen AN, Parr CL (2016) The underestimated biodiversity of tropical grassy biomes. Philos Trans R Soc B Biol Sci 371:20150319
Nadeau CP, Urban MC, Bridle JR (2017) Climates past, present, and yet-to-come shape climate change vulnerabilities. Trends Ecol Evol 32:786–800
Nascimento G, Câmara T, Arnan X (2022) Critical thermal limits in ants and their implications under climate change. Biol Rev 97:1287–1305
Nowrouzi S, Andersen AN, Bishop TR, Robson SK (2018) Is thermal limitation the primary driver of elevational distributions? Not for montane rainforest ants in the Australian Wet Tropics. Oecologia 188:333–342
Oliveira BF, Scheffers BR (2019) Vertical stratification influences global patterns of biodiversity. Ecography 42:249–249
Otero LM, Huey RB, Gorman GC (2015) A few meters matter: local habitats drive reproductive cycles in a tropical lizard. Am Nat 186:E72–E80
Parkash R, Rajpurohit S, Ramniwas S (2008) Changes in body melanisation and desiccation resistance in highland vs. lowland populations of D. melanogaster. J Insect Physiol 54:1050–1056
Parr CL, Bishop TR (2022) The response of ants to climate change. Glob Chang Biol 28:3188–3205
Paul KI, Polglase PJ, Smethurst PJ, O'Connell AM, Carlyle CJ, Khanna PK (2004) Soil temperature under forests: a simple model for predicting soil temperature under a range of forest types. Agric For Meteorol 121:167–182
Penick CA, Tschinkel WR (2008) Thermoregulatory brood transport in the fire ant, Solenopsis invicta. Insect Soc 55:176–182
Pfeiffer M, Linsenmair KE (2000) Contributions to the life history of the Malaysian giant ant Camponotus gigas (Hymenoptera, Formicidae). Insect Soc 47:123–132
Pincebourde S, Sinoquet H, Combes D, Casas J (2007) Regional climate modulates the canopy mosaic of favourable and risky microclimates for insects. J Anim Ecol 76:424–438
Pintanel P, Tejedo M, Ron SR, Llorente GA, Merino-Viteri A (2019) Elevational and microclimatic drivers of thermal tolerance in Andean Pristimantis frogs. J Biogeogr 46:1664–1675
Plowman NS, Hood AS, Moses J, Redmond C, Novotny V, Klimes P, Fayle TM (2017) Network reorganization and breakdown of an ant–plant protection mutualism with elevation. Proc R Soc B 284
Plowman NS, Mottl O, Novotny V, Idigel C, Philip FJ, Rimandai M, Klimes P (2020) Nest microhabitats and tree size mediate shifts in ant community structure across elevation in tropical rainforest canopies. Ecography 43:431–442
Rodríguez-Trelles F, Tarrío R, Santos M (2013) Genome-wide evolutionary response to a heat wave in Drosophila. Biol Lett 9:20130228
Roeder KA, Roeder DV, Bujan J (2021) Ant thermal tolerance: a review of methods, hypotheses, and sources of variation. Ann Entomol Soc Am 114:459

Román-Palacios C, Wiens JJ (2020) Recent responses to climate change reveal the drivers of species extinction and survival. Proc Natl Acad Sci 117:4211–4217

Sanders NJ, Lessard JP, Fitzpatrick MC, Dunn RR (2007) Temperature, but not productivity or geometry, predicts elevational diversity gradients in ants across spatial grains. Glob Ecol Biogeogr 16:640–649

Sangma JSA, Prasad SB (2021) Population and nesting behaviour of weaver ants, *Oecophylla smaragdina* from Meghalaya, India. Sociobiology 68:e7204–e7204

Saupe EE, Qiao HJ, Hendricks JR, Portell RW, Hunter SJ, Soberon J, Lieberman BS (2015) Niche breadth and geographic range size as determinants of species survival on geological time scales. Glob Ecol Biogeogr 24:1159–1169

Scheffers BR, Williams SE (2018) Tropical mountain passes are out of reach – but not for arboreal species. Front Ecol Environ 16:101–108

Scheffers BR, Phillips BL, Laurance WF, Sodhi NS, Diesmos A, Williams SE (2013) Increasing arboreality with altitude: a novel biogeographic dimension. Proc R Soc B Biol Sci 280:20131581

Scheffers BR, Evans TA, Williams SE, Edwards DP (2014) Microhabitats in the tropics buffer temperature in a globally coherent manner. Biol Lett 10:20140819

Scheffers BR, Edwards DP, Macdonald SL, Senior RA, Andriamahohatra LR, Roslan N, Rogers AM, Haugaasen T, Wright P, Williams SE (2017a) Extreme thermal heterogeneity in structurally complex tropical rain forests. Biotropica 49:35–44

Scheffers BR, Shoo L, Phillips B, Macdonald SL, Anderson A, VanDerWal J, Storlie C, Gourret A, Williams SE (2017b) Vertical (arboreality) and horizontal (dispersal) movement increase the resilience of vertebrates to climatic instability. Glob Ecol Biogeogr 26:787–798

Seifert B, Fiedler P, Schultz R (2017) Escape to the high canopy—thermal deficiency causes niche expansion in a forest-floor ant. Insect Sci 24:699–707

Soare TW, Tully SI, Willson SK, Kronauer DJC, O'Donnell S (2011) Choice of nest site protects army ant colonies from environmental extremes in tropical montane forest. Insect Soc 58:299–308

Sosiak CE, Barden P (2021) Multidimensional trait morphology predicts ecology across ant lineages. Funct Ecol 35:139–152

Southwood T (1988) Tactics, strategies and templets. Oikos 52:3–18

Spicer ME, Stark AY, Adams BJ, Kneale R, Kaspari M, Yanoviak SP (2017) Thermal constraints on foraging of tropical canopy ants. Oecologia 183:1007–1017

Stark AY, Adams BJ, Fredley JL, Yanoviak SP (2017) Out on a limb: thermal microenvironments in the tropical forest canopy and their relevance to ants. J Therm Biol 69:32–38

Stott P (2016) How climate change affects extreme weather events. Science 352:1517–1518

Sunday JM, Bates AE, Dulvy NK (2011) Global analysis of thermal tolerance and latitude in ectotherms. Proc R Soc Lond B Biol Sci 278:1823–1830

Sunday JM, Bates AE, Kearney MR, Colwell RK, Dulvy NK, Longino JT, Huey RB (2014) Thermal-safety margins and the necessity of thermoregulatory behavior across latitude and elevation. Proc Natl Acad Sci 111:5610–5615

Szewczyk T, McCain CM (2016) A systematic review of global drivers of ant elevational diversity. PLoS One 11:e0155404

Tanaka HO, Yamane S, Itioka T (2010) Within-tree distribution of nest sites and foraging areas of ants on canopy trees in a tropical rainforest in Borneo. Popul Ecol 52:147

Thuiller W, Lavorel S, Araújo MB (2005) Niche properties and geographical extent as predictors of species sensitivity to climate change. Glob Ecol Biogeogr 14:347–357

Urban MC (2018) Escalator to extinction. Proc Natl Acad Sci 115:11871–11873

Vasconcelos HL, Neves KC, Andersen AN (2023) Vertical stratification of ant assemblages varies along a latitudinal gradient in Brazilian savanna. J Biogeogr 50:1331–1340

Westoby M, Yates L, Holland B, Halliwell B (2023) Phylogenetically conservative trait correlation: quantification and interpretation. J Ecol 111:2105–2117
Wetterer JK (1990) Diel changes in forager size, activity, and load selectivity in a tropical leaf-cutting ant, Atta cephalotes. Ecol Entomol 15:97–104
Williams SE, Shoo LP, Isaac JL, Hoffmann AA, Langham G (2008) Towards an integrated framework for assessing the vulnerability of species to climate change. PLoS Biol 6:e325
Wong MK, Guénard B (2017) Subterranean ants: summary and perspectives on field sampling methods, with notes on diversity and ecology (Hymenoptera: Formicidae). Myrmecol News 25:1–16
Wright SJ, Muller-landau HC, Schipper J (2009) The future of tropical species on a warmer planet. Conserv Biol 23:1418–1426
Xing S, Leahy L, Ashton LA, Kitching RL, Bonebrake TC, Scheffers BR (2023) Ecological patterns and processes in the vertical dimension of terrestrial ecosystems. J Anim Ecol 92:538–551
Yanoviak SP, Kaspari M (2000) Community structure and the habitat templet: ants in the tropical forest canopy and litter. Oikos 89:259–266
Yanoviak SP, Berghoff SM, Linsenmair KE, Zotz G (2011) Effects of an epiphytic orchid on arboreal ant community structure in Panama. Biotropica 43:731–737
Yusah KM, Foster WA, Reynolds G, Fayle TM (2018) Ant mosaics in Bornean primary rain forest high canopy depend on spatial scale, time of day, and sampling method. PeerJ 6:e4231

Part V
Plant-Ant Interactions and Associated Ecosystem Services

Chapter 13
Climate-Change Impacts on Ant-Plant Mutualistic Networks: Insights from Current Studies and Future Prospects

Talita Câmara, Inara R. Leal, and Xavier Arnan

13.1 Introduction

One of the greatest challenges of the future is preserving biodiversity in the face of climate change. Many species are predicted to adjust their distributions globally through phenotypic plasticity or genetic adaptation (Valladares et al. 2014; King et al. 2018). However, species unable to cope with new climate conditions may face extinction (Urban et al. 2015). Beyond the risks of species redistribution or loss, a critical concern is the disruption of ecological interactions that are essential for maintaining ecosystem stability and function (Kaiser-Bunbury et al. 2010). Mutualistic relationships—critical for ecosystem resilience—are particularly sensitive to changes in climate-related variables such as temperature (Jones et al. 2008; Cruz et al. 2023) and precipitation (Leal and Peixoto 2017). These factors can drive partner-switching due to physiological stress (Fitzpatrick et al. 2014; Oliveira et al. 2021; Câmara et al. 2025), significantly altering the specialization of mutualistic interactions across spatial (Câmara et al. 2018) and temporal scales (Câmara et al. 2019).

T. Câmara (✉)
Universidade de Pernambuco—Campus Garanhuns, Garanhuns, PE, Brazil

Programa de Pós-Graduação em Biologia Vegetal, Universidade Federal de Pernambuco, Recife, PE, Brazil

Programa de Pós-Graduação em Ecologia e Evolução, Universidade Estadual de Feira de Santana, Feira de Santana, BA, Brazil
e-mail: camara.talita@yahoo.com.br

I. R. Leal
Departamento de Botânica, Universidade Federal de Pernambuco, Recife, Pernambuco, Brazil
e-mail: inara.leal@ufpe.br

X. Arnan
Universidade de Pernambuco – Campus Garanhuns, Garanhuns, Pernambuco, Brazil
e-mail: xavier.arnan@upe.br

I. R. Leal et al. (eds.), *Ant Responses to Climate Change*, Entomology in Focus 10,
https://doi.org/10.1007/978-3-032-21127-9_13

Mutualisms have traditionally been categorized as facultative (where species can survive and reproduce without their partners) or obligate (where species depend on them for survival and reproduction), alternatively as generalist (interacting with multiple partners) or specialist interactions (interacting with only a few partners) (Chomicki et al. 2020). However, this classification is increasingly seen as outdated, because generalist/specialist exists along a continuum rather than as a strict dichotomy (Bronstein 2015). Understanding the vulnerability of mutualisms to climate change requires considering their broader community context as *they directly influence the stability and resilience of ecological communities* (Vidal et al. 2021). To address this, network ecologists have developed metrics that assess the degree of specialization of ecological interactions at the whole-of-community level (Blüthgen et al. 2006). In highly generalized networks, species interact with many partners, with interactions more evenly distributed among partners, leading to high resource overlap and lower niche partitioning. Such networks are often hypothesized to be more resilient to climate change. If one partner species declines or goes extinct, other species can potentially compensate, maintaining the mutualistic function (Nagaishi and Takemoto 2018). In contrast, highly specialized networks are characterized by species interacting with only a few partners or disproportionately with specific partners, resulting in low-resource overlap and high niche partitioning (Blüthgen et al. 2006). While potentially offering high efficiency in stable environments, highly specialized networks are potentially thought to be less adaptable to environmental fluctuations (Takimoto and Suzuki 2016; Vidal et al. 2021). The loss or decline of even a single key partner can disrupt or collapse the entire interaction, leading to cascading effects throughout the community. Therefore, understanding how climate change alters network specialization is critical for predicting the fate of mutualisms and the ecosystems they support.

Despite their importance, empirical evidence of climate-driven changes in mutualistic networks remains limited (Schleuning et al. 2012; Dalsgaard et al. 2011; Sebastián-Gonzalez et al. 2015; Jones et al. 2023). In this chapter, we focus on the potential impacts of climate change on the degree of specialization in ant-plant mutualisms, using ecological networks as a tool. One predictive approach involves space-for-time studies, which infer ecological responses to climate change by examining current patterns across climate gradients (Fukami and Wardle 2005, Rustad 2008; see Chap. 8, 9, 10). This provides insights into potential temporal trajectories of ecological change (Lovell et al. 2023; Kharouba and Williams 2024). This framework is particularly relevant for understanding how changes in climatic variables such as temperature and precipitation may influence ant-plant mutualisms, given the rapid responses of ant communities to environmental changes (Caddy-Retalic et al. 2019; Parr and Bishop 2022). Additionally, we employed modelling techniques to project future scenarios of ant-plant network specialization, assessing the resilience of these interactions under changing climatic conditions.

Ant-plant interactions have significantly advanced our understanding of mutualisms (Bronstein 1998). Ants play a direct role in plant evolution, shaping traits related to defence, seed dispersal, and, to a lesser extent, pollination (Rico-Gray and Oliveria 2007). These interactions occur along a continuum from generalized to highly specialized associations. Here, we focus on three key types of ant-plant mutualisms: those mediated by extrafloral nectaries (EFNs), domatia, and diaspores. EFN-mediated mutualisms involve plants secreting carbohydrate-rich nectar, which attracts a diverse array of opportunistic ants that, in turn, defend plants against herbivores (Bentley and Elias 1983; see Chap. 14). Although ants generally exhibit low fidelity to specific plants, competition for nectar resources among attendant ant species can drive niche partitioning (Blüthgen and Fiedler 2004) and influence community-level specialization (Blüthgen et al. 2004). In domatia-mediated mutualisms, plants provide structural resources such as hollow stems or swollen thorns for ant nesting, often supplemented with lipid-rich food bodies. In return, resident ants offer consistent protection against herbivores (Rico-Gray and Oliveria 2007). However, some domatia can also be occupied by opportunistic ant species, which may influence the degree of specialization as well as plant survival and reproduction (Emer et al. 2013; Moraes and Vasconcelos 2009). Myrmecochory, or ant-mediated seed dispersal, occurs when ants transport seeds bearing lipid-rich elaiosomes, benefiting numerous plant species (Leal et al. 2007; see Chap. 16). However, ant seed dispersal is not exclusive to myrmecochorous plants; ants also collect fallen fleshy fruits and remove seeds from them (Christianini et al. 2012). Competition for seeds among ants may similarly drive niche partitioning (Davidson 1977) and influence the degree of specialization within communities.

Our study addresses the following key questions: (1) How does climate influence the network structures of ant-plant mutualisms? (2) How does the degree of mutualistic specialization vary along climatic gradients? (3) How do different types of ant-plant mutualisms—such as ant-EFN, ant-domatia, and ant-seed dispersal—differ in their specialization responses to future climate change? To answer these questions, we conducted a multi-faced analysis. First, we performed a comprehensive review of the existing evidence on how climate-related variables—including temperature, precipitation, elevation, latitude, and future climatic scenarios—affect the structures of ant-plant networks. Second, we quantitatively evaluated the association between current climatic variables and specialization in overall ant-plant mutualisms, specifically focusing on ant-domatia, ant-seed, and ant-EFN networks. Finally, we modelled potential shifts in network specialization for each type of ant--plant mutualism—ant-domatia, ant-seed, and ant-EFN networks—under future climate scenarios.

13.2 Effects of Climate-Related Variables on the Network Structure of Ant-Plant Mutualisms

We performed a literature review of studies that have investigated the influence of climate on ant-plant networks. Our searches in Web of Science and Google Scholar used the following keywords: "ant-plant network*" AND "climat*" OR "climate chan*" OR "elevat*" OR "latit*" OR "precipitat*" OR "temperatu*". This search yielded 26 papers, of which only six met our criteria: studies that explicitly addressed the impacts of climate on ant-plant interaction networks either using direct measures of precipitation or temperature or elevation or latitude as climate proxies. Six of the studies were conducted in the Neotropics, one involved myrmecochory, three on myrmecophilous interactions, and one on obligate mutualisms involving ant--gardeners. Additionally, there is a single study from Papua New Guinea that includes myrmecophytic plants. For this synthesis, we focus on network metrics (e.g. connectance, interaction diversity, interaction evenness, nestedness, niche specialization) of community-level interactions.

Temperature Two of the studies examined temperature. In Mexico, an ant-plant mutualism mediated by floral and extrafloral nectar and honeydew showed increased nestedness with rising temperatures. This pattern was attributed to reduced ant activity under cooler temperatures (Rico-Gray et al. 2012). In Brazilian savanna, the temperature did not affect nestedness or niche-based specialization (H_2') in seed-ant networks (Anjos et al. 2019).

Precipitation *Three of the studies examined precipitation.* In the above Mexican study, network nestedness increased with increasing precipitation (Rico-Gray et al. 2012). In Brazilian Caatinga, precipitation was found to increase niche-based specialization H_2' and reduce interaction evenness in ant-EFN networks (Câmara et al. 2018). In myrmecochorous networks in Brazil, there was no relationship between precipitation and either H_2' or nestedness (Anjos et al. 2019).

Elevation *One study examined elevation.* In Papua New Guinea, ant-domatia networks showed decreased specialization (H_2') and modularity at higher elevations, driven by reduced partner and increased connectance due to the dominance of *Anonychomyrma* sp. (Plowman et al. 2017).

Latitude *One study examined latitude.* A study of tree-nesting ant-plant networks in a Neotropical savanna found increasing interaction diversity with increasing latitude, but there was no relationship for specialization (H_2'), nestedness, or modularity (Dáttilo and Vasconcelos 2019).

Future Climate Scenarios *There is only one study that examined future climate scenarios, that of ant-gardener networks* in the Neotropics (Morales-Lineares et al. 2021). These networks showed a 50% decrease in nestedness under future scenarios, while modularity and specialization (H_2') remained unchanged. Despite the changes in nestedness, ant-gardener networks were found to be robust against climate change in terms of overall structure (Morales-Lineares et al. 2021).

13.3 Climate Impacts on Ant-Plant Mutualism in the Neotropics

13.3.1 Influence of Climatic Variables

To assess the influence of precipitation and temperature on ant-plant mutualistic networks, we compiled data from 48 quantitative ant-plant networks reported in Fayle et al. (2024), along with two additional networks provided by our research (Câmara et al. 2018, 2025) and one provided by the author (Alencar et al. 2023). We

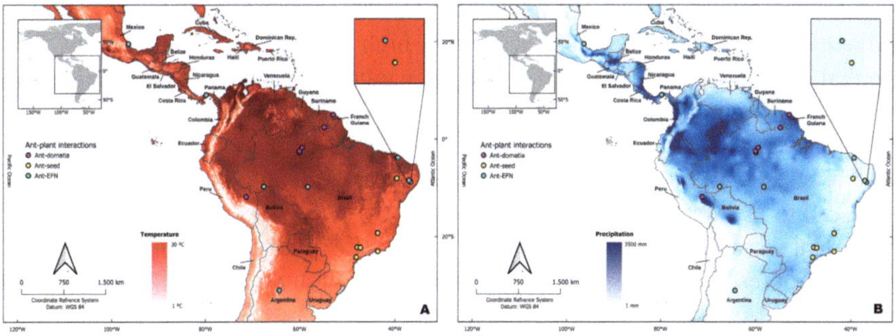

Fig. 13.1 America map highlighting the study sites and its climatic conditions, with (**a**) mean annual temperature and (**b**) mean annual precipitation. Areas with higher intensity of red (**a**) represent regions with higher mean annual temperature, while areas with higher intensity of blue (**b**) indicate regions of higher mean annual precipitation. The dots on the maps represent the 21 mutualistic networks, with different colours indicating the type of interactions between ants and plants: purple for ant-domatia, yellow for ant-seed, and green for ant-extrafloral nectar (EFN) networks. Two overlapping dots are highlighted to indicate studies conducted at the same site. The data for these networks were sourced from Fayle et al. (2024), which compiled ant-plant interaction data from the following studies: ant-domatia networks (Davidson 1989; Dáttilo et al. 2013; Dejean et al. 2018; Emer et al. 2013; Fonseca and Ganade 1996; Passmore et al. 2012), ant-seed interactions (Arruda et al. 2020; Christianini et al. 2012; da Silva Laviski et al. 2021; Leal et al. 1998, 2014; Oliveira et al. 2019, Pizo and Oliveira 2001), and ant-EFN networks (Díaz-Castelazo et al. 2020; Dáttilo et al. 2014; Miranda et al. 2022; Schemske 1982). Ant-EFN networks (Alencar et al. 2023; Câmara et al. 2018; Câmara et al. 2025) were additionally incorporated into this dataset

selected studies that specifically documented ants attending EFNs, nesting in domatia, or interacting with seeds, resulting in a total of 30 networks. Given that some studies were represented by multiple subplots, we aggregated the subplots within each local study into a single network. While this aggregation may obscure fine--scale spatial variation, it enables the examination of broader patterns at a larger spatial scale, which aligns with the objectives of this study. Nearly 70% of these studies were conducted in the Neotropics (Fig. 13.1), and so we considered only these as a coherent biota with a shared evolutionary history. In total, our study included six ant-domatia networks, seven ant-seed networks, and eight ant-EFN networks, for a combined total of 21 ant-plant interaction networks. Ant-EFN and ant-seed interactions were documented across a broad range of climatic environments, whereas ant-domatia networks were observed predominantly in ecosystems characterized by consistently high temperatures and year-round precipitation (Fig. 13.2). Lou et al. (2023) found that the richness of domatia-bearing plants is positively correlated with precipitation levels and lower seasonality.

We selected 19 temperature- and precipitation-related bioclimatic variables from WorldClim v.2.1 (Fick and Hijmans 2017), representing the 1970–2000 historical period, at a spatial resolution of 1 km^2. To reduce dimensionality and eliminate correlations among variables, we performed a principal component analysis (PCA). This analysis utilized data from 21 coordinates where ant-plant networks were sampled; bioclimatic variables were extracted and standardized before analysis. The same 19 bioclimatic variables used to assess current conditions were also applied to project future climates. Two future climate scenarios for 2041–2060 were considered: SSP2-4.5, representing a stabilization pathway following moderate mitigation efforts, and SSP5-8.5, assuming high greenhouse gas emissions with minimal mitigation (O'Neill et al. 2016). Separate PCA analyses were conducted for each scenario to evaluate potential impacts on ant-plant interactions.

Our findings revealed that temperature variables played a more important role in explaining the variation in different types of ant-plant networks in the Neotropics compared to precipitation (Fig. 13.2). The first two PCA axes accounted for 77.8% of the total variability in ant-plant networks. PCA1 explained 60% of the total variance and represented a gradient of decreasing temperature, including mean annual temperature (loading = −0.92), isothermality (−0.85), temperature seasonality (0.90), minimum temperature of the coldest month (−0.95), temperature annual range (0.88), mean temperature of the driest quarter (−0.92), mean temperature of warmest quarter (−0.85), and mean temperature of the coldest quarter (−0.94). PCA1 thus represents a gradient transitioning from warmer and stable temperatures to cooler and more variable ones. PCA2 captured precipitation gradients, explaining an additional 17.8% of the variance. The variables driving this axis included annual precipitation (loading = −0.56), precipitation of the driest month (−0.60), precipitation seasonality (0.64), precipitation of the driest quarter (−0.60), and precipitation of the warmest quarter (−0.65).

When projecting ant-plant interaction networks under future climate scenarios, thermal gradients continued to play a dominant role in determining the arrangement of different types of ant-plant networks within climate space. In the optimistic scenario, the first PCA axis (thermal gradient) explained 59.8% of the total variance, while the second axis (precipitation gradient) accounted for 18.2%. In the pessimistic scenario, the first axis explained 59% of the variance, and the second axis 18.9%.

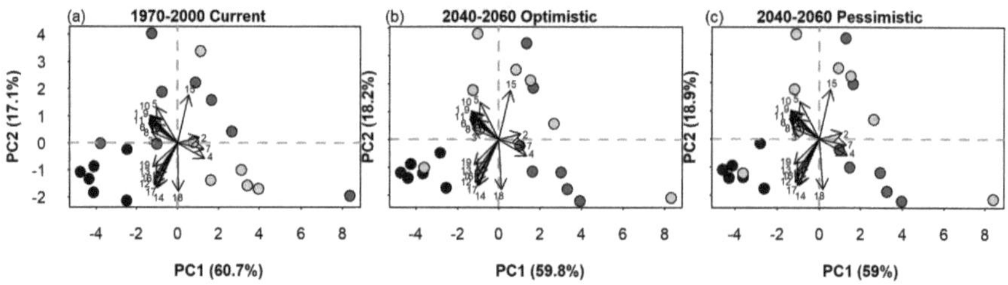

Fig. 13.2 Principal component analysis (PCA) of 19 climatic variables related to temperature and precipitation was extracted from the locations where 21 ant-plant mutualistic networks were sampled. The PCA includes current climatic conditions (**a**) and two future climate scenarios for 2040–2060: SSP2-4.5 (optimistic) (**b**) and SSP5-8.5 (pessimistic) (**c**). Dot colours represent different types of ant-plant interactions: purple for ant-domatia, yellow for ant-seed, and green for ant--extrafloral nectar (EFN) networks. Each number represents a biclimatic: 1, annual mean temperature; 2, mean diurnal range; 3, isothermality (2/7); 4, temperature seasonality; 5, max temperature of warmest month; 6, min temperature of coldest month; 7, temperature annual range (5–6); 8, mean temperature of wettest quarter; 9, mean temperature of driest quarter; 10, mean temperature of warmest quarter; 11, mean temperature of coldest quarter; 12, annual precipitation; 13, precipitation of wettest month; 14, precipitation of driest month; 15, precipitation seasonality; 16, precipitation of wettest quarter; 17, precipitation of driest quarter; 18, precipitation of warmest quarter; 19, precipitation of coldest quarter

13.3.2 Current Climate and Specialization of Ant-Plant Interactions

We assessed the specialization of the 21 ant-plant mutualistic networks, using the frequency-based index H_2' (Blüthgen et al. 2006). The H_2' measures the exclusiveness of interactions (niche partitioning) at the network-level and is standardized by the marginal totals, ranging from 0.0 (indicating high generalization) to 1.0 (indicating high specialization). This metric is particularly effective for comparing network-wide patterns across mutualisms, as it reflects the overall distribution and exclusivity of interactions, independent of total abundance and diversity (Blüthgen et al. 2008). Ant-domatia mutualism exhibited the highest mean specialization (mean ± SD; $H_2' = 0.85 \pm 0.06$), followed by ant-EFN ($H_2' = 0.27 \pm 0.11$) and ant-seed ($H_2' = 0.25 \pm 0.17$) networks. We used a generalized additive model (GAM) to predict specialization (response variable) along a thermal gradient, represented by the first axis of the PCA previously conducted (explanatory variable). To facilitate interpretation, we multiplied the values of the first PCA axis by −1, so that higher values represent warmer conditions and lower values cooler conditions. The GAM allows us to capture non-linear relationships between specialization and climate.

We observed that *as PC1 increased, ant-plant mutualism specialization also increased*, suggesting that *warmer and stable temperatures play a key role in shaping specialized interactions within these networks* (GAM: $R^2_{adj} = 0.67$, $t = 11.67$, $P < 0.001$). Ant-domatia mutualisms, producing the most specialized group networks, are found particularly in consistently warmer environments (Fig. 13.3a), which could be shaping the observed patterns of specialization along thermal gradients. Thus, to assess the impact of ant-domatia networks, we repeated the analysis excluding them. Even after their exclusion, the thermal gradient continued to be significantly correlated with ant-plant specialization, with specialization increasing more linearly along the thermal gradient (Fig. 13.3b; GAM, $R^2_{adj} = 0.15$; $t = 7.45$; $P < 0.001$).

Our findings suggest that specialization in ant-plant mutualisms increases in warmer, more thermally stable environments, likely driven by two non-exclusive hypotheses.

Evolutionary Stability Hypothesis: Long-term ecological stability in warm climates has facilitated the co-evolution of highly specialized ant-plant interactions, particularly for domatia-bearing plants that provide critical nesting resources. Domatia-bearing plants have diversified primarily in tropical environments (Chomicki and Renner 2015), especially in the Amazon rainforest, where they exhibit high species richness (Lou et al. 2023). Arboreal habitats expose ants to increased physiological challenges, such as desiccation, and higher predation pressure (Nelsen et al. 2018). Nesting space is often a limiting resource for arboreal ants, intensifying competition for stable and protective nesting sites. In this context, domatia-bearing plants provide a crucial advantage, fostering highly specialized and co-evolved relationships. The size and availability of domatia likely co-evolved with colony structure, a process favoured by long-term ecological stability (Fonseca 1999).

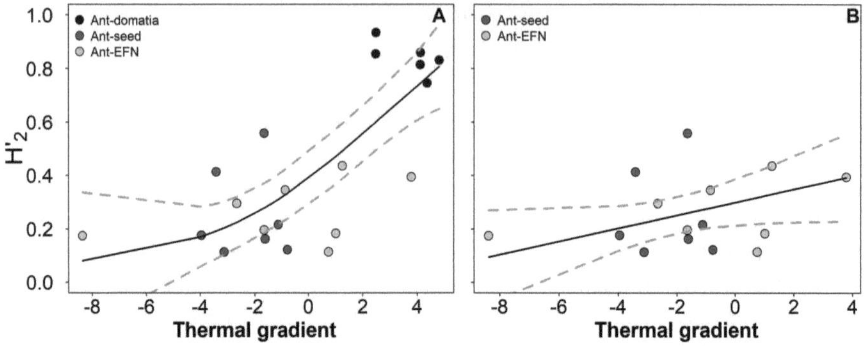

Fig. 13.3 Relationship between specialization (H_2') of ant-plant networks and the current thermal gradient represented by the first PCA axis. Figure (**a**) presents all ant-plant interactions, while Figure (**b**) displays the results after excluding the highly specialized ant-domatia interactions. Dot colours indicate different types of ant-plant interactions: purple for ant-domatia, yellow for ant--seed, and green for ant-extrafloral nectar (EFN) networks. The solid black line represents the fitted model from a generalized additive model (GAM), while the dashed lines represent the 95% confidence interval. The generalized additive model was weighted according to the sampling intensity $= \frac{\sqrt{N_i}}{size_i}$, where N_i is the number of interactions in network *i* and $size_i$ is the product of the number of plant species and the number of animal species in network *i* (Schleuning et al. 2012)

Resource Availability and Competition Hypothesis: In warmer environments, increased production of plant rewards intensifies competition, driving dominant ants to adopt stronger monopolization strategies and reinforcing specialization in ant-plant interactions. Warmer, more stable conditions accelerate organ initiation in plants, enhancing the production of extrafloral nectar, fruits, and seeds (Morison and Lawlor 1999). These environments also support high ant species richness, particularly in tropical rainforests, further increasing competitive pressure (Blüthgen et al. 2000). Stronger monopolization of high-quality EFNs, fruits, and seeds by dominant ants may reinforce competitive hierarchies, promoting greater niche partitioning in ant-plant interactions at the community level.

13.3.3 Ant-Plant Network Specialization Under Future Climate Scenarios

Our above findings raise the question: could mutualisms mediated by EFN and diaspores be more resilient to climate change than mutualism mediated by domatia? To try to answer this relevant question, we predicted the specialization of ant-plant networks under future climate change scenarios. We used the same two future climate scenarios for the years 2040–2060: a more optimistic scenario (SSP2-4.5) and a more pessimistic scenario (SSP5-8.5), both used on subtopic 3.1. To estimate the specialization under each scenario, we first fitted the GAM using the current network specialization as the responsible variable and the first axis of PCA (thermal gradient) as the predictor variable. We then used this fitted model to predict future specialization values by replacing current PCA values with the first PCA axis of future climatic conditions from the optimistic and pessimistic scenarios, respectively. By comparing these predicted values to current specialization levels, we can assess the potential impact of climate change on the specialization of ant-plant mutualistic networks across different scenarios of climate change.

Under both optimistic and pessimistic scenarios, specialization of ant-domatia showed a consistent decrease for all networks (Fig. 13.4a, b), with a mean reduction of 0.06 under optimistic scenario and pessimistic scenario (paired *T*-test: optimistic scenario vs current scenario, $t = 2.83$, df = 5, $P = 0.04$; pessimistic scenario vs current scenario, $t = 3.40$; df = 5; $P = 0.02$; Fig. 13.4c, f). For ant-seed networks, we observed varying trends: five out of seven networks exhibited an increase in specialization, while two networks showed a decrease under both scenarios of climate change (Fig. 13.4a, b). These opposing patterns led to no significant differences in mean specialization overall (paired *T*-test: optimistic scenario vs current scenario, $t = -0.16$; df = 6; $P = 0.88$; pessimistic scenario vs current scenario, $t = -0.10$; df = 6; $P = 0.92$; Fig. 13.4d,g). Similarly, for ant-EFN interactions, we observed an increase of specialization for four networks and decreased in three, with one showing no effect (Fig. 13.4a, b). This resulted in no change overall (paired *T*-test: optimistic scenario vs current scenario, $t = -1.02$; df = 7; $P = 0.34$; pessimistic scenario vs current scenario, $t = -1.09$; df = 7; $P = 0.31$; Fig. 13.4e, h).

Our findings reveal a nuanced response of ant-plant mutualisms to future climate scenarios. While highly specialized ant-domatia interactions are consistently predicted to become more generalized, ant-seed and ant-EFN mutualisms exhibit variable responses, potentially experiencing either increases or decreases in specialization. This underscores the potential for climate change to disrupt mutualistic interactions across the specialization spectrum. Specifically, species reliant on highly specialized partnerships, such as domatia-bearing plants and their associated ants, are likely to face significant disruptions as their interactions become more generalized. However, it is also important to recognize that shifts towards generalization in traditionally generalized interactions, such as those involving ant-seed and ant-EFN mutualisms, can also lead to increased vulnerability. This highlights the complexity of climate change impacts, as specialization changes, in either direction, can disrupt established mutualistic relationships and potentially destabilize ecological networks.

This vulnerability of specialized mutualisms carries significant conservation implications, particularly in a biodiversity hotspot like the Amazon rainforest. The Amazon rainforest, home to a vast diversity of ant-plant mutualisms, including highly specialized domatia-bearing plants, is under considerable threat from "savannization" and global warming (Bottino et al. 2024). The breakdown of specialized ant-plant interactions can accelerate the process of biodiversity loss. Youngsteadt et al. (2024) demonstrated the influence of environmental change on ant-plant mutualisms by examining the impact of urbanization on *Cordia nodosa* myrmecophytes. They observed a shift from specialized mutualistic ants in forests to

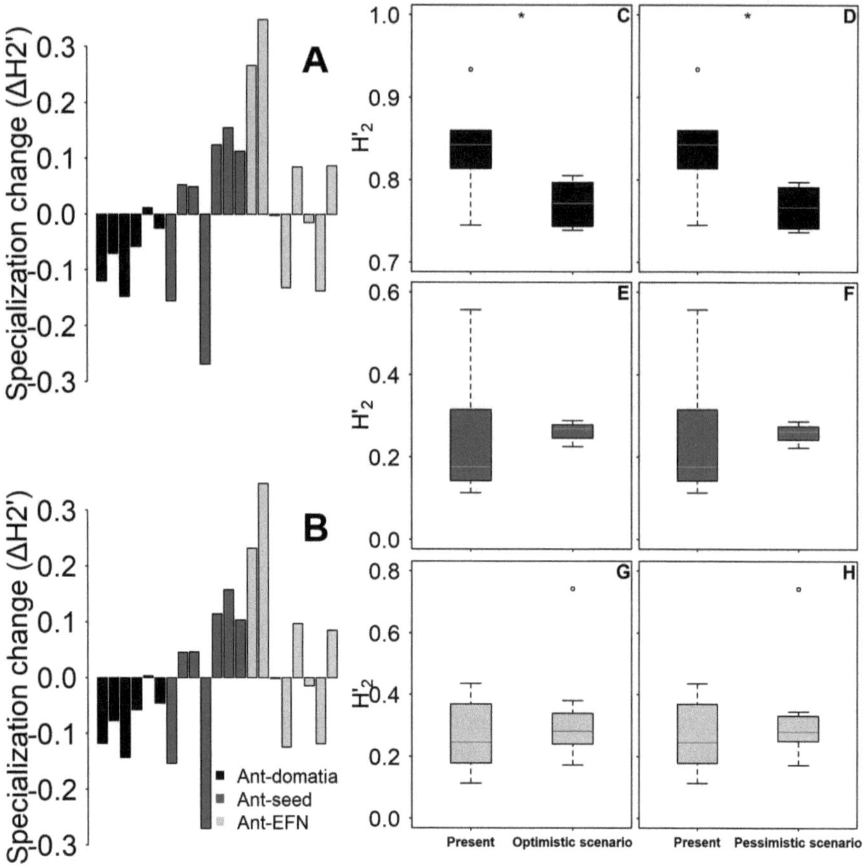

Fig. 13.4 Specialization change ($H_2' - H_{2future}$) in ant-plant networks under two climate change scenarios for 2040–2060: SSP2-4.5 (optimistic) and SSP5-8.5 (pessimistic). (**a**) Barplot showing the specialization change for each of 21 ant-plant networks, categorized by interaction type, under the optimistic scenario and (**b**) pessimistic scenario. Boxplots (**c**–**e**) show paired comparisons between current and future specialization for each ant-plant interaction type under the optimistic scenario and (**f**–**h**) for the pessimistic scenario. Colours represent different ant-plant interactions: purple for ant-domatia, yellow for ant-seed, and green for ant-extrafloral nectar (EFN) networks

opportunistic ants in urban areas, accompanied by an increase in the critical thermal maximum (CT_{max}) of urban ants. This finding raises a crucial question: could thermal traits, such as CT_{max}, act as a biological filter under climate change scenarios? Indeed, research indicates that greater heat tolerance is positively correlated with ant abundance, activity, and, ultimately, distribution in warmer environments (Roeder et al. 2021, Nascimento et al. 2022). This is particularly relevant for specialized arboreal ant species, whose distributions may be strongly influenced by their thermal tolerances in a warming climate.

The varied responses of more generalized ant-plant mutualisms to climate change underscore the necessity of considering ecosystem-specific dynamics when predicting ecological outcomes. For instance, studies have documented a shift from less heat-tolerant to more heat-tolerant ant species in association with EFN-bearing plants and seeds, as higher temperatures tend to favour ants with elevated critical thermal maxima (CT_{max}) and greater tolerance for foraging on warmer leaf and ground surfaces (Fritzpatrick et al. 2014, Stuble et al. 2014, Câmara et al. 2025). This observation raises pertinent questions: could the subset of heat-tolerant ant species experience reduced competition among themselves, leading to more generalized interaction networks? Alternatively, might seeds and EFNs become valuable resources in future scenarios, intensifying competition even among the few heat-tolerant species? Such variations highlight the complexity of predicting interaction network responses to climate change. Global patterns may obscure significant variations at finer scales, emphasizing the importance of integrating detailed studies to understand and mitigate the impacts of climate change on ant-plant mutualisms.

13.4 Conclusion

In this chapter, we have aimed to dissect the complex relationship between climate and ant-plant mutualism networks, with a particular focus on how climatic variables influence network specialization. Our comprehensive analysis, integrating a literature review and quantitative modelling, revealed that temperature, precipitation, and elevation significantly shape these ecological interactions, while latitude exhibited a comparatively weaker effect. Notably, temperature emerged as a pivotal driver of network specialization (H_2'), with stable, warmer environments fostering increased niche partitioning and consequently, higher specialization. Our findings raise significant concerns regarding the resilience of ant-plant mutualisms in the face of projected climate change. Specialized networks, particularly those involving ant-domatia interactions, are likely to experience a reduction in specialization, potentially transitioning towards more generalized interactions. The responses of other mutualistic interactions, such as ant-seed dispersal and ant-extrafloral nectary (EFN) interactions, remain less clear, with our models suggesting the possibility of either increased or decreased specialization depending on the specific climate change scenario.

However, we must acknowledge the inherent limitations of modelling studies. While our analysis provides a valuable framework for understanding large-scale patterns, it represents only a piece of the complex puzzle. To fully understand the impacts of climate on ant-plant mutualisms, future research must delve into the mechanistic underpinnings of these interactions. Specifically, we need to investigate how temperature influences key functional traits, such as domatia abundance, EFN production, nectar secretion, and fruit/seed production. Furthermore, understanding the thermal tolerances (see Chap. 4) and foraging behaviours (see Chap. 2) of ants, including their ability to defend plants and disperse seeds across varying temperatures, is crucial for accurate climate change impact predictions.

Moreover, our analysis did not account for phenotypic plasticity and acclimatization (see Chap. 4), both of which could play significant roles in buffering the effects of climate change. Plants and ants may exhibit considerable flexibility in their traits and behaviours, allowing them to adapt to changing environmental conditions. Additionally, the reliance on historical air temperature data in our models necessitates a consideration of microclimatic effects. Microclimates, which can significantly deviate from ambient temperatures, likely play a vital role in mediating the impacts of climate change on ant-plant interactions, offering refuge from extreme conditions. Therefore, we advocate for a multi-faceted approach to future research. Large-scale studies are essential for predicting long-term evolutionary responses and assessing the adaptive capacity of these networks. However, these studies must be complemented by local-scale investigations that examine the quantitative and qualitative shifts in interactions across diverse microclimates. Only through such integrated efforts can we truly understand the vulnerability and resilience of ant-plant mutualisms in the face of rapid climate change.

References

Alencar CLS, Nogueira A, Vicente RE, Coutinho IAC (2023) Plant species with larger nectaries produce better quality nectar when needed and interact with the best ant partners. J Exp Bot 74:4613–4623. https://doi.org/10.1093/jxb/erad160

Anjos D, Luna P, Borges CCR, Dáttilo W, Del-Claro K (2019) Structural changes over time in individual-based networks involving a harvester ant, seeds, and invertebrates. Ecol Entomol 44(6):753–761. https://doi.org/10.1111/een.12768

Arruda AJ, Costa FV, Guerra TJ, Junqueira PA, Dayrell RLC, Messeder JVS, Rodrigues HTS, Buisson E, Silveira FAO (2020) Topsoil disturbance reshapes diaspore interactions with ground-foraging animals in a megadiverse grassland. J Veg Sci 31(6):1039–1052. https://doi.org/10.1111/jvs.12866

Bentley BL, Elias TS (eds) (1983) The biology of nectaries. Cary Arboretum, New York Botanical Garden, New York

Blüthgen N, Verhaagh M, Goitía W, Jaffé K, Morawetz W, Barthlott W (2000) How plants shape the ant community in the Amazonian rainforest canopy: the key role of extrafloral nectaries and homopteran honeydew. Oecologia 125(2):229–240. http://www.jstor.org/stable/4222767

Blüthgen N, Stork NE, Fiedler K (2004) Bottom-up control and co-occurrence in complex communities: honeydew and nectar determine a rainforest ant mosaic. Oikos 106(2):344–358. http://www.jstor.org/stable/3548119

Blüthgen N, Menzel F, Blüthgen N (2006) Measuring specialization in species interaction networks. BMC Ecol 6:9. https://doi.org/10.1186/1472-6785-6-9

Blüthgen N, Fründ J, Vázquez DP, Menzel F (2008) What do interaction network metric tell us about specialization and biological traits? Ecology 89(12):3387–3399. https://doi.org/10.1890/07-2121.1

Bottino MJ, Nobre P, Juniror S, Capistrano VB, Malagutti M, Tamoki JN, Oliveira BFA, Nobre CA (2024) Amazon savannization and climate change are projected to increase dry season length and temperature extremes over Brazil. Sci Rep 14:5131. https://doi.org/10.1038/s41598-024-55176-5

Bronstein J (1998) The contributions of ant-plant protection to our understanding of mutualism. Biotropica 30(2):150–161. https://doi.org/10.1111/j.1744-7429.1998.tb00050.x
Bronstein J (2015) Mutualism. Oxford University Press, Oxford
Caddy-Retalic S, Hoffmann BD, Guerin GR, Andersen AN, Wardle GM, McInerney FA, Lowe AJ (2019) Plant and ant assemblages predicted to decouple under climate change. Divers Distrib 25(4):551–567. https://doi.org/10.1111/ddi.12858
Câmara T, Leal IR, Blüthgen N, Oliveira FMP, de Queiroz RT, Arnan X (2018) Effects of chronic anthropogenic disturbance and rainfall on the specialization of ant-plant mutualistic networks in the Caatinga, a Brazilian dry forest. J Anim Ecol 87(4):1022–1033. https://doi.org/10.1111/1365-2656.12820
Câmara T, Leal I, Blüthgen N, Oliveira FMP, Arnan X (2019) Anthropogenic disturbance and rainfall variation threaten the stability of plant-ant interactions in the Brazilian Caatinga. Ecography 42(11):1960–1972. https://doi.org/10.1111/ecog.04531
Câmara T, Cavalcante NT, Falcão HM, Santana E, Teixeira GSS, Arnan X (2025) Increased temperatures could heighten vulnerability of ant-plant mutualism. Oecologia 207:8. https://doi.org/10.1007/s00442-024-05646-4
Chomicki G, Renner SS (2015) Phylogenetics and molecular clocks reveal the repeated evolution of ant-plants after the late Miocene in Africa and the early Miocene in Australasia and the Neotropics. New Phytol 207(2):411–424. https://doi.org/10.1111/nph.13271
Chomicki G, Kiers T, Renner SS (2020) The evolution of mutualistic dependence. Annu Rev Ecol Evol Syst 51:409–432. https://doi.org/10.1146/annurev-ecolsys-110218-024629
Christianini AV, Mayhé-Nunes A, Oliveira PS (2012) Exploitation of fallen diaspores by ants: are there ant-plant partner choices? Biotropica 44(3):360–367. https://doi.org/10.1111/j.1744-7429.2011.00822.x
Cruz AR, Davidowitz G, Moore CM, Bronstein JL (2023) Mutualisms in a warming world. Ecol Lett 26(8):1432–1451. https://doi.org/10.1111/ele.14264
Da Silva Laviski BF, Mayhé-Nunes AJ, Nunes-Freitas AF (2021) Structure of ant-diaspore networks and their functional outcomes in a Brazilian Atlantic Forest. Sociobiology 68(3):e7104. https://doi.org/10.13102/sociobiology.v68i3.7104
Dalsgaard B, Trojelsgaard K, González AMM, Nogués-Bravo D, Ollerton J, Petanidou T, Sandel B, Svenning J (2011) Specialization in plant-hummingbird networks is associated with species richness, contemporary precipitation and quaternary climate-change velocity. PLoS One 6(10):e25891. https://doi.org/10.1371/journal.pone.0025891
Dáttilo W, Vasconcelos HL (2019) Macroecological patterns and correlates of ant–tree interaction networks in Neotropical savannas. Glob Ecol Biogeogr 28(9):1283–1294. https://doi.org/10.1111/geb.12932
Dáttilo W, Izzo TJ, Vasconcelos HL, Rico-Gray V (2013) Strength of the modular pattern in Amazonian symbiotic ant–plant networks. Arthropod Plant Interact 7:455–461. https://doi.org/10.1007/s11829-013-9256-1
Dáttilo W, Díaz-Castelazo C, Rico-Gray V (2014) Ant dominance hierarchy determines the nested pattern in ant–plant networks. Biol J Linn Soc 113(2):405–414. https://doi.org/10.1111/bij.12350
Davidson DW (1977) Foraging ecology and community organization in desert seed-eating ants. Ecology 58(4):725–737. https://doi.org/10.2307/1936209
Davidson DW, Snelling RR, Longino JT (1989) Competition among ants for myrmecophytes and the significance of plant trichomes. Biotropica 21(1):64–73. https://doi.org/10.2307/2388444
Dejean A, Azémar F, Petitclerc F, Delabie JH, Corbara B, Leroy C, Compin A (2018) Highly modular pattern in ant-plant interactions involving specialized and non-specialized myrmecophytes. Sci Nat 105:43. https://doi.org/10.1007/s00114-018-1570-0
Díaz-Castelazo C, Martínez-Adriano CA, Dáttilo W, Rico-Gray V (2020) Relative contribution of ecological and biological attributes in the fine-grain structure of ant-plant networks. PeerJ 8:e8314. https://doi.org/10.7717/peerj.8314

Emer C, Venticinque EM, Fonseca CR (2013) Effects of dam-induced landscape fragmentation on Amazonian ant-plant mutualistic networks. Conserv Biol 27(4):763–773. https://doi.org/10.1111/cobi.12045

Fayle TM, Jorge LR, Kilmes P, Dáttilo W, Lima PF, Christianini A, Nunes-Freitas AF, da Silva Laviski BF, Wang B, Emer C, Lara C, Guevera CC et al (2024) Data from: ant-plant specialization influenced more by network types than by disturbance, elevation, or latitude [Dataset]. Dryad. https://doi.org/10.5061/dryad.vdncjsz17

Fick SE, Hijmans RJ (2017) WorldClim 2: new 1-km spatial resolution climate surfaces for global land areas. Int J Climatol 37(12):4302–4315. https://doi.org/10.1002/joc.5086

Fitzpatrick G, Lanan MC, Bronstein JL (2014) Thermal tolerance affects mutualist attendance in ant-plant protection mutualism. Oecologia 176:129–138. https://doi.org/10.1007/s00442-014-3005-8

Fonseca CR (1999) Amazonian ant plant interactions and the nesting space limitation hypothesis. J Trop Ecol 15(6):807–825. http://www.jstor.org/stable/2560045

Fonseca CR, Ganade G (1996) Asymmetries, compartments and null interactions in an Amazonian ant-plant community. J Anim Ecol 65(3):339–347. https://doi.org/10.2307/5880

Fukami T, Wardle DA (2005) Long-term ecological dynamics: reciprocal insights from natural and anthropogenic gradients. Proc R Soc B 272:2105–2115. https://doi.org/10.1098/rspb.2005.3277

Jones AM, Berkelmans R, van Oppen MJH, Mieog JC, Sinclair W (2008) A community change in the algal endosymbionts of a scleractinian coral following a natural bleaching event: field evidence of acclimatization. Proc R Soc B 275:1359–1365. https://doi.org/10.1098/rspb.2008.0069

Jones C, Zurell D, Wiesner K (2023) Novel analytic methods for predicting extinctions in ecological networks. Ecol Monogr 94(2):e1601. https://doi.org/10.1002/ecm.1601

Kaiser-Bunbury C, Traveset A, Hansen DM (2010) Conservation and restoration of plant-animal mutualisms on oceanic islands. Perspect Plant Ecol Evol Syst 12(2):131–143. https://doi.org/10.1016/j.ppees.2009.10.002

Kharouba HM, Williams JL (2024) Forecasting species' responses to climate change using space-for-time substitution. Trends Ecol Evol 39(8):716–725. https://doi.org/10.1016/j.tree.2024.03.009

King NG, McKeown NJ, Smale DA, Moore PJ (2018) The importance of phenotypic plasticity and local adaptation in driving intraspecific variability in thermal niches of marine macrophytes. Ecography 41:1469–1484. https://doi.org/10.1111/ecog.03186

Leal IR, Oliveira PS (1998) Interactions between fungus-growing ants (Attini), fruits and seeds in Cerrado vegetation in Southeast Brazil. Biotropica 30(2):170–178. https://doi.org/10.1111/j.1744-7429.1998.tb00052.x

Leal IR, Wirth R, Tabarelli T (2007) Seed dispersal by ants in the semi-arid Caatinga of north-east Brazil. Ann Bot 99(5):885–894. https://doi.org/10.1093/aob/mcm017

Leal LC, Peixoto PEC (2017) Decreasing water availability across the globe improves the effectiveness of protective ant-plant mutualisms: a meta-analysis. Biol Rev 92(3):1785–1794. https://doi.org/10.1111/brv.12307

Leal LC, Andersen AN, Leal IR (2014) Anthropogenic disturbance reduces seed dispersal services for myrmecochorus plants in the Brazilian Caatinga. Oecologia 174:173–181. https://doi.org/10.1007/s00442-013-2740-6

Lovell RSL, Collins S, Martin SH, Pigot AL, Phillimore AB (2023) Space-for-time substitutions in climate change ecology and evolution. Biol Rev 98(6):2243–2270. https://doi.org/10.1111/brv.13004

Luo Y, Taylor A, Weigelt P, Guénard B, Economo EP, Nowak A, Inderjit, Kreft H (2023) Climate and ant diversity explain the global distribution of ant-plant mutualisms. Ecography 11:e06841. https://doi.org/10.1111/ecog.06841

Miranda PN, Ribeiro JELS, Corro EJ, Brasil I, Delabie JHC, Dáttilo W (2022) Structural stability of ant-plant mutualistic networks mediated by extrafloral nectaries: looking at the effects of forest fragmentation in the Brazilian Amazon. Sociobiology 69(3):e8261. https://doi.org/10.13102/sociobiology.v69i3.8261

Moraes SC, Vasconcelos HL (2009) Long-term persistence of a Neotropical ant-plant population in the absence of obligate plant-ants. Ecology 90:2375–2383. https://doi.org/10.1890/08-1274.1
Morales-Linares J, Flores-Palacios A, Corona-López A, Toledo-Hernández VH (2021) Structure and robustness of the Neotropical ant-gardens network under climate change. Insect Conserv Divers 14(5):635–646. https://doi.org/10.1111/icad.12497
Morison JIL, Lawlor DW (1999) Interactions between increasing CO_2 concentration and temperature on plant growth. Plant Cell Environ 22(6):659–682. https://doi.org/10.1046/j.1365-3040.1999.00443.x
Nagaishi E, Takemoto K (2018) Network resilience of mutualistic ecosystems and environmental changes: an empirical study. R Soc Open Sci 5:180706. https://doi.org/10.1098/rsos.180706
Nascimento G, Câmara T, Arnan X (2022) Critical thermal limits in ants and their implications under climate change. Biol Rev 97(4):1287–1305. https://doi.org/10.1111/brv.12843
Nelsen MP, Ree RH, Moreau CS (2018) Ant-plant interactions evolved through increasing interdependence. Proc Natl Acad Sci USA 115(48):12253–12258. https://doi.org/10.1073/pnas.179794115
O'Neill BC, Tebaldi C, van Vuuren DP, Eyring V, Friedlingstein P, Hurtt G, Knutti R, Kriegler E, Lamarque JF, Lowe J, Meehl GA, Moss R, Riahi K, Sanderson BM (2016) The Scenario Model Intercomparison Project (ScenarioMIP) for CMIP6 Geoscientific Model Development 9(9):3461–3482. https://doi.org/10.5194/gmd-9-3461-2016
Oliveira FMP, Andersen AN, Arnan X, Ribeiro-Neto JD, Arcoverde GB, Leal IR (2019) Effects of increasing aridity and chronic anthropogenic disturbance on seed dispersal by ants in Brazilian Caatinga. J Anim Ecol 88(6):870–880. https://doi.org/10.1111/1365-2656.12979
Oliveira FMP, Câmara T, Durval JIF, Oliveira CLS, Arnan X, Andersen AN, Ribeiro EMS, Leal IR (2021) Plant protection services mediated by extrafloral nectaries decline with aridity but are not influenced by chronic anthropogenic disturbance in Brazilian Caatinga. J Ecol 109(1):260–272. https://doi.org/10.1111/1365-2745.13469
Parr CL, Bishop TR (2022) The response of ants to climate change. Glob Change Biol 28(10):3188–3205. https://doi.org/10.1111/gcb.16140
Passmore HA, Bruna EM, Heredia SM, Vasconcelos HL (2012) Resilient networks of ant-plant mutualists in Amazonian forest fragments. PLoS One 7(8):e40803. https://doi.org/10.1371/journal.pone.0040803
Pizo MA, Oliveira PS (2001) Size and lipid content of nonmyrmecochorous diaspores: effects on the interaction with litter-foraging ants in the Atlantic rain forest of Brazil. Plant Ecol 157(1):37–52. https://doi.org/10.1023/A:1013735305100
Plowman NS, Hood ASC, Moses J, Redmond C, Novotny V, Klimes P, Fayle TM (2017) Network reorganization and breakdown of ant-plant protection mutualism with elevation. Proc R Soc B 284:20162564. https://doi.org/10.1098/rspb.2016.2564
Rico-Gray V, Oliveira P (2007) The ecology and evolution of ant–plant interactions. University of Chicago Press, Chicago
Rico-Gray V, Díaz-Castelazo C, Ramírez-Hernández A, Guimarães PR, Holland N (2012) Abiotic factors shape temporal variation in the structure of ant-plant network. Arthropod Plant Interact 6:289–295. https://doi.org/10.1007/s11829-011-9170-3
Roeder KA, Roeder DV, Bujan J (2021) Ant thermal tolerance: a review of methods, hypotheses, and sources of variation. Ann Entomol Soc Am 114(4):459–469. https://doi.org/10.1093/aesa/saab018
Rustad LE (2008) The response of terrestrial ecosystem to global climate change: towards an integrated approach. Sci Total Environ 404:222–235. https://doi.org/10.1016/j.scitotenv.2008.04.050
Schemske DW (1982) Ecological correlates of a Neotropical mutualism: ant assemblages at Costus extrafloral nectaries. Ecology 63(4):932–941. https://doi.org/10.2307/1937233
Schleuning M, Fründ J, Klein A, Abrahamczyk S, Alarcón R, Albrecht M, Andersson GKS, Bazaian S, Böhning-Gaese K, Bommarco R, Dalsgaard B, Dehling DM et al (2012) Specialization of

mutualistic interaction networks decreases toward tropical latitudes. Curr Biol 22:1925–1931. https://doi.org/10.1016/j.cub.2012.08.015

Sebastián-González E, Dalsgaard B, Sandel B, Guimarães PR (2015) Macroecological trends in nestedness and modularity of seed-dispersal networks: human impacts matter. Glob Ecol Biogeogr 24(3):293–303. https://doi.org/10.1111/geb.12270

Stuble KL, Patterson CM, Rodriguez-Cabal MA, Ribbons RR, Dunn RR, Sanders NJ (2014) Ant-mediated seed dispersal in a warmed world. PeerJ 2:e286. https://doi.org/10.7717/peerj.286

Takimoto G, Suzuki K (2016) Global stability of obligate mutualism modules with facultative mutualists. Oikos 125(4):535–540. https://doi.org/10.1111/oik.02741

Urban MC (2015) Accelerating extinction risk from climate change. Science 348:571. https://doi.org/10.1126/science.aaa4984

Valladares F et al (2014) The effects of phenotypic plasticity and local adaptation on forecasts of species range shifts under climate change. Ecol Lett 17(11):1351–1364. https://doi.org/10.1111/ele.12348

Vidal MC, Annenberg TJ, Curé AE, Althoff DM, Segraves KA (2021) The variable effects of global change on insect mutualisms. Curr Opin Insect Sci 47:46–52. https://doi.org/10.1016/j.cois.2021.03.002

Youngsteadt E, Prado SG, Aquino AKD, Valdeiglesias JP, Ojeda TG, Quispe JSG (2024) Urbanization drives partner switching and loss of mutualism in an ant-plant symbiosis. Ecology 105(11):e4449. https://doi.org/10.1002/ecy.4449

Chapter 14
Ant-Plant Protection Mutualisms Mediated by Extrafloral Nectar: A Macroecological Analysis of Outcomes in Relation to Precipitation

Laura C. Leal, Amanda V. da Silva, Bruno Melati, Luana S. Prochazka, Douglas D. de Souza, and Anselmo Nogueira

14.1 Introduction

The magnitude of both costs and benefits of mutualistic interactions can vary markedly in time and space. Indeed, variation in biotic and abiotic conditions can turn a mutualistic interaction into a neutral or even antagonistic one (see Bronstein 2015; Chamberlain et al. 2014, and references therein). In recent decades, researchers have focused on identifying factors and mechanisms underlying the variation in mutualistic outcomes (e.g. Bertness and Callaway 1994; Billick and Tonkel 2003; Jorge and Howe 2009; Thrall et al. 2007). However, we are still far from unravelling such mechanisms because so many species are involved in so many interactions in all habitats around the globe. Filling this knowledge gap has become an urgent task in the current climatic crisis, which is expected to progressively modify the ecological conditions under which mutualisms are embedded (Vidal et al. 2021).

Supplementary Information The online version contains supplementary material available at https://doi.org/10.1007/978-3-032-21127-9_14.

L. C. Leal (✉)
Departamento de Zoologia, Universidade Federal do Paraná, Curitiba, PR, Brazil

Departamento de Ecologia e Biologia Evolutiva, Universidade Federal de São Paulo, Diadema, SP, Brazil
e-mail: laura.leal@ufpr.br

A. V. da Silva · B. Melati · D. D. de Souza · A. Nogueira
Centro de Ciências Naturais e Humanas, Universidade Federal do ABC, Alameda da Universidade, São Bernardo do Campo, SP, Brazil
e-mail: a.nogueira@ufabc.edu.br

L. S. Prochazka
Programa de Pós-Graduação em Biodiversidade Vegetal e Meio Ambiente, Instituto de Pesquisas Ambientais, Avenida Miguel Estefno, São Paulo, SP, Brazil

I. R. Leal et al. (eds.), *Ant Responses to Climate Change*, Entomology in Focus 10,
https://doi.org/10.1007/978-3-032-21127-9_14

According to climatic models, Earth will become progressively warmer, subject to more frequent and severe drought events (Dai et al. 2018; Karl and Trenberth 2003). Such changes are expected to affect the outcome of mutualistic interactions. This is the case of mutualistic interactions involving plants offering nutritional rewards to animals that, in return, provide benefits such as protection against herbivores (e.g. Leal et al. 2023, Melati et al. 2018) or transport of gametes or propagules (e.g. Howe and Smallwood 1982; Leal and Koski 2024). These nutritional rewards are energetically costly for plants to produce (Xu and Chen 2015), especially in environmental conditions that constrain photosynthetic rates (e.g. Proietti et al. 2000). Given that water availability is one of the main factors regulating plant photosynthetic rates (Wang et al. 2018), it is expected that the drier the environment, the more costly it is for plants to engage in mutualistic interactions and the lower the net benefits obtained from animal partners. Such expectations have been supported by studies reporting a predictable decline in the net benefit for plants interacting with animals along precipitation gradients and the disruption of the mutualism in most plant populations (e.g. Chalcoff et al. 2012; Nogueira et al. 2015; Pringle et al. 2013). However, this evidence remains preliminary and scattered across a few locations, impairing our ability to predict the magnitude and direction of future impacts resulting from the increase in Earth's surface aridity on the dynamics of animal-plant mutualistic interactions.

Ant-plant protection mutualisms are particularly threatened by increased aridity (Aljbory and Chen 2018; Mooney et al. 2019; Parr and Bishop 2022; Warren II and Bradford 2014). This mutualism between protective ants and plants is usually mediated by extrafloral nectaries (EFNs)—glands producing nectar unrelated to pollinators attraction (Bentley 1977a). The secretion of extrafloral nectar attracts and rewards ants whose foraging protects the plants from herbivores (Bentley 1977a, b). This anti-herbivory defence occurs in more than 4000 vascular plant species worldwide across more than 20% of plant families (Keeler et al. 2024; Weber and Keeler 2013). Since extrafloral nectar is a carbohydrate-based resource produced by plants, its relative energetic cost is expected to vary according to photosynthetic rates, which can be constrained by lower water availability (McDowell and Sevanto 2010).

Under dry conditions, plant photosynthesis and carbon limitation can lead ant-plant mutualisms to divergent outcomes. On the one hand, reduced water availability may decrease plant investment in partner rewards, thereby reducing ant attendance and its protection against herbivores (McDowell and Sevanto 2010). On the other hand, carbon becomes more limiting with increased aridity, which increases the relative cost of leaf losses caused by herbivores; this might lead to increased investment in anti-herbivory defences, including biotic defence by ants (Pringle 2013). Current evidence points to the second scenario. For instance, a macroecological meta-analysis found that the magnitude of the protective effects of ants on plants with EFNs decreases slightly as mean precipitation increases, with ants having a mostly negative impact on plants in moister habitats (Leal and Peixoto 2017). This slight decline was associated with a reduced frequency of plant interactions with behaviourally dominant (i.e. higher-quality protector) ant species (Leal and Peixoto 2017), which could be mediated by variations in the relative value of extrafloral nectar secreted by plants along the precipitation gradient (Alencar et al. 2023; Dáttilo et al. 2015; Leal and Peixoto 2017). In this context, water availability may have altered the dynamics between ants, herbivores, and plants, slightly modifying the protective outcomes of these interactions for plants. However, this meta-analysis combined metrics describing the ant effects on both herbivory and plant performance descriptors. Although such effects are commonly connected, often there is a stronger positive effect on plant anti-herbivory defence than on plant reproductive performance (see Leal et al. 2023 and references therein), meaning that these two types of ant effects may respond differently to variation in water availability.

This chapter assesses how four types of ant effects on plants bearing EFNs vary along a macroecological precipitation gradient. To do so, we combine the most comprehensive dataset available to date—including data from several previous meta-analyses—with findings from more recent studies. We examined the ant effects on (i) herbivore occurrence on plants, (ii) levels of herbivory, and (iii) vegetative and (iv) reproductive plant performance. This approach offers new insights into the patterns and mechanisms by which climatic change may mediate the outcomes of biotic defence in plants across different habitats worldwide.

14.2 Effect of Protective Ants on Plants with EFNs Along the Precipitation Gradient

To determine the effect of precipitation on plant defence by ants, we combined data from four previous meta-analyses evaluating the defensive role of ants on plants with EFNs: Chamberlain and Holland (2009), Rosumek et al. (2009), Trager et al. (2010), and Leal et al. (2023). These meta-analyses used datasets compiling information from published studies evaluating the effect of ants on herbivores, herbivory, and vegetative and reproductive performance of plants bearing EFNs. Given that each study assessed different types of ant effects—some focusing only on herbivores or herbivory, while others included both herbivory and plant performance—all were considered here to compile the most up-to-date dataset available for assessing the protective service provided by ants to plants with EFNs.

To ensure maximum coverage, we conducted an additional literature search to identity studies that had not been included in previous meta-analyses (Sect. 14.1; Electronic Supplementary Material). This resulted in a total of 59 suitable studies for our analysis. Of these, 46 had already been included in earlier quantitative reviews, while 13 were newly added and had not been used in any prior meta-analysis. These new studies are mostly from America and cover a wide range of mean annual precipitation from 160 to 3609 mm (Fig. 14.1). From each of the 59 studies, we extracted the mean, standard deviation (SD), and sample size for metrics describing different types of ant effects on plants bearing EFNs. When studies reported the standard error (SE) instead of SD, we converted SE to SD by

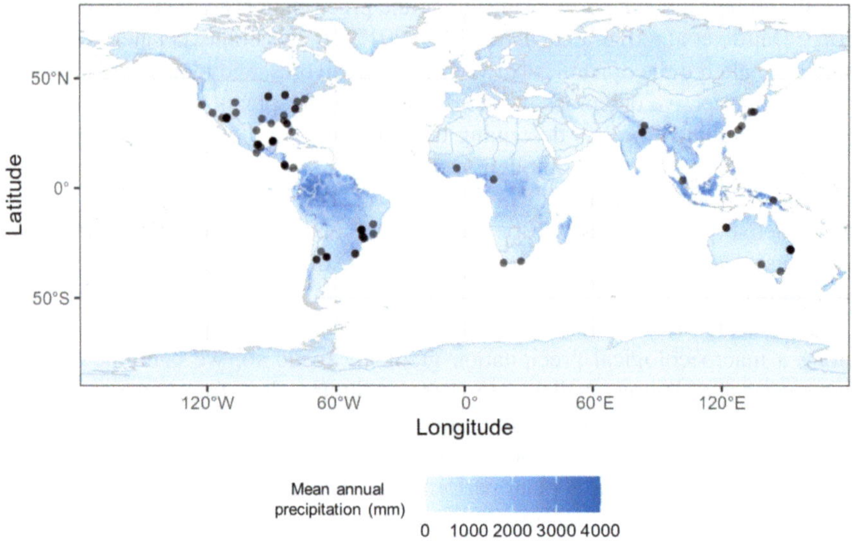

Fig. 14.1 Mean annual precipitation (mm) across the globe showing the study sites where the studies evaluating the role of ants on plant defence included in our meta-analysis were carried out

multiplying the SE by the square root of the sample size. If the means, SD, or SE were available in the figures, we extracted their values using the metaDigitise package v. 1.0.1 (Pick et al. 2018) in R v. 4.1.3 (R Core Team 2023). We then divided the metrics into four categories, including ant effect on herbivores, herbivory, vegetative, and reproductive plant performance (Table 14.1), to evaluate the magnitude of direct and indirect ant effects on plants with EFNs. The effects of ants in deterring or preying on herbivores and reducing plant tissue loss were considered direct and immediate protective effects on plants. In contrast, the effects on variables describing plant performance—whether vegetative or reproductive—were considered indirect effects, representing potential long-term consequences of ant protection on plants.

We used Hedges' g as the metric describing effect size of ants on plants with EFNs (see Sect. 14.1 in the Electronic Supplementary Material for more details). This metric represents the standardized difference between the means of two groups: plants with and without ant access to their EFNs. We calculated the Hedges' g using the *escalc* function from the metafor package v. 2.4.0 (Viechtbauer 2010). For effect sizes describing the ant effect on herbivores and herbivory, we calculated the Hedges' g, subtracting the mean values for the group of plants in which ants were present on plants from those in which ants were excluded from plants. For the effect sizes describing ant effects on plant performance, we did the opposite, subtracting the mean values of plants in which ants were absent from those in which ants were present on the plants. We used this strategy to favour the biological interpretation of

Table 14.1 Classification of metrics of ant effects on plants with EFNs in the studies included in our dataset

Ant effect type	Variables measured in the original studies
Herbivores	Number of eggs, number of larvae, number of offspring, number of herbivores per census, mean number of herbivores per census per stem, herbivores abundance, herbivore abundance on male and/or female inflorescence, number of herbivores per plant, herbivore richness, percentage of egg survival, time spent on plants
Herbivory	Percentage of leaf area consumed, rate of leaf area consumed, percentage of consumed leaf area/branch, percentage of leaf area consumed on male and/or female plants, daily percentage of leaf loss, change in leaf damage score per leaf, proportion of area damaged, proportion of perforation
Plant vegetative performance	Number of new leaves produced per branch, number of new leaves, foliar area, produced leaf area per branch, plant height, plant growth, growth rate, and plant biomass
Plant reproductive performance	Number of buds, number of flowers per plant, number of female flowers per plant, the proportion of flowers producing fruits, number of fruits, number of buds turning into fruits, number of fruits per branch, number of unripe and/or ripe fruits per plant, number of fruits per cladode, the proportion of fruits ripening, the proportion of fruits per bud, total pods, mature pods, mature pods per total pods, number of seeds per branch, number of seeds per plant, number of seeds per inflorescence, mean seed mass, seed to ovules ratio per plant, eight of seed pods, number of ovules

We grouped the metrics into four effect types to evaluate the magnitude of direct and indirect ant effects on plants separately

Hedges' g values. Regardless of the category of ant effect sizes, the positive and higher the values of Hedges' g, the stronger and more beneficial the direct effect of ants on reducing herbivory and/or indirect benefit of ants increasing the plant performance.

We obtained 292 effect sizes from 59 studies, including 67 effect sizes related to herbivore occurrence on plants, 90 on herbivory, and 22 and 113 on plant vegetative and reproductive performance, respectively. Our dataset included 60 species of plants distributed across 42 genera and 23 families. The three most commonly studied plant families comprised 64% of total effect sizes, represented by Fabaceae (32.2%), Euphorbiaceae (20.2%), and Malvaceae (11.6%). Interestingly, these three families represent 2.7% of the plant families in which EFNs occur (Weber and Keller 2013), although they have a high abundance of species with extrafloral nectaries, unlike others with few species interacting with ants.

To evaluate the differences of the direction and magnitude of ant effects, we built multilevel meta-analytical models using the ant effect types as moderator, along with several random effects (model 1). These random variables included: (i) study identity, to account for potential methodological biases; (ii) plant species, as some were evaluated in multiple studies; and (iii) phylogenetic relatedness, to control for similarities due to shared ancestry (see Sect. 14.1 in the Electronic Supplementary

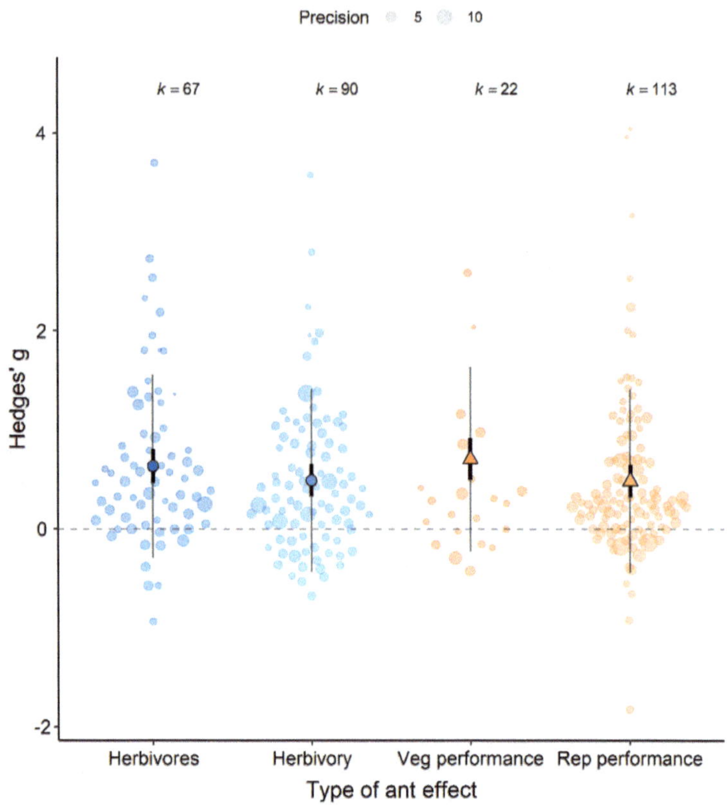

Fig. 14.2 Variation of the mean effect size of ants (Hedges' g) on herbivores, herbivory, and vegetative and reproductive performance of plants with extrafloral nectaries. Colours and symbols group the metrics of the ant's direct (blue circles) and indirect (orange triangles) effects on plants with extrafloral nectaries. *Veg performance* and *rep performance* indicate the mean effect of ants on the vegetative and reproductive performance of plants with EFNs, respectively. The vertical thicker line is the 95% CI, and the vertical thinner line is the precision interval. The dashed horizontal line represents no ant effect on plants bearing extrafloral nectaries (Hedges' $g = 0$). Each dot represents an effect size, and the size of the dot represents its precision (the inverse of standard error). *K* is the number of effect sizes (dots)

Material for details). Although we found a similarly positive effect of ants on both plant defence metrics (i.e. herbivore occurrence and herbivory intensity) and plant performance metrics (i.e. vegetative and reproductive performance) (Fig. 14.2), the magnitude of these effects varied slightly among the four effect categories used in our analysis (Model 1: $Q_{typeofanteffect} = 12.40$; df = 3; $p = 0.006$). Overall, ants conferred a positive mean benefit to plants across all effect types.

To create the precipitation gradient, we extracted the mean annual precipitation of each study site using Chelsea v. 1.2 (Karger et al. 2017; Karger et al. 2018). Chelsea has a resolution of 30 arcsecs (~1 km) and corrects for orographic variables affecting precipitation, such as topography and wind field. To assess how mean annual precipitation influences the direction and magnitude of ant effects on plants with EFNs, we built two additional multilevel meta-analytical models using precipitation and categories of ant effects as moderators, along with the same random effects described earlier (Model 2 and 3). As a result, we observed that precipitation had a different effect on each type of ant effect considered (Model 2: $Q_{precipitation*typeofanteffect} = 24.18$; df = 7; $p = 0.001$, Fig. 14.3). Ant effect on the herbivore occurrence and herbivory of plants with extrafloral nectaries was similar along the precipitation gradient (herbivore occurrence, intercept = 0.82; p-intercept <0.0001; slope = −0.0002; p-slope = 0.22; herbivory, intercept = 0.57; p-intercept = 0.11; slope = −0.0001; p-slope = 0.35; Fig. 14.3). Meanwhile, increasing in mean

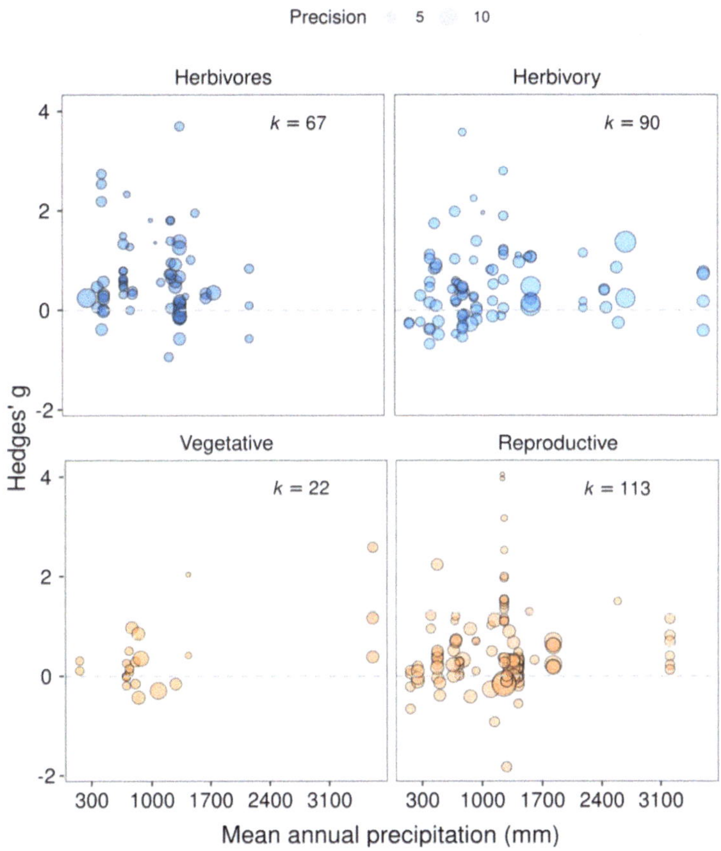

Fig. 14.3 Precipitation do not affect the mean positive effect of ant protection on the magnitude of ant effects on herbivores occurrence, herbivory, reproductive, and vegetative performance of plants with extrafloral nectaries (Model 3 with the exclusion of extreme values above 2500 mm of precipitation). Colours are grouping the metrics of ant's direct (blue) and indirect (orange) effects on plants with extrafloral nectaries. The dashed horizontal line represents no ant effect on plants. Each dot represents an effect size, and the size of the dot represents its precision (the inverse of standard error). *K* indicates the number of effect sizes (dots)

precipitation positively affected the magnitude of ant effect on plant's vegetative performance (intercept = 0.43; p-intercept = 0.04; slope = 0.0002; p-slope = 0.01; Fig. 14.3) but played no rule on the ant effects on plant's reproductive performance (intercept = 0.48; p-intercept = 0.05; slope = 0; p-slope = 0.24; Fig. 14.3). After running our models, we noticed that our effects sizes were not evenly distributed along our precipitation gradients, mainly due to lacking studies in sites above 2500 mm. We had only three effect sizes of ants on plant's vegetative performance, six on plant reproductive performance, and four on herbivory in areas with mean annual precipitation above 2500 mm. Such asymmetric distribution of effect sizes along our gradient could bias our results, and because of that, we performed a new analysis removing these extreme values. After removing these values, precipitation did not influence the magnitude of the ant effect on the plant's vegetative performance (Model 3, $Q_{precipitation:typeofanteffect}$ = 9.14; df = 7; p = 0.24; intercept = 0.78; slope = −0.0002; p-slope = 0.96). In summary, the protection service mediated by ants was, on average, consistently positive and similar for plants across the precipitation gradient, regardless of whether the protection metric focused on herbivores or plant performance. Additionally, more effect sizes were retrieved from the literature in drier regions of the gradient, showing higher variability in interaction outcomes. Despite the increased variability, the average effect remained similar to that observed in the other areas across the precipitation gradient.

14.3 Publication Biases and Heterogeneity

Meta-analytical models are usually affected by two different types of biases: (1) outcome report bias, in which non-significant results are more challenging to publish and, consequently, less frequent on meta-analytical datasets, and (2) temporal biases, which acknowledges that methods improve over time, increasing the measurement's precision, and the probability of neutral or negative effects (Gurevitch et al. 2013). We evaluated these biases in our meta-analytical dataset following the methods described in the Electronic Supplementary Material.

We found evidence for publication bias in our dataset ($Q_{outcome\ bias}$ = 108.02; df = 1; $p < 0.01$). Less precise studies (i.e. lower sample sizes and/or large variability in the effects reported) were more likely to be published the higher the magnitude of the effect size of ants on plants with EFNs (Fig. 14.4a). Conversely, we did not find evidence for temporal bias ($Q_{temporal\ bias}$ = 1.03; df = 1; $p = 0.31$; Fig. 14.4b), the precision of the ant effects sizes on plants with EFNs did not vary predictably over time. Finally, our models had a high level of heterogeneity, mainly explained by differences between studies and plant species (Table 14.2).

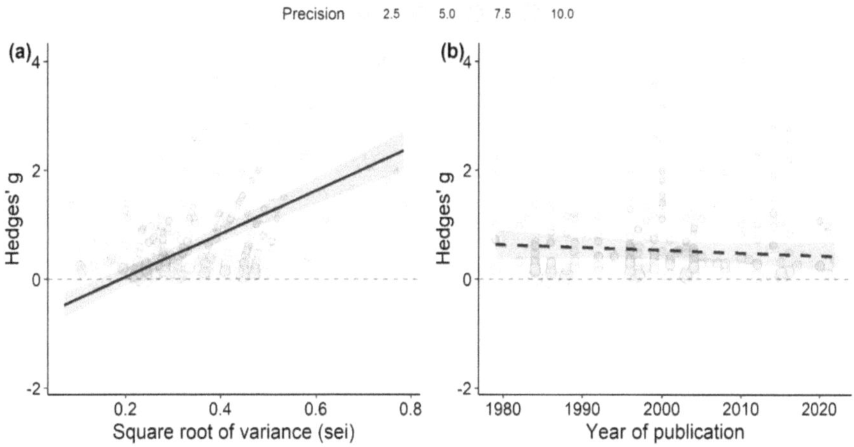

Fig. 14.4 Publication bias (**a**) and temporal bias (**b**). Publication bias refers to the publication of non-significant results, while temporal bias refers to the tendency of effect sizes to reduce or be more precise through time

14.4 Discussion

14.4.1 Variation of Herbivory and Ant Effects on Plants Along Precipitation Gradients

We investigated how the outcome of EFN-mediated ant-plant mutualisms varies along a macroecological precipitation gradient. We found that ants effectively defend plants by reducing herbivore abundance and damage, resulting in an overall positive impact on plant vegetative and reproductive performance. Contrary to our expectation, these positive effects remained consistent across the entire precipitation gradient. This result was unexpected given that along the precipitation gradient, plant physiology, and EFN functioning can vary strongly (e.g. Nogueira et al. 2015; Sandel et al. 2010), as do the assemblages of ants and herbivores on plants (e.g. Câmara et al. 2019; Moreira et al. 2015; Nogueira et al. 2015; Weissflog et al. 2018). Nevertheless, from a plant perspective, these differences were not decisive for the overall outcome of ant-mediated protective mutualisms.

The finding that ants were equally effective at reducing herbivory in both drier and wetter habitats contrasts with a previous meta-analysis that reported a negative relationship between precipitation and the protective effects of ants on plants with EFNs, with effective herbivore deterrence observed only in drier environments (Leal and Peixoto 2017). The ability of plants with EFNs to sustain mutualistic interactions with protective ants across a broad precipitation gradient likely stems from the inherently low cost and strategic regulation of extrafloral nectar production. Although studies on induction of extrafloral nectar secretion in response to herbivory are still limited—covering only about 16 plant species to date (see

Table 14.2 Heterogeneity estimates obtained for random effects, including study identity, plant species, and plant phylogeny, across three different meta-analytical models investigating the ant effects on plants with EFNs

	I^2 (95% CI)	σ^2 (95% CI)
Model 1: Type of ant effect metric		
Study ID	35.30 (33.39, 37.22)	0.32 (0.19, 0.47)
Plant species	38.61 (36.70, 40.53)	0.33 (0.23, 0.47)
Plant phylogeny	1.38 (−0.54, 3.30)	0.06 (0.00, 0.45)
Total	75.29 (73.37, 77.21)	–
Model 2: Precipitation * type of ant effect metric		
Study ID	34.27 (35.35, 36.19)	0.32 (0.19, 0.47)
Plant species	38.72 (36.80, 40.64)	0.34 (0.22, 0.48)
Plant phylogeny	2.90 (0.97, 4.81)	0.09 (0.00, 0.51)
Total	75.89 (73.96, 77.80)	–
Model 3: Precipitation * type of ant effect metric (performance with annual mean precipitation < 2500 mm)		
Study ID	33.22 (31.30, 35.14)	0.32 (0.18, 0.48)
Plant species	38.92 (37.00, 40.84)	0.35 (0.23, 0.49)
Plant phylogeny	4.89 (2.97, 6.1)	0.12 (0.00, 0.55)
Total	77.03 (75.11, 78.95)	–

The first model did not include moderators other than the type of ant effect, whereas the other two also included precipitation. The two key estimates (I^2 and σ^2) describe heterogeneity for each random effect in the meta-analysis, where I^2 indicates the percentage of total variation across studies that is due to real differences rather than chance, with higher values reflecting greater heterogeneity, and σ^2 represents the estimated variance of the true effects among studies, capturing how much effect sizes vary due to random factors. In both cases, the 95% confidence interval (CI) reflects the uncertainty surrounding each estimate

Table 4 in Chinarelli et al. 2022, and Alencar et al. 2023)—the results consistently indicate that plants can increase extrafloral nectar production and/or ant visitation when attacked, demonstrating some level of regulation of extrafloral nectary function. This targeted secretion allows plants to allocate biotic defences more effectively during herbivore attacks, minimizing energetic expenditure even under restrictive conditions (Alencar et al. 2023). Moreover, in *Ochroma pyramidale* (Bombacaceae), the sugar content of extrafloral nectar represented only about 1% of the total energy allocated to leaves (O'Dowd 1979). If such a proportion is consistent across plant species with EFNs, it suggests that even under moderate resource constraints plants can maintain EFN functionality.

The synchronization between producing new plant tissues (e.g. leaves) and the seasonal increase in environmental resources (e.g. water, light) can enhance EFN functionality (e.g. Nogueira et al. 2020). This alignment reduces the relative cost of extrafloral nectar production of plants with EFNs across climatic gradients. In habitats with restricted rainfall, where precipitation is concentrated within a short period, new leaves and branches production often coincides with EFN activation and nectar secretion (e.g. Nogueira et al. 2015). As a result, even in water-limited environments, biotic defence can align with sudden increases in water availability and the need for greater protection of newly produced tissues against herbivores (Calixto et al. 2020). Conversely, plants may experience less water stress over time in high-rainfall environments like forest systems but contend with reduced light availability, particularly in shaded understories. During the driest part of the year—when water remains sufficient—greater canopy light penetration occurs, promoting new plant tissue production and higher EFN activity, enhancing biotic defences (Nogueira et al. 2020). Thus, this phenological synchronization can underpin the broadly positive effects of biotic defences observed across precipitation gradient.

Finally, the ubiquity of potential ant partners from various genera across tropical and subtropical ecosystems (Luo et al. 2023) underscores the widespread availability of numerous ant species capable of protecting plants against herbivores. From a plant perspective, ants are an abundant resource (Schultheiss et al. 2022), and, given the relatively low cost of nectar production required to attract them, they provide an efficient and economical means of recruitment for defence (Leal et al. 2023). Therefore, as long as plants can maintain extrafloral nectar secretion, plant species bearing EFNs will, on average, benefit from ant protection along precipitation gradients.

14.4.2 Influence of Climatic Change on Ant-Plant Mutualisms

Although our meta-analysis shows that the protective outcome of EFN-mediated mutualisms remains consistently positive across the global precipitation gradient, the magnitude and resilience of these effects are likely to vary among biomes under climate change. In particular, wetter regions—such as tropical rainforests—may face a higher risk of disruption. In these environments, the interacting partners (plants, herbivores, and ants) typically exhibit fewer physiological adaptations to cope with drought. As extreme drought and flooding events become more frequent and spatially variable (Jenkins et al. 2011; Szewczyk and McCain 2016), local population dynamics and community structures will likely be the first to respond (Caddy-Retalic et al. 2019; Klapwijk et al. 2013; Parr and Bishop 2022). These disruptions are more likely to compromise the persistence of mutualistic networks in wetter habitats, despite the overall positive role that ants play in plant defence across different precipitation regimes.

Although we could not directly evaluate species turnover along the precipitation gradient, existing studies provide important insights. Drought intensification is expected to reduce ant richness and alter community composition from wet tropical areas to deserts (Almeida et al. 2023, Gibb et al. 2019; see Chap. 11). Plant communities will also shift, but at a slower pace—estimated to be 7.5 times less sensitive to climate projections than ants (Caddy-Retalic et al. 2019). This mismatch may decouple ant-plant-herbivore interactions, representing a key mechanism through which climate change threatens ecological interactions (Rafferty et al. 2015). Even among species that persist, EFN-mediated defence may be challenged by increased costs under water stress. Rising temperatures and altered rainfall are expected to increase herbivory (Hamann et al. 2021), heightening the adaptive value of anti-herbivory defences. However, water deprivation reduces carbon assimilation in plants, increasing the relative cost of maintaining anti-herbivory defences and raising the risk of failing to express them (McDowell and Sevanto 2010), particularly in wetter regions where plants are less physiologically adapted to endure drought (Allen et al. 2010). By contrast, EFN-bearing species from drier environments, already adapted to warm and arid conditions, are more likely to maintain EFN production and thus sustain mutualistic defence.

While most research on climate impacts on ant-plant interactions has emphasized changes in species composition, it is equally important to consider how mutualistic outcomes are affected among species that do survive. For instance, carbohydrate-rich EFNs may enhance thermal tolerance in ants, as colonies with higher carbohydrate intake have been shown to exhibit greater heat tolerance (Bujan and Kaspari 2017; Freires et al. 2023). This suggests that sustained investment in EFN production could improve ant survival under harsher climatic conditions. This may foster mutualistic persistence or even intensification in seasonally dry or arid environments, which could emerge as future hotspots for these interactions. Nevertheless, these hypotheses require further investigation into the physiological strategies of plants with EFNs across precipitation gradients, including drought-sensitive species. Moreover, the increasing frequency of unpredictable extreme climatic events (Stott 2016) may exceed the ability of mutualists to acclimate to new conditions, changing patterns of diversity and dominance among species (e.g. Rodrigues-Filho et al. 2024) and potentially leading to local extinctions (Wiens et al. 2016). Future work should focus on the potential recovery of ant-plant mutualisms following such disturbances and on identifying thresholds beyond which these interactions collapse.

14.5 Conclusion

Our macroecological study demonstrates that EFN-mediated ant-plant mutualisms are generally resilient to shifts in precipitation. By incorporating 13 new studies and 13 additional plant species with EFNs —a 27% increase—we raised the number of effect sizes from 144 to 292, yet our database still covers only 60 of the 3941 known EFN-bearing species (Weber and Keeler 2013), highlighting both the early stage of research in this field and the divergence among existing meta-analyses. Further studies are particularly needed in underrepresented environments with annual rainfall below 400 mm or above 2500 mm to better understand these interactions in deserts and humid rainforests. Most available data used in our meta-analysis come from seasonally dry regions (500–1500 mm/year), where biotic defence tends to enhance plant fitness (Leal and Peixoto 2017), although considerable variability exists. Rare but informative cases of failure or overprotection—such as those reported in *Anemopaegma album* (Bignoniaceae) and *A. scabriuscula* under similar ecological conditions (Nogueira et al. 2012, 2015)—highlight how shifts in ant assemblages or plant resource allocation can disrupt mutualism outcomes. Moreover, plant species from wetter habitats—underrepresented in our dataset—generally exhibit fewer drought-adaptive traits, making them more vulnerable to future climatic stress. In such conditions, these plants may reduce investment in EFN secretion, which could lead to a decline in ant-plant mutualisms in wet regions. As a result, these interactions may become increasingly restricted to arid environments, where species are better adapted to drought. Given that biotic defences are more prevalent in wetter environments (Luo et al. 2023), their potential collapse under climate-change impacts could trigger widespread ecological consequences across multiple levels of biological organization. Addressing these knowledge gaps will require further research on how increasing temperatures and drought severity affect EFN physiology and whether other anti-herbivory strategies can compensate across environmental gradients.

References

Alencar CLS, Nogueira A, Vicente RE, Coutinho IAC (2023) Plant species with larger extrafloral nectaries produce better quality nectar when needed and interact with the best ant partners. J Exp Bot 74:4613–4627. https://doi.org/10.1093/jxb/erad160

Aljbory Z, Chen M (2018) Indirect plant defense against herbivores: a review. Ins Sci 25:2–23. https://doi.org/10.1111/1744-7917.12436

Allen CD, Macalady AK, Chenchouni H, Bachelet D, McDowell N, Vennetier M, Kitzberger T, Rigling A, Breshears DD, Hogg EH, Gonzalez P, Fensham R, Zhang Z, Castro J, Demidova N, Hwan-Lim J, Allard G, Running SW, Semerci A, Cobb N (2010) A global overview of drought and heat-induced tree mortality reveals emerging climate change risks for forests. For Ecol Manag 259:660–684. https://doi.org/10.1016/j.foreco.2009.09.001

Almeida RP, Silva RR, da Costa AC, Ferreira LV, Meir P, Elison AM (2023) Induced drought strongly affects richness and composition of ground dwelling ants in eastern Amazon. Oecologia 201:299–309. https://doi.org/10.1007/s00442-023-05316-x

Bentley BL (1977a) Extrafloral nectaries and protection by pugnacious bodyguards. Ann Rev Ecol Evol Syst 8:407–427

Bentley BL (1977b) The protective function of ants visiting extrafloral nectaries in *Bixa Orellana* (Bixaceae). J Ecol 65:27–38

Bertness MD, Callaway R (1994) Positive interactions in communities. Trends Ecol Evol 9:191–193. https://doi.org/10.1016/0169-5347(94)90088-4

Billick I, Tonkel K (2003) The relative importance of spatial vs. temporal variability in generating a conditional mutualism. Ecology 84:289–295. https://doi.org/10.1890/0012-9658(2003)084[0289:TRIOSV]2.0.CO;2

Bronstein JL (2015) The study of mutualism. In: Bronstein JL (ed) Mutualism. Oxford University Press, Oxford, pp 3–19

Bujan J, Kaspari M (2017) Nutrition modifies critical thermal maximum of a dominant canopy ant. J Insect Physiol 102:1–6. https://doi.org/10.1016/j.jinsphys.2017.08.007

Caddy-Retalic S, Hoffman BD, Guerin GR, Andersen AN, Wardle GM, McInerney FA, Lowe AJ (2019) Plant and ant assemblages predicted to decouple under climate change. Divers Distrib 25:551–567. https://doi.org/10.1111/ddi.12858

Calixto ES, Novaes LR, Santos DFB, Lange D, Moreira X, Del-Claro K (2020) Climate seasonality drives ant–plant–herbivore interactions via plant phenology in an extrafloral nectary-bearing plant community. J Ecol 109:639–651. https://doi.org/10.1111/1365-2745.13492

Câmara T, Leal IR, Blüthgen N, Oliveira FMP, Arnan X (2019) Anthropogenic disturbance and rainfall variation threaten the stability of plant–ant interactions in the Brazilian Caatinga. Ecography 42:1960–1972. https://doi.org/10.1111/ecog.04531
Chalcoff VR, Aizen MA, Ezcurra C (2012) Erosion of a pollination mutualism along an environmental gradient in a south Andean treelet, *Embothrium coccineum* (Proteaceae). Oikos 121:471–480. https://doi.org/10.1111/j.1600-0706.2011.19663.x
Chamberlain SA, Holland JN (2009) Quantitative synthesis of context dependency in ant–plant protection mutualisms. Ecology 90:2384–2392. https://doi.org/10.1890/08-1490.1
Chamberlain SA, Bronstein JL, Rudgers JA (2014) How context dependent are species interactions? Ecol Lett 17:881–890. https://doi.org/10.1111/ele.12279
Chinarelli HD, Nogueira A, Leal LC (2022) Extrafloral nectar production induced by simulated herbivory does not improve ant bodyguard attendance and ultimately plant defence. Biol J Linn Soc 135:429–446. https://doi.org/10.1093/biolinnean/blab159
Dai A, Zhao T, Chen J (2018) Climate change and drought: a precipitation and evaporation perspective. Curr Clim Chang Rep 4:301–312. https://doi.org/10.1007/s40641-018-0101-6
Dáttilo W, Aguirre A, Flores-Flores RV, Fagundes R, Lange D, García-Chávez J, Del-Claro K, Rico-Gray V (2015) Secretory activity of extrafloral nectaries shaping multitrophic ant-plant-herbivore interactions in an arid environment. J Arid Environ 114:104–109. https://doi.org/10.1016/j.jaridenv.2014.12.001
Freires ASA, Ferreira CAS, Nascimento G, Arnan X (2023) Carbohydrate-rich diet increases critical thermal maximum in ants. Entomol Exp et Appl 171:240–247. https://doi.org/10.1111/eea.13268
Gibb H, Grossman BF, Dickman CR, Decker O, Wardle GM (2019) Long-term responses of desert ant assemblages to climate. J Anim Ecol 88:1549–1563. https://doi.org/10.1111/1365-2656.13052
Gurevitch J, Koricheva J, Mengersen K (2013) Handbook of meta-analysis in ecology and evolution. Princeton University Press, Princeton
Hamann E, Blevins C, Franks SJ, Jameel MI, Anderson JT (2021) Climate change alters plant–herbivore interactions. New Phytol 229:1894–1910. https://doi.org/10.1111/nph.17036
Howe HF, Smallwood J (1982) Ecology of seed dispersal. Annu Rev Ecol Evol Syst 11:201–228. https://doi.org/10.1146/annurev.es.13.110182.001221
Jenkins CN, Sanders NJ, Andersen AN, Arnan X, Brühl CA, Cerda X, Ellison AM, Fisher BL, Fitzpatrick MC, Gotelli NJ, Gove AD, Guénard B, Lattke JE, Lessard JP, McGlynn TP, Menke SB, Parr CL, Philpott SM, Vasconcelos HL, Weiser MD, Dunn RR (2011) Global diversity in light of climate change: the case of ants. Divers Distrib 17:652–662. https://doi.org/10.1111/j.1472-4642.2011.00770.x
Jorge MLSP, Howe HF (2009) Can forest fragmentation disrupt a conditional mutualism? A case from Central Amazon. Oecologia 161:709–718. https://doi.org/10.1007/s00442-009-1417-7
Karger DN, Conrad O, Böhner J, Kawohl T, Kreft H, Soria-Auza RW, Zimmermann NE, Linder P, Kessler M (2017) Climatologies at high resolution for the Earth land surface areas. Sci Data 4:170122. https://doi.org/10.1038/sdata.2017.122
Karger DN, Conrad O, Böhner J, Kawohl T, Kreft H, Soria-Auza RW, Zimmermann NE, Linder HP, Kessler M (2018) Data from: Climatologies at high resolution for the Earth's land surface areas. EnviDat. https://doi.org/10.16904/envidat.228.v2.1
Karl TR, Trenberth KE (2003) Modern global climate change. Science 302:1719–1723. https://doi.org/10.1126/science.1090228
Keeler KH, Porturas LD, Weber MG (2024) World list of plants with extrafloral nectaries. www.extrafloralnectaries.org
Klapwijk MJ, Csóka G, Hirka A, Bjorkman C (2013) Forest insects and climate change: long term trends in herbivory damage. Ecol Evol 3:4183–4196. https://doi.org/10.1002/ece3.717
Leal LC, Koski MH (2024) Linking pollen limitation to seed dispersal effectiveness. Ecol Lett 27:e14347. https://doi.org/10.1111/ele.14347

Leal LC, Peixoto PEC (2017) Decreasing water availability across the globe improves the effectiveness of protective ant–plant mutualisms: a meta-analysis. Biol Rev 92:1785–1794. https://doi.org/10.1111/brv.12307

Leal LC, Nogueira A, Peixoto PEC (2023) Which traits optimize plant benefits? Meta-analysis on the effect of partner traits on the outcome of an ant-plant protective mutualism. J Ecol 111:263–275. https://doi.org/10.1111/1365-2745.14031

Luo Y, Taylor A, Weigelt P, Guénard B, Economo EP, Nowak A, Inderjit, Kreft H (2023) Climate and ant diversity explain the global distribution of ant-plant mutualisms. Ecography 11:e06841. https://doi.org/10.1111/ecog.06841

McDowell NG, Sevanto S (2010) The mechanisms of carbon starvation: how, when, or does it even occur at all? New Phytol:264–266. https://doi.org/10.1111/j.1469-8137.2010.03232.x

Melati BG, Leal LC (2018) Aggressive bodyguards are not always the best: preferential interaction with more aggressive ant species reduces the reproductive success of plants bearing extrafloral nectaries. PLOSOne 13:e0199764

Mooney E, Davidson B, Uyl JD, Mullins M, Medina E, Nguyen P, Owens J (2019) Elevated temperatures alter an ant-aphid mutualism. Entomol Exp Appl 167:891–905. https://doi.org/10.1111/eea.12839

Moreira X, Abdala-Roberts L, Parra-Tabla V, Mooney KA (2015) Latitudinal variation in herbivory: influences of climatic drivers, herbivore identity and natural enemies. Oikos 124:1444–1452. https://doi.org/10.1111/oik.02040

Nogueira A, Guimarães E, Machado SR, Lohmann LG (2012) Do extrafloral nectaries present a defensive role against herbivores in two species of the family Bignoniaceae in a Neotropical savanna? Plant Ecol 213:289–301. https://doi.org/10.1007/s11258-011-9974-3

Nogueira A, Rey PJ, Alcántara JM, Feitosa RM, Lohmann LG (2015) Geographic mosaic of plant evolution: extrafloral nectary variation mediated by ant and herbivore assemblages. PLoS One 10:e0123806. https://doi.org/10.1371/journal.pone.0123806

Nogueira A, Baccaro FB, Leal LC, Rey PJ, Lohmann LG, Bronstein JL (2020) Variation in the production of plant tissues bearing extrafloral nectaries explains temporal patterns of ant attendance in Amazonian understorey plants. J Ecol 108:1578–1591. https://doi.org/10.1111/1365-2745.13340

O'Dowd DJ (1979) Foliar nectar production and ant activity on a Neotropical tree, *Ochroma pyramidale*. Oecologia 43:233–248. https://doi.org/10.1007/BF00344773

Parr CL, Bishop TR (2022) The response of ants to climate change. Glob Chang Biol 28:3188–3205. https://doi.org/10.1111/gcb.16140

Pick JL, Nakagawa S, Noble DWA (2018) Reproducible, flexible and high-throughput data extraction from primary literature: the metaDigitise R package. Met Ecol Evol 10:426–431. https://doi.org/10.1111/2041-210X.13118

Pringle EG, Akçay E, Raab TK, Dirzo R, Gordon DM (2013) Water stress strengthens mutualism among ants, trees and scale insects. PLoS One 11:e1001705. https://doi.org/10.1371/journal.pbio.1001705

Proietti P, Palliottia A, Famiania F, Antognozzi E, Ferranti F, Andreutti R, Frenguelli G (2000) Influence of leaf position, fruit and light availability on photosynthesis of two chestnut genotypes. Sci Hortic 85:63–73. https://doi.org/10.1016/S0304-4238(99)00129-6

R Core Team (2023) R: a language and environment for statistical computing. R Foundation for Statistical Computing: https://www.R-project.org/

Rafferty NE, CaraDonna PJ, Bronstein JL (2015) Phenological shifts and the fate of mutualisms. Oikos 124:14–21. https://doi.org/10.1111/oik.01523

Rodrigues-Filho CAS, Costa FRC, Schietti J, Nogueira A, Leitão RF, Menger J, Borba G, Gerolamo CS, Avilla SS, Emilio T, Castilho CV, Bastos DA, Rocha EX, Fernandes IO, Cornellius C, Zuanon J, Souza JLP, Utta ACS, Baccaro FB (2024) Multi-taxa responses to climate change in the Amazon Forest. Glob Chang Biol 30:e17598. https://doi.org/10.1111/gcb.17598

Rosumek F, Silveira FAO, Neves FS, Barbosa NPU, Diniz L, Oki Y, Pezzini F, Fernandes GW, Cornelissen T (2009) Ants on plants: a meta-analysis of the role of ants as plant biotic defenses. Oecologia 160:537–549. https://doi.org/10.1007/s00442-009-1309-x

Sandel B, Goldstein LJ, Kraft NJB, Okie JG, Shuldman MI, Ackerly DD, Cleland EE, Suding KN (2010) Contrasting trait responses in plant communities to experimental and geographic variation in precipitation. New Phytol 188:565–575. https://doi.org/10.1111/j.1469-8137.2010.03382.x

Schultheiss P, Nooten SS, Wang R, Guénard B (2022) The abundance, biomass, and distribution of ants on Earth. PNAS 119:e2201550119. https://doi.org/10.1073/pnas.2201550119

Stott P (2016) How climate change affects extreme weather events. Science 352:1517–1518. https://doi.org/10.1126/science.aaf72

Szewczyk T, McCain CM (2016) A systematic review of global drivers of ant elevational gradient. PLoS One 11:e0155404. https://doi.org/10.1371/journal.pone.0155404

Thrall PH, Slattery JF, Broadhurst LM, Bickford S (2007) Geographic patterns of symbiont abundance and adaptation in native Australian *Acacia*-rhizobia interactions. J Ecol 95:1110–1122. https://doi.org/10.1111/j.1365-2745.2007.01278.x

Trager M, Bhotika S, Hostetler JA, Andrade GV, Rodriguez-Cabal MA, McKeon S, Osenberg CW, Bolker BM (2010) Benefits for plants in ant-plant protective mutualisms: a meta-analysis. PloS One 5:e14308. https://doi.org/10.1371/journal.pone.0014308

Vidal MC, Anneberg TJ, Curé AE, Althoff DM, Segraves KA (2021) The variable effects of global change on insect mutualisms. Curr Opin Insect Sci 47:46–52. https://doi.org/10.1016/j.cois.2021.03.002

Viechtbauer W (2010) Conducting meta-analyses in R with the metafor package. J Stat Softw 36:1–48. https://doi.org/10.18637/jss.v036.i03

Wang Z, Li G, Sun H, Ma L, Guo Y, Zhao Z, Gao H, Mei L (2018) Effects of drought stress on photosynthesis and photosynthetic electron transport chain in young apple tree leaves. Biol Open 7:bio035279. https://doi.org/10.1242/bio.035279

Warren RJ II, Bradford MA (2014) Mutualism fails when climate response differs between interacting species. Glob Chang Biol 20:466–474. https://doi.org/10.1111/gcb.12407

Weber MG, Keeler KH (2013) The phylogenetic distribution of extrafloral nectaries in plants. Ann Bot 111:1251–1261. https://doi.org/10.1093/aob/mcs225

Weissflog A, Markesteijn L, Lewis OT, Comita LS, Engelbrecht BMJ (2018) Contrasting patterns of insect herbivory and predation pressure across a tropical rainfall gradient. Biotropica 50:302–311. https://doi.org/10.1111/btp.12513

Wiens JJ (2016) Climate-related local extinctions are already widespread among plant and animal species. PLoS Biol 14:e2001104. https://doi.org/10.1371/journal.pbio.2001104

Xu FF, Chen J (2015) Extrafloral nectar secretion is mainly determined by carbon fixation under herbivore-free condition in the tropical shrub *Clerodendrum philippinum* var. *simplex*. Flora 217:10–13. https://doi.org/10.1016/j.flora.2015.09.004

Chapter 15
Climate Change and the Provision of Ant-Mediated Ecosystem Services

Luciana Elizalde, María Natalia Lescano, Mariano A. Rodriguez-Cabal, and Gabriela I. Pirk

15.1 Introduction

The concept of ecosystem services (ES) provides a valuable framework for recognizing the immense value of nature, as it encompasses the broad spectrum of benefits that humans derive from natural ecosystems and biodiversity (Constanza et al. 2017). Ecosystem services can be categorized into three main groups—provisioning services (e.g. food, water, timbre), regulation and maintenance services (e.g. climate regulation, water purification, flood and pest control, soil formation, nutrient cycling), and cultural services (e.g. recreation, aesthetic enjoyment)—also referred to as material, regulating, and non-material, respectively (Díaz et al. 2018). Since climate change is altering the provision of ES (Pörtner et al. 2021), there is an increasing need to better understand how to conserve biodiversity in order to maintain ES and identify keystone taxa that are crucial for their provision. Recent research has emphasized the functional relationships between biodiversity and ES, highlighting that functional diversity is an important attribute of biodiversity that

L. Elizalde (✉) · M. N. Lescano · G. I. Pirk
LIHO—INIBIOMA, Universidad Nacional del Comahue, CONICET, San Carlos de Bariloche, Río Negro, Argentina
e-mail: lelizalde@comahue-conicet.gob.ar; natilescano@comahue-conicet.gob.ar; pirkgi@comahue-conicet.gob.ar

M. A. Rodriguez-Cabal
Rubenstein School of Environment and Natural Resources, University of Vermont, Burlington, Vermont, USA

Grupo de Ecología de Invasiones - INIBIOMA, Universidad Nacional del Comahue, CONICET, San Carlos de Bariloche, Río Negro, Argentina

I. R. Leal et al. (eds.), *Ant Responses to Climate Change*, Entomology in Focus 10,
https://doi.org/10.1007/978-3-032-21127-9_15

influences ES by affecting the underlying ecosystem processes (de Bello et al. 2010). As climate change increasingly disrupts these processes, global efforts to assess its magnitude and pace (e.g. Intergovernmental Panel on Climate Change) have been closely linked with initiatives to understand the consequences for ES and the parallel declines in biodiversity (Intergovernmental Platform on Biodiversity and Ecosystem Services).

Ants are widespread inhabitants of terrestrial ecosystems, contributing a substantial proportion of global faunal biomass and playing a pivotal role in maintaining the web of ES that sustain human well-being (Del Toro et al. 2012, Elizalde et al. 2020). Their remarkable adaptability, complex social structures, and diverse ecological interactions make them key players in ensuring the health and resilience of ecosystems (Elizalde et al. 2020). The diverse feeding and nesting strategies of >18,000 ant species contribute to a variety of ES, including biological control, nutrient cycling, soil bioturbation, and seed dispersal. Aside from bees, which attract attention due to their commercial importance as pollinators, ants are the most studied insect group in terms of ES provided, with at least 15 ES attributed to them (Noriega et al. 2018, Table 15.1). However, ants can also generate disservices, with negative effects on human productive systems or well-being, such as those caused by becoming pests in agroecosystems or urban areas (Del Toro et al. 2012). Furthermore, although ants as social insects are generally expected to be relatively resilient, some groups that carry out relevant ES can be highly sensitive to a changing climate (Parr and Bishop 2022). This underscores the need to consider the potential impacts of climate change on the diversity and functioning of ant communities, which could in turn affect the vital ES they provide.

In this chapter, we review the literature on ES provided by ants to summarize the current knowledge and assess how climate change may affect these services, with a focus on warming and changing precipitation patterns, for which there is some existing evidence. We first highlight relevant attributes of ants that make them important ES providers and explore how these attributes might be affected by climate change (Sect. 15.2). Then, we consider the role of ants in ES that have been most extensively researched, such as seed dispersal, pollination, biological control, and soil services (Sect. 15.3), as well as their potential disservices (Sect. 15.4). In doing so, we explore and discuss different actions and management strategies that could help sustain and enhance the ES provided by ants, while also reducing any negative disservices, in the light of climate change. Finally, we highlight key research questions and identify knowledge gaps that need to be addressed in order to more precisely understand and quantify the impacts of climate change on ES involving ants, and we suggest how ES provided by ants can help mitigate the negative effects of climate change on human societies (Sect. 15.5). By synthesizing the current understanding and identifying knowledge gaps, we aim to provide valuable information for researchers and land managers to help them maximize conservation value, enhance multiple ES, and minimize disservices involving ants.

Table 15.1 Ecosystem services provided by ants (modified from Del Toro et al. 2012), including the code classification of the Common International Classification of Ecosystem Services (CICES, Haines-Young and Potschin 2018). References for literature reviews or meta-analyses of particular services are included (if no reference is given, it means that no review of that particular ES in ants was published according to our knowledge). Services in bold are those treated in detail in this chapter (Sect. 15.3)

Category	Service	CICES code	References
Provisioning/material	Food	1.1.6.1	Rastogi (2011)
	Biomedical/pharmaceutical	1.1.6.2	Rastogi (2011)
Regulation and maintenance	**Seed dispersal**	2.2.2.2	
	Pollination	2.2.2.1	Das and Das (2023)
	Animal community regulation	2.2.3.1	Cerdá and Dejean (2011)
	Biological control	2.2.3.1	Anjos et al. (2022)
	Nutrient cycling	2.2.4.1	Farji Brener and Werenkraut (2017)
	Soil formation	2.2.4.1	Folgarait (1998)
	Decomposition	2.2.4.2	
	Carbon cycling	2.1.1.2	
Cultural/non-material	Religion/symbology	3.2.1.2	
	Literature	3.2.1.3	
	New World traditions	3.1.2.2	
	Western culture	3.1.2.2	
	Biological indication[a]	3.1.2.1	

[a]Biological indication was included by Del Toro et al. (2012) as a supporting service; however, it is best classified as a cultural service according to CICES

15.2 Ant Attributes Relevant for ES Provision in the Light of Climate Change

Several ant attributes make them key players in maintaining ecosystem health and resilience and, consequently, the ES that they provide. Ant abundance and species richness are crucial to ecosystem functioning and are likely to be impacted by climate change (see Parr and Bishop (2022) for a review). The success of ants as a highly abundant and diverse group is largely attributed to their unique social behaviours, such as chemical communication, coupled with their heavy reliance on sugary resources (Hölldobler and Wilson 1990). Furthermore, certain ant species are stress-tolerant and relatively easy to manage, making them valuable for enhancing ES (Elizalde et al. 2020). In the context of climate change, understanding how ants interact with their environment and how their biological traits may determine their adaptive capacity is critical for predicting climate change impacts on ecosystem functions.

15.2.1 Abundance and Species Richness

Ants are remarkable both in terms of abundance and biomass. Recent estimates suggest that the number of ants on Earth is approximately 20 quadrillion individuals, with a total dry biomass of 12 metric tons of carbon, exceeding the combined dry biomass of all wild birds and mammals, and representing approximately 20% of total human dry biomass (Schultheiss et al. 2022). Higher ant abundances are linked to high-quality seed dispersal (Ortiz et al. 2021), efficient removal of food resources and prey (Griffiths et al. 2018, Elizalde et al. 2022), stronger influence within food webs (Moya-Laraño and Wise 2007), and significant improvements in soil properties (Tuma et al. 2022). Thus, higher ant abundances can enhance the provisioning of ES, including seed dispersal, pest control, decomposition of dead matter, reduction of bacterial contamination, and soil enrichment. However, climate change poses a threat to these services, as declines in abundant, highly efficient species could lead to large-scale reductions in ecosystem performance and stability.

While ant abundance directly affects ES provision, ant species richness plays an important role in maintaining functional redundancy within ecosystems under changing conditions. Recent evidence suggests that ant species richness can be a good predictor of functional redundancy in some assemblages and time frames (Bihn et al. 2010, Houadria et al. 2016, Arnan et al. 2019). Species-rich communities may exhibit greater resilience to environmental changes, as seen in Mediterranean habitats where species-poor communities in oak forests were most vulnerable, while species-rich communities in open habitats demonstrated higher resilience in food resource exploitation (Arnan et al. 2019). However, the relationship between species richness and ES provision is complex and not always straightforward. For instance, higher ant species richness does not necessarily correlate with increased foraging rates or seed removal in some ecosystems (Gove and Majer 2006, Thompson and McLachlan 2007, Gove et al. 2009), particularly when a stress-tolerant keystone species dominates the ecosystem function (Andersen 2019).

Although high abundance and species richness are important factors that can, in some cases, buffer the effects of environmental fluctuations, some ants remain highly sensitive to climate change (see review in Parr and Bishop (2022)). Temperature strongly influences ant survival, reproduction, physiology, and behaviour (Sanders et al. 2007, Pelini et al. 2011b, Diamond et al. 2012, Warren and Chick 2013, Stuble et al. 2014, Warren and Bradford 2014). Ant foraging activity is particularly affected by temperature, especially in species with lower critical thermal maxima (CT_{max}) (see Chap. 4). While changes in rainfall patterns due to climate change might also impact ants, the evidence is much scarcer compared to that concerning temperature effects. In arid areas, increases in long-term rainfall have been shown to boost the activity of dominant ant species, although the impacts on subordinate ants remain less clear (Gibb et al. 2019). Consequently, rising temperatures and shifting precipitation regimes could reduce foraging windows, potentially diminishing ES related to ant foraging activities, such as seed dispersal, pollination, and biological control. However, the social nature of ants may help mitigate some of these impacts, as their collective responses to environmental stressors could enhance the resilience and adaptability of ant colonies (Parr and Bishop 2022), which might ensure the continued provision of ES in the face of climate change, especially in temperate regions.

15.2.2 Social Behaviours

One key to the success of ants as social insects is their highly coordinated and organized behaviour, which allows them to exploit resources more efficiently and ultimately provide vital ES to humanity (Elizalde et al. 2020). For example, the ability to respond quickly to changes in both the spatial abundance and density of food sources via recruitment systems that enhance functional responses (Beckers et al. 1989) makes ants important predators of pests (Morris et al. 2015, Risch and Carroll, 1982). Trail pheromones in ants allow the massive recruitment of foragers to a food source (Morgan 2008), thus enhancing the efficiency of ES related to foraging activities (e.g. seed dispersal, pollination, and biological control), and can be manipulated to control their effects (e.g. Sunamura 2018). Climate change might impact ant communication due to the effects of temperature on trail pheromones, such as through pheromone evaporation (van Oudenhove et al. 2012; see Chap. 2), which in turn could have negative implications for the ES they provide.

Coordination of cooperative behaviour and worker caste organization increases colony cohesion and efficiency. However, it was recently proposed that such attributes might make ants sensitive to detrimental impacts of changing climate, extreme weather, invasive species, and land conversion (Fisher et al. 2019). In such scenarios, chemical coordination of cooperative behaviour might be affected by homogenization of cuticular hydrocarbons—mainly due to reduced diversity of resources exploited by colonies—that are fundamental in nestmate recognition. This predicted convergence of chemical profiles among conspecific social insect colonies will erode the nestmate recognition system and interfere with the dynamic response of workers to their environment, for example by weakening colony defences and increasing exploitation by social parasites (Fisher et al. 2019). These predictions require further research and evidence, in particular, experimental information on how sociality of ant colonies will help in providing flexibility or will limit ants' ES in future scenarios of changing climate.

Fig. 15.1 Mutualistic relationships between ants and plants that provide key sugary resources to ES-provisioning ants, mediated by extrafloral nectar (left) or ant-tended hemipteran (right). In north-western Patagonia, Argentina, the native ant species, *Dorymyrmex tener*, frequently visits sugary sources provided by vegetable crops, as shown here feeding on an extrafloral nectary of *Vicia faba* (left, photo credit: L.N. González-García) or feeding on aphid honeydew on the exotic plant *Carduus thoermeri* (right, photo credit: M.N. Lescano)

15.2.3 Use of Sugary Resources

Natural sugary sources, such as floral and extrafloral nectaries and honeydew produced by hemipterans (ant-EFN/H mutualisms, Fig. 15.1), are known to structure ant communities (Blüthgen and Fiedler 2004, Aranda-Rickert et al. 2014) and contribute to the ecological dominance and success of ants (O'Dowd et al. 2003, Savage et al. 2011), thus helping sustain and enhance the ES that ants provide. The preference for nectar by ants facilitates floral visitation, which can involve them in pollination, leading to both positive and negative interactions with plants (Das and Das 2023, Sect. 15.3.2). The exploitation of extrafloral nectaries and honeydew by dominant ants, which actively patrol and defend the plant from non-sap-sucking herbivores and deter the natural enemies of hemipterans, underscores their significance in biological control efforts (Clough et al. 2017, Sect. 15.3.3). Furthermore, ES provided by ants that do not directly depend on sugary sources, such as seed dispersal and soil-related services, may be indirectly influenced by the availability of these resources since sugars help subsidize large ant colonies (Davidson and Cook 2008). Their attraction to sugary food sources also makes ants susceptible to manipulation, either to control them when they become urban nuisances (Rust and Su 2012, Sect. 15.4) or to disrupt their mutualisms when ant-tended hemipterans are considered crop pests (Nagy et al. 2013, Wäckers et al. 2017). Lastly, the consumption of sugary resources may play a crucial role in providing resilience to ant populations under climate change scenarios (Bujan and Kaspari 2017, Freires et al. 2023), thereby sustaining the ES they provide.

Predicting how ant interactions with nectar and EFN/H mutualisms might respond to climate change is complex due to the diverse characteristics of each partner, such as nutritional status, growth, reproduction, and foraging patterns that respond differently and simultaneously to different abiotic factors (see Chaps. 13, 14, and 17). For example, higher temperatures could enhance foraging activity in ants with greater thermal tolerance, leading to stronger mutualisms and increased protection for plants and hemipterans (Devegili et al. 2020). However, the same conditions might negatively impact aphid populations by reducing their fecundity (Wang et al. 2001, Dampc et al. 2021). Climate change could also alter the quantity and quality of nectar and honeydew, which are critical to these mutualisms. Elevated temperatures have been shown to reduce nectar volume and alter sugar content in plants (Takkis et al. 2018), while increasing honeydew excretion rates in hemipterans (Zhou et al. 2017). Higher CO_2 levels might increase the carbon-to-nutrient ratio in plants (Cernusak et al. 2019), diluting the nutritional content of these resources and potentially reducing their attractiveness to ants (Grasso et al. 2015). However, hemipterans may produce more honeydew in response to diluted sap nutrients, potentially reinforcing ant-hemipteran mutualisms under these conditions (Kremer et al. 2018, Blanchard et al. 2022). The wide variety of possible responses to climate change, both at the level of individual partners in ant-EFN/H mutualisms and their consequences at the community level, underscores the need for further study to better understand the potential impacts of climate change on the maintenance of the services provided by ants.

15.3 Assessing and Managing Ant-Mediated Services in a Changing World

As a first step to analyse the type and amount of research carried out on ants in relation to their role in ecosystems, we performed a search in WoS (April 17, 2024), including research articles with "ants" *or* "Formicidae" *and* "ecosystem" as terms in the Title, Keywords, and Abstract. This search yielded 3406 articles, 2848 of which focused on ants and their roles in ecosystems. Based on these, we generated a bibliometric map using VOSviewer (Van Eck and Waltman 2010) to represent relationships among the topics covered.

The bibliometric map reveals the diverse roles that ants play in ecosystems and how these functions connect to potential ES (Fig. 15.2). The cluster "seed" focuses on research related to seed dispersal and granivory. Ants assist in seed dispersal, ensuring that seeds are spread to favourable locations. This role, linked to supporting services, is particularly vital where ants are important dispersal agents, such as

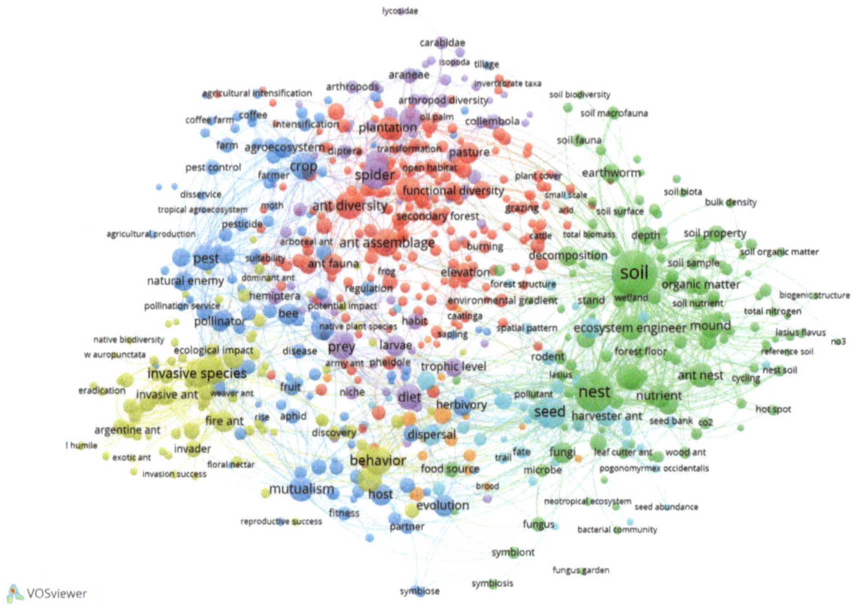

Fig. 15.2 Bibliometric network of research on ants and their role in ecosystems. Each circle represents a key term extracted from the search and is proportional in size to the number of occurrences. Links between circles (lines) represent the co-occurrence between them, with the intensity of the line being proportional to its strength. Circles are grouped into clusters (different colours) according to their co-occurrence interaction's strength. The light blue cluster ("seed") combines research terms related to seed dispersal. The cluster in blue ("crop") is related to ant functions in agroecosystems, while the fuzzy purple cluster encompasses studies of ants' trophic interaction with other animals. The red cluster ("diversity") in the centre merges studies on changes in ant diversity. The cluster in green ("soil-nest") is related to the functions that ants perform in the soil ecosystem, while the yellow ("invasive species") cluster encompasses research related to invasive ants

open arid habitats of the Mediterranean climate in Australia and South Africa (Lengyel et al. 2010). In contrast, some ants can also be major granivores, particularly in desert ecosystems (Brown et al. 1979), and they might contribute to weed control in agricultural systems by preying on naturally dispersed weed seeds, reducing weed abundance and germination (Evans and Gleeson 2016). By influencing plant recruitment, both positively and negatively, ants contribute to vegetation dynamics, the provision of food and habitat for other organisms, and therefore ecosystem resilience. The "crop" cluster relates to ant involvement in agroecosystems. Ants act as biological control agents by preying on crop pests, but their mutualisms with hemipterans can also indirectly cause problems for producers by increasing hemipteran pest populations. Their role in pollination also ties them to provisioning services within agricultural landscapes. The "diet" cluster reflects the complex food webs ants are part of, interacting as both prey and predators with a range of organisms. These interactions influence various supporting and regulating services. The "diversity" includes studies on ant biodiversity, species richness, and community composition across various habitats and environmental gradients, influencing the range of ES provided across different environments. This cluster underscores how ants contribute to ecosystem stability and resilience under changing environmental conditions, such as climate change. One of the largest clusters emphasizes the "soil-nest relationship", including studies evaluating the impacts of ants on the soil and as ecosystem engineers. By constructing nests and mounds, ants enhance soil aeration, water infiltration, and nutrient cycling. These activities have direct implications for soil fertility and structure, contributing to regulating services like improved soil health and carbon sequestration. Ant nests also create microhabitats that increase soil biodiversity, indirectly supporting other species and thereby enhancing overall ecosystem resilience. Lastly, the "invasive species" shows how research has focused on the impact of invasive ants, like the Argentine ant (*Linepithema humile*) and fire ants (*Solenopsis* spp.), which can disrupt native ecosystems, alter species interactions, and affect ES negatively. While the bibliographic search was designed to capture studies directly linking ants to ecosystem impacts, functions, and services, we acknowledge that its focus on the term "ecosystem" may have underrepresented certain ecological interactions, such as those involving extrafloral nectaries and other mutualistic interactions, which are important for understanding some ant-mediated ES (as mentioned in Sect. 15.2.3).

In the following sections, we analyse in more detail the role of ants for ES relating to seed dispersal, pollination, biological control, and soil-related services. For each of these services, we evaluate how climate change may impact the contribution of ants, highlight key research questions, and identify knowledge gaps that need to be addressed to better understand the effect of climate change. We also explore management strategies to sustain and enhance the ES in a changing climate scenario, providing examples of how ant-mediated ES can potentially help mitigate its negative impacts on human societies.

15.3.1 Seed Dispersal

Myrmecochory, or seed dispersal by ants, is a widespread dispersal mechanism, documented in approximately 4.5% of plant species (Beattie and Hughes 2002; Rico-Gray and Oliveira 2007; Lengyel et al. 2010). Globally, more than a hundred ant species provide effective seed dispersal services to plants (Gove et al. 2007, Warren and Giladi 2014). Myrmecochorous plant species produce seeds with a lipid-rich fleshy appendage known as elaiosome, which attracts ants (see Chap. 16). The quality of dispersal services provided varies significantly and is influenced by behavioural and morphological differences among ant species (Gove et al. 2007, Ness et al. 2009). Generally, large-bodied, solitary-foraging ants are considered superior dispersers because individual workers can collect seeds, transport them over relatively long distances, and deposit them singly or in small groups in external nest refuse piles (Ness et al. 2004, Gove et al. 2007, Leal et al. 2014b). In contrast, small-bodied, recruit-foraging ants are often regarded as low-quality dispersers because they tend to consume the elaiosome in situ without transporting the diaspore, or they transport seeds over very short distances, depositing high numbers of seeds in nest refuse (Andersen and Morrison 1998, Leal et al. 2014b).

The reliance of many ant-dispersed plant species on a single ant species (or species complex) for seed dispersal may make them particularly vulnerable for disruption by ongoing climatic change (Gove et al. 2007, Ness et al. 2009, Warren et al. 2011, Warren and Bradford 2014). For example, in deciduous forests of North America, ~30% of understory herb species are ant-dispersed, with a single species – *Aphaenogaster rudis*—responsible for a majority of that dispersal (Beattie and Culver 1981, Zelikova et al. 2008, Ness et al. 2009). While some studies in these forests suggest that experimental warming may not significantly affect seed removal rates (Stuble et al. 2014), others have reported reductions in various foraging activities, including seed removal, following even slight temperature increases (Pelini et al. 2011a). Ants are particularly sensitive to ground-surface temperature, which influences foraging activity (Stuble et al. 2014). This suggests that while chronic experimental warming may not directly influence seed dispersal rates, temperature does correlate with rates of seed removal. In the Brazilian Caatinga, Oliveira et al. (2019) studied the combined effects of anthropogenic disturbance and a precipitation gradient on myrmecochory. Through field surveys and experiments, they found that high-quality disperser composition varied only with aridity, while ant-seed interactions, seed removal rates, and mean removal distance all declined with increasing aridity. These responses were mainly driven by the reduced abundance of the dominant disperser *Dinoponera quadriceps* (Oliveira et al. 2019). These findings highlight the vulnerability of seed dispersal services under a warmer, drier climate, likely due to low functional redundancy in high-quality dispersers.

Phenological timing is critical to the success of many mutualisms, and one of the primary impacts of climate change on ant-mediated seed dispersal is phenological mismatch between plants and ants. Changes in temperature can desynchronize the timing of plant flowering, seed production, and ant activity, leading to reduced seed dispersal efficiency. Extreme weather events, such as drought and storms, can further disrupt seed dispersal by affecting ant foraging behaviour and seed transport dynamics. Myrmecochorous plants in the Northeastern United States release their seeds relatively early in the spring in accordance with peak foraging by *A. rudis* at that time (Warren and Giladi 2014). Seed dispersal outcomes for early- vs. late-flowering plant species can differ depending on whether cold- or warm-adapted ant species are foraging. If myrmecochorous plants and ants respond synchronously to global warming, then these critical dispersal mutualisms should be relatively robust to the detrimental impacts of climate change. For example, Warren et al. (2011) explored the influence of changes in temperature, light, and moisture on the synchrony of spring fruiting phenology of two myrmecochorous plants—*Hexastylis arifolia* and *Hepatica nobilis*—and the foraging phenology of ants in the deciduous forests of the Eastern United States. The authors found that warming temperatures act as the primary phenological cue for plant fruiting but that the key seed disperser, *A. rudis*, fails to function as a seed disperser during early fruit set. Additionally, in a novel transplant experiment, Warren and Bradford (2014) found that early-blooming plant species were more at risk from asynchronous phenological timing because they were dependent on cold-adapted species alone, whereas later-blooming species were dispersed by both warm- and cold-adapted ant species, which enhanced effective seed dispersal.

Seed dispersal by ants might represent a valuable tool for aiding in the restoration of degraded areas and enhancing the resilience of ecosystems to ongoing environmental changes. In degraded areas with harsh environmental conditions and poor soils, the transport of seeds to ants' nests, where soils are enriched, could improve the chances of seed germination and seedling establishment. This interaction could be artificially induced by attaching attractive baits to seeds of non-myrmecochorous plants (Henao-Gallego et al. 2011, Ortiz et al. 2022, Fig. 15.3), a promising tool still under research (Ortiz et al. 2022). These artificially baited seeds could be designed to attract stress-tolerant ant species, which are usually abundant in areas with high anthropic disturbance (e.g. quarries, Pereda-Gomez et al. 2019) to enhance the populations of non-myrmecochorous plants that are important for restoration and resilient to climate change.

Climate change poses multifaceted challenges to ant-mediated seed dispersal, with potential implications for ecosystem structure, function, and resilience. Future

Fig. 15.3 *Dorymyrmex tener* carrying a non-myrmecochorous seed with an artificial bait (light square with seeds embedded) developed to help in the restoration of degraded areas. (Photo credits: D.P. Ortiz)

research should aim to elucidate the mechanisms underlying the effects of climate change on ant-mediated seed dispersal across a broader range of ecosystems and geographical regions. By advancing our understanding of these complex interactions, we can better predict and mitigate the ecological consequences of climate change on seed dispersal networks. Ultimately, preserving the critical role of ants in seed dispersal is essential for maintaining biodiversity and ecosystem stability in a rapidly changing world.

15.3.2 Pollination

Pollinators encompass a diverse array of species, ranging from vertebrates like birds, bats, mice, and lizards (Willmer 2011) to insects such as bees, flies, butterflies, moths, wasps, beetles, and ants (Kevan and Baker 1983). Traditionally, ants have been regarded as nectar robbers and disruptors of pollination due to their small size, winglessness, potential production of antibiotic secretions that can inhibit pollen viability, and aggression towards more effective pollinators (Rico-Gray and Oliveira 2007; Torezan-Silingardi and Del-Claro 2024). However, increasing evidence shows that ants play a more significant positive role in pollination than previously thought: not only are they effective pollinators of certain plant species (Das and Das 2023) but also in some cases their presence in flowers increases cross-pollination or may repel less effective pollinators or nectar robbers (Ness 2006; Wielgoss et al. 2014; Gonzálvez et al. 2015). For example, certain plants in dry ecosystems and savannas rely heavily on ants for pollination, showcasing a specialized interaction termed "ant-pollination syndrome" (Torezan-Silingardi and Del-Claro 2024).

In the light of climate change, since ants that pollinate flowers (mostly species of *Camponotus*, *Crematogaster*, *Dorymyrmex* (Fig. 15.4), *Formica*, *Pheidole*, *Solenopsis*, and *Tapinoma*, Das and Das 2023) also exploit alternative resources, such as insects, honeydew, and extrafloral nectaries, they are resilient to changes in plant abundance or phenology. So, in these cases, pollination services provided by ants might not be severely affected. This adaptability seems to align with broader patterns of resilience: a recent study found no evidence of global declines in ant richness, in contrast to declines observed in other Hymenoptera families, including bees (Zattara and Aizen 2021).

Nevertheless, the positive contributions of ants to pollination remain limited to a relatively small subset of plant species (e.g. 41 species across 23 families, Das and Das, 2023). Furthermore, the role of ants as pollinators often involves trade-offs, as their negative impacts on plants can outweigh their positive contributions. Consequently, efforts to mitigate pollination declines should focus on the entire pollinator assemblage rather than solely on ants. While many flower-visiting ants negatively affect pollination, there are instances where ants enhance pollination but simultaneously pose challenges, such as by attacking beehives (Sarwar 2016). Additionally, some ants tend aphids, which can reduce plant fitness (Devegili et al.

Fig. 15.4 A *Dorymyrmex tener* worker visiting a raspberry flower in Patagonia, Argentina. Its role in pollination is currently under study. (Photo credits: V.R. Chalcoff)

2021), resulting in an overall neutral or negative impact on the plants they visit. Among actions to prevent pollination decline under climate change are landscape management to achieve connectivity, which allows species to migrate (Chen et al. 2011); translocation of pollinators into new locations, where they may have a better chance of survival (Seddon et al. 2014); and keeping high biodiversity levels to ensure plant-pollinator phenological synchrony and thus pollination function (Brittain et al. 2013, Bartomeus et al. 2013).

Given the scarcity of studies explicitly addressing the balance of positive vs. negative effects of ants on pollination, future research should aim to quantify these interactions under changing environmental conditions. While ants may play a role in compensating for the loss of other pollinators, their contributions will not replace the efficiency of specialized pollinators such as bees. Instead, ants' involvement should be seen as a complementary component of broader ecosystem resilience, particularly in degraded or resource-limited systems.

15.3.3 Animal Community Regulation and Biological Control

The regulation of animal communities, particularly through the control of pests, pathogens, and disease vectors, is a major challenge for sustainable global food security and human health in the context of climate change (Gregory et al. 2009, Campbell-Lendrum et al. 2015, Skendži et al. 2021). Biological control, a method of pest management that leverages natural enemies like parasitoids and predators, provides huge benefits to human societies. For instance, in the United States, the value of biological control in agriculture alone is estimated at US$13.6 billion (Losey and Vaughan 2006). In the context of climate change, where both pests and diseases vectorized by insects as well as crop losses due to pests are predicted to increase with rising temperatures (Deutsch et al. 2018, Skendži et al. 2021), biological control gains much relevance as a service for human societies (Lamichhane et al. 2015). The growing relevance of biological control under climate change underscores the need for innovative approaches to address the predicted increases in pest outbreaks and the potential breakdown of current biocontrol agents (Thurman et al. 2017). One promising strategy is to focus on generalist predators already present in the environment (i.e. conservation biological control), as these species are well-adapted to preying on pests that expand their ranges due to climate change and are less likely to experience phenological mismatches (Thurman et al. 2017). Among these generalist predators, ants stand out as a key group with significant potential to contribute to pest regulation in changing climates.

Predatory and omnivorous ants are important generalist providers of biological control services for managing insect pests in both agricultural and urban settings (Offenberg 2015, Perfecto and Philpott 2023). Over 26 ant species from 16 genera have been evaluated regarding its biocontrol service, with *Pheidole pallidula* (Myrmicinae) and *Lasius niger* and *L. grandis* (Formicinae) being the most studied. These ants prey on at least 30 pest species, most of them honeydew-producing species, from 13 different crops, including citrus, mango, apple, and cocoa (Anjos et al. 2022), which are mainly tropical crops. Ants contribute to reducing the abundance of pests, thereby decreasing crop damage and increasing yields (Anjos et al. 2022). Additionally, ants play a role in controlling termite pests in both agricultural fields and urban wooden structures (Tuma et al. 2020), and specialized army ants in the Neotropics clean out cockroaches, fleas, and rats when their foraging columns enter into the houses (Ellison and Gotelli 2021). Ants can also aid in carrion decomposition by directly feeding on dead corpses and by preying on fly larvae (Barton and Evans 2017), which has important implications for controlling fly disease vectors, while simultaneously accelerating the removal of dead matter in urban and rural landscapes. However, ants might also decrease carrion decomposition rates when their predation on fly larvae is too high (Wei et al. 2023), and they can reduce the abundance of natural enemies—mainly generalist predators—and increase the abundance of honeydew-producing pests, such as mealybugs, soft scales, aphids, psyllids, and whiteflies (Anjos et al. 2022, Sect. 15.5).

Despite the critical role ants play in animal community regulation, no study has directly evaluated how ant-based biological control might be influenced by climate change. Most ant species relevant in pest control are stress-tolerant and opportunistic, characteristics that may allow them to tolerate the harsh conditions present in agroecosystems and cities, enabling them to adapt successfully to changing climatic conditions (Uno et al. 2010, Menke et al. 2011). Thus, as global temperatures rise, ants might play an increasingly important role in biological control. Their ability to respond rapidly to changes in food availability and their large colony sizes position them as potentially key biocontrol agents in future scenarios of increased pest outbreaks (Thomson et al. 2010, Diehl et al. 2013). In this context, ants' characteristics such as their ability to quickly respond to changes in both the spatial abundance and density of food sources via recruitment systems, as well as the great variation in colony size, which is positively correlated with the number of foragers available (Risch and Carroll 1982, Morris et al. 2015), deserve further research in the context of climate change. For example, when comparing the predatory abilities of two cohabiting *Dorymyrmex* species on live lepidopteran larvae, the species with more foragers was more successful and faster in the key steps of predation (i.e. discovery, recruitment, and removal of prey) than the species with fewer foragers, even when the caterpillars were at experimental high densities to mimic a pest outbreak (Elizalde et al. 2022, Fig. 15.5).

Nevertheless, two groups of ants that play important roles as arthropod community regulators and biocontrol agents might be at risk due to climate change. Ants in the family Ponerinae are important predators in tropical habitats (Hanisch et al. 2020) and have been recently identified at risk due to climate warming (Parr and

Fig. 15.5 *Dorymyrmex tener* recruiting in high numbers to predate on four alive caterpillars in a field experiment in the Patagonian steppe. (Photo credits: L. Elizalde)

Bishop 2022), mainly based on traits expected to suffer higher stress from increasing temperatures, such as large body size and dark colouration. Another vulnerable group includes aggressive arboreal ants, such as *Oecophylla* and *Azteca* species, which are crucial for biological control in tropical agroecosystems (Van Mele 2008, Morris et al. 2018). These canopy ants might face a greater risk from warming climate due to their limited access to thermal microhabitat refuges, which might make them more directly exposed to rising temperatures than their soil-dwelling counterparts (Parr and Bishop 2022). In the Caatinga, the largest seasonally dry tropical forest, ant protection services mediated by extrafloral nectar declined with increasing aridity due to shifts in ant species composition, highlighting the vulnerability of nectar-bearing plants to climate change through reduced herbivore predation (Oliveira et al. 2021). Thus, efforts should concentrate on managing tropical ant communities, in particular agroecosystems, where these species are keystone providers.

From a management perspective, there is a growing interest in how omnivorous ants change their feeding preferences and aggression towards prey when sugar sources are added to their diets. Manipulating ants in this context is a highly promising tool for increasing their predatory potential and altering their mutualistic associations with honeydew-producing pests in agroecosystems (e.g. Nagy et al. 2013, Wäckers et al. 2017). In parallel, management for habitat diversity increases ant species diversity, which might improve ant predation levels on pest species, benefiting biological control, at least in coffee and cacao agroforestry systems (Philpott and Armbrecht 2006). Crop shading, organic agriculture, and agroforestry can enhance the pest control provided by ants in some systems (Wittwer et al. 2021, Anjos et al. 2022) but not in others. For example, in organic vegetable crops, tree cover and cropped areas in the landscape reduced ant predation but increased the predation by other predators, as most ant species in this system typically inhabit open habitats (Frizzo et al. 2020). Techniques such as applying mulch, reducing tillage practices, and minimizing insecticide use are employed by managers to increase ant species diversity in agroecosystems, thereby enhancing their biocontrol service (Drummond and Choate 2011), even of weeds (Evans and Gleeson 2016). Finally, direct management of ant colonies, such as moving colonies into the crop areas and providing food and/or nests for their survival, as practised with *Oecophylla* ants in many locations worldwide (Thurman et al. 2019), should be tested with other relevant pest predator ant species. Given that some ant species are as or even more effective than pesticides and at lower costs, they could be used within integrated pest management programmes (Offenberg 2015). Despite these advances, further research is needed to fully understand how to manage ants to maximize their benefits as pest and pathogen regulators in the diverse contexts that climate change will present in agricultural and urban settings worldwide.

15.3.4 Soil-Related Services

Soils deliver fundamental ES, principally due to a range of biogeochemical processes that occur within and the functionality of its biodiversity (Lavelle et al. 2006). Soils store more carbon (C) than the atmosphere and terrestrial vegetation combined (Ciais et al. 2013), making C storage an important ES of soils. Since C storage in soils can reduce atmospheric CO_2 concentrations, soils are key in regulating climate (Whitmore et al. 2015). Additionally, soil organic matter enhances water purification and water holding capacity, provides protection against erosion risk, and supports food and fibre provision through improved soil fertility (Smith et al. 2015). Thus, C sequestration in soils not only regulates climate but also bolsters other soil-derived ES, such as agricultural production, clean water supply, and biodiversity (Lal 2004, Smith et al. 2020).

The effects of climate change on soils are complex, as they depend on highly interconnected processes. Carbon enters the soil mainly as above- or belowground plant litter and exudates, and litter decomposition is the main driver of nutrient and C cycling process in terrestrial ecosystems, influencing nutrient availability and soil C dynamics (Aerts 1997). Soil biota play a key role in modifying these inputs and also through bioturbation (i.e. the movement and mixing of soil by organisms), translocating the organic matter through the soil profile and mixing it with mineral particles, thus contributing to C stabilization (Frouz 2018). Since soil organisms responsible for decomposition increase their activity with increasing temperatures, climate change is expected to enhance decomposition rates, potentiating the release of CO_2 to the atmosphere, speeding up global warming through a positive feedback mechanism (i.e. carbon-climate feedback, Lu et al. 2013). Conversely, soil porosity, which is important for water holding capacity and infiltration under predicted increased flooding scenarios, is influenced by organic C and biotic activity (Kirkham 2014). Understanding how soil biota will respond to climate change is key to predicting future soil functioning (Bradford et al. 2016).

Ants that live in soils are key components of the macrofauna, participating in nutrient and C cycling, decomposition, and bioturbation (Folgarait 1998, Frouz and Jilková 2008, Del Toro et al. 2012). Ant nests have been shown to increase soil nutrient content (nitrogen, phosphorus, and potassium, including C) and cations (Ca, Na, Al, Mg, Cu, Fe, Mn, Zn), which in turn improve plant growth and reproduction and increase performance and fitness of plants (Farji-Brener and Werenkraut 2017). These effects are observed across various species and latitudes, mainly in species that build large nests and house populous ant colonies (Farji-Brener and Werenkraut 2017). However, ant species with non-apparent nests can also enrich the soil and benefit plant growth (e.g. Pirk et al. 2020). Ants can enhance decomposition of organic matter by up to 170% (Luo et al. 2023), by directly shredding leaves and organic material for feeding or nesting activities or indirectly influencing soil community composition or their microenvironment (Hölldobler and Wilson 1990, Del Toro et al. 2015). This indirect pathway might result in a positive effect on decomposition rate, for example, by providing more O_2 to decomposer microorganisms, or it might reduce decomposition rates via top-down effects, by predating on decomposer community (Parr et al. 2016, Walker et al. 2022, but see Cates et al. 2021). Finally, soil bioturbation by global ant population is estimated in circa 7.5–10 Gt of sediment per year, comparable to river-borne sediment delivery to oceans, although much lower than the 31 Gt estimated for human activities (Viles and Coombes 2022). Bioturbation by ants enhances water infiltration in the soil, both through nest entrances at the surface and as a result of increased soil macroporosity due to their tunnelling activities. The effect of ant nests on infiltration can be long-lasting: 2 years after ant nests were removed, there was three times more water infiltration in areas where ants nested than in areas without ant nests (Evans et al. 2011). In addition, high densities of nests of *Aphaenogaster barbigula* in Australia increased water infiltration about four times than in zones without nests (Eldridge 1993).

Most studies of the effects of ants on soils were carried out in natural environments, although evidence of the provision of soil-related ES by ants in agroecosystems is increasing. In an annual cropping agroecosystem, a species of *Pheidole* is estimated to add to the soil 2.44, 0.071, and 0.628 kg/ha per year of C, phosphorus, and nitrogen, respectively, contributing to soil nutrient and crop yield enhancement (Shukla et al. 2013, 2018). In a recent review, it was concluded that ant nests in agroecosystems show less cation content than adjacent, non-nest soils, probably due to the high anthropogenic nutrient input in agroecosystems—although that conclusion is based on just four studies (Farji-Brener and Werenkraut 2017). In warm and dry agricultural habitats in Australia, where earthworms are absent, ants and termites were found to increase wheat yield by 36% through improved soil water infiltration and nitrogen levels (Evans et al. 2011).

Climate change impacts on ant-mediated soil ES are complex and variable, with evidence often revealing contradictory responses across different contexts. Experimental evidence under warming scenarios suggests that ant nests might further improve soil nutrient pools and plant growth (Liao et al. 2023). However, nitrogen concentration and decomposition can be negatively affected, as evidenced in a mesocosm experiment with *Formica subsericea* (Toro et al. 2015). It is possible that the highly simplified environment in this mesocosm restricted the microbial nest soil community, affecting nutrient cycling and decomposition rates. In fact, a recent field warming experiment showed that ants were more active in decomposing organic matter under warmed conditions than other soil organisms (Luo et al. 2023). Bioturbation by ants is also expected to increase as climate warms, as ants increase their activity as a behavioural response to escape to the higher temperatures, excavating deeper nests to move their brood to optimum temperature conditions within the soil. For example, *F. subsericea* displaced significantly more soil and built deeper nests in warmer treatments than in the controls in the aforementioned warming mesocosm experiment (Del Toro et al. 2015). Another mesocosm experiment showed that colonies of *Lasius niger* exposed to medium and high temperatures excavated larger and deeper nests than those exposed to mild temperature (García Ibarra et al. 2024). However, temperature increase was not linearly related to the increase in bioturbation, since nests excavated under high and medium temperatures had the same depth (García Ibarra et al. 2024). These examples illustrate the complex, non-linear relationships between climate warming and ants' ES on soils, mediated by the interaction with the associated microbial and faunal communities.

Ant activity introduces additional complexity to soil CO_2 emissions. In boreal forests, large colonies of *Formica rufa* group have been identified as CO_2 hotspots, contributing to nearly 1% of total ecosystem emissions (Ohashi et al. 2005, Risch et al. 2005, Domisch et al. 2006). Similarly, some leaf-cutting ants, with their large colonies and fungus cultivar, are associated with elevated CO_2 emissions (see Chap. 18). Increased bioturbation activity under warming scenarios may also amplify the effects of ants on CO_2 fluxes (Del Toro et al. 2015). However, emissions may primarily reflect a redistribution of CO_2, as ants concentrate C from foraged organic material within their nests, replacing the more diffuse emissions from decomposition processes occurring across the landscape (Mehring et al. 2021). While this redistribution may not indicate a net increase in ecosystem emissions, it underscores the need to investigate how ant nests influence C dynamics compared to scattered emissions across the landscape. Accurately quantifying these dynamics will be essential for integrating ant-mediated processes into biogeochemical models (Bradford et al. 2016).

While the evidence underscores the significance of ants in soil nutrient enhancement, C dynamics, decomposition, and bioturbation under changing climatic conditions, it also reveals the complexity and variability of these interactions across different ecosystems. The patchiness of ant effects, the interplay with microbial communities, and the differential impacts across biomes highlight critical areas for future research. Particularly, more studies are needed in agroecosystems and urban areas, as well as across diverse biomes, to understand how generalizable these findings are. Future models should incorporate the relationships between ants, soil processes, and other soil community components, considering both direct and indirect effects at various spatial and temporal scales. Additionally, the geomorphological impacts of ants on vegetation cover, soil properties, and biogeochemical cycling warrant further investigation in the context of a changing climate. As we deepen our understanding of these complex interactions, we will gain a clearer appreciation of the full potential of ant-mediated ES in helping soils adapt to the impacts of climate change.

15.4 Ecosystem Disservices

Many ant activities can also have detrimental effects on humans and the environment, referred to as "disservices". Among the most known disservices are the ones that arise in urban areas and productive systems, where humans and ants coexist closely, and those resulting from the ecological impacts of invasive ant species. These disservices depend on the presence, abundance, activities, and species composition of ants, all of which are likely to be influenced by the effects of climate change.

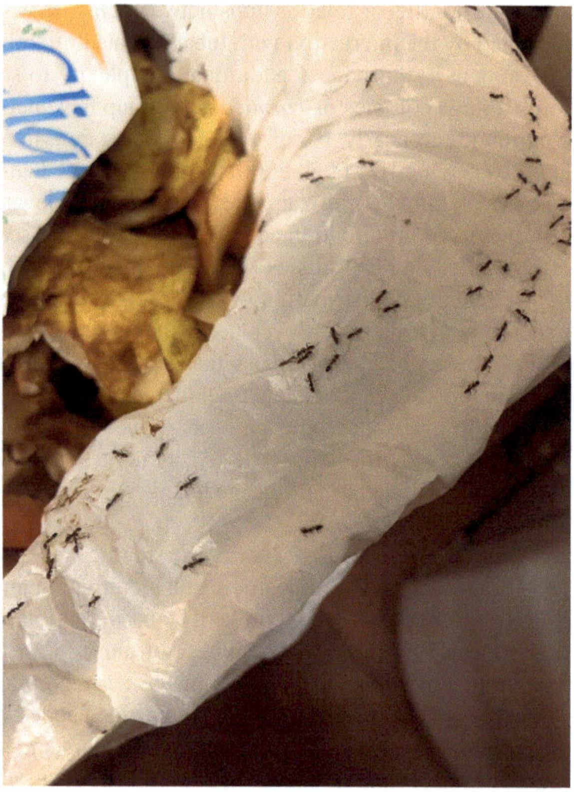

Fig. 15.6 *Linepithema humile* exploiting a trash bin in an urban apartment, in Bariloche, Argentina. The picture was taken by a citizen during a citizen science initiative, which allowed researchers to estimate the extent of *Linepithema humile*'s invasion in the city, just a few days after its discovery

In urban areas, although ants provide multiple ES (Perfecto and Philpott 2023), they can become nuisance pests when they turn into frequent home visitors or dwellers. Urban ants, either native or invasive, often nest within houses causing structural problems (e.g. *Camponotus* spp., Akre et al. 1995) and even colonize electrical equipment causing enormous damage (e.g. *Solenopsis invicta*, Vinson and MacKay 2019). The sting of some urban species can provoke anaphylactic reactions (e.g. *Solenopsis invicta* (Solley et al. 2002), *Myrmecia pilosula* (Brown et al. 2004)), and it has been found that ants can carry pathogens (Klotz et al. 1995), with the risk of spreading diseases, which is a serious concern especially in hospitals (Chadee and Le Maitre 1990). Finally, ants within homes can exploit human food and waste, and their mere presence can be a hassle to people (Fig. 15.6).

Regarding the effects of climate change on ants inhabiting urban areas, it has been suggested that as cities are under the heat island effect (i.e. they experience warmer temperatures than surrounding areas), urban-adapted ants may be particularly successful in tracking climate change. A study in North Carolina, USA, has shown that species found in urbanized areas have greater tolerance to warmer temperatures and dry conditions than species found in more natural forested sites, so they may be more successful at moving with further human expansion and at tracking future climate change (Menke et al. 2011). In fact, some studies that measured temperature tolerance of species in urban vs. natural areas found a higher tolerance of urban ants to higher temperatures (Angilletta et al. 2007). Also, *Temnothorax curvispinosus* in urban habitats exhibited greater running speed at warm test temperatures compared to those from rural areas, which could enhance resource acquisition in urban environments and help compensate for the increased metabolic demands of urban living (Chick et al. 2021). Cities may actually facilitate movement of species, acting as corridors for the continued expansion of both native and non-native species favoured by warmer temperatures. For example, *Tetramorium immigrans* in southeastern France can persist in the north of its range under harsher climates by colonizing highly urbanized areas (Cordonnier et al. 2020). Likewise, the characteristics of the cities allow *L. humile* to inhabit areas otherwise not climatically suitable (López-Collar and Cabrero-Sañudo 2021; Werenkraut et al. 2025). The availability of resources and nesting microhabitats, plus thermal and moisture conditions, makes cities a favourable environment for many invasive and native species. So, cities could become "reservoirs" from which invasive species could colonize new regions and eventually invade surrounding areas when conditions become more adequate (Borden and Flory 2021). Thus, it seems that ants, both native and invasive, will persist and even find suitable refuges in cities under climate change.

More than 200 ant species have been successful in invading new ranges (Suarez et al. 2010) and 5 of them are listed among the 100 of the World's Worst Invasive Alien Species (Global Invasive Species Database 2024). When they invade natural areas, they can establish strong interactions (i.e. competition or predation) with other species, producing compositional changes, declines, and even the local extinction of other arthropods, birds, reptiles, amphibians, and mammals (Holway et al. 2002, Angulo et al. 2024). They can also disrupt interactions such as seed dispersal (Rodriguez-Cabal et al. 2009) or change decomposition rates and nutrient cycles, affecting soil and plants (Dunham and Mikheyev 2010).

It has been assumed that climate change would worsen the threat of ant invasions as many species would expand their ranges facing less-limiting temperatures (see Chap. 7). Bertelsmeier et al. (2015) shed some light on these ideas by modelling suitable areas for 15 of the worst invasive ant species, both currently and with predicted climate change. Their results were surprising, as the potential distribution of only five species was predicted to increase, especially in tropical and subtropical regions, and most would decline. However, as we have discussed, invasive species could still reach new areas by moving among the microclimate created by urbanized and modified landscapes from where they may invade natural areas. Therefore, the invasion of these species will remain a huge concern in the future.

Another major disservice of ants is their potential to disrupt agricultural production. Leaf-cutting ant species (genera *Atta* and *Acromyrmex*) can cause huge damage to crops, especially in Neotropical areas (Montoya-Lerma et al. 2012). Some of these species are favoured in anthropogenically modified landscapes, such as croplands and deforested areas, where they increase their nest densities and attacks on plants (Leal et al. 2014a). They are considered pests of many Neotropical crops (e.g. cocoa, cotton, citrus, maize) and cause important losses (Montoya-Lerma et al. 2012). Other ants can cause damage to crops by benefitting pests. Ants, especially invasive species, which tend honeydew-excreting hemipteran (e.g. scales, aphids, mealybugs), might favour these insect populations, thus decreasing crop quality. For example, the presence of tending species such as the invasive *L. humile* is associated with mealybug outbreaks in vineyards and orchards worldwide (Daane et al. 2007) and can interfere with its biological control (Mgocheki and Addison 2009).

Climate change will likely have effects on the impact of leaf-cutting ants in agricultural systems. Higher herbivory rates are expected due to higher consumption rates caused by elevated temperatures and the need of herbivores to compensate for the expected decline in leaf nutritive quality in elevated CO_2 environments (Coley 1998). Also, the distribution of some leafcutters may change with climate change. A recent study using species distribution models showed that under climate change scenarios, two of the three leaf-cutting ants inhabiting Mexico would decrease their ranges, but *Atta cephalotes* would expand its range by 30% (Gómez-Díaz et al. 2023). Regarding Hemiptera-tending ant species in orchards, as many of them are invasive and some invasive species are likely going to extend their range (Bertelsmeier et al. 2015), it is likely that more productive areas will suffer their effects. However, as the interaction between hosting aphids and ants can result in a net positive effect for plants (see Chap. 17), this may not necessarily lead to a negative effect on productive areas.

In conclusion, the disservices provided by ants are not expected to diminish and may even increase due to climate change. Moreover, climate change will act synergistically with other global change drivers, such land use change and biological invasions, further favouring problematic ant species. This presents a significant challenge: managing the problems caused by some ants, while preserving the benefits they provide. In many cases, the same species contributes to both services and disservices (e.g. leaf-cutting ants enhance nutrient cycling but defoliate crops; generalist species protect aphids yet control other pests or pollinate crops). Detailed studies will be crucial to inform control or management actions that balance these opposing effects.

Prevention remains a cornerstone in addressing these challenges. Early detection of problematic ants is increasingly critical under climate change scenarios, as shifting environmental conditions may facilitate new incursions and expansions of invasive species. Predictive models are instrumental in identifying species that pose the greatest threats and pinpointing areas of potential introduction. Combined with monitoring efforts, these models can guide prevention campaigns and allocate resources effectively. Citizen science initiatives in urban areas are particularly valuable, as these environments often serve as entry points for propagules (Fig. 15.6, Sheard et al. 2020). Engaging urban citizens, producers, and wildlife managers in surveillance efforts can significantly enhance detection capabilities (e.g. Werenkraut et al. 2025).

Preventing the arrival of propagules is another essential strategy. Public education campaigns aimed at reducing the transport of ants via plants, soil, wood, cars, or personal belongings can lower propagule pressure (Ward et al. 2006). For example, climate change may increase the movement of goods and people across regions (Seebens et al. 2015), which could inadvertently aid the spread of invasive ants.

Integrating prevention campaigns with educational initiatives can mitigate these risks.

When pest or invasive ants are already established, control measures should prioritize environmentally friendly practices to minimize unintended ecological impacts. For example, push-pull methods, which show promise in managing leaf-cutting ants (Perri et al. 2021), require further development and adaptation to address climate-related challenges. Toxic baits are effective for localized control in urban areas, but coordinated, long-term actions across infested regions are essential to prevent reinfestation and range expansion (Josens et al. 2014).

To address the growing challenges of ant disservices under climate change, research and control efforts should focus on (1) identifying problematic species that are likely to expand their range under warming scenarios, (2) predicting areas of potential introduction using climate models to guide surveillance and mitigation strategies, (3) informing the population to prevent the arrival and establishment of propagules through public awareness campaigns, and (4) developing and refining environmentally friendly control strategies tailored to climate-resilient ant populations. These targeted approaches align with broader climate change mitigation efforts. By reducing crop losses caused by ant pests, they can discourage deforestation and the expansion of agricultural areas. Sustainable management of ant disservices will not only support ecosystem health but also contribute to long-term climate resilience.

15.5 Conclusions and Future Directions

Ants provide critical ES to humans and will continue to do so under anticipated climate change scenarios, owing to their resilience and adaptability. For instance, seed dispersal by ants could enhance plant biodiversity in recovering areas, contributing to climate resilience. Some ant species, given their potential resilience to climate impacts, may also help sustain processes such as pollination and biological control in systems where other providers, like bees or classical biocontrol agents, are displaced due to climate change. Furthermore, ant activity in soil processes, such as organic matter decomposition processes (Luo et al. 2023) and silicate weathering (Dorn 2014; Vicca et al. 2021), could contribute to nutrient cycling and carbon sequestration. Ant bioturbation enhances water infiltration, potentially mitigating the effects of increased precipitation, especially in urban areas (Perfecto and Philpott 2023). However, the ability of ants to improve water infiltration may be limited under intense rainfall (Cerdà and Jurgensen 2011). While these functions show promise, further research is needed to confirm their scope and effectiveness, as certain services, such as specialized seed dispersal interactions or predation by groups like army ants, ponerines, and tree-nesting ants, may be negatively impacted by future climate change scenarios.

Throughout this chapter, we have identified several knowledge gaps that need to be addressed to advance the quantification and valuation of ant-mediated ES in the context of climate change. These gaps include the impact of climate change on the relationship between ant abundance and species richness in ES provision, the spatial influence of ants on ES, and the development of management strategies to enhance positive contributions, while minimizing negative effects. Addressing these gaps is essential to develop effective solutions to the challenges posed by climate change. We propose two main research pathways. The first involves identifying and managing keystone ant species that play pivotal roles in maintaining ecosystem functionality and supporting multiple ES (Hevia et al. 2017). Targeting the conservation and management of these species may be the most effective strategy for maintaining ES resilience (Biggs et al. 2012). Integrating keystone species into predictive models will also improve our ability to forecast shifts in ecosystem dynamics and service provision under climate change (Wills and Landis 2018). The second pathway emphasizes the development of trait-based approaches, which provide a robust framework for generalizing biodiversity assembly mechanisms and predicting how species and communities respond to climate change (de Bello et al. 2021). For example, body size is a trait that encapsulates numerous physiological, life history, and behavioural processes. While body size can be useful in predicting range shifts and ES provision, its explanatory power is often constrained by trait covariations, which introduce complexity (Roeder et al. 2021; Gibb et al. 2023).

Balancing the benefits of ant-provided ES with their potential disservices is equally important. While ants can play a significant role in pest control and soil improvement, effective management strategies should focus on enhancing their positive contributions, while reducing potential conflicts, particularly in agricultural and urban settings. Studying ants and their ES in the context of climate change exemplifies the evolution of ecology into a more predictive and solution-oriented discipline. By addressing the knowledge gaps and focusing on the proposed research pathways, we can better understand and harness the contributions of ants to ecosystem resilience and climate-change adaptation.

Acknowledgements We would like to thank the editors for inviting us to participate in this book and for their helpful suggestions to improve quality and clarity. This research was partially funded by the Agencia Nacional de Promoción Científica y Tecnológica (PICT 2020-2302) and Universidad Nacional del Comahue (04/B259, 2023-2026). The authors would like to thank CONICET (Consejo Nacional de Investigaciones Científicas y Técnicas, Argentina). Katie Stuble contributed to the first draft of this chapter, and Vanina Chalcoff, Daniela Ortiz, and Lucas González-García provided photos to illustrate this chapter.

References

Aerts R (1997) Climate, leaf litter chemistry and leaf litter decomposition in terrestrial ecosystems: a triangular relationship. Oikos 79:439. https://doi.org/10.2307/3546886

Akre RD, Hansen LD, Myhre EA (1995) Insect life: my house or yours? The biology of carpenter ants. Am Entomol 41:221–226. https://doi.org/10.1093/AE/41.4.221

Andersen AN (2019) Responses of ant communities to disturbance: five principles for understanding the disturbance dynamics of a globally dominant faunal group. J Anim Ecol 88:350–362. https://doi.org/10.1111/1365-2656.12907

Andersen AN, Morrison SC (1998) Myrmecochory in Australia's seasonal tropics: effects of disturbance on distance dispersal. Austral Ecol 23:483–491. https://doi.org/10.1111/j.1442-9993.1998.tb00756.x

Angilletta MJ, Wilson RS, Niehaus AC et al (2007) Urban physiology: City ants possess high heat tolerance. PLoS One 2:e258. https://doi.org/10.1371/JOURNAL.PONE.0000258

Angulo E, Guénard B, Balzani P et al (2024) The Argentine ant, *Linepithema humile*: natural history, ecology and impact of a successful invader. Entomol Gen 44:41–61. https://doi.org/10.1127/ENTOMOLOGIA/2023/2187

Anjos DV, Tena A, Viana-Junior AB et al (2022) The effects of ants on pest control: a meta-analysis. Proc R Soc B Biol Sci 289:20221316. https://doi.org/10.1098/rspb.2022.1316

Aranda-Rickert A, Diez P, Marazzi B (2014) Extrafloral nectar fuels ant life in deserts. AoB Plants 6. https://doi.org/10.1093/aobpla/plu068

Arnan X, Molowny-Horas R, Blüthgen N (2019) Food resource exploitation and functional resilience in ant communities found in common Mediterranean habitats. Sci Total Environ 684:126–135. https://doi.org/10.1016/j.scitotenv.2019.05.260

Bartomeus I, Ascher JS, Gibbs J et al (2013) Historical changes in northeastern US bee pollinators related to shared ecological traits. Proc Natl Acad Sci USA 110:4656–4660. https://doi.org/10.1073/pnas.1218503110

Barton PS, Evans MJ (2017) Insect biodiversity meets ecosystem function: differential effects of habitat and insects on carrion decomposition. Ecol Entomol 42:364–374. https://doi.org/10.1111/EEN.12395

Beattie AJ, Culver DC (1981) The guild of myrmecochores in the herbaceous flora of West Virginia forests. Ecology 62:107–115. https://doi.org/10.2307/1936674

Beattie AJ, Hughes L (2002) Ant–plant interactions. In: Herrera CM, Pellmyr O (eds) Plant–animal interactions: an evolutionary approach. Wiley, pp 211–235

Beckers R, Goss S, Deneubourg J-L, Pasteels J (1989) Colony size, communication, and ant foraging strategy. Psyche (Stuttg) 96:239–256

Bertelsmeier C, Luque GM, Hoffmann BD, Courchamp F (2015) Worldwide ant invasions under climate change. Biodivers Conserv 24. https://doi.org/10.1007/s10531-014-0794-3

Biggs R, Schlüter M, Biggs D et al (2012) Toward principles for enhancing the resilience of ecosystem services. Annu Rev Environ Resour 37:421–448. https://doi.org/10.1146/ANNUREV-ENVIRON-051211-123836/1

Bihn JH, Gebauer G, Brandl R (2010) Loss of functional diversity of ant assemblages in secondary tropical forests. Ecology 91:782–792. https://doi.org/10.1890/08-1276.1

Blanchard S, Verheggen F, Van De Vreken I et al (2022) Combined elevation of temperature and CO_2 impacts the production and sugar composition of aphid honeydew. J Chem Ecol 48:772–781. https://doi.org/10.1007/s10886-022-01385-z

Blüthgen N, Fiedler K (2004) Competition for composition: lessons from nectar-feeding ant communities. Ecology 85:1479–1485. https://doi.org/10.1890/03-0430

Borden JB, Flory SL (2021) Urban evolution of invasive species. Front Ecol Environ 19:184–191. https://doi.org/10.1002/FEE.2295

Bradford MA, Wieder WR, Bonan GB et al (2016) Managing uncertainty in soil carbon feedbacks to climate change. Nat Clim Chang 68(6):751–758. https://doi.org/10.1038/nclimate3071

Brittain C, Williams N, Kremen C et al (2013) Synergistic effects of non-*Apis* bees and honey bees for pollination services. Proc R Soc B 280(1754):20122767

Brown JH, Reichman OJ, Davidson DW (1979) Granivory in desert ecosystems. Annu Rev Ecol Evol Syst 10:201–227. https://doi.org/10.1146/ANNUREV.ES.10.110179.001221

Brown SGA, Haas MA, Black JA et al (2004) In vitro testing to diagnose venom allergy and monitor immunotherapy: a placebo-controlled, crossover trial. Clin Exp Allergy 34:792–800. https://doi.org/10.1111/j.1365-2222.2004.01949.x

Bujan J, Kaspari M (2017) Nutrition modifies critical thermal maximum of a dominant canopy ant. J Insect Physiol 102:1–6. https://doi.org/10.1016/J.JINSPHYS.2017.08.007

Campbell-Lendrum D, Manga L, Bagayoko M, Sommerfeld J (2015) Climate change and vector-borne diseases: what are the implications for public health research and policy? Philos Trans R Soc B Biol Sci 370:1–8. https://doi.org/10.1098/RSTB.2013.0552

Cates AM, Wills BD, Kim TN et al (2021) No evidence of top-down effects by ants on litter decomposition in a temperate grassland. Ecosphere 12:e03638. https://doi.org/10.1002/ECS2.3638

Cerdá X, Dejean A (2011) Predation by ants on arthropods and other animals. In: Predation in the hymenoptera: an evolutionary perspective, pp 39–78

Cerdà A, Jurgensen MF (2011) Ant mounds as a source of sediment on citrus orchard plantations in eastern Spain. A three-scale rainfall simulation approach. Catena 85:231–236. https://doi.org/10.1016/J.CATENA.2011.01.008

Cernusak LA, Haverd V, Brendel O et al (2019) Robust response of terrestrial plants to rising CO_2. Trends Plant Sci 24:578

Chadee DD, Le Maitre A (1990) Ants: potential mechanical vectors of hospital infections in Trinidad. Trans R Soc Trop Med Hyg 84:297

Chick LD, Waters JS, Diamond SE (2021) Pedal to the metal: cities power evolutionary divergence by accelerating metabolic rate and locomotor performance. Evol Appl 14:36–52. https://doi.org/10.1111/eva.13083

Ciais P, Sabine C, Bala G et al (2013) Carbon and other biogeochemical cycles. In: Climate Change 2013: The physical science basis. Contribution of Working Group I to the Fifth Assessment Report of the Intergovernmental Panel on Climate Change Change, pp 465–570

Clough Y, Philpott S, Tscharntke T (2017) Services and disservices of ant communities in tropical cacao and coffee agroforestry systems. In: Ant-plant interactions: impacts of humans on terrestrial ecosystems

Coley PD (1998) Possible effects of climate change on plant/herbivore interactions in moist tropical forests. Clim Chang 39:455

Cordonnier M, Bellec A, Escarguel G, Kaufmann B (2020) Effects of urbanization–climate interactions on range expansion in the invasive European pavement ant. Basic Appl Ecol 44:46–54. https://doi.org/10.1016/J.BAAE.2020.02.003

Costanza R, De Groot R, Braat L, Kubiszewski I, Fioramonti L, Sutton, P, Farber S, Grasso M (2017) Twenty years of ecosystem services: how far have we come and how far do we still need to go? Ecosyst Serv 28:1–16

Daane KM, Sime KR, Fallon J, Cooper ML (2007) Impacts of Argentine ants on mealybugs and their natural enemies in California's coastal vineyards. Ecol Entomol 32:583–596. https://doi.org/10.1111/j.1365-2311.2007.00910.x

Dampc J, Mołoń M, Durak T, Durak R (2021) Changes in aphid–plant interactions under increased temperature. Biology (Basel) 10:480. https://doi.org/10.3390/biology10060480

Das S, Das A (2023) Ants are more than just curious bystanders to some flowers—they act as significant pollinators. Front Insect Sci 3:1–18

Davidson DW, Cook SC (2008) Tropical arboreal ants: linking nutrition to roles in rainforest ecosystems. In: Tropical Forest Community Ecology

de Bello F, Lavorel S, Díaz S et al (2010) Towards an assessment of multiple ecosystem processes and services via functional traits. Biodivers Conserv 19:2873–2893. https://doi.org/10.1007/S10531-010-9850-9/TABLES/2

de Bello F, Carmona CP, Dias ATC et al (2021) Handbook of trait–based ecology: from theory to R tools. Cambridge University Press, Cambridge

Del Toro I, Ribbons RR, Pelini S (2012) The little things that run the world revisited: a review of ant-mediated ecosystem services and disservices (hymenoptera: Formicidae). Myrmecological News 17:133–146

Del Toro I, Ribbons RR, Ellison AM (2015) Ant-mediated ecosystem functions on a warmer planet: effects on soil movement, decomposition and nutrient cycling. J Anim Ecol 84:1233–1241. https://doi.org/10.1111/1365-2656.12367

Deutsch CA, Tewksbury JJ, Tigchelaar M (2018) Increase in crop losses to insect pests in a warming climate. Science (80-1) 919:916–919

Devegili AM, Lescano MN, Gianoli E, Farji-Brener AG (2020) Defence variation within a guild of aphid-tending ants explains aphid population growth. Ecol Entomol 45:1180–1189. https://doi.org/10.1111/EEN.12904

Devegili AM, Lescano MN, Gianoli E, Farji-Brener AG (2021) Evidence of indirect biotic resistance: native ants decrease invasive plant fitness by enhancing aphid infestation. Oecologia 196:607–618. https://doi.org/10.1007/s00442-021-04874-2

Diamond SE, Sorger DM, Hulcr J et al (2012) Who likes it hot? A global analysis of the climatic, ecological, and evolutionary determinants of warming tolerance in ants. Glob Chang Biol 18:448–456. https://doi.org/10.1111/j.1365-2486.2011.02542.x

Díaz S, Pascual U, Stenseke M et al (2018) Assessing nature's contributions to people. Science (80-1) 359. https://doi.org/10.1126/science.aap8826

Diehl E, Sereda E, Wolters V, Birkhofer K (2013) Effects of predator specialization, host plant and climate on biological control of aphids by natural enemies: a meta-analysis. J Appl Ecol 50:262–270. https://doi.org/10.1111/1365-2664.12032

Domisch T, Finér L, Ohashi M et al (2006) Contribution of red wood ant mounds to forest floor CO2 efflux in boreal coniferous forests. Soil Biol Biochem 38:2425–2433. https://doi.org/10.1016/J.SOILBIO.2006.03.004

Dorn RI (2014) Ants as a powerful biotic agent of olivine and plagioclase dissolution. Geology 42:771–774. https://doi.org/10.1130/G35825.1

Drummond F, Choate B (2011) Ants as biological control agents in agricultural cropping systems. Terr Arthropod Rev 4:157–180. https://doi.org/10.1163/187498311x571979

Dunham AE, Mikheyev AS (2010) Influence of an invasive ant on grazing and detrital communities and nutrient fluxes in a tropical forest. Divers Distrib 16:33–42. https://doi.org/10.1111/J.1472-4642.2009.00620.X

Eldridge DJ (1993) Effect of ants on sandy soils in semi-arid eastern Australia—local distribution of nest entrances and their effect on infiltration of water. Soil Res 31:509–518. https://doi.org/10.1071/SR9930509

Elizalde L, Arbetman M, Arnan X et al (2020) The ecosystem services provided by social insects: traits, management tools and knowledge gaps. Biol Rev 95:1418–1441. https://doi.org/10.1111/brv.12616

Elizalde L, Lescano MN, Werenkraut V, Pirk GI (2022) Numbers matter: predatory ability increases with forager group size in omnivorous ant species with similar predatory traits. Ecol Entomol 47:930–940. https://doi.org/10.1111/een.13181

Ellison AM, Gotelli NJ (2021) Ants (hymenoptera: Formicidae) and humans: from inspiration and metaphor to 21st -century symbiont. Myrmecol News 31:225–240. https://doi.org/10.25849/myrmecol.news

Evans TA, Gleeson PV (2016) Direct measurement of ant predation of weed seeds in wheat cropping. J Appl Ecol 53:1177–1185. https://doi.org/10.1111/1365-2664.12640

Evans TA, Dawes TZ, Ward PR, Lo N (2011) Ants and termites increase crop yield in a dry climate. Nat Commun 2:262. https://doi.org/10.1038/ncomms1257

Farji-Brener AG, Werenkraut V (2017) The effects of ant nests on soil fertility and plant performance: a meta-analysis. J Anim Ecol 86:866–877. https://doi.org/10.1111/1365-2656.12672

Fisher K, West M, Lomeli AM et al (2019) Are societies resilient? Challenges faced by social insects in a changing world. Insect Soc 66:5–13. https://doi.org/10.1007/S00040-018-0663-2/TABLES/2

Folgarait PJ (1998) Ant biodiversity and its relationship to ecosystem functioning: a review. Biodivers Conserv 7:1221–1244

Freires ASA, Ferreira CAS, Nascimento G, Arnan X (2023) Carbohydrate-rich diet increases critical thermal maximum in ants. Entomol Exp Appl 171:240–247. https://doi.org/10.1111/eea.13268

Frizzo TLM, Souza LM, Sujii ER, Togni PHB (2020) Ants provide biological control on tropical organic farms influenced by local and landscape factors. Biol Control 151:104378. https://doi.org/10.1016/j.biocontrol.2020.104378
Frouz J (2018) Effects of soil macro- and mesofauna on litter decomposition and soil organic matter stabilization. Geoderma 332:161–172. https://doi.org/10.1016/J.GEODERMA.2017.08.039
Frouz J, Jilková V (2008) The effect of ants on soil properties and processes (hymenoptera: Formicidae). Myrmecol News 11:191–199
García Ibarra F, Jouquet P, Bottinelli N et al (2024) Experimental evidence that increased surface temperature affects bioturbation by ants. J Anim Ecol 93:319–332. https://doi.org/10.1111/1365-2656.14040
Gibb H, Grossman BF, Dickman CR et al (2019) Long-term responses of desert ant assemblages to climate. J Anim Ecol 88:1549–1563. https://doi.org/10.1111/1365-2656.13052
Gibb H, Bishop TR, Leahy L et al (2023) Ecological strategies of (pl)ants: towards a world-wide worker economic spectrum for ants. Funct Ecol 37:13–25. https://doi.org/10.1111/1365-2435.14135
Global Invasive Species Database (2024). http://www.iucngisd.org/gisd/100_worst.php on 14-05-2024. Accessed 20 Jun 2024
Gómez-Díaz JA, Baena ML, González-Zamora A, Delfín-Alfonso CA (2023) Potential present and future distributions of the genus *Atta* of Mexico. PLoS One 18:e0292072. https://doi.org/10.1371/JOURNAL.PONE.0292072
Chen IC, Hill JK, Ohlemüller R, Roy DB, and & Thomas CD (2011) Rapid range shifts of species associated with high levels of climate warming. Science, 333(6045)1024-1026
Gonzálvez FG, Chen J, Rodríguez-Gironés MA (2015) The function of ant repellence by flowers: testing the "nectar protection" and "pollinator protection" hypotheses. Evol Ecol 29:391–403. https://doi.org/10.1007/s10682-014-9742-7
Gove AD, Majer JD (2006) Do isolated trees encourage arboreal ant foraging at ground-level? Quantification of ant activity and the influence of season, in Veracruz, Mexico. Agric Ecosyst Environ 113:272–276. https://doi.org/10.1016/J.AGEE.2005.09.019
Gove AD, Majer JD, Dunn RR (2007) A keystone ant species promotes seed dispersal in a "diffuse" mutualism. Community Ecol:687–697. https://doi.org/10.1007/s00442-007-0756-5
Gove AD, Majer JD, Rico-Gray V (2009) Ant assemblages in isolated trees are more sensitive to species loss and replacement than their woodland counterparts. Basic Appl Ecol 10:187–195. https://doi.org/10.1016/j.baae.2008.02.005
Grasso DA, Pandolfi C, Bazihizina N et al (2015) Extrafloral-nectar-based partner manipulation in plant–ant relationships. AoB Plants 7. https://doi.org/10.1093/AOBPLA/PLV002
Gregory PJ, Johnson SN, Newton AC, Ingram JSI (2009) Integrating pests and pathogens into the climate change/food security debate. J Exp Bot 60:2827–2838. https://doi.org/10.1093/JXB/ERP080
Griffiths HM, Ashton LA, Walker AE et al (2018) Ants are the major agents of resource removal from tropical rainforests. J Anim Ecol 87:293–300. https://doi.org/10.1111/1365-2656.12728
Haines-Young R, Potschin MB (2018) Common international classification of ecosystem services (CICES) V5.1: guidance on the application of the revised structure. www.cices.eu. Accessed 15 May 2024
Hanisch PE, Drager K, Yang WH et al (2020) Intra- and interspecific variation in trophic ecology of 'predatory' ants in the subfamily Ponerinae. Ecol Entomol 45:444–455. https://doi.org/10.1111/EEN.12817
Henao-Gallego N, Escobar-Ramírez S, Calle Z et al (2011) An artificial aril designed to induce seed hauling by ants for ecological rehabilitation purposes. Restor Ecol 20:550–560. https://doi.org/10.1111/j.1526-100X.2011.00852.x
Hevia V, Sara BM, Marina P et al (2017) Trait based approaches to analyze links between the drivers of change and ecosystem services: synthesizing existing evidence and future challenges. Ecol Evol 7:831–844. https://doi.org/10.1002/ece3.2692
Hölldobler B, Wilson EO (1990) The ants. Harvard University Press, Cambridge

Holway DA, Lach L, Suarez AV et al (2002) The causes and consequences of ant invasions. Annu Rev Ecol Syst 33:181–233. https://doi.org/10.1146/annurev.ecolsys.33.010802.150444
Houadria M, Blüthgen N, Salas-Lopez A et al (2016) The relation between circadian asynchrony, functional redundancy, and trophic performance in tropical ant communities. Ecology 97:225–235. https://doi.org/10.1890/14-2466.1
Josens R, Sola FJ, Marchisio N et al (2014) Knowing the enemy: ant behavior and control in a pediatric hospital of Buenos Aires. Springerplus 3:1–13. https://doi.org/10.1186/2193-1801-3-229
Kevan PG, Baker HG (1983) Insects as flower visitors and pollinators. Annu Rev Entomol 28. https://doi.org/10.1146/annurev.en.28.010183.002203
Kirkham MB (2014) Principles of soil and plant water relations. Academic Press, San Diego, CA
Klotz JH, Mangold JR, Vail KM et al (1995) A survey of the urban pest ants (hymenoptera: Formicidae) of peninsular Florida. Florida Entomol 78. https://doi.org/10.2307/3495674
Kremer JMM, Nooten SS, Cook JM et al (2018) Elevated atmospheric carbon dioxide concentrations promote ant tending of aphids. J Anim Ecol 87:1475–1483. https://doi.org/10.1111/1365-2656.12842
Lal R (2004) Soil carbon sequestration to mitigate climate change. Geoderma 123:1–22. https://doi.org/10.1016/J.GEODERMA.2004.01.032
Lamichhane JR, Barzman M, Booij K et al (2015) Robust cropping systems to tackle pests under climate change. A review. Agron Sustain Dev 35:443–459. https://doi.org/10.1007/S13593-014-0275-9/FIGURES/4
Lavelle P, Decaens T, Aubert M et al (2006) Soil invertebrates and ecosystem services. Eur J Soil Biol 42:S3–S15. https://doi.org/10.1016/j.ejsobi.2006.10.002
Leal IR, Wirth R, Tabarelli M (2014a) The multiple impacts of leaf-cutting ants and their novel ecological role in human-modified Neotropical forests. Biotropica 46:516–528. https://doi.org/10.1111/BTP.12126
Leal LC, Andersen AN, Leal IR (2014b) Anthropogenic disturbance reduces seed-dispersal services for myrmecochorous plants in the Brazilian Caatinga. Oecologia 174:173–181. https://doi.org/10.1007/s00442-013-2740-6
Lengyel S, Gove AD, Latimer AM et al (2010) Convergent evolution of seed dispersal by ants, and phylogeny and biogeography in flowering plants: a global survey. Perspect Plant Ecol Evol Syst 12:43–55. https://doi.org/10.1016/J.PPEES.2009.08.001
Liao J, Lu M, Gu H et al (2023) Warming-induced shifts on Tibetan plateau: the overlooked ants and their ecological impacts. Landsc Ecol 38:3999–4008. https://doi.org/10.1007/s10980-023-01770-3
López-Collar D, Cabrero-Sañudo FJ (2021) Update on the invasion status of the Argentine ant, *Linepithema humile* (Mayr, 1868), in Madrid, a large city in the interior of the Iberian Peninsula. J Hymenopt Res 85:161
Losey JE, Vaughan M (2006) The economic value of ecological services provided by insects. Bioscience 56:311–323
Lu M, Zhou X, Yang Q et al (2013) Responses of ecosystem carbon cycle to experimental warming: a meta-analysis. Ecology 94:726–738. https://doi.org/10.1890/12-0279.1
Luo B, Huang M, Wang W et al (2023) Ant nests increase litter decomposition to mitigate the negative effect of warming in an alpine grassland ecosystem. Proc R Soc B 290:20230613
Mehring AS, Martin RM, Delavaux CS et al (2021) Leaf-cutting ant (*Atta cephalotes*) nests may be hotspots of methane and carbon dioxide emissions in tropical forests. Pedobiologia (Jena) 87:150754. https://doi.org/10.1016/J.PEDOBI.2021.150754
Menke SB, Guénard B, Sexton JO et al (2011) Urban areas may serve as habitat and corridors for dry-adapted, heat tolerant species; an example from ants. Urban Ecosyst 14:135–163. https://doi.org/10.1007/s11252-010-0150-7
Mgocheki N, Addison P (2009) Interference of ants (hymenoptera: Formicidae) with biological control of the vine mealybug Planococcus ficus (Signoret) (Hemiptera: Pseudococcidae). Biol Control 49:180–185. https://doi.org/10.1016/J.BIOCONTROL.2009.02.001

Montoya-Lerma J, Giraldo-Echeverri C, Armbrecht I et al (2012) Leaf-cutting ants revisited: towards rational management and control. Int J Pest Manag 58:225–247
Morgan ED (2008) Chemical sorcery for sociality: exocrine secretions of ants (hymenoptera: Formicidae). Myrmecol News 11:79–90
Morris JR, Vandermeer J, Perfecto I (2015) A keystone ant species provides robust biological control of the coffee berry borer under varying pest densities. PLoS One 10:1–15. https://doi.org/10.1371/journal.pone.0142850
Morris JR, Jiménez-Soto E, Philpott SM, Perfecto I (2018) Ant-mediated (hymenoptera: Formicidae) biological control of the coffee berry borer: diversity, ecological complexity, and conservation biocontrol. Myrmecol News 26:1–17
Moya-Laraño J, Wise DH (2007) Direct and indirect effects of ants on a forest-floor food web. Ecology 88:1454–1465. https://doi.org/10.1890/05-1474
Nagy C, Cross JV, Markó V (2013) Sugar feeding of the common black ant, *Lasius Niger* (L.), as a possible indirect method for reducing aphid populations on apple by disturbing ant-aphid mutualism. Biol Control 65:24–36. https://doi.org/10.1016/j.biocontrol.2013.01.005
Ness JH (2006) A mutualism's indirect costs: the most aggressive plant bodyguards also deter pollinators. Oikos 113:506–514. https://doi.org/10.1111/j.2006.0030-1299.14143.x
Ness JH, Bronstein JL, Andersen AN, Holland JN (2004) Ant body size predicts dispersal distance of ant-adapted seeds: implications of small-ant invasions. Ecology 85:1244–1250. https://doi.org/10.1890/03-0364
Ness JH, Morin DF, Giladi I (2009) Uncommon specialization in a mutualism between a temperate herbaceous plant guild and an ant: are *Aphaenogaster* ants keystone mutualists? Oikos 118:1793–1804. https://doi.org/10.1111/j.1600-0706.2009.17430.x
Noriega JA, Hortal J, Azcárate FM et al (2018) Research trends in ecosystem services provided by insects. Basic Appl Ecol 26:8–23. https://doi.org/10.1016/j.baae.2017.09.006
O'Dowd DJ, Green PT, Lake PS (2003) Invasional "meltdown" on an oceanic Island. Ecol Lett 6. https://doi.org/10.1046/j.1461-0248.2003.00512.x
Offenberg J (2015) Ants as tools in sustainable agriculture. J Appl Ecol 52:1197–1205. https://doi.org/10.1111/1365-2664.12496
Ohashi M, Finér L, Domisch T et al (2005) CO_2 efflux from a red wood ant mound in a boreal forest. Agric For Meteorol 130:131–136. https://doi.org/10.1016/J.AGRFORMET.2005.03.002
Oliveira FMP, Andersen AN, Arnan X et al (2019) Effects of increasing aridity and chronic anthropogenic disturbance on seed dispersal by ants in Brazilian Caatinga. J Anim Ecol 88:1365–2656.12979. https://doi.org/10.1111/1365-2656.12979
Oliveira FMP, Câmara T, Durval JIF et al (2021) Plant protection services mediated by extrafloral nectaries decline with aridity but are not influenced by chronic anthropogenic disturbance in Brazilian Caatinga. J Ecol 109:260–272. https://doi.org/10.1111/1365-2745.13469
Ortiz DP, Elizalde L, Pirk GI (2021) Role of ants as dispersers of native and exotic seeds in an understudied dryland. Ecol Entomol 46:626–636. https://doi.org/10.1111/een.13010
Ortiz DP, Elizalde L, Pirk GI (2022) How to induce ant–plant interactions: seed baiting to enhance its dispersal as a tool for restoration. J Appl Entomol 146:1353–1360. https://doi.org/10.1111/JEN.13065
Parr CL, Bishop TR (2022) The response of ants to climate change. Glob Chang Biol 28:3188–3205. https://doi.org/10.1111/gcb.16140
Parr CL, Eggleton P, Davies AB et al (2016) Suppression of savanna ants alters invertebrate composition and influences key ecosystem processes. Ecology 97:1611–1617. https://doi.org/10.1890/15-1713.1
Pelini SL, Boudreau M, McCoy N et al (2011a) Effects of short-term warming on low and high latitude forest ant communities. Ecosphere 2:1–12. https://doi.org/10.1890/ES11-00097.1
Pelini SL, Bowles FP, Ellison AM et al (2011b) Heating up the forest: open-top chamber warming manipulation of arthropod communities at Harvard and Duke forests. Methods Ecol Evol 2. https://doi.org/10.1111/j.2041-210X.2011.00100.x

Pereda-Gomez ME, Pessacq P, Elizalde L (2019) Stress-tolerant ants and the impact of quarries on an ant community in Patagonia. J Arid Environ:104017. https://doi.org/10.1016/j.jaridenv.2019.104017
Perfecto I, Philpott SM (2023) Ants (hymenoptera: Formicidae) and ecosystem functions and services in urban areas: a reflection on a diverse literature. Myrmecol News 33:103–122. https://doi.org/10.25849/myrmecol.news
Perri DV, Gorosito NB, Schilman PE et al (2021) Push-pull to manage leaf-cutting ants: an effective strategy in forestry plantations. Pest Manag Sci 77:432–439. https://doi.org/10.1002/ps.6036
Philpott SM, Armbrecht I (2006) Biodiversity in tropical agroforests and the ecological role of ants and ant diversity in predatory function. Ecol Entomol 31:369–377
Pirk GI, Elizalde L, Lescano MN, Werenkraut V (2020) Essential but invisible: non-apparent but widespread ant nests favour soil nutrients and plant growth in semi-arid areas. Ecol Entomol 45:1408–1417. https://doi.org/10.1111/EEN.12925
Pörtner H-O, Scholes RJ, Agard J, et al (2021) IPBES-IPCC co-sponsored workshop report on biodiversity and climate change. 1818, WIMEK,
Rastogi N (2011) Provisioning services from ants: food and pharmaceuticals. Asian Myrmecol 4:103–120
Rico-Gray V, Oliveira PS (2007) The ecology and evolution of ant-plant interactions. University of Chicago Press
Risch SJ, Carroll CR (1982) The ecological role of ants in two Mexican agroecosystems. Oecologia 55:114–119. https://doi.org/10.1007/BF00386726
Risch AC, Jurgensen MF, Schütz M, Page-Dumroese DS (2005) The contribution of red wood ants to soil C and N pools and CO2 emissions in subalpine forests. Ecology 86:419–430. https://doi.org/10.1890/04-0159
Rodriguez-Cabal MA, Stuble KL, Nuñez MA, Sanders NJ (2009) Quantitative analysis of the effects of the exotic Argentine ant on seed-dispersal mutualisms. Biol Lett 5:499–502. https://doi.org/10.1098/rsbl.2009.0297
Roeder KA, Roeder DV, Bujan J (2021) Ant thermal tolerance: a review of methods, hypotheses, and sources of variation. Ann Entomol Soc Am 114:459–469. https://doi.org/10.1093/aesa/saab018
Rust MK, Su N-Y (2012) Managing social insects of urban importance. Annu Rev Entomol 57:355–375. https://doi.org/10.1146/annurev-ento-120710-100634
Sanders NJ, Lessard J-P, Fitzpatrick MC, Dunn RR (2007) Temperature, but not productivity or geometry, predicts elevational diversity gradients in ants across spatial grains. Glob Ecol Biogeogr 16:640–649
Sarwar M (2016) Insect pests of honey bees and choosing of the right management strategic plan. Int J Entomol Res 1:16–22
Savage AM, Johnson SD, Whitney KD, Rudgers JA (2011) Do invasive ants respond more strongly to carbohydrate availability than co-occurring non-invasive ants? A test along an active Anoplolepis gracilipes invasion front. Austral Ecol 36. https://doi.org/10.1111/j.1442-9993.2010.02152.x
Schultheiss P, Nooten SS, Wang R et al (2022) The abundance, biomass, and distribution of ants on Earth. Proc Natl Acad Sci USA 119:1–9. https://doi.org/10.1073/pnas.2201550119
Seddon PJ, Griffiths CJ, Soorae PS, Armstrong DP (2014) Reversing defaunation: restoring species in a changing world. Science 345:406–412. https://doi.org/10.1126/science.12518
Seebens H, Essl F, Dawson W et al (2015) Global trade will accelerate plant invasions in emerging economies under climate change. Glob Change Biol 2:4128–4140. https://doi.org/10.1111/gcb.13021
Sheard JK, Sanders NJ, Gundlach C, Schär S, Larsen RS (2020) Monitoring the influx of new species through citizen science: the first introduced ant in Denmark. PeerJ 8:e8850. https://doi.org/10.7717/peerj.8850

Shukla RK, Singh H, Rastogi N, Agarwal VM (2013) Impact of abundant *Pheidole* ant species on soil nutrients in relation to the food biology of the species. Appl Soil Ecol 71:15–23. https://doi.org/10.1016/J.APSOIL.2013.05.002

Shukla RK, Rastogi N, Singh H (2018) Contribution of the nutrient-enriched ant nest debris soil to growth and yield of Kalmegh (*Andrographis paniculata*) under natural and experimental field conditions. Biol Agric Hortic 34:173–185. https://doi.org/10.1080/01448765.2017.1408492

Skendži S, Zovko M, Pajă I (2021) The impact of climate change on agricultural insect pests. Insects 12:440

Smith P, Cotrufo MF, Rumpel C et al (2015) Biogeochemical cycles and biodiversity as key drivers of ecosystem services provided by soils. Soil 1:665–685. https://doi.org/10.5194/SOIL-1-665-2015

Smith P, Soussana JF, Angers D et al (2020) How to measure, report and verify soil carbon change to realize the potential of soil carbon sequestration for atmospheric greenhouse gas removal. Glob Chang Biol 26:219–241. https://doi.org/10.1111/gcb.14815

Solley GO, Vanderwoude C, Knight GK (2002) Anaphylaxis due to red imported fire ant sting. Med J Aust 176. https://doi.org/10.5694/j.1326-5377.2002.tb04548.x

Stuble KL, Patterson CM, Dunn RR, Sanders NJ (2014) Ant-mediated seed dispersal in a warmed world. PeerJ:1–15. https://doi.org/10.7717/peerj.286

Suarez AV, McGlynn TP, Tsutsui ND (2010) Biogeographic and taxonomic patterns of introduced ants. Ant Ecology 233–244

Sunamura E (2018) Application of trail pheromones to management of pest ants. In: Chemical ecology of insects. CRC Press, Taylor & Francis Group, 2017, Boca Raton, pp 159–169

Takkis K, Tscheulin T, Petanidou T (2018) Differential effects of climate warming on the nectar secretion of early-and late-flowering Mediterranean plants. Front Plant Sci 9. https://doi.org/10.3389/fpls.2018.00874

Thompson B, McLachlan S (2007) The effects of urbanization on ant communities and myrmecochory in Manitoba, Canada. Urban Ecosyst 10:43–52. https://doi.org/10.1007/s11252-006-0013-4

Thomson LJ, Macfadyen S, Hoffmann AA (2010) Predicting the effects of climate change on natural enemies of agricultural pests. Biol Control 52:296–306. https://doi.org/10.1016/j.biocontrol.2009.01.022

Thurman JH, Crowder DW, Northfield TD (2017) Biological control agents in the Anthropocene: current risks and future options. Curr Opin Insect Sci 23:59–64. https://doi.org/10.1016/j.cois.2017.07.006

Thurman JH, Northfied TD, Snyder WE (2019) Weaver ants provide ecosystem services to tropical tree crops. Front Ecol Evol 7:1–9. https://doi.org/10.3389/fevo.2019.00120

Torezan-Silingardi HM, Del-Claro K (2024) Ant-pollination: conditional outcomes depending on environmental features and species interactions. J App Entomol. https://doi.org/10.1111/jen.13360

Tuma J, Eggleton P, Fayle TM (2020) Ant-termite interactions: an important but under-explored ecological linkage. Biol Rev 95:555–572. https://doi.org/10.1111/brv.12577

Tuma J, Frouz J, Veselá H et al (2022) The impacts of tropical mound-building social insects on soil properties vary between taxa and with anthropogenic habitat change. Appl Soil Ecol 179:104576. https://doi.org/10.1016/J.APSOIL.2022.104576

Uno S, Cotton J, Philpott SM (2010) Diversity, abundance, and species composition of ants in urban green spaces. Urban Ecosyst 13:425–441. https://doi.org/10.1007/s11252-010-0136-5

Van Eck NJ, Waltman L (2010) Software survey: VOSviewer, a computer program for bibliometric mapping. Scientometrics 84:523–538

Van Mele P (2008) A historical review of research on the weaver ant *Oecophylla* in biological control. Agric For Entomol 10:13–22

van Oudenhove L, Boulay R, Lenoir A et al (2012) Substrate temperature constrains recruitment and trail following behavior in ants. J Chem Ecol 38:802–809. https://doi.org/10.1007/S10886-012-0130-X/TABLES/2

Vicca S, Goll DS, Hagens M et al (2021) Is the climate change mitigation effect of enhanced silicate weathering governed by biological processes? Glob Chang Biol 28:1–16. https://doi.org/10.1111/gcb.15993

Viles H, Coombes M (2022) Geomorphology biogeomorphology in the Anthropocene: a hierarchical, traits-based approach. Geomorphology 417:108446. https://doi.org/10.1016/j.geomorph.2022.108446

Vinson SB, MacKay WP (2019) Effects of the fire ant, *Solenopsis invicta*, on electrical circuits and equipment. In: Applied myrmecology. CRC Press, pp 496–503

Wäckers FL, Alberola JS, Garcia-Marí F, Pekas A (2017) Attract and distract: manipulation of a food-mediated protective mutualism enhances natural pest control. Agric Ecosyst Environ 246:168–174. https://doi.org/10.1016/j.agee.2017.05.037

Walker AEL, Robertson MP, Eggleton P et al (2022) Indirect control of decomposition by an invertebrate predator. Funct Ecol 36:2943–2954. https://doi.org/10.1111/1365-2435.14198

Wang C, Strazanac JS, Butler L (2001) Association between ants (hymenoptera: Formicidae) and habitat characteristics in oak-dominated mixed forests. Environ Entomol 30:842–848. https://doi.org/10.1603/0046-225X-30.5.842

Ward DF, Beggs JR, Clout MN et al (2006) The diversity and origin of exotic ants arriving in New Zealand via human-mediated dispersal. Divers Distrib 12:601–609. https://doi.org/10.1111/j.1366-9516.2006.00270.x

Warren RJ, Bradford MA (2014) Mutualism fails when climate response differs between interacting species. Glob Chang Biol 20:466–474. https://doi.org/10.1111/gcb.12407

Warren RJ, Chick L (2013) Upward ant distribution shift corresponds with minimum, not maximum, temperature tolerance. Glob Chang Biol 19:2082–2088. https://doi.org/10.1111/gcb.12169

Warren RJ, Giladi I (2014) Ant-mediated seed dispersal: a few ant species (hymenoptera: Formicidae) benefit many plants. Myrmecol News 20:129–140

Warren RJ, Bahn V, Bradford MA (2011) Temperature cues phenological synchrony in ant-mediated seed dispersal. Glob Chang Biol 17:2444–2454. https://doi.org/10.1111/j.1365-2486.2010.02386.x

Wei C, Sun W, Hu X et al (2023) Territorial ants delay carrion decomposition by depressing Diptera larvae in a Tibetan alpine meadow. Ecol Entomol 48:127–133. https://doi.org/10.1111/een.13207

Werenkraut V, Lescano MN, Elizalde L, Pirk GI (2025). Citizen science quickly reveals the Argentine ant Linepithema humile (Mayr, 1868) distribution in an invaded urban area and provides unexpected findings. Neotrop Entomol 54:133. https://doi.org/10.1007/s13744-025-01348-5

Whitmore AP, Kirk GJD, Rawlins BG (2015) Technologies for increasing carbon storage in soil to mitigate climate change. Soil Use Manag 31:62–71. https://doi.org/10.1111/SUM.12115

Wielgoss A, Tscharntke T, Rumede A et al (2014) Interaction complexity matters: disentangling services and disservices of ant communities driving yield in tropical agroecosystems. Proc R Soc B 281(1775):20132144

Willmer P (2011) Pollination and floral ecology. Princeton University Press

Wills DA, Landis DA (2018) The role of ants in north temperate grasslands: a review. Oecologia 186:323–338. https://doi.org/10.1007/s00442-017-4007-0

Wittwer RA, Bender SF, Hartman K et al (2021) Organic and conservation agriculture promote ecosystem multifunctionality. Sci Adv 7:eabg6995. https://doi.org/10.1126/SCIADV.ABG6995/SUPPL_FILE/SCIADV.ABG6995_SM.PDF

Zattara EE, Aizen MA (2021) Worldwide occurrence records suggest a global decline in bee species richness. One Earth 4:114–123. https://doi.org/10.1016/j.oneear.2020.12.005

Zelikova TJ, Dunn RR, Sanders NJ (2008) Variation in seed dispersal along an elevational gradient in Great Smoky Mountains National Park. Acta Oecol 34:155–162. https://doi.org/10.1016/J.ACTAO.2008.05.002

Zhou A, Qu X, Shan L, Wang X (2017) Temperature warming strengthens the mutualism between ghost ants and invasive mealybugs. Sci Rep 7:959. https://doi.org/10.1038/s41598-017-01137-0

Chapter 16
Climate Change Impacts on Ant-Mediated Seed Dispersal

Fernanda M. P. Oliveira, Diego Centeno-Alvarado, Byanca N. T. Cavalcante, Paulo S. Oliveira, and Inara R. Leal

16.1 Introduction

Human-driven climate change poses an unprecedented threat to global biodiversity, with its effects accelerating at an alarming rate (IPCC 2023). Shifts in atmospheric CO_2 levels, temperature, and rainfall regimes are already reshaping ecosystems worldwide (Araújo and Rahbek 2006; Damschen et al. 2010; IPCC 2023). In response to such changes, organisms will need to adapt, migrate, or face population declines and ultimately extinction (Walther et al. 2002; Parmesan 2006; Hoffmann and Sgrò 2011). However, accurately predicting the ecological consequences of climate change requires understanding not only its direct effects on individual organisms but also its indirect effects mediated through interactions with other species (Kiers et al. 2010). Mutualisms—interactions between species in which both partners benefit (Bronstein 2009)—are fundamental to the origin and maintenance of biodiversity and play a critical role in ecosystem functioning (Strauss and Irwin

Supplementary Information The online version contains supplementary material available at https://doi.org/10.1007/978-3-032-21127-9_16.

F. M. P. Oliveira (✉) · I. R. Leal
Departamento de Botânica, Universidade Federal de Pernambuco, Recife, Pernambuco, Brazil
e-mail: fmp.oliveira@ufpe.br; inara.leal@ufpe.br

D. Centeno-Alvarado · B. N. T. Cavalcante
Programa de Pós-Graduação em Biologia Vegetal, Universidade Federal de Pernambuco, Recife, PE, Brazil
e-mail: diego.centenoa@ufpe.br; byanca.cavalcante@ufpe.br

P. S. Oliveira
Departamento de Biologia Animal, Universidade Estadual de Campinas, Campinas, SP, Brazil
e-mail: pso@unicamp.br

I. R. Leal et al. (eds.), *Ant Responses to Climate Change*, Entomology in Focus 10,
https://doi.org/10.1007/978-3-032-21127-9_16

2004; Bronstein 2015; Andresen et al. 2018). These interactions often form intricate networks, meaning that their disruption can amplify the effects of climate change, accelerating biodiversity loss and impairing ecosystem services (Tylianakis et al. 2008; Kiers et al. 2010; Bascompte et al. 2019).

Seed dispersal is a key ecosystem service that plays a fundamental role in plant recruitment, which in turn affects the distribution of species and diversity and composition of plant communities (Howe and Smallwood 1982; Jordano 1995; McConkey et al. 2012). By moving seeds away from parent plants, dispersal reduces parent-offspring competition, seed predation, and sibling competition, thus increasing the likelihood of successful seedling establishment. It also facilitates the colonization of new areas—processes that are critical for ecosystem resilience in a rapidly changing world (Howe and Smallwood 1982; Nathan and Muller-Landau 2000; Passos and Oliveira 2002; Teixido et al. 2022). Many seed dispersal mechanisms involve mutualistic interactions with animals (Jordano 2014; Schowalter 2022). Among these, myrmecochory—seed dispersal by ants—is a globally important syndrome found in approximately 4.5% of plant species (Lengyel et al. 2009; Rico-Gray and Oliveira 2007). However, seed dispersal by ants will potentially be disrupted by future changes in the composition and abundance of seed-dispersing ant species and by geographic or phenological mismatches between plant species and their ant dispersers (Parr and Bishop 2022; see Chap. 15).

The effectiveness of seed dispersal depends on both quantitative and qualitative components. The quantitative component includes the number of seeds dispersed, the rates at which they are dispersed, and dispersal distance. The qualitative component relates to seed fate, incorporating post-dispersal processes that influence the likelihood of germination and seedling survival (see Schupp et al. 2010, 2017). In ant-mediated seed dispersal, both quantitative and qualitative components are strongly influenced by the species composition of ant partners. For instance, large-bodied ant species often provide the highest-quality seed dispersal services due to their efficiency in collecting seeds and transporting them over longer distances (Andersen and Morrison 1998; Gove et al. 2007; Christianini and Oliveira 2010; Leal et al. 2014a, b; Warren II and Giladi 2014; Oliveira et al. 2017; Vidal et al. 2021).

Climate change will potentially alter ant species composition, leading to cascading effects on the effectiveness of seed dispersal by ants. Smaller insects are often assumed to be more vulnerable to high temperatures due to desiccation risks (Baudier et al. 2015; Kühsel et al. 2017). However, large- rather than small-bodied ants might actually be most sensitive to climate change (Andrew et al. 2019; Gibb et al. 2018; Oliveira et al. 2019) because of their greater resource requirements and longer maturation periods, which limit their adaptive capacity (Gibb et al. 2018; McCain and King 2014; Savage et al. 2004). Consequently, climate-induced declines in large-bodied ant populations could reduce the quality of seed dispersal services. While many ants have generalist diets and can persist without specific plant partners, plants that rely on ant-mediated dispersal face a higher risk of seedling establishment failure if key ant dispersers decline or shift their ranges (Christian 2001).

In this chapter, we first provide an overview of ant-mediated seed dispersal. We then review studies addressing how variation in temperature and precipitation along geographic gradients affects the composition of seed dispersing ants and ant-mediated dispersal services. Finally, we present a case study that models the current distributional overlap between a keystone large-bodied ant disperser and its associated diaspores in a highly threatened ecosystem, projecting potential shifts in overlap and thus dispersal effectiveness under future climate scenarios.

16.2 Ant-Mediated Seed Dispersal

Myrmecochory is a globally important seed dispersal syndrome found in approximately 11,000 angiosperm species from 77 plant families (Lengyel et al. 2009). This mutualism is particularly prominent in Australia's sclerophyllous shrublands, South Africa's fynbos, the Mediterranean region, Brazil's semi-arid Caatinga, and North America's eastern deciduous forests (Beattie and Culver 1981; Westoby and Rice 1981; Bond and Slingsby 1984; Leal et al. 2007; Rico-Gray and Oliveira 2007). The seeds of myrmecochorous plants possess elaiosomes—lipid-rich appendages that attract ants and facilitate seed transport (Beattie 1985, Fig. 16.1a–d). Typically, ants carry the whole diaspore to their nests, remove the elaiosome, and discard the intact seed in nest galleries or external refuse piles, where the seeds can germinate and establish (Hughes and Westoby 1992; Manzaneda and Rey 2012). In this mutualism, ants receive a food reward, while plants can receive several potential benefits (see Beattie 1985, for a review), including (1) moving the seed away from the parent plant, reducing density dependence, seed predation, parent-offspring competition, and sibling competition (Handel 1978; Bond and Breytenbach 1985; Smith et al. 1989); (2) discarding the seed into nutrient-rich microsites, such as ant nests or refuse piles (Culver and Beattie 1980; Davidson and Morton 1981); and (3) burying the seed and protecting it from disturbances such as fire (Berg 1975; Bond and Slingsby 1983). Myrmecochory occurs almost exclusively in plants of small stature, which suggests that distance dispersal is the primary benefit (Leal et al. 2015).

Although this chapter focuses on myrmecochory, ants can also opportunistically disperse a variety of fleshy diaspores after being attracted to fruit pulp (Fig. 16.1e–h, Böhning-Gaese et al. 1999; Passos and Oliveira 2002, 2004). By 'cleaning' fleshy diaspores, ants can enhance seed germination (Leal and Oliveira 1998). Their role as secondary dispersers may positively affect the fate of seeds primarily adapted to vertebrate dispersal, ultimately complementing and amplifying the benefits provided by vertebrates (Leal and Oliveira 1998; Pizo and Oliveira 1998; Christianini et al. 2012; Levey and Byrne 1993; Passos and Oliveira 2002, 2004; Camargo et al. 2016; Christianini and Oliveira 2009; Magalhães et al. 2018). In some species, for example, birds facilitate long-distance dispersal, while ants relocate diaspores to microsites that enhance establishment (Christianini and Oliveira 2010; Falcon et al. 2024). Seed dispersal by ants can also occur incidentally by granivorous species

Fig. 16.1 Examples of diaspores of myrmechocorous plants (**a**, **b**, and **c**—*Jatropha mutabilis*, Euphorbiaceae; **d**—*J. ribifolia*, Euphorbiaceae) and non-myrmechocorous plants (**e**—*Simaba fer*

through the dropping of seeds during transport to nests (Mull 2003; Martyn et al. 2022; Azcárate et al. 2005) or by the mistaken discarding on refuse piles (Retana et al. 2004).

Mutualist species differ in traits that influence the fitness of their interacting partners (Palmer et al. 2010). In myrmecochory, ant species vary in morphology and behavior, affecting both the quantitative and qualitative components of dispersal effectiveness (Andersen and Morrison 1998; Gove et al. 2007; Warren II and Giladi 2014; Vidal et al. 2021). High-quality seed dispersers are typically medium-to-large ants (body length >5.0 mm) that transport diaspores over relatively long distances (>2 m) and deposit them either individually or in small groups in nests or nest refuse piles (Fig. 16.1a, b, e, f, and h). Conversely, low-quality dispersers are smaller ants that either consume diaspores in situ without removal or transport them short distances, often depositing them in large groups in nest refuse piles (Fig. 16.1c–e, Leal et al. 2014a, b; Oliveira et al. 2019). Because many small ant species cannot readily transport seeds, they often act as seed "cleaners," consuming the lipid-rich elaiosome in situ and preventing effective seed dispersal (e.g., Beaumont et al. 2011). Despite their large size and ability to transport seeds over long distances, leaf-cutting ants are classified as low-quality dispersers because they frequently cut or bury seedlings growing near their nests, hindering seedling establishment (Leal and Oliveira 1998; Leal et al. 2017).

16.3 Effects of Climate Change on Ant-Mediated Seed Dispersal

To understand the potential effects of climate change on myrmecochory, we systematically reviewed research on the impacts of climatic variation on ant-mediated seed dispersal, searching the Web of Science (Clarivate Analytics) and Scopus (Elsevier), using the keywords and Boolean operators: ("ant" OR "Formicidae" OR ant-plant mutualis*) AND ("seed dispersal" OR "seed removal" OR myrmecochor*) AND ("climate change" OR climat*) (Sect. 16.1; Electronic Supplementary Material). The resulting 498 papers were then filtered to yield 18 papers that directly or indirectly analyze the effects of climatic variables on any component of myrmecochory (i.e., seed-dispersing ant community, seed removal rate, seed dispersal distances, phenological synchrony (timing between seed release and ant foraging) and

Fig. 16.1 (continued) *ruginea*, Simaroubaceae; **f**—*Sideroxylon obtusifolium*, Sapotaceae; **g**—*Pilosocereus gounellei*, Cactaceae; and **h**—*Commiphora leptophloeos*, Burseraceae) dispersed by high-quality seed dispersers (**a**, **e**, **f**, and **h**—*Dinoponera quadriceps*, Ponerinae; **b**—*Ectatomma muticum*, Ectatomminae) and low-quality dispersers (**c**—*Solenopsis tridens*, Myrmicinae; **d**—*Pheidole* sp., Myrmicinae; and **g**—*Dorymyrmex thoracicus*, Dolichoderinae) in Caatinga dry forest, Brazil. (Photo credits: Haymée Alencar (**a**), Mikael Castro (**b** and **h**) and Fernanda Oliveira (**c**, **d**, **e**, **f**, and **g**))

post-dispersal services such as seedling establishment. Studies that merely referenced climatic factors without specific analysis were excluded (Sect. 16.1; Electronic Supplementary Material).

The 18 papers reported a total of 108 cases of relevant responses to either mean temperature (58) or mean precipitation (50, Table 16.1; Electronic Supplementary Material). Each case represents an independent analytical unit reported within individual studies. Two analytical units, for example, would be represented by a single study that reported findings on how precipitation affected (1) seed removal rates and (2) seed dispersal distance. Most (83%) considered dispersal distance, and others considered rates of removal (7%), the structure and/or composition of seed-dispersing ant communities (6%), phenological synchrony (2%), and post-dispersal processes (2%) (Table 16.1; Electronic Supplementary Material). Separately, we identified three studies investigating other climatic parameters (soil moisture: Warren II et al. 2010, 2012; air pressure deficit: Christianini and Oliveira 2013), which were excluded from our core analysis due to insufficient comparable data but are noted here for ecological context.

The studies are distributed across a wide variety of contexts (Fig. 16.2, Table 16.1; Electronic Supplementary Material). For example, in terms of climate, the mean annual temperature across the study areas spans from 4.5 to 27.4 °C, while the mean annual precipitation varies between 232 mm and 2585 mm. Geographically, 32.4% of the cases are from the Nearctic, 22.2% from the Neotropics, 21.3% from Australasia, 16.6% from the Palearctic, 3.7% from Indomalaya, and 2.8% from the Afrotropic realm (Fig. 16.2). When classified by ecoregions, 33% of the cases focus on temperate broadleaf and mixed forests, 24% on Mediterranean forests, woodlands, and scrub, 13% on tropical and subtropical grasslands, savannas, and shrublands, 11% on tropical and subtropical moist broadleaf forests, 9% on temperate grasslands, savannas, and shrublands, 6% on deserts and xeric shrublands, and 4% on tropical and subtropical dry broadleaf forests (Fig. 16.2).

16.3.1 Composition of Seed-Dispersing Species

As described in previous chapters of this volume (see Chaps. 8, 9, 10, 11, and 12), ant diversity and composition vary markedly with variation in both temperature and precipitation. Such variation applies to the diversity and composition of seed-dispersing species. However, both broad climatic gradients and local conditions interact to shape ant seed-disperser communities. For example, lower temperatures associated with higher elevations can reduce the richness and abundance of seed-dispersing ants, potentially limiting their contribution to seed dispersal (Zelikova et al. 2008). However, other factors associated with increased altitude, such as a reduction in the availability of suitable ant microhabitats, also play a significant role in shaping these communities (Joseph et al. 2019; Lasmar et al. 2020). In contrast, cooler soil temperatures—often resulting from increased canopy cover—can positively affect the diversity and activity of seed-dispersing ants. For example, at

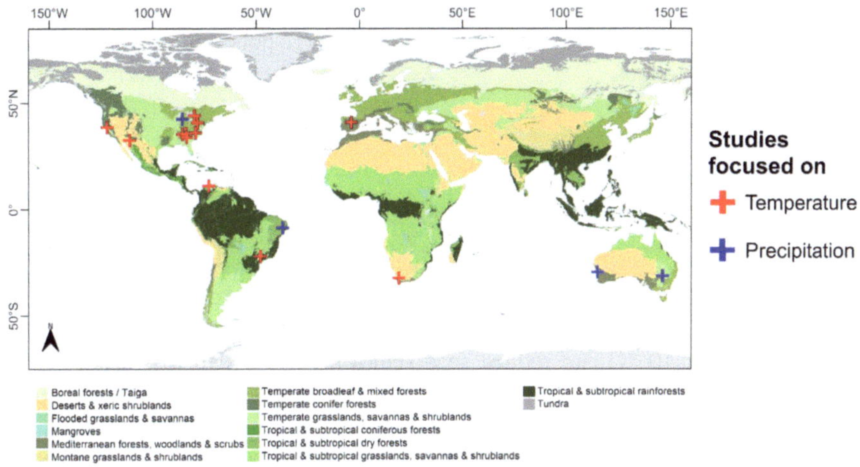

Fig. 16.2 Global distribution of studies on the effects of climate on ant-mediated seed dispersal across ecoregions (*sensu* Dinerstein et al. 2017). Even though the analysis included 18 studies, only 16 symbols are displayed because some of the studies were carried out in the same locations and are only represented by one marker on the map

open-pit mining sites in the tropical dry regions of Colombia, both the richness of seed-dispersing ants and the rate of seed removal increased with restoration age and improved microclimatic conditions (Dominguez-Haydar and Armbrecht 2010). Elevational gradients have been widely used to explore climatic effects. However, elevation often correlates with other environmental factors such as vegetation types, available area, and microclimatic variations (Joseph et al. 2019). These variations complicate the interpretation of isolated climatic effects. Disentangling the impacts of altitude and climate on ecological processes requires a careful approach to separate their distinct influences (Lasmar et al. 2020). The complex relationship between climate and environmental factors can lead to varying ecological responses along gradients, depending on the interactions between biotic and abiotic factors (see Chaps. 9 and 10).

Variation in temperature and precipitation can drive shifts in the composition of ant seed-disperser communities, as species occupy distinct habitats and exhibit contrasting physiological tolerances and behavioral responses (Warren II et al. 2012). For instance, ant species composition changes along elevation gradients as a result of temperature variation (Botes et al. 2005). Keystone seed dispersers such as *Aphaenogaster* spp. (Myrmicinae), for example, are unable to occupy the wettest habitats in deciduous forests in eastern North America (Warren II et al. 2012). In these environments, low-quality seed dispersers tend to dominate, potentially diminishing the overall quality of seed dispersal services (Warren II et al. 2012). Likewise, in Brazilian semi-arid Caatinga, increased aridity is associated with a significant decline in the abundance of the highest-quality seed disperser *Dinoponera quadriceps* (Ponerinae) (Oliveira et al. 2019). Similarly, in Brazilian Cerrado (tropical savanna), Christianini and Oliveira (2013) showed that increased landscape fragmentation and the resulting creation of edges can alter the composition of seed-dispersing ant communities. Notably, key disperser large poneromorph ants become less frequent near edges.

16.3.2 Rates of Seed Removal

Seed removal rates can be influenced by changes in ant species composition, abundance, or foraging activity, all of which may be affected by climatic variation, as noted previously. For example, in mixed conifer forests of North America, temperature increases of 10 °C can reduce seed removal rates by up to sixfold due to decreased ant activity (Ness and Bressmer 2005). Conversely, in deciduous forests of North American and Mediterranean grasslands, lower and cold temperatures, often associated with higher elevations, can also reduce seed removal rates by limiting ant foraging hours (Warren II et al. 2010; Silvestre et al. 2019). Moderate temperatures, on the other hand, may improve ant foraging efficiency and thus increase seed removal rates (Dominguez-Haydar and Armbrecht 2010). The overall impact of temperature on seed removal can also depend on the ants' maximum thermal tolerance (CT_{max}), with species possessing higher CT_{max} continuing to remove seeds effectively even at higher temperatures (Stuble et al. 2014). Additionally, increased aridity or reduced precipitation can decrease seed removal rates (Oliveira et al. 2019), with this effect mediated by a reduction in the abundance of key seed dispersers. Climate-driven shifts in the abundance of key seed-dispersing ants—such as *Aphaenogaster rudis* in North American deciduous forests and *Rhytidoponera* spp. in Australian shrublands—can exert a stronger influence on seed removal rates than broader changes in ant diversity or overall foraging levels (Zelikova et al. 2008; Gove et al. 2007). Temperature, in particular, can also play a critical role in shaping ant-mediated seed removal through its effects on the physical and chemical traits of elaiosomes (Drees and Shea 2023).

16.3.3 Seed Dispersal Distance

As with rates of seed removal, climatic variation can influence seed dispersal distances through its effects on both ant behavior and the identity or abundance of disperser species. Dispersal distances tend to be lower under low mean annual temperature and higher under higher maximum temperatures during the warmest month and higher minimum temperatures during the coldest month (Gómez and Espadaler 2013). Precipitation seems to have different effects on seed dispersal distances depending on the location and temporal scale. In the oak-hickory forests of North America, short-term events of unusually high precipitation reduce both the frequency and duration of ant foraging, along with increasing slug activity and attractivity to elaiosomes of myrmecochorous seeds, which in turn decrease seed attractiveness to ants (Kalisz et al. 1999). Conversely, in seasonally dry tropical forests of South America, chronically low precipitation levels—characteristic of the region's climate—may limit seed dispersal distances due to declines in the abundance of *Dinoponera quadriceps*, a key seed-dispersing ant species (Oliveira et al. 2019). Similar to seed removal rates, fluctuations in the populations of dominant disperser species caused by climate variation—like *Aphaenogaster rudis* in North American deciduous forests and *Rhytidoponera* spp. in Australian shrublands—often play a more decisive role than community-wide shifts in species richness or activity patterns (Zelikova et al. 2008; Gove et al. 2007).

16.3.4 Phenological Synchrony: Seed Release and Ant Foraging Activity

Species-specific responses to climate change can disrupt the phenological synchrony between seed release and ant foraging in myrmecochorous systems, as warming temperatures serve as key cues for both plant phenology and ant activity. Shifts in temperature regimes and microclimatic conditions can result in mismatches in the timing of these events (Warren II et al. 2010; Gordon et al. 2019). For example, in temperate forest understories of the eastern United States, *Hepatica nobilis* released its seeds earlier than usual due to warming, before the emergence of its main ant disperser, *Aphaenogaster rudis*, resulting in poor seed dispersal. In contrast, *Hexastylis arifolia*, which fruits later, benefited from warming temperatures that aligned its seed release with peak ant foraging, ensuring successful dispersal (Warren II et al. 2010). These results suggest that climate-driven phenological asynchrony may create winners and losers within mutualistic networks. In another study conducted in a deciduous forest in southern Ontario, Canada, seed removal by *Aphaenogaster rudis* was tightly correlated with temperature, peaking early in the season. However, fruit dehiscence in five myrmecochorous plant species occurred at discrete intervals, with no overlap between peak seed removal and fruit dehiscence, highlighting the potential for phenological mismatches (Gordon et al. 2019). Beyond reducing seed dispersal effectiveness, such mismatches may also expose seeds to increased risks: in southern Spain, seed release in *Helleborus foetidus* is tightly synchronized with the foraging activity of effective ant dispersers, while seeds released outside this window suffer high predation by rodents (Boulay et al. 2007). Together, these studies highlight how the erosion of temporal synchrony can undermine mutualistic interactions and reshape plant community dynamics.

16.3.5 Seedling Establishment

Seedling establishment is strongly influenced by climatic variables, especially rainfall. However, evidence on how climatic variation mediates the effects of ants on seedling establishment remains limited, and general principles are difficult to establish. In particular, increased temperature, solar radiation, and humidity, along with fewer wet days per year, have been positively correlated with the likelihood of successful seed burial, which protects them from seed predation by *Pogonomyrmex subdentatus* (Pan et al. 2021), potentially contributing to seed establishment. Conversely, Warren II et al. (2010) observed that mean rates of ant-mediated seed removal declined at higher soil moisture levels, while seedling recruitment remained relatively stable and adult plants were highly clustered in wetter areas. This pattern suggests a potential mismatch between where seeds are dispersed and the conditions that support long-term adult survival. Additionally, cold-adapted ants restore effective plant establishment by dispersing seeds of early-blooming plants that fail to synchronize with warm-adapted ants, reducing clumping and improving seedling survival (Warren II and Bradford 2014).

16.4 Modeling Climate Change Impacts on Ant-Mediated Seed Dispersal by *Dinoponera quadriceps*

Arid and semi-arid environments are often hotspots for ant-mediated seed dispersal (Dunn et al. 2007; Lengyel et al. 2010; Leal et al. 2015; Munguía-Rosas and Álvarez-Espino 2022). Here we focus on *Dinoponera quadriceps*, a dominant and highly effective seed disperser in Brazilian Caatinga, the largest seasonally dry tropical forest region in the Americas (Silva et al. 2017). This ant is the highest-quality disperser in the region (Leal et al. 2007, 2014a, b, 2017) and has been documented to transport seeds up to 27.5 m (Oliveira et al. 2019; see also Fourcassié and Oliveira 2002, on *D. gigantea*)—far exceeding the global average of 2 m (Gómez and Espadaler 2013). Moreover, compared with control areas, post-dispersal seed predation is lower in areas near the nest entrances of *D. quadriceps* where dispersed seeds are deposited, and seedling abundance is more than twice as high as in surrounding areas (Leal et al. 2017). We modeled the current climatic niches of this ant and the myrmechocorous Euphorbiaceae species (*Cnidoscolus quercifolius*, *C. urens*, *C. bahianus*, *Croton argyophylloides*, *C. campestris*, *C. heliotropiifolius*, *C. sonderianus*, *Jatropha mollissima*, *J. mutabilis*, and *J. ribifolia*) dispersed by it, all of which are the only myrmecochorous species documented in the literature to date as being dispersed by this ant (Leal et al. 2007, 2014a, b, Oliveira et al. 2019, 2024, Sect. 16.2, Table 16.2 and Table 16.3; Electronic Supplementary Material). We also projected potential future (i.e., 2081–2100) climatic shifts under optimistic (RCP 4.5) and pessimistic (RCP 8.5) greenhouse gas emission scenarios (IPCC 2023) (Sect. 16.2; Electronic Supplementary Material). Subsequently, we modeled current and future niche overlap (Schoener's D metric) between *D. quadriceps* and

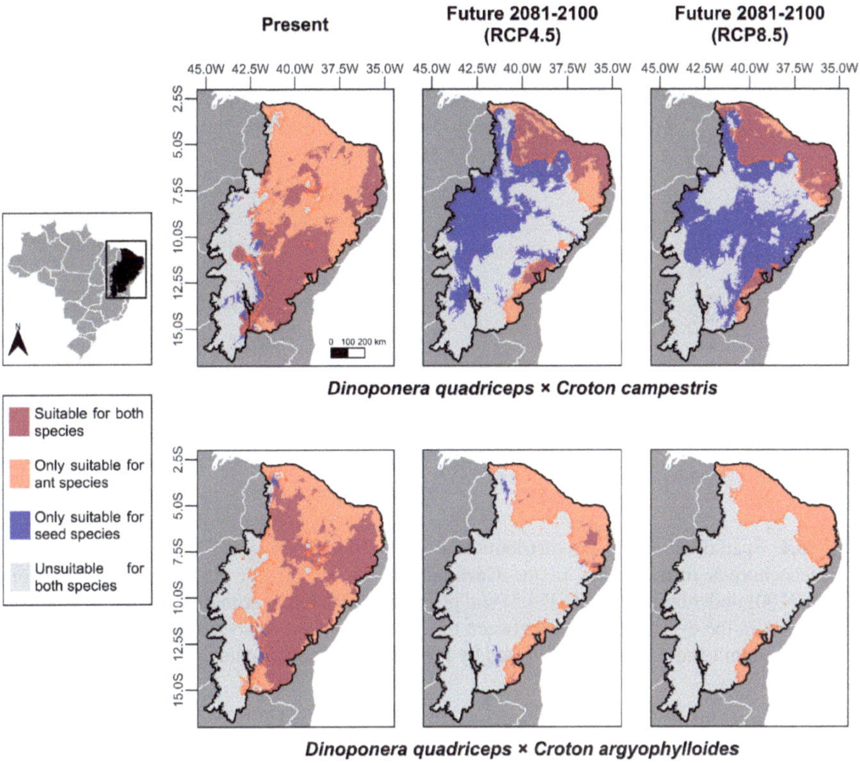

Fig. 16.3 Spatial overlap between *Dinoponera quadriceps* and the seed species it disperses within the Caatinga dry forest: current distribution and future projections (2081–2100) under optimistic (RCP 4.5) and pessimistic (RCP 8.5) scenarios. The maps illustrate two seed species representing contrasting patterns: *Croton campestris*, which shows the smallest reduction in spatial overlap with *D. quadriceps* despite an expansion in its distribution area, and *C. argyophylloides*, which exhibits the greatest reduction in spatial overlap accompanied by a contraction in its distribution

the myrmecochores (e.g., Tsafack et al. 2021) (Fig. 16.3, Sect. 16.2; Electronic Supplementary Material). Schoener's D is a niche overlap metric (range: 0–1) that measures the similarity of two species' environmental requirements based on relative abundance-weighted suitability scores, reflecting their probability of co-occurrence under shared ecological conditions (Schoener 1970; Warren et al. 2008, 2010, 2021; Badali and Zilman 2020), with applications in assessing niche shifts, species interactions, and climate change impacts (Zhang et al. 2014; Ikeda et al. 2016; Kurnaz and Yousefkhani 2019; Pili et al. 2020; Püts et al. 2020; Atwater and Barney 2021; Sadeghi et al. 2024).

When comparing niche overlap between *D. quadriceps* and each seed species across present and different future scenarios, the results show a clear trend of reduction in all cases (Fig. 16.4, Table 16.4; Electronic Supplementary Material). On average, about 64% (i.e., based on all 10 seed species combined) of the myrmecochore current niches overlap with that of *D. quadriceps*, and this is projected to

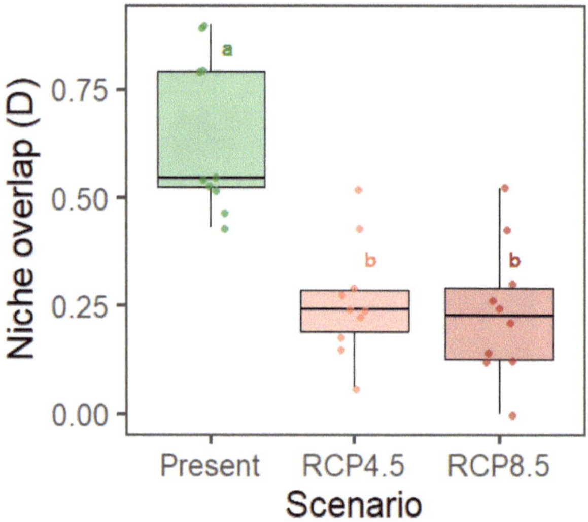

Fig. 16.4 Spatial overlap in the distribution areas of *Dinoponera quadriceps* and its dispersed myrmecochorous plant species in the Caatinga dry forest: present versus future projections (2081–2100) under optimistic (RCP 4.5) and pessimistic (RCP 8.5) scenarios. Differences in spatial overlap in the distribution areas between time scenarios were analyzed using post-hoc Tukey tests, with significant differences indicated by lowercase letters above the bars

decline to about 26% in future optimistic scenario and 23% in future pessimistic scenario. The decrease from the present to the future optimistic scenario ranges from 32.56% to 88.68%, and from 30.23% to 99.98% for the future pessimistic scenario (Table 16.4; Electronic Supplementary Material). However, the potential implications of niche overlap reductions may vary among species. For *Croton argyrophylloides*, *C. heliotropiifolius*, *C. sonderianus*, *Jatropha mutabilis*, and *J. ribifolia*, both suitable distribution areas and niche overlap with *D. quadriceps* are projected to decline, indicating a possible disruption of the interaction due to the local extinction of both partners. In contrast, species such as *Cnidoscolus quercifolius*, *C. urens*, *Croton campestris*, and *Jatropha mollissima* are predicted to experience an expansion of suitable habitat, yet their niche overlap with *D. quadriceps* still decreases. In these cases, plants may persist but could face reduced access to effective seed dispersal, potentially limiting recruitment, population connectivity, and long-term persistence. Interestingly, for *Cnidoscolus bahianus*, the outcome appears to depend on the greenhouse gas emission scenario: under the pessimistic scenario, both partners may face local extinctions, while under the optimistic scenario, the species' suitable area may expand despite a reduction in niche overlap.

While other ant species (e.g., *Ectatomma muticum*) contribute to seed dispersal in areas beyond the range of *D. quadriceps*, differences in dispersal distance and deposition patterns in zones of co-occurrence could still influence seed survival and recruitment. Although many myrmecochorous species persist without *D. quadriceps* across large portions of their ranges, the reduced spatial overlap in areas where co-occurrence does happen may constrain seed removal opportunities, potentially diminishing the contribution of this effective disperser. This is particularly relevant because disruptions to the ant-mediated seed dispersal service—even if localized—can still affect plant recruitment and community dynamics in the Caatinga (Leal et al. 2007). Moreover, our results suggest that the ongoing recovery of this ecological service in the Caatinga dry forest (Oliveira et al. 2024) could be compromised in areas where *D. quadriceps* and key myrmecochores currently co-occur. Although co-occurrence is not a prerequisite for population persistence (as these species clearly survive in many areas without *D. quadriceps*), reductions in dispersal by this ant could increase seed vulnerability to granivores and potentially impact plant fitness and local recruitment dynamics (Heithaus 1981; VanderWall et al. 2005). In addition, these Euphorbiaceae species are all diplocorous (Leal et al. 2007), and in areas where *D. quadriceps* is absent, they may rely exclusively on ballistic dispersal. However, ballistic dispersal is much more limited in range compared to ant-mediated dispersal (Leal et al. 2007).

16.5 Conclusion

Despite the growing recognition of the importance of seed dispersal services provided by ants, our systematic review underscores the limited studies that directly assess the impacts of climatic variation on myrmecochory. Moreover, most studies have concentrated on mean temperature and precipitation, while other potentially important factors, such as climatic variability, seasonality, and extreme weather events, remain underexplored. Further, there is limited geographic representation, with few studies conducted in myrmecochory hotspots such as the sclerophyllous shrublands of Australia and South Africa's fynbos and the Brazilian Caatinga (Lengyel et al. 2009; Leal et al. 2015). The limited available research suggests that climate change may affect multiple aspects of the dispersal process, ranging from the structure of ant disperser communities to post-dispersal plant establishment.

Research on ant-mediated seed dispersal has focused predominantly on ant disperser community structure and the quantitative components of seed dispersal effectiveness (e.g., seed removal). Few studies have addressed the qualitative aspects (e.g., post-dispersal processes), which are critical for understanding the overall outcome of this mutualism and its role as an ecosystem service. Dispersal distance is crucial in determining how far seeds move from the parent plant, which can influence plant population dynamics and the genetic diversity of plant communities. Similarly, post-dispersal processes, which include seedling predation, germination, seedling recruitment, and plant establishment, are essential for understanding the success of this mutualism. Without examining these qualitative aspects, we miss important insights into the long-term ecological outcomes of ant-mediated seed dispersal. For instance, even if seeds are efficiently removed by ants, their success in establishing new plants may be hindered by factors such as seed predation or unsuitable microhabitats. Understanding both the quantitative and qualitative components of seed dispersal is necessary to fully grasp the role of ants as ecosystem engineers, facilitating not only seed movement but also the long-term survival and establishment of plant communities.

A recurring pattern is the role of species composition, particularly the presence of high-quality seed dispersers, which mediates seed dispersal efficiency. While their presence alone does not determine efficiency, these species are highly sensitive to climate change. An example of such a key disperser is *Dinoponera quadriceps*. Our projections suggest that climate change could lead to a significant reduction in the overlap between *D. quadriceps* and its associated diaspores, potentially reducing myrmecochory and altering plant community composition and jeopardizing ecosystem resilience. Given the widespread occurrence of key ant dispersers in other biomes, similar vulnerabilities may extend beyond the Caatinga.

To advance the field, future research should prioritize expanding the geographic scope of studies, incorporating additional climatic variables such as extreme weather events and soil moisture fluctuations, and investigating the long-term consequences of climate change on post-dispersal seed fate. It is also crucial to disentangle the effects of climate change from those of other environmental gradients, such as altitude, to better isolate the drivers of observed patterns. Additionally, it is crucial to consider plant traits, such as the quantity and quality of seed and fruit rewards (i.e., elaiosome, aril, and pulp) and plant phenology, since these features significantly influence seed dispersal effectiveness. Understanding how climate-driven changes in plant attributes affect interactions with ant dispersers will provide a more comprehensive understanding of this mutualism. Furthermore, a broader community-level perspective of ant dispersers is needed to assess all components of dispersal effectiveness, including seed removal, dispersal distance, and post-dispersal processes. A more thorough understanding of these dynamics is essential for predicting biodiversity responses and designing conservation strategies to enhance ecosystem resilience in the face of climate change.

References

Andersen AN, Morrison SC (1998) Myrmecochory in Australia's seasonal tropics: effects of disturbance on distance dispersal. Aust J Ecol 23(5):483–491. https://doi.org/10.1111/j.1442-9993.1998.tb00756.x

Andresen E, Arroyo-Rodríguez V, Escobar F (2018) Tropical biodiversity: the importance of biotic interactions for its origin, maintenance, function, and conservation. In: Dáttilo W, Rico-Gray V (eds) Ecological networks in the tropics: an integrative overview of species interactions from some of the most species-rich habitats on earth. Springer International Publishing, Cham, pp 1–13

Andrew NR, Miller C, Hall G et al (2019) Aridity and land use negatively influence a dominant species' upper critical thermal limits. PeerJ 6:e6252. https://doi.org/10.7717/peerj.6252

Araújo MB, Rahbek C (2006) How does climate change affect biodiversity? Science 313(5792):1396–1397. https://doi.org/10.1126/science.1131758

Atwater DZ, Barney JN (2021) Climatic niche shifts in 815 introduced plant species affect their predicted distributions. Glob Chang Biol 30(8):1671–1684. https://doi.org/10.1111/geb.13342

Azcárate FM, Arqueros L, Sánchez AM et al (2005) Seed and fruit selection by harvester ants, Messor barbarus, in Mediterranean grassland and scrubland. Funct Ecol 19(2):273–283. https://doi.org/10.1111/j.0269-8463.2005.00956.x

Badali M, Zilman A (2020) Effects of niche overlap on coexistence, fixation and invasion in a population of two interacting species. R Soc Open Sci 7(2):192181. https://doi.org/10.1098/rsos.192181

Bascompte J, García MB, Ortega R et al (2019) Mutualistic interactions reshuffle the effects of climate change on plants across the tree of life. Sci Adv 5(5):eaav2539. https://doi.org/10.1126/sciadv.aav2539

Baudier KM, Mudd AE, Erickson SC et al (2015) Microhabitat and body size effects on heat tolerance: implications for responses to climate change (army ants: Formicidae, Ecitoninae). J Anim Ecol 84(5):1322–1330. https://doi.org/10.1111/1365-2656.12388

Beattie AJ (1985) The evolutionary ecology of ant-plant mutualisms. Cambridge University Press, Cambridge

Beattie AJ, Culver DC (1981) The guild of myrmecochores in the herbaceous flora of West Virginia forests. Ecology 62(1):107–115. https://doi.org/10.2307/1936674

Beaumont KP, Mackay DA, Whalen MA (2011) Interactions between ants and seeds of two myrmecochorous plant species in recently burnt and long-unburnt forest sites. Austral Ecol 36:767–778. https://doi.org/10.1111/j.1442-9993.2010.02215.x

Berg RY (1975) Myrmecochorous plants in Australia and their dispersal by ants. Aust J Bot 23(3):475–508. https://doi.org/10.1071/BT9750475

Böhning-Gaese K, Gaese BH, Rabemanantsoa SB (1999) Importance of primary and secondary seed dispersal in the Malagasy tree *Commiphora guillaumini*. Ecology 80(3):821–832. https://doi.org/10.1890/0012-9658(1999)080[0821:IOPASS]2.0.CO;2

Bond WJ, Slingsby P (1983) Seed dispersal by ants in shrublands of the Cape Province and its evolutionary implications. S Afr J Sci 79:231–233.

Bond WJ, Breytenbach GJ (1985) Ants, rodents and seed predation in Proteaceae. S Afr J Zool 20(3):150–154. https://doi.org/10.1080/02541858.1985.11447928

Bond WJ, Slingsby P (1984) Collapse of an ant-plant mutualism: the Argentine ant (*Iridomyrmex humilis)* and myrmecochorous Proteaceae. Ecology 65(4):1031–1037. https://doi.org/10.2307/1938311

Botes A, McGeoch MA, Robertson HG et al (2005) Ants, altitude and change in the northern cape floristic region. J Biogeogr 33:71–90. https://doi.org/10.1111/j.1365-2699.2005.01336.x

Boulay R, Coll-Toledano J, Manzaneda AJ, Cerda X (2007) Geographic variations in seed dispersal by ants: are plant and seed traits decisive? Naturwissenschaften 94:242–246. https://doi.org/10.1007/s00114-006-0185-z

Bronstein JL (2009) The evolution of facilitation and mutualism. J Ecol 97(6):1160–1170. https://doi.org/10.1111/j.1365-2745.2009.01566.x

Bronstein JL (ed) (2015) Mutualism. Oxford University Press, Oxford

Camargo PH, Martins MM, Feitosa RM et al (2016) Bird and ant synergy increases the seed dispersal effectiveness of an ornithochoric shrub. Oecologia 181:507–518. https://doi.org/10.1007/s00442-016-3571-z

Christian CE (2001) Consequences of a biological invasion reveal the importance of mutualism for plant communities. Nature 413:635–639. https://doi.org/10.1038/35098093

Christianini AV, Oliveira PS (2009) The relevance of ants as seed rescuers of a primarily bird-dispersed tree in the Neotropical Cerrado savanna. Oecologia 160(4):735–745. https://doi.org/10.1007/s00442-009-1349-2

Christianini AV, Oliveira PS (2010) Birds and ants provide complementary seed dispersal in a neotropical savanna. J Ecol 98(3):573–582. https://doi.org/10.1111/j.1365-2745.2010.01653.x

Christianini AV, Oliveira PS (2013) Edge effects decrease ant-derived benefits to seedlings in a neotropical savanna. Arthropod Plant Interact 7:191–199. https://doi.org/10.1007/s11829-012-9229-9

Christianini AV, Mayhé-Nunes AJ, Oliveira PS (2012) Exploitation of fallen diaspores by ants: are there ant–plant partner choices? Biotropica 44(3):360–367. https://doi.org/10.1111/j.1744-7429.2011.00822.x

Culver DC, Beattie AJ (1980) The fate of *Viola* seeds dispersed by ants. Am J Bot 67(5):710–714. https://doi.org/10.2307/2442664

Damschen EI, Harrison S, Grace JB (2010) Climate change effects on an endemic-rich edaphic flora: resurveying Robert H. Whittaker's Siskiyou sites (Oregon, USA). Ecology 91(12):3609–3619. https://doi.org/10.1890/09-1057.1

Davidson DW, Morton SR (1981) Competition for dispersal in ant-dispersed plants. Science 213(4513):1259–1261. https://doi.org/10.1126/science.213.4513.1259

Dominguez-Haydar Y, Armbrecht I (2010) Response of ants and their seed removal in rehabilitation areas and forests at El Cerrejón coal mine in Colombia. Restor Ecol 19:178–184. https://doi.org/10.1111/j.1526-100X.2010.00735.x

Drees TH, Shea K (2023) Climate warming increases insect-driven seed removal of two elaiosome-bearing invasive thistle species. Ecology 105:e4233. https://doi.org/10.1002/ecy.4223

Dinerstein E, Olson D, Joshi A, Vynne C, Burgess ND, Wikramanayake E, Hahn N, Palminteri S, Hedao P, Noss R, Hansen M, Locke H, Ellis EC, Jones B, Barber CV, Hayes R, Kormos C, Martin V, Crist E, Sechrest W, Price L, Baillie JEM, Weeden D, Suckling K, Davis C, Sizer N, Moore R, Thau D, Birch T, Potapov P, Turubanova S, Tyukavina A, Souza N, Pintea L, Brito JC, Llewellyn OA, Miller AG, Patzelt A, Ghazanfar SA, Timberlake J, Klöser H, Shennan-Farpón Y, Kindt R, Lillesø J-PB, Breugel P, Graudal L, Voge M, Al-Shammari KF, Saleem M (2017) An ecoregion-based approach to protecting half the terrestrial realm. BioScience 67(6):534–545. https://doi.org/10.1093/biosci/bix014

Dunn RR, Gove A, Barraclough TG et al (2007) Convergent evolution of an ant-plant mutualism across plant families, continents and time. Evol Ecol Res 9:1349–1362

Falcon JE, Schoereder JH, Ribeiro VS et al (2024) How do birds and ants contribute to the recruitment of a tropical tree? Biotropica 56(6):e13372. https://doi.org/10.1111/btp.13372

Fourcassié V, Oliveira PS (2002) Foraging ecology of the giant Amazonian ant *Dinoponera gigantea* (Hymenoptera, Formicidae, Ponerinae): activity schedule, diet, and spatial foraging patterns. J Nat Hist 36:2211–2227. https://doi.org/10.1080/00222930110097149

Gibb H, Sanders NJ, Dunn RR et al (2018) Habitat disturbance selects against both small and large species across varying climates. Ecography 41(7):1184–1193. https://doi.org/10.1111/ecog.03244

Gómez C, Espadaler X (2013) An update of the world survey of myrmecochorous dispersal distances. Ecography 36:1193–1201. https://doi.org/10.1111/j.1600-0587.2013.00289.x

Gordon SCC, Meadley-Dunphy SA, Prior KM et al (2019) Asynchrony between ant seed dispersal activity and fruit dehiscence of myrmecochorous plants. Am J Bot 106:71–80. https://doi.org/10.1002/ajb2.1214

Gove AD, Majer JD, Dunn RR (2007) A keystone ant species promotes seed dispersal in a "diffuse" mutualism. Oecologia 153:687–697. https://doi.org/10.1007/s00442-007-0756-5

Handel SN (1978) The competitive relationship of three woodland sedges and its bearing on the evolution of ant-dispersal of *Carex pedunculata*. Evolution 32:151–163. https://doi.org/10.1111/j.1558-5646.1978.tb01105.x

Heithaus E (1981) Seed predation by rodents on three ant-dispersed plants. Ecology 62:136–145. https://doi.org/10.2307/1936677

Hoffmann AA, Sgrò CM (2011) Climate change and evolutionary adaptation. Nature 470(7335):479–485. https://doi.org/10.1038/nature09670

Howe HF, Smallwood J (1982) Ecology of seed dispersal. Annu Rev Ecol Syst 13:201–228. https://doi.org/10.1146/annurev.es.13.110182.001221

Hughes L, Westoby M (1992) Fate of seeds adapted for dispersal by ants in Australian sclerophyll vegetation. Ecology 73(4):1285–1299. https://doi.org/10.2307/1940676

Ikeda DH, Max, TL, Allan GJ, Lau MK, Shuster SM, Whitham TG (2016) Genetically informed ecological niche models improve climate change predictions. Glob Chang Biol 23:164–176. https://doi.org/10.1111/gcb.13470
IPCC (Intergovernmental Panel on Climate Change) (2023) Climate change 2023: synthesis report: contribution of working group I, II and III to the sixth assessment report of the intergovernmental panel on climate change. Intergovernmental Panel on Climate Change, Geneva
Jordano P (1995) Frugivore-mediated selection on fruit and seed size: birds and St. Lucie's cherry, Prunus mahaleb. Ecology 76(8):2627–2639. https://doi.org/10.2307/2265833
Jordano P (2014) Fruits and frugivory. In: Fenner M (ed) Seeds: the ecology of regeneration in plant communities, 2nd edn. Wallingford Cabi, Wallingford, pp 18–61
Joseph GS, Muluvhahothe MM, Seymour CL et al (2019) Stability of Afromontane ant diversity decreases across an elevation gradient. Glob Ecol Conserv 17:e00596. https://doi.org/10.1016/j.gecco.2019.e00596
Kalisz S, Hanzawa FM, Tonsor SJ et al (1999) Ant-mediated seed dispersal alters the pattern of relatedness in a population of Trillium grandiflorum. Ecology 80(11):2620–2634. https://doi.org/10.1890/0012-9658(1999)080[2620:AMSDAP]2.0.CO;2
Kiers ET, Palmer TM, Ives AR et al (2010) Mutualisms in a changing world: an evolutionary perspective. Ecol Lett 13(12):1459–1474. https://doi.org/10.1111/j.1461-0248.2010.01598.x
Kühsel S, Brückner A, Schmelzle S et al (2017) Surface area–volume ratios in insects. Insect Sci 24(5):829–841. https://doi.org/10.1111/1744-7917.12362
Kurnaz M, Yousefkhani SSH (2020) Ecological niche divergence between Darevskia rudis and D. bithynica (Lacertidae) in Turkey. Biologia 75(9):1307–1312. https://doi.org/10.2478/s11756-019-00374-0
Lasmar CJ, Ribas CR, Louzada J et al (2020) Disentangling elevational and vegetational effects on ant diversity patterns. Acta Oecol 102:103489. https://doi.org/10.1016/j.actao.2019.103489
Leal IR, Oliveira PS (1998) Interaction between fungus-growing ants (Attini), fruits, and seeds in Cerrado vegetation in Southeast Brazil. Biotropica 30:170–178. https://doi.org/10.1111/j.1744-7429.1998.tb00052.x
Leal IR, Wirth R, Tabarelli M (2007) Seed dispersal by ants in the semi-arid Caatinga of North-East Brazil. Ann Bot 99(5):885–894. https://doi.org/10.1093/aob/mcm017
Leal LC, Andersen AN, Leal IR (2014a) Anthropogenic disturbance reduces seed dispersal services for myrmecochorous plants in the Brazilian Caatinga. Oecologia 174:173–181. https://doi.org/10.1007/s00442-013-2740-6
Leal LC, Neto MCL, Oliveira AFM et al (2014b) Myrmecochores can target high-quality disperser ants: variation in elaiosome traits and ant preferences for myrmecochorous Euphorbiaceae in Brazilian Caatinga. Oecologia 174:493–500. https://doi.org/10.1007/s00442-013-2789-2
Leal IR, Leal LC, Andersen AN (2015) The benefits of myrmecochory: a matter of stature. Biotropica 47:281–285. https://doi.org/10.1111/btp.12213
Leal IR, Leal LC, Oliveira FMP et al (2017) Effects of human disturbance and climate change on myrmecochory in Brazilian Caatinga. In: Oliveira PS, Koptur S (eds) Ant-plant interactions: impacts of humans on terrestrial ecosystems. Cambridge University Press, Cambridge, pp 112–132
Lengyel S, Gove AD, Latimer AM et al (2009) Ants sow the seeds of global diversification in flowering plants. PLoS One 4(5):e5480. https://doi.org/10.1371/journal.pone.0005480
Lengyel S, Gove AD, Latimer AM et al (2010) Convergent evolution of seed dispersal by ants, and phylogeny and biogeography in flowering plants: a global survey. Perspect Plant Ecol Evol Syst 12:43–55. https://doi.org/10.1016/j.ppees.2009.08.001
Levey DJ, Byrne MM (1993) Complex ant-plant interactions: rain-forest ants as secondary dispersers and post-dispersal seed predators. Ecology 74(6):1802–1812. https://doi.org/10.2307/1939938
Magalhães VB, Santo NBE, Salles LFP, Soares H Jr, Oliveira PS (2018) Secondary seed dispersal by ants in Neotropical cerrado savanna: species-specific effects on seeds and seedlings

of *Siparuna guianensis* (Siparunaceae). Ecol Entomol 43:665–674. https://doi.org/10.1111/een.12640
Manzaneda AJ, Rey PJ (2012) Geographical and interspecific variation and the nutrient-enrichment hypothesis as an adaptive advantage of myrmecochory. Ecography 35(4):322–332. https://doi.org/10.1111/j.1600-0587.2011.06923.x
Martyn T, Kwapich C, Kline A et al (2022) Granivorous ants prefer small and unprotected seeds – implications for restoration in arid ecosystems. Restor Ecol 31(2):e13759. https://doi.org/10.1111/rec.13759
McCain CM, King SR (2014) Body size and activity times mediate mammalian responses to climate change. Glob Chang Biol 20(6):1760–1769. https://doi.org/10.1111/gcb.12499
McConkey KR, Prasad S, Corlett RT et al (2012) Seed dispersal in changing landscapes. Biol Conserv 146(1):1–13. https://doi.org/10.1016/j.biocon.2011.09.018
Mull JF (2003) Dispersal of sagebrush-steppe seeds by the western harvester ant (*Pogomyrmex occidentalis*). West N Am Nat 63(3):358–362
Munguía-Rosas MA, Álvarez-Espino RX (2022) What are elaiosomes for? Effects of elaiosomes on ant attraction, seed removal and germination in wild chaya (*Cnidoscolus aconitifolius*). J Arid Environ 205:104826. https://doi.org/10.1016/j.jaridenv.2022.104826
Nathan R, Muller-Landau HC (2000) Spatial patterns of seed dispersal, their determinants and consequences for recruitment. Trends Ecol Evol 15(7):278–285. https://doi.org/10.1016/S0169-5347(00)01874-7
Ness JH, Bressmer K (2005) Abiotic influences on the behaviour of rodents, ants, and plants affect an ant-seed mutualism. Ecoscience 12:76–81. https://doi.org/10.2980/i1195-6860-12-1-76.1
Oliveira FMP, Andersen AN, Arnan X et al (2019) Effects of increasing aridity and chronic anthropogenic disturbance on seed dispersal by ants in Brazilian Caatinga. J Anim Ecol 88(6):870–880. https://doi.org/10.1111/1365-2656.12979
Oliveira FR, Oliveira FMP, Centeno-Alvarado D et al (2024) Rapid recovery of ant-mediated seed dispersal service along secondary succession in a Caatinga dry forest. For Ecol Manag 554:121670. https://doi.org/10.1016/j.foreco.2023.121670
Palmer TM, Doak DF, Stanton ML et al (2010) Synergy of multiple partners, including freeloaders, increases host fitness in a multispecies mutualism. Proc Natl Acad Sci 107(40):17234–17239. https://doi.org/10.1073/pnas.1006872107
Pan VS, McMunn M, Karban R et al (2021) Mucilage binding to ground protects seeds of many plants from harvester ants: a functional investigation. Funct Ecol 35:2448–2460. https://doi.org/10.1111/1365-2435.13881
Parmesan C (2006) Ecological and evolutionary responses to recent climate change. Annu Rev Ecol Evol Syst 37(1):637–669. https://doi.org/10.1146/annurev.ecolsys.37.091305.110100
Parr CL, Bishop TR (2022) The response of ants to climate change. Glob Chang Biol 28(10):3188–3205. https://doi.org/10.1111/gcb.16140
Passos L, Oliveira PS (2002) Ants affect the distribution and performance of *Clusia criuva* seedlings, a primarily bird-dispersed rainforest tree. J Ecol 90:517–528. https://doi.org/10.1046/j.1365-2745.2002.00687.x
Passos L, Oliveira PS (2004) Interaction between ants and fruits of *Guapira opposita* (Nyctaginaceae) in a Brazilian sandy plain rainforest: ant effects on seeds and seedlings. Oecologia 139:376–382. https://doi.org/10.1007/s00442-004-1531-5
Pili AN, Tingley R, Sy EY, Diesmos MLL, Diesmos AC (2020) Niche shifts and environmental non-equilibrium undermine the usefulness of ecological niche models for invasion risk assessments. Sci Rep 10:7972. https://doi.org/10.1038/s41598-020-64568-2
Pizo MA, Oliveira PS (1998) Interaction between ants and seeds of a nonmyrmecochorous neotropical tree, Cabralea canjerana (Meliaceae), in the Atlantic forest of southeast Brazil. Am J Bot 85(5):669–674. https://doi.org/10.2307/2446536
Püts M, Taylor MH, Núñez-Riboni I, Steenbeek J, Stäbler M, Möllmann C, Kempf A (2020) Insights on integrating habitat preferences in process-oriented ecological models – a case study of the southern North Sea. Ecol Model 431:109189. https://doi.org/10.1016/j.ecolmodel.2020.109189

Retana J, Picó X, Rodrigo A (2004) Dual role of harvesting ants as seed dispersers of a non-myrmecochorous Mediterranean perennial herb. Oikos 105(2):377–385. https://doi.org/10.1111/j.0030-1299.2004.12854.x
Rico-Gray V, Oliveira PS (2007) The ecology of ant-plant interactions. The University of Chicago Press, Chicago
Sadegui M, Malekian M, Esfahani MT (2024) Interspecific niche overlap and climatic associations of native Quercus species in the Zagros forests of Iran. Global Ecol Conserv 51:e02878. https://doi.org/10.1016/j.gecco.2024.e02878
Savage VM, Gillooly JF, Brown JH et al (2004) Effects of body size and temperature on population growth. Am Nat 163(3):429–441. https://doi.org/10.1086/381872
Schoener TW (1970) Nonsynchronous spatial overlap of lizards in patchy habitats. Ecology 51:408–418. https://doi.org/10.2307/1935376
Schowalter TD (2022) Insect ecology: an ecosystem approach. Academic
Schupp EW, Jordano P, Gómez JM (2010) Seed dispersal effectiveness revisited: a conceptual review. New Phytol 188(2):333–353. https://doi.org/10.1111/j.1469-8137.2010.03402.x
Schupp EW, Jordano P, Gómez JM (2017) A general framework for effectiveness concepts in mutualisms. Ecol Lett 20(5):577–590. https://doi.org/10.1111/ele.12764
Silva JMC, Leal IR, Tabarelli M. (Eds.) (2017) Caatinga: the largest tropical dry forest region in South America. Springer International Publishing Cham. https://doi.org/10.1007/978-3-319-68339-3
Silvestre M, Aguilar A, Seoane J et al (2019) Responses of seed size, ant worker size, and seed removal rate to elevation in Mediterranean grasslands. Oecologia 189:781–793. https://doi.org/10.1007/s00442-019-04356-6
Smith BH, Forman PD, Boyd AE (1989) Spatial patterns of seed dispersal and predation of two myrmecochorous forest herbs. Ecology 70(6):1649–1656
Strauss SY, Irwin RE (2004) Ecological and evolutionary consequences of multispecies plant-animal interactions. Annu Rev Ecol Evol Syst 35(1):435–466. https://doi.org/10.1146/annurev.ecolsys.35.112202.130215
Stuble KL, Patterson CM, Rodríguez-Cabal MA et al (2014) Ant-mediated seed dispersal in a warmed world. PeerJ 2:e286. https://doi.org/10.7717/peerj.286
Teixido AL, Fuzessy LF, Souza CS et al (2022) Anthropogenic impacts on plant-animal mutualisms: a global synthesis for pollination and seed dispersal. Biol Conserv 266:109461. https://doi.org/10.1016/j.biocon.2022.109461
Tsafack N, Wang X, Xie Y, Fattorini S (2021) Niche overlap and species co-occurrence patterns in carabid communities of the northern Chinese steppes. In: Spence J, Casale A, Assmann T et al (eds) Systematic zoology and biodiversity science: a tribute to Terry Erwin (1940–2020). ZooKeys, pp 929–949
Tylianakis JM, Didham RK, Bascompte J et al (2008) Global change and species interactions in terrestrial ecosystems. Ecol Lett 11(12):1351–1363. https://doi.org/10.1111/j.1461-0248.2008.01250.x
VanderWall SB, Kuhn KM, Beck MJ (2005) Seed removal, seed predation, and secondary dispersal. Ecology 86:801–806. https://doi.org/10.1890/04-0847
Vidal MC, Anneberg TJ, Curé AE et al (2021) The variable effects of global change on insect mutualisms. Curr Opin Insect Sci 47:46–52. https://doi.org/10.1016/j.cois.2021.03.002
Walther GR, Post E, Convey P et al (2002) Ecological responses to recent climate change. Nature 416(6879):389–395. https://doi.org/10.1038/416389a
Warren RJ II, Bradford MA (2014) Mutualism fails when climate response differs between interacting species. Glob Chang Biol 20:466–474. https://doi.org/10.1111/gcb.12407
Warren DL, Glor RE, Turelli M (2008) Environmental niche equivalency versus conservatism: quantitative approaches to niche evolution. Evolution 62(11):2868–2883. https://doi.org/10.1111/evo.2008.62.issue-11 10.1111/j.1558-5646.2008.00482.x
Warren DL, Glor RE, Turelli M (2010) ENMTools: a toolbox for comparative studies of environmental niche models. Ecography 33(3):607–611. https://doi.org/10.1111/j.1600-0587.2009.06142.x

Warren RJ II, Giladi I (2014) Ant-mediated seed dispersal: a few ant species (Hymenoptera: Formicidae) benefit many plants. Myrmecol News 20:129–140

Warren RJ II, Giladi I, Bradford MA (2010) Ant-mediated seed dispersal does not facilitate niche expansion. J Ecol 98:1178–1185. https://doi.org/10.1111/j.1365-2745.2010.01694.x

Warren RJ II, Giladi I, Bradford MA (2012) Environmental heterogeneity and interspecific interactions influence nest occupancy by key seed-dispersing ants. Environ Entomol 41:463–468. https://doi.org/10.1603/EN12027

Warren DL, Matzke NJ, Cardillo M, Baumgartner JB, Beaumont LJ, Turelli M, Glor RE, Huron NA, Simões M, Iglesias TL, Piquet JC, Dinnage R (2021) ENMTools 1.0: an R package for comparative ecological biogeography. Ecography 44(4):504–511. https://doi.org/10.1111/ecog.05485

Westoby M, Rice B (1981) A note on combining two methods of dispersal-for-distance. Aust J Ecol 6(2):189–192

Zelikova TJ, Dunn RR, Sanders NJ (2008) Variation in seed dispersal along an elevational gradient in Great Smoky Mountains National Park. Acta Oecol 34:155–162. https://doi.org/10.1016/j.actao.2008.05.002

Zhang Y, Chen C, Li L, Zhao C, Chen W, Huang Y (2014) Insights from ecological niche modeling on the taxonomic distinction and niche differentiation between the black-spotted and red-spotted tokay geckoes (Gekko gecko). Ecol Evol 4(17):3383–3394. https://doi.org/10.1002/ece3.1183

Chapter 17
What Future for Ant-Aphid Mutualism Under Climate Change?

Solène Blanchard and Claire Detrain

17.1 Introduction

Mutualistic interactions between two or more species are essential for the stability and functioning of ecosystems. Services provided by mutualisms take many forms, including assistance for reproduction, nutrition, protection, hygiene, and dispersal. Hemipteran insects belonging to the suborder Sternorrhyncha feed on phloem sap and trade the honeydew they excrete to get protection from ants against natural enemies and improved hygiene of their colony (Doebeli and Knowlton 1998; Stadler and Dixon 2005). These sap-sucking insects include aphids, mealybugs, planthoppers, psyllids, scale insects, and whiteflies. Mutualistic relationships between honeydew-harvesting ants and sap-feeding insects have contributed to the ecological success of both partners, with far-reaching ecological consequences for natural and agricultural ecosystems (Stadler and Dixon 2005; Styrsky and Eubanks 2007; Van Emden and Harrington 2017).

Aphids are the most comprehensively studied ant mutualists and are the focus of our chapter. They are likely to reflect the impacts of climate change on ant-hemipteran mutualisms more generally. Myrmecophily, i.e. an external symbiotic association with ants, has evolved and disappeared multiple times in aphid subfamilies and even within the same genus (Shingleton and Stern 2003; Stadler and Dixon 2005; Depa et al. 2020). Ant-aphid associations range from close (obligate mutualism) to occasional (facultative mutualism) (Stadler and Dixon 2005; Way 1963). In non-myrmecophilous aphids, the excreted honeydew is simply expelled and falls from the anus. Conversely, myrmecophilous aphids are usually stimulated by ants,

S. Blanchard
Research Centre for Biodiversity and Environment, University of Toulouse III, Toulouse, France

C. Detrain (✉)
Unit of Social Ecology, University of Brussels, Brussels, Belgium
e-mail: claire.detrain@ulb.be

I. R. Leal et al. (eds.), *Ant Responses to Climate Change*, Entomology in Focus 10,
https://doi.org/10.1007/978-3-032-21127-9_17

Fig. 17.1 Interaction between a *Lasius niger* ant forager and its *Aphis fabae* aphid partners, both in adult and larval stages. (©Solène Blanchard)

through antennal contacts, in order to emit honeydew droplets that are kept over the perianal cuticle until being ingested by ants. The most studied aphid-tending ants belong to the genera *Formica, Lasius* and *Myrmica* and are known to be involved in mutualistic interactions with almost 60% of all aphid species, most of which belong to the subfamily Aphidinae (Shingleton et al. 2005) (Fig. 17.1).

Although several studies have described aphid-ant mutualism as a fair interaction (El-Ziady and Kennedy 1956; Way 1963), the ant-aphid relationship is probably best conceptualised as one of reciprocal exploitation, in which each partner incurs costs such as energy or time expenditure, physiological costs or even morphological adaptations (Doebeli and Knowlton 1998; Stadler and Dixon 1998; Herre et al. 1999; Shingleton et al. 2005; Stadler and Dixon 2005; Ness et al. 2010; Yao 2014).

Ant attendance can be costly for aphids as honeydew must be produced regularly throughout the season, in sufficient quantity and quality to attract ants at the level required for effective protection and hygiene (Katayama and Suzuki 2002; Katayama et al. 2013). For ants, there are also several time and energy costs associated with aphid tending. Ant foragers spend a considerable amount of time contacting and stimulating aphids before successfully inducing the release of honeydew droplets. Time is also required to clean sticky honeydew from aphid bodies or leaves, to remove corpses of dead aphids as well as to protect the aphids against predators like ladybugs, hoverfly larvae (Jiggins et al. 1993; Detrain et al. 2017), or parasitoids (Müller and Godfray 1999). Any significant reduction in the net benefit received by one partner, or any new constraint that disrupts ant-aphid interactions, makes facultative mutualists susceptible to partner switching or shifts to antagonistic relationships (Rafferty et al. 2015). The stability of the ant-aphid relationship depends on the phenotypic plasticity of both partners and their ability to cope with biotic and abiotic changes in their environment. Obligate mutualists are expected to be particularly affected by climate change because the loss of ant partners can threaten their local existence (Kiers et al. 2010; Barton and Ives 2014).

Concerns over climate change have increased over the last few decades, with scientific reports indicating major changes due to human activities (IPCC 2023; see Lee et al. 2023). Changes in temperature, precipitation patterns, droughts, and concentration of atmospheric gases such as carbon dioxide (CO_2) or ozone (O_3) are likely to alter the functioning and stability of ecosystems at all levels, from individuals to communities, and to modify networks of interactions between organisms, including mutualistic ones (Barnosky et al. 2011). At the level of ecological communities, we expect major changes in species diversity, geographic distribution and phenological synchronisation between trophic levels. In this regard, how climate change may alter the phenotypes of ants and aphids as well as their interactions, with cascading effects on aphid natural enemies (Van Baaren et al. 2010; Boullis et al. 2015; Rosenblatt and Schmitz 2016), has to be critically examined. Furthermore, due to the high dependence of aphid host plants on local environmental conditions, ant-aphid mutualism is likely to be particularly sensitive to climate change, through induced changes in plant sap and consequently in aphid honeydew. From the early establishment of the mutualistic interaction to its persistence in time and space, abiotic factors may thus modify the way ants and aphids interact.

Our chapter focuses on observed or expected effects of increases in ambient temperature and CO_2 on ant-aphid mutualism as this represents the core of published studies (e.g. Barton and Ives 2014; Kremer et al. 2018; Blanchard et al. 2019; Mooney et al. 2019; Blanchard et al. 2021), with occasional references to other sap-sucking hemipterans. Almost all these studies have been laboratory-based, examining the influence of a single abiotic stressor, such as increasing temperatures under controlled humidity and CO_2 levels. The effects of drought on ants (Couper et al. 2021; Ord 2023), aphid populations, host plants and parasitoids have also been documented (for a review, Johnson et al. 2011, Leybourne et al. 2021) but, strangely, little is known on how drought affects mutualistic interactions between sap-sucking insects and ants (Pringle et al. 2013). In this chapter, we first consider how climate change may affect chemically mediated interactions between ants and aphids, from the localisation of partners to the establishment and maintenance of their relationship. Particular attention is given to the production and composition of honeydew, as it is the keystone upon which this mutualism is built. We then examine the extent of climate-driven changes in populations of myrmecophilous aphids, as well as in the level of ant commitment to aphid care. The key value of aphid digestive microflora in the context of interactions with ants and higher trophic levels is also discussed. Finally, we consider alternative carbohydrate sources competing for ant visitation, such as plant nectaries, which may contribute to a remodelling of ant nutritional ecology under global change.

17.2 Chemical Communication Under Climate Change

Chemical communication has a crucial role at every step of the ant-aphid partnership. Aphid-associated volatiles and cuticular hydrocarbons help the ants to detect aphids and influence whether they display aggressive or tending behaviour. The trail pheromone emitted by ant foragers is another critical compound as it determines how many workers are mobilised to care for and protect aphid colonies (Fig. 17.2). The effects of increasing temperatures or CO_2 levels on these semiochemicals can

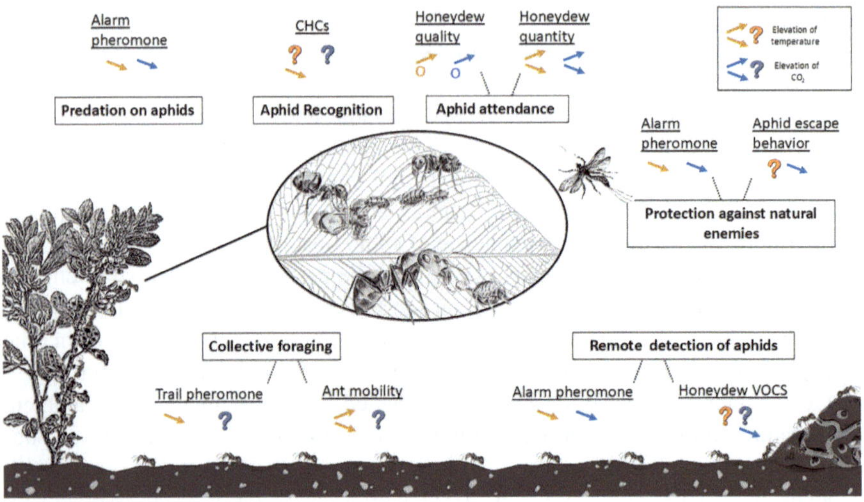

Fig. 17.2 Graphical synthesis of the expected effects of an elevation of temperature (orange arrows) and atmospheric CO_2 concentration (blue arrows) on the chemical compounds mediating each step of the ant-aphid mutualism. The effects are addressed from the ant perspective. Downward and upward arrows indicate respectively negative and positive effects on the emission/detection of semiochemicals, aphid honeydew or insect behaviour. "?"means that scientific data are currently lacking. CHCs Cuticular Hydrocarbon Chains; VOCs Volatile Organic Compounds. (Detailed explanations are given in Section 17.2)

thus have a profound impact on this mutualism, from the detection up to the long-term exploitation of aphids by ants.

17.2.1 Detection of Aphid Colonies: Alarm Pheromone

In response to attacks by natural enemies, aphids secrete an alarm pheromone, with (E)-β-farnesene (EβF) as its major compound, that warns conspecifics of dangerous situations and triggers an escape response (Bowers et al. 1972; Nault et al. 1973). At low concentrations, the alarm pheromone can act as a chemical cue used by ant foragers to remotely locate the aphid colonies (Verheggen et al. 2012). At higher concentrations, aphid alarm pheromone increases the number of ant workers that display aggressive behaviour, thereby reinforcing aphid protection against natural enemies (Katayama and Suzuki 2002) and potentially strengthening their mutualistic interaction. However, above a certain concentration of alarm pheromone, ant workers can become so aggressive that they attack and prey on the aphids (Nault et al. 1976).

A rise of ambient temperatures can impact the emission and detection of pheromones in insects by increasing their volatility. Highly volatile compounds such as EβF may therefore be harder to detect by ants, and the aphid colonies more difficult to locate from a distance (Verheggen et al. 2012), although this has still to be demonstrated (Boullis and Verheggen 2016) (Fig. 17.2). These lower levels of alarm pheromone may also dampen aggressive behaviour and predation exerted by ants. Concerning CO_2 elevation, it was found in one study to reduce the emission of EβF by 12% and 35% following a predator attack (Boullis et al. 2017) and attenuate the escape behaviour of aphids exposed to natural doses of alarm pheromone (Mondor et al. 2004; Sun et al. 2010; Hentley et al. 2014; Boullis et al. 2017) (Fig. 17.2). This reduced emission of alarm pheromone coupled a weaker escape response of aphids is assumed to make them less responsive to natural enemies and more dependent on ants for protection under a CO_2 enriched environment. The lower mobility and reactivity of aphids to external stressors is also expected to down-regulate ant agressivity leading to unwanted predation. This may promote a higher stability of aphid colonies and hence to facilitate their visitation and attendance by ants under elevated CO_2 conditions.

17.2.2 *Detection of Aphid Colonies: Honeydew Volatile Compounds*

Several volatile organic compounds (VOCs) are associated with fresh honeydew droplets. These volatile compounds are typically fermentation by-products, as honeydew sugars and amino acids provide an excellent growth medium for microorganisms. The digestive tract of aphids harbours a diverse endosymbiotic microflora, some of which is acquired during feeding on the host plant and eventually transferred to the honeydew. The honeydew-associated volatiles can act as kairomones (a compound sensed by and beneficial to another species than the emitter), used by natural enemies and ant partners to locate aphid colonies (Leroy et al. 2011; Fischer et al. 2015). Ants can even discriminate between two aphid species at a distance based on the VOCs associated with honeydew (Fischer et al. 2017).

The impact of increased temperatures on the VOCs' signature of honeydew remains unexplored. Only one study has investigated how CO_2 concentration influences the production and diversity of honeydew volatiles (Boullis et al. 2018), showing that an elevation of CO_2 concentration lowers VOCs' diversity. These changes are probably due to a remodelling of the bacterial community associated with aphid honeydew. A possible explanation is that an increase in CO_2 level alters the metabolism of the host plants causing changes in phloem sap and aphid honeydew, particularly in their content of amino acids and sugars that act as precursors for bacterial metabolism. To date, there are too few studies investigating the effects of climate change on VOC signature to make reliable predictions, but there will be a possible reduction of its complexity (Boullis et al. 2018), which may impair the detection and discrimination of aphid colonies by the ants.

17.2.3 Recognition of Tended Aphid Colonies: Cuticular Hydrocarbon Chains

The ant cuticle is covered by a thin layer of cuticular hydrocarbon chains (CHCs) whose composition is highly specific, enabling ants to distinguish nestmates from foreign workers, reproductive individuals from sterile workers, and workers according to their tasks (Greene and Gordon 2003; Martin and Drijfhout 2009; Blomquist and Bagnères 2010; Sprenger and Menzel 2020). CHCs can also be involved in short-range ant-aphid interactions. Indeed, when ants interact with aphids, they transfer some of their cuticular hydrocarbons to the aphid bodies, which makes the aphids recognisable to other ant foragers (Endo and Itino 2012). Ant foragers are then able to discriminate between recently tended and unattended aphids and to classify colonies as potential trophobionts or prey (Sakata 1994; Lang and Menzel 2011).

As insects are prone to desiccation due to their high surface-to-volume ratio, another important role of the CHCs is to minimise water loss (Gibbs and Rajpurohit 2010; see Chap. 5). Under temperature variation and/or moisture stress, the plasticity of CHCs profile facilitates the adaptation of some species, or even some behavioural caste of workers, to changing environmental conditions (Wagner et al. 2001; Menzel et al. 2017, 2018). After exposure to higher temperatures, CHCs that show tight molecular alignment (n-alkanes) increase in relative abundance, while compounds that disrupt molecular alignment (dimethyl and trimethyl alkanes, alkadienes) decrease. In addition to improving cuticle waterproofing (Gibbs and Mousseau 1994; Menzel et al. 2018), these changes increase the viscosity of the CHC layer (Sprenger et al. 2018), which influences the diffusion of recognition cues across the body surface (Cooper et al. 2009). Temperature changes could thus affect CHC diversity and, to some extent, the recognition processes of ant workers (Villalta et al. 2020) and ant-attended aphids (Fig. 17.2). As for the possible effects of a CO_2-enriched environment on the CHCs of ants, they remain unexplored. Regardless of climate-induced changes in the chemical composition of CHC layer, the high learning abilities of ants (Czaczkes and Kumar 2020) make them likely to adapt and still be able to discriminate between aphids recently tended or not. Conversely, for the few aphid species producing de novo compounds similar to ant cuticular hydrocarbons to avoid being preyed by ant foragers (Endo and Itino 2013), they may lose their ability to escape predation by ants because temperature-induced changes in CHCs may be too rapid for such chemical mimicry to evolve.

17.2.4 Exploitation of Aphid Colonies: The Regulating and Guiding Role of the Trail Pheromone

Once aphids have been discovered by ants, the collective exploitation of honeydew relies on the release of a recruitment signal by ant foragers. Ant individuals that have ingested a honeydew droplet typically lay a chemical trail on their way back to the nest that triggers the nest exit of workers and guides them to the plant where aphids are located. The trail-laying intensity, and hence the level of aphid attendance, typically increases with the amount of honeydew produced and its content in sugars such as sucrose and melezitose that are highly attractive to the ants (Detrain et al. 2010).

When ants are exposed to higher ambient temperatures, they walk more rapidly, with increased flows of foragers that reinforce the trail and thereby the collective exploitation of aphids. On the other hand, ants show a reduced ability to follow a chemical trail (Van Oudenhove et al. 2011, 2012; see Chap. 2) (Fig. 17.2). This may be the result of an accelerated decay of the trail due to increased evaporation or degradation of its components. The concentration of trail pheromone deposited on the ground may become too low for the self-amplifying process of nestmate recruitment to occur and hence for a collective exploitation of aphid colonies to emerge (Detrain and Deneubourg 2006). Increased trail evaporation at elevated temperatures could particularly affect the collective exploitation of aphid colonies with a small population size or producing honeydew of low quality/amount since, in these cases, the amount of trail pheromone laid by ant scouts is already quite limited. Similarly, short-lived trails consisting of low molecular weight compounds are more sensitive to high temperatures and are more likely to fall below the threshold amount that triggers the recruitment of ant foragers. Elevated temperatures could therefore have varying degrees of impact on trail-based foraging by ants, depending on the physico-chemical properties and amount of the trail pheromone.

The impact of elevated CO_2 concentration on ant trail pheromones has not been studied (Fig. 17.2). Yet, it is likely that CO_2 interference with the volatile components of the trail pheromone will be more limited than that caused by an increase in ambient temperature.

Climate change may therefore impact ants and their associated hemipterans at multiple levels in their chemical communication, altering or strengthening their interactions depending on the climate factor studied. Another keystone upon which interactions between ants and aphids are built is the sugar-enriched honeydew, of which the quality and production rate is tightly linked to the aphid host plants and therefore might be highly susceptible to climate change.

17.3 Aphid Honeydew Under Climate Change

Climate change can alter aphid honeydew, with major implications for ant-aphid mutualism, as this carbohydrate source is the cement of their relationship. Aphid honeydew is a mixture of water-soluble carbohydrates, amino acids, proteins, minerals, and vitamins (Way 1963; Ali et al. 2024), derived from the plant phloem that is sucked by the insect. Its composition is highly variable depending on the host plant (Hendrix et al. 1992; Völkl et al. 1999; Fischer and Shingleton 2001; Pringle et al. 2014), the aphid species or clone (Hendrix et al. 1992; Völkl et al. 1999; Vantaux et al. 2011), and even the symbiotic bacteria associated with aphids

(Wilkinson et al. 1997). Its production also varies according to the health and number of aphids, as well as their feeding behaviour (Auclair 1963). The emission rate/volume of honeydew droplets, coupled with their sugar concentration (Woodring et al. 2004; Detrain and Prieur 2014), determines the feeding behaviour of ants as well as the intensity of their collective recruitment to food (e.g. Detrain et al. 1999; Völkl et al. 1999; Mailleux et al. 2003; Detrain et al. 2010). The energetic value associated with aphid honeydew is crucial for the initiation and maintenance of the ant-aphid relationship (Way 1963; Völkl et al. 1999; Stadler and Dixon 2005; Pringle et al. 2014). This partly explains why the level of ant visitation can vary between aphids feeding on the same host plant and even on the same plant organ, some being always attended, while others being never visited by ants (Stadler et al. 2003).

17.3.1 Quantity of Honeydew

If aphids are stimulated by ants to excrete honeydew droplets at a higher rate, the honeydew is enriched in amino acids, which may lead to a shortage of nitrogen that would otherwise be available for aphid growth and reproduction (Fischer and Shingleton 2001; Yao and Akimoto 2002). To compensate for this nitrogen deficit, aphids must assimilate even larger quantities of phloem sap and allocate more time to food intake. For the ants, the amount of honeydew produced determines how many foragers succeed in filling their crop, recruiting additional nestmates and possibly developing a fidelity to this carbohydrate source (Mailleux et al. 2003).

Raising the ambient temperature up to 23 °C does not seem to alter the ability of aphids to produce honeydew, as the number of droplets released by each aphid and the number of ant antennal contacts required to trigger their emission do not change significantly in the case of *Lasius niger-Aphis fabae* partnership (Blanchard et al. 2021, 2022). On the other hand, ant metabolism increases with ambient temperature, inducing faster and more intense flows of ants heading towards the aphid-infested plant. This increased foraging activity at moderately elevated temperatures results in higher total amount of honeydew retrieved by ants, with potential enhancing effects on their mutualistic interactions with aphids (Fig. 17.2). The maintenance of this partnership over prolonged periods of warmer conditions depends on the ability of aphids to produce enough honeydew to satisfy the higher flows of foragers. It is also worth noting that a further increase in temperature can lead to the cessation of honeydew harvesting by ants, as foragers' flows decrease and even become null when critical thermal values are approached (Cerdá et al. 1998, Blanchard et al. 2021; see Chap. 4).

Concurrently, the building of a collective trail can be seriously impaired under high temperature conditions, as seen earlier in this chapter. Moreover, feeding performances of aphids are negatively affected on plants that are water-stressed following a prolonged exposure to high temperatures (Nachappa et al. 2016; Bari et al. 2018; Kansman et al. 2020). Drought associated with high temperatures thus reduces honeydew production what may further jeopardise trophobiotic aphid interactions with ants (Fig. 17.2). Temperature seems to exert the same influence on other sap-sucking hemipterans, with an increase in honeydew production at moderately elevated temperatures and a reversed impact when approaching their critical thermal values (Sagata and Gibb 2016).

Under elevated CO_2, the N content of phloem decreases, which sap-sucking herbivores can compensate for by ingesting more phloem sap (Stiling and Cornelissen 2007; Sun et al. 2009; Sun and Ge 2011; Sun et al. 2016; Kremer et al. 2018). As a result of a higher feeding rate, under elevated CO_2, aphids can release up to three times more honeydew (Kremer et al. 2018), which in turn can increase the number of aphid-tending ants (Fig. 17.2). However, other studies found no effect (Blanchard et al. 2022) or even a negative effect (Boullis et al. 2018 on non-myrmecophilous species) of CO_2 concentration on the amount of honeydew released by aphids.

17.3.2 Composition of Honeydew

Several honeydew sugars, such as fructose, glucose, sucrose, or xylose, are not synthesised by the aphids but come directly from the phloem sap (Hendrix et al. 1992; Ashford et al. 2000; Fischer and Shingleton 2001), whereas others, such as melezitose, are biosynthesised by aphids (Bacon and Dickinson 1957). Under elevated temperatures, aphids produce honeydew with higher concentrations of sucrose, trehalose and melezitose (Mooney et al. 2019, Seeburger et al. 2022, but see Blanchard et al. 2022). This is expected to enhance collective ant foraging on aphid colonies, as melezitose and sucrose are the most attractive sugars for the ants, trigger their feeding response and promote the laying of a trail pheromone (Detrain et al. 2010; Detrain and Prieur 2014) (Fig. 17.2). Furthermore, when considering drought associated with higher temperatures, myrmecophilous aphids produce higher amounts of melezitose in order to osmoregulate the higher osmolality of the phloem sap of water-stressed plants (Seeburger et al. 2022). Drought could therefore further enhance ant-aphid mutualism at higher temperatures, provided these environmental conditions stay below critical values for both partners. Likewise, under water stress, the increased sugar amounts in the honeydew of scale insects enhance ant tending on the colony (Pringle et al. 2013). In contrast, under conditions of high relative ambient humidity, the presence of melezitose in aphid honeydew could be lowered (Seeburger et al. 2022), leading to a decrease of interest from ants in the honeydew resources and possibly a switch towards other sugar producers.

An elevation of atmospheric CO_2 induces no change in carbohydrate concentrations in the phloem sap of some plant species, while fructose and glucose contents significantly increase in others (Oehme et al. 2013). Although the carbohydrate content of plant tissues can be modified, the relative amount of each sugar in the released honeydew seems to be not significantly altered by an elevation of CO_2 alone (Blanchard et al. 2022). Likewise, amino acids in honeydew (Sun et al. 2009; Kremer et al. 2018) were not altered under elevated CO_2 concentrations, although decreasing amounts of amino acids have been observed in foliar tissues of plants (De Lucia et al. 2012). Thus, a CO_2–enrichment of atmosphere alone has no significant impact on honeydew composition (Fig. 17.2). Interestingly, when combined with an increase in temperature, high CO_2 levels then increase the amount of sucrose and melezitose in aphid honeydew (Blanchard et al. 2022) and may thus reinforce ant-aphid interactions.

Thus, while atmospheric CO_2 levels have no clear-cut impact on aphid honeydew, an increase in temperatures may either reinforce ant-aphid mutualism for moderate thermic elevation or may dampen it for extreme heat waves, through changes in honeydew composition and production. However, the species-specificity of honeydew changes highlights the need for further field and laboratory studies that consider the full complexity of honeydew-mediated interactions between ants and aphids, namely by assessing the combined effects of elevated temperature and CO_2.

17.4 Populations of Aphids and Tending Ants Under Climate Change

Beyond the impact of elevated temperature or CO_2 concentration on the semiochemicals and honeydew involved in the ant-aphid partnership, the following section reports on our current knowledge of higher-level effects of climate change on the aphid populations, particularly myrmecophilous ones, and on the global involvement of ants in the care of sap-sucking hemipterans.

17.4.1 Impact on Aphid Populations

Temperature is probably the most important factor affecting aphid behaviour, spatio-temporal distribution, development, or reproduction (Kocmánková et al. 2010). Likewise, in other honeydew-producing hemipterans such as mealybugs, an elevation of ambient temperature can strongly influence their abundance and geographical distribution (Wei et al. 2020). Drought, which is often associated with heatwaves, also affects aphid populations, with typically negative impacts due to reduced vigour of their host plant and increased chemical defences in water-stressed plants (Leybourne et al. 2021). The effects of increased ambient temperature are less clear-cut: they can be both direct and indirect through changes in the host plant of sap-sucking insects, their natural enemies or protective mutualists (Diamond et al. 2012, 2013; Robinson et al. 2017; Hamann et al. 2021). Studies have found mostly a positive effect on aphid population growth of increasing temperature up to 28 °C (Fig. 17.3a), but above 30 °C, population growth, development time and reproductive capacity are lowered to varying degrees depending on the thermal tolerance of the aphid species (Fig. 17.3a) (Asin and Pons 2001; Ma et al. 2004, Ma

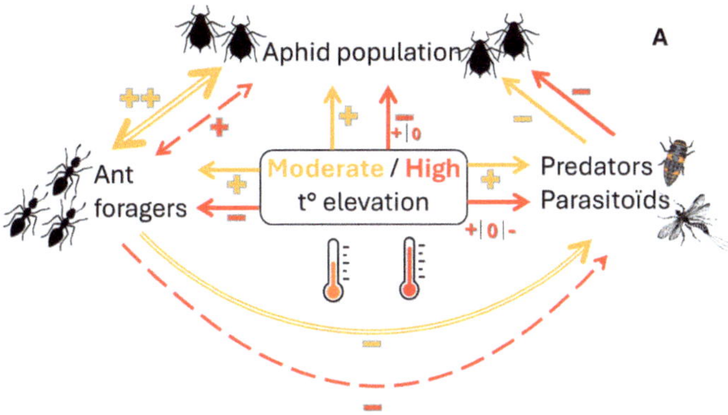

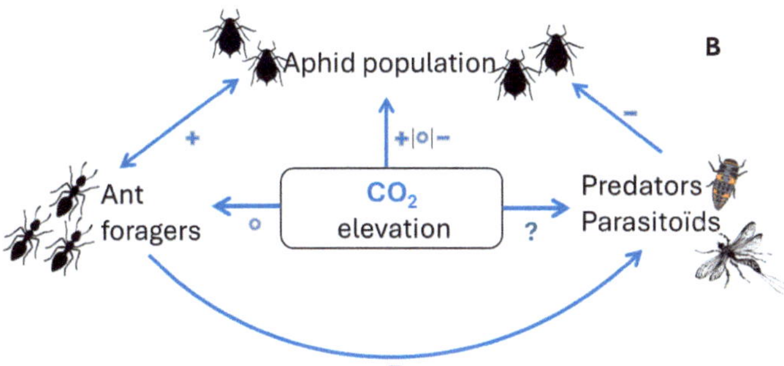

Fig. 17.3 Overview of the effects of Elevated temperatures (**a**) or atmospheric CO_2 concentration (**b**) on the population of aphids, ant foragers and aphid natural enemies as well as on their interspecific relationships. "+, –, 0" respectively account for a positive, negative or lack of effects of the abiotic factor on insect populations. "?" means that no scientific data are currently available in literature. For interspecific relationships, "+, -" signs put close to arrows connecting insect species respectively indicate mutualistic or antagonistic relationships. For these interspecific links, double arrows and dashed arrows respectively represent an expected reinforcement or weakening of the relationships existing between species compared to the current situation. Plain single arrows indicate that no clear-cut impact of climate change on interspecific relationships has been reported so far. Concerning changes in temperature, the effects of a moderate elevation (in orange) and extreme warm episodes (in red) are considered separately. (For detailed explanations, please refer to Sects. 17.4.2 and 17.4.3)

et al. 2015; Barton and Ives 2014). Aphid species with complex life histories that obligately alternate between host plants and have time-sensitive movements between these plants should face even greater risks of decline under climate change (Crossley et al. 2021). Indeed, host hopping exposes aphids to phenological mismatches, such as when their phenology is more accelerated by an increase in temperature than that of their host plant. Such a phenological mismatch can impair the settlement of aphid colonies or at least limit their growth to such an extent that mutualistic interactions with ants cannot be established or maintained. Furthermore, higher temperatures may prolong the yearly period of parthenogenetic reproduction, consequently reducing the genetic variability within the aphid population. Such a reduced genetic variability may lead to contrasting local patterns of interactions, with some sites having aphid clones producing high quality of honeydew and being intensively tended by ants (Vantaux et al. 2011) and others having clones left unattended due to the poor nutritional value of the honeydew produced. Predicting the indirect effects of temperatures on aphid natural enemies is also challenging. Higher temperatures usually increase the metabolism, activity, and attack rates of predators/ parasitoids (Fig. 17.3a) but, at the same time, lower their performance due to enhanced intraguild predation or interference (Rall et al. 2010; Laubmeier et al. 2023). Furthermore, phenotypic plasticity enables parasitoids to adjust life-history traits in responses to changes in temperature, including host-foraging behaviours (Jerbi-Elayed et al. 2021, 2022). The effects of increasing temperatures on aphid-parasitoids interactions are therefore highly idiosyncratic and difficult to predict.

The impact of CO_2 enrichment is primarily indirect through its influence on plants (Urban 2003), with variable effects on aphid population growth (Fig. 17.3b). Some studies report an increase in aphid populations under CO_2 enrichment (Hullé et al. 2010; Ryan et al. 2015), whereas others found no impact (Salt et al. 1996; Docherty et al. 1997; Awmack et al. 2004) or even negative effects (Newman et al. 1999; Awmack and Harrington 2000; Adler et al. 2007) due to prolonged development of aphids and reduced fecundity (Moreno-Delafuente et al. 2021; Carreras Navarro et al. 2020; Chen et al. 2019). The effects of elevated CO_2 on aphid predators and parasitoids are even more uncertain (Awmack et al. 1997, 2004; Sun et al. 2011).

17.4.2 Does Myrmecophily Reinforce or Dampen Climate-Induced Changes in Aphid Populations?

Several studies have investigated how the populations of different aphid species, being myrmecophilous or not, are affected by an elevation of temperature. For example, frequent events of extremely high temperatures (34 °C to 38 °C) decrease the abundance of non-myrmecophilous aphids, such as *Sitobion avenae* and *Schizaphis graminum*, whereas they have no effect or even positive ones on *Rhopalosiphum padi*, a myrmecophilous species with higher heat tolerance (Ma et al. 2015). However, due to the small number of studies, it is unclear whether myrmecophilous aphids are more heat tolerant than non-myrmecophilous ones. Ants are usually thought to favour the growth of myrmecophilous aphid colonies (Way 1963) but in some cases, may have a negative effect on their development times, body size or lifetime fecundity (Yao 2014). Knowing that tending ants have varying, sometimes opposite, effects on aphid growth, it is difficult to predict whether the presence of ant foragers will attenuate or increase climate-induced changes in populations of myrmecophilous aphid species. For instance, Barton and Ives (2014) have shown that a 5 °C increase in ambient temperature enhances the growth of *Rhopalosiphum maidis* populations, but this positive effect is attenuated by the presence of tending ants, which are less abundant and less aggressive towards predators when temperature increases. In this case, warming tends to weaken the ant-aphid mutualism and counter-intuitively reduces the abundance of this aphid pest species (Barton and Ives 2014).

As seen earlier, the rise of CO_2 concentration has variable and sometimes opposite effects on the growth, fecundity, and performance of aphid populations. Globally, studies that have observed negative effects of CO_2 were done on myrmecophilous aphid species (Newman et al. 1999; Chen et al. 2019; Carreras Navarro et al. 2020; Moreno-Delafuente et al. 2021), while those that found none or a positive effect were considering both myrmecophilous and non-myrmecophilous species (Salt et al. 1996 for myrmecophilous species, Docherty et al. 1997, Awmack and Harrington 2000 for non-myrmecophilous species).

17.4.3 Impact on Population of Aphid-Tending Ants

Several studies have investigated the impact of climate change on ant morphology (Bishop et al. 2016; see Chap. 3), physiology (Diamond et al. 2012, 2013; Roeder et al. 2021; Nascimento et al. 2022; see Chaps. 4 and 5), behaviour (van Oudenhove et al. 2012; see Chap. 2) and population dynamics (Couper et al. 2021; see Chaps. 6 and 7). At the community level, differences in thermal tolerance among ant species (Roeder et al. 2021; see Chap. 4) generate within- and between-habitat variation in the composition of ant assemblages (Bishop et al. 2017; Roeder et al. 2018; Couper et al. 2021; Parr and Bishop 2022), but also determine patterns of resource use (Cerdá et al. 1998; Roeder et al. 2018).

In terms of temperature elevation, ant attendance of myrmecophilous aphids is likely to be differentially affected by moderate temperature increases or extreme warm episodes (Fig. 17.3a). As long as the temperature elevation remains moderate, it will accelerate aphid development and make them available to ants earlier in the spring in temperate ecosystems. This may facilitate the emergence of loyalty of ant foragers to this type of sugary resource and may extend the period during which ants are tending aphids. Furthermore, as also seen in ant-mealybug mutualism (Zhou et al. 2017), the foraging behaviour of ants is stimulated by ambient warming: foragers perform more antennal contacts on aphids and run at higher speed with more intense flows towards the aphid-infested plant (Blanchard et al. 2021). This increased foraging activity of ants at moderately elevated temperatures results in higher total amount of honeydew collected by ant colonies and in enhanced benefits derived from ant interactions for aphid colonies. Combined with a higher growth of aphid populations around their thermal optima (Asin and Pons 2001; Barton and Ives 2014), a moderate increase in mean ambient temperature is thus expected to strengthen ant-aphid mutualism by increasing the number of aphids producing honeydew and the population of tending ants (Fig. 17.3a). In contrast, prolonged heat waves and drought in summer can impair aphid development (Mooney et al. 2019), induce a drop in their population size and reduce honeydew production (Sagata and Gibb 2016). Under daily and seasonal temperature variations that include several heat waves, ant interest in aphid colonies decreases globally (Mooney et al. 2019; Robinson et al. 2017) as well as their aggressivity towards predators (Barton and Ives 2014) resulting in aphids being less protected against natural enemies. Coupled with drought-induced relocation of ant colonies (Ord 2023), extreme heat episodes may break down ant-aphid mutualism and may thus promote a welcome and counterintuitive control of some aphid pest species (Fig. 17.3a).

As for the impact of CO_2 enrichment, it is worth noting that ant workers are used to being exposed inside the nest to very high CO_2 levels (Römer et al. 2018). No significant alteration in ant behaviour is thus expected within the range of atmospheric CO_2 concentrations predicted for the future (Fig. 17.3b). Concerning the level of aphid attendance by ants, it is, however, expected to increase (Kremer et al. 2018) (Fig. 17.3b), mainly due to changes in the quality and quantity of honeydew provided to the ant partner, as seen earlier in this chapter (Hendrix et al. 1992; Völkl et al. 1999; Yao and Akimoto 2002; Pringle et al. 2014).

Overall, further work is needed to investigate thermal tolerance and plasticity of ant species (see Chap. 4), in particular those associated with sap-sucking insects. In this respect, honeydew carbohydrates not only fuel the activity of ant workers but can also increase their thermal tolerance, as shown for sucrose, which increases the maximum critical temperature of ants, enabling them to forage at warmer temperatures (Bujan and Kaspari 2017; Freires et al. 2023). This demonstrates that the thermal tolerance of an ant species is linked not only to the mean and extreme temperatures of its habitat but also to its partnership with aphids. Such a thermal plasticity is very important in predicting the resilience of aphid-ant interactions in a warming environment, as it determines the ability of one species to cope with heat waves and changes in the average temperature of its environment (Rodrigues and Beldade 2020).

17.4.4 The Unsuspected Role of Microorganisms

Some obligate microorganisms are required for the survival and reproductive success of sap-sucking insects that feed upon an unbalanced diet (Baumann 2005; Douglas 2009; Ferrari and Vavre 2011). For instance, *Buchnera* spp. bacteria are typically obligate symbionts that provide essential amino acids to their aphid host (Douglas 1998, 2009). Aphids also harbour a wide range of facultative symbionts that can be maternally transmitted or acquired from the environment, including when feeding on host plants (Moran et al. 2008; Oliver et al. 2010; Pons et al. 2019; Renoz et al. 2019; Kaszyca-Taszakowska and Depa 2022). Heat-tolerant genotypes of the obligate symbiont *Buchnera aphidicola* and some facultative symbionts (*Rickettsia* sp., *S. symbiotica*) can improve or maintain aphid fitness under heat stress (Montllor et al. 2002; Heyworth et al. 2020; Tougeron and Iltis 2022). On the other hand, if obligate symbionts endure a rapid decline due to heat stress, this can lead to premature death, sterility and collapse of hemipteran populations (Kikuchi et al. 2016; Zhang et al. 2019). The microbiome is thus a key actor of the adaptation of aphid communities to climate change, with some specific co-evolving bacterial strains that improve their thermal tolerance. The microbiota of sap-sucking insects can also influence higher trophic levels (Oliver et al. 2010; Álvarez-Pérez et al. 2024) by having a significant impact on the rate of parasitism and the fitness of aphid natural enemies (Vorburger et al. 2010; Attia et al. 2022). Microorganisms in honeydew namely produce VOCs that act as a kairomone for the location of the aphid host and the initiation of oviposition (Leroy et al. 2011). Facultative symbionts can also be transmitted from the aphids to the tending ants (Renoz et al. 2019), with possible cascading effects on the fitness of the ant partner that have still to be examined. Finally, the presence of bacteria in aphid gut and/or honeydew may strengthen or dampen mutualistic interactions between aphids and ants (Depa et al. 2020; Jossart et al. 2025). For instance, the presence of *Staphylococcus xylosus* bacteria can make aphids more attractive to the ants (Fischer et al. 2015), probably due to induced changes in the honeydew volatilome. The microbial ecology involving hemipterans, honeydew, natural enemies and ants thus requires to be scrutinised to uncover whether microorganisms can shape heat-tolerant communities of sap-sucking insects and influence their relationships with ant partners under a scenario of climate change.

17.5 Impact on the Nutritional Ecology and Partnerships of Ants

How climate change will affect ant-aphid mutualism depends on the ability of each partner to counter, attenuate or even benefit from these new environmental conditions. Most mutualists are not locked into one interaction and many possess traits that allow them to endure a prolonged absence of their partner (Kiers et al. 2010). In the case of aphids, approximately a third of all species are obligate ant mutualists and another third are facultative ones (Depa et al. 2020). Aphid species that did not co-evolve with ants are more likely to overcome the loss of ant partnership, namely because they are usually less susceptible to predators and parasitoids of which the efficacy can be concurrently reduced by climate change (Boullis et al. 2015; Hentley and Wade 2016). Conversely, strictly myrmecophilous aphids may greatly suffer from the disruption of their interaction with ants since they are more limited in the alternative adaptive ways.

Ants are expected to be less susceptible than aphids to extreme heat episodes, drought, or elevation of CO_2 concentration. First, ants can attenuate the effects of unfavourable conditions by adapting their foraging behaviour, relocating their nest, or rearranging the distribution of workers inside the anthill (see Chap. 2). Second, in the case of a climate-induced decline of their usual aphid partner, ants are able to switch to new aphid species or clones that are less sensitive and/or best producers of honeydew (Addicott 1978; Cushman and Addicott 1989) or can even partner with mealybugs. Indeed, the honeydew profile of mealybugs is similar to that of aphids but often enriched in sucrose and thereby more attractive to the ants (Zhou et al. 2015, 2017). As these honeydew-producing hemipterans share the same ecological niches and temperature optima as the aphids, mealybugs are likely to compete for ant care. It is, however, unknown whether mealybugs may accelerate some disruption of ants' relationship with aphids by becoming their preferential partner under changing climatic conditions.

When considering possible switches in mutualistic partnerships, aphids may also have to compete with floral and extrafloral nectaries (EFN) of plants (Engel et al. 2001, Blüthgen et al. 2004; see Chap. 14). Plant nectar releases volatiles that attract insects such as ant foragers (Kessler et al. 2008). Carbohydrate-rich extrafloral nectar also provides a significant amount of energy which may distract ants from honeydew and thereby act as an indirect defence of plants against myrmecophilous aphids. When competing for ant attendance, EFNs are usually more attractive due to their higher sugar content, although large aphid colonies may outcompete them by producing huge amounts of honeydew (Katayama et al. 2013). Furthermore, global warming is expected to promote the invasion of many EFN-producing plants in northern ecosystems. This may result in significant changes in the diet of ant communities in boreal and subarctic forests (Holopainen et al. 2020) that would prefer foraging on EFN richer in sugars and easier to collect than aphid honeydew (Becerra and Venable 1989). Ultimately, these new food opportunities may weaken ant-aphid mutualism or even tip the balance towards ant predation on aphids under a scenario of global warming (Sakata and Hashimoto 2000).

17.6 Concluding Remarks

Understanding the ability of species to adapt and co-evolve with their different partners or antagonists under environmental stressors is of crucial importance to map the evolution of ecosystems in the coming decades. Ants have evolved various mutualistic interactions with insects or plants, thereby contributing to shaping the local community structure, including the population of phytophagous insects. Predicted average values of ambient temperatures and atmospheric CO_2 concentration are likely to favour the growth and development of many insect pests, such as aphids, and thus increase occurrences of ant-aphid interactions. However, as shown in this review, complexity arises when closely examining the impact of increasing temperatures or CO_2 levels on intraspecific and interspecific communication as well as on honeydew-mediated interactions between ants and sap-sucking hemipterans. This explains why the expected effects of climate change on aphid-ant mutualism are usually ambiguous and often have to be considered on a case-by-case basis. Despite this, we can draw four general trajectories:

1. Deep changes in the chemically mediated interactions between ants and their aphid partners are expected. A moderate elevation of temperature may increase

flows of ant foragers that reinforce the recruitment trail and strengthen a collective exploitation of aphid colonies. The emergence of a foraging trail may, however, be delayed or prevented during extreme heat waves due to the accelerated decay of this pheromone. Temperature-induced changes in CHCs may also alter recognition of tended aphids, although these effects may be mitigated by the high learning capacities of ants to quickly recognise new chemical patterns. Under elevated CO_2 conditions, the reduced release of alarm pheromone by aphids, while possibly affecting their remote detection by ant foragers, can make the aphid colonies more stable in time and space and facilitate their visitation by ants.
2. An increase in temperatures seems to favour honeydew production by hemipterans, which may, however, decline at too high temperatures and/or when the host plants and aphids become water stressed. A moderate temperature elevation may also have a positive effect on honeydew sugar content, thereby increasing its attractiveness to the ants, triggering the laying of a foraging trail, and promoting a collective exploitation of aphid colonies. A CO_2-enriched atmosphere has no clear-cut effect on honeydew production, which varies between aphid species. Interestingly, while the elevation of atmospheric CO_2 has no significant impact on honeydew composition, a concurrent elevation of both temperature and CO_2 concentration increases its sugar content. This highlights the need to study not only the influence of a single abiotic condition on ant-aphid mutualism, but also the combined effects of multiple environmental factors.
3. An increase in temperatures is expected to favour aphid population growth, provided that temperatures remain within their tolerance range and do not cause phenological mismatches with their host plant. The effects of elevated CO_2 are highly variable across aphid species. The few existing studies make it difficult to predict whether myrmecophilous aphid species will be positively or negatively affected by climate change, and whether these changes can be attenuated or enhanced by the presence of tending ants.
4. Aphid species that are more tolerant of high temperatures may be favoured over others that are more susceptible to sudden variations in their environment. In this respect, aphid microflora deserves to be scrutinised since it can influence the thermal tolerance of hemipterans and hence may shape the future of ant-aphid mutualism. This literature review has also confirmed the high degree of species-specificity in the responses of ants and aphids to climate change.

To conclude, climate change is clearly a driver for the adaptation and selection of aphid populations, species, or clones and thus indirectly affects ant-aphid mutualisms. An understanding of climate-change impacts requires an ecosystemic approach that considers alternative partnerships and resources as well as cascading effects in the trophic chain. For instance, negative impacts of climate change may be reduced by functional redundancy involving the replacement of one host species by another. Several hemipteran species, but also plant nectaries, are functionally redundant within the guild of sugar producers and can lead to the reinforcement or even the emergence of new partnerships for the ants. In addition to these expected changes in sugar producers, ants will probably have to face new competing species taking advantage of milder weather to spread outside their usual ecological niches (see Chap. 7). Combined with major changes in plant physiology, the climate-driven evolution of ant-aphid mutualisms is likely to deeply impact natural and agricultural ecosystems worldwide, probably requiring new management strategies to control aphid pests for the decades to come.

References

Addicott JF (1978) Competition for mutualists: aphids and ants. Can J Zool 56:2093–2096. https://doi.org/10.1139/z78-283

Adler LS, de Valpine P, Harte J, Call J (2007) Effects of long-term experimental warming on aphid density in the field. J Kansas Entomol Soc 80:156–168. https://doi.org/10.2317/0022-8567(2007)80

Ali J, Abbas A, Abbas S, Ji Y, Khan KA, Ghramh HA, Mahamood M, Chen R (2024) Honeydew: a keystone in insect–plant interactions, current insights and future perspectives. J Appl Entomol. https://doi.org/10.1111/JEN.13269

Álvarez-Pérez S, Lievens B, de Vega C (2024) Floral nectar and honeydew microbial diversity and their role in biocontrol of insect pests and pollination. Curr Opin Insect Sci 61:101138. https://doi.org/10.1016/J.COIS.2023.101138

Ashford DA, Smith WA, Douglas AE (2000) Living on a high sugar diet: the fate of sucrose ingested by a phloem-feeding insect, the pea aphid *Acyrthosiphon pisum*. J Insect Physiol 46:335–341. https://doi.org/10.1016/S0022-1910(99)00186-9

Asin L, Pons X (2001) Effect of high temperature on the growth and reproduction of corn aphids (Homoptera: Aphididae) and implications for their population dynamics on the northeastern Iberian Peninsula. Environ Entomol 30:1127–1134. https://doi.org/10.1603/0046-225X-30.6.1127

Attia S, Renoz F, Pons I, Louâpre P, Foray V, Piedra JM, Sanané I, Le Goff G, Lognay G, Hance T (2022) The aphid facultative symbiont *Serratia symbiotica* influences the foraging behaviors and the life-history traits of the parasitoid *Aphidius ervi*. Entomol Gen 42:21–33. https://doi.org/10.1127/entomologia/2021/1274

Auclair JL (1963) Aphid feeding and nutrition. Annu Rev Entomol 8:439–490. https://doi.org/10.1146/ANNUREV.EN.08.010163.002255

Awmack CS, Harrington R (2000) Elevated CO2 affects the interactions between aphid pests and host plant flowering. Agric For Entomol 2:57–61. https://doi.org/10.1046/J.1461-9563.2000.00050.X

Awmack CS, Woodcock CM, Harrington R (1997) Climate change may increase vulnerability of aphids to natural enemies. Ecol Entomol 22:366–368. https://doi.org/10.1046/j.1365-2311.1997.00069.x

Awmack CS, Harrington R, Lindroth RL (2004) Aphid individual performance may not predict population responses to elevated CO2 or O3. Glob Chang Biol 10:1414–1423. https://doi.org/10.1111/j.1365-2486.2004.00800.x

Bacon J, Dickinson B (1957) The origin of melezitose: a biochemical relationship between the lime tree (*Tilia* spp.) and an aphis (*Eucallipterus tiliae* L.). Biochem J 66:289

Bari GMM, Ahmad M, Rahman MR (2018) Determination of feeding performance of mustard aphid, *Lipaphis erysimi* (Kalt.) on different mustard varieties at water stress condition. Fundam Appl Agric 3:467–473–467–473. https://doi.org/10.5455/FAA.290184

Barnosky AD, Matzke N, Tomiya S, Wogan GOU, Swartz B, Quental TB, Marshall C, McGuire JL, Lindsey EL, Maguire KC, Mersey B, Ferrer EA (2011) Has the Earth's sixth mass extinction already arrived? Nature 471:51–57. https://doi.org/10.1038/nature09678

Barton BT, Ives AR (2014) Direct and indirect effects of warming on aphids, their predators and ant mutualists. Ecology 95:587–593. https://doi.org/10.1890/13-1977.1

Baumann P (2005) Biology of bacteriocyte-associated endosymbionts of plant sap-sucking insects. Ann Rev Microbiol 59:155–189. https://doi.org/10.1146/ANNUREV.MICRO.59.030804.121041

Becerra JXI, Venable DL (1989) Extrafloral nectaries: a defense against ant-homoptera mutualisms? Oikos 55:276. https://doi.org/10.2307/3565432

Bishop TR, Robertson MP, Gibb H, van Rensburg BJ, Braschler B, Chown SL, Foord SH, Munyai TC, Okey I, Tshivhandekano PG, Werenkraut V, Parr CL (2016) Ant assemblages have darker and larger members in cold environments. Glob Ecol Biogeogr 25:1489–1499. https://doi.org/10.1111/GEB.12516

Bishop TR, Robertson MP, Van Rensburg BJ, Parr CL (2017) Coping with the cold: minimum temperatures and thermal tolerances dominate the ecology of mountain ants. Ecol Entomol 42:105–114. https://doi.org/10.1111/EEN.12364

Blanchard S, Lognay G, Verheggen F, Detrain C (2019) Today and tomorrow: impact of climate change on aphid biology and potential consequences on their mutualism with ants. Physiol Entomol 44:77–86. https://doi.org/10.1111/phen.12275

Blanchard S, Van Offelen J, Verheggen F, Detrain C (2021) Towards more intimacy: moderate elevation of temperature drives increases in foraging and mutualistic interactions between *Lasius niger* and *Aphis fabae*. Ecol Entomol 46:406–418. https://doi.org/10.1111/EEN.12982

Blanchard S, Verheggen F, Van De Vreken I, Richel A, Detrain C (2022) Combined elevation of temperature and CO2 impacts the production and sugar composition of aphid honeydew. J Chem Ecol 48:772–781. https://doi.org/10.1007/S10886-022-01385-Z

Blomquist GJ, Bagnères AG (2010) Structure and analysis of insect hydrocarbons. In: Insect hydrocarbons: biology, biochemistry and chemical ecology. Cambridge University Press, Cambridge, pp 19–34

Blüthgen N, Stork NE, Fiedler K (2004) Bottom-up control and co-occurrence in complex communities: honeydew and nectar determine a rainforest ant mosaic. Oikos 106:344–358. https://doi.org/10.1111/J.0030-1299.2004.12687.X

Boullis A, Verheggen FJ (2016) Chemical ecology of aphids (Hemiptera: Aphididae). In: Biology and ecology of aphids. CRC Press, Boca Ranton, pp 171–198

Boullis A, Francis F, Verheggen FJ (2015) Climate change and tritrophic interactions: will modifications to greenhouse gas emissions increase the vulnerability of herbivorous insects to natural enemies? Environ Entomol 44:277–286. https://doi.org/10.1093/ee/nvu019

Boullis A, Fassotte B, Sarles L, Lognay G, Heuskin S, Vanderplanck M, Bartram S, Haubruge E, Francis F, Verheggen FJ (2017) Elevated carbon dioxide concentration reduces alarm signaling in aphids. J Chem Ecol 43:164–171. https://doi.org/10.1007/s10886-017-0818-z

Boullis A, Blanchard S, Francis F, Verheggen F (2018) Elevated CO2 concentrations impact the semiochemistry of aphid honeydew without having a cascade effect on an aphid predator. Insects 9:47. https://doi.org/10.3390/INSECTS9020047

Bowers WS, Nault LR, Webb RE, Dutky SR (1972) Aphid alarm pheromone: isolation, identification, synthesis. Science (80-) 177:1121–1122. https://doi.org/10.1126/SCIENCE.177.4054.1121

Bujan J, Kaspari M (2017) Nutrition modifies critical thermal maximum of a dominant canopy ant. J Insect Physiol 102:1–6. https://doi.org/10.1016/J.JINSPHYS.2017.08.007

Carreras Navarro E, Lam SK, Trębicki P (2020) Elevated carbon dioxide and nitrogen impact wheat and its aphid pest. Front Plant Sci 11:605337. https://doi.org/10.3389/FPLS.2020.605337

Cerdá X, Retana J, Cros S (1998) Critical thermal limits in Mediterranean ant species: trade-off between mortality risk and foraging performance. Funct Ecol 12:45–55. https://doi.org/10.1046/j.1365-2435.1998.00160.x

Chen Y, Serteyn L, Wang Z, He KL, Francis F (2019) Reduction of plant suitability for corn leaf aphid (Hemiptera: Aphididae) under elevated carbon dioxide condition. Environ Entomol 48:935–944. https://doi.org/10.1093/EE/NVZ045

Cooper R, Lee H, González JM, Butler J, Vinson SB, Liang H (2009) Lubrication and surface properties of roach cuticle. J Tribol 131:1–4. https://doi.org/10.1115/1.3002327/468682

Couper LI, Sanders NJ, Heller NE, Gordon DM (2021) Multiyear drought exacerbates long-term effects of climate on an invasive ant species. Ecology 102:e03476. https://doi.org/10.1002/ECY.3476

Crossley MS, Smith OM, Davis TS, Eigenbrode SD, Hartman GL, Lagos-Kutz D, Halbert SE, Voegtlin DJ, Moran MD, Snyder WE (2021) Complex life histories predispose aphids to recent abundance declines. Glob Chang Biol 27:4283–4293. https://doi.org/10.1111/GCB.15739

Cushman JH, Addicott JF (1989) Intra- and interspecific competition for mutualists. Oecologia 79:315–321. https://doi.org/10.1007/BF00384310

Czaczkes TJ, Kumar P (2020) Very rapid multi-odour discrimination learning in the ant *Lasius niger.* Insect Soc 67:541–545. https://doi.org/10.1007/S00040-020-00787-0

De Lucia E, Nabity P, Zavala J, Berenbaum M (2012) Climate change: resetting plant-insect interactions. Plant Physiol 160:1677–1685. https://doi.org/10.1104/pp.112.204750

Depa Ł, Kaszyca-Taszakowska N, Taszakowski A, Kanturski M (2020) Ant-induced evolutionary patterns in aphids. Biol Rev 95:1574–1589. https://doi.org/10.1111/BRV.12629

Detrain C, Deneubourg JL (2006) Self-organized structures in a superorganism: do ants "behave" like molecules? Phys Life Rev 3:162–187. https://doi.org/10.1016/j.plrev.2006.07.001

Detrain C, Prieur J (2014) Sensitivity and feeding efficiency of the black garden ant Lasius niger to sugar resources. J Insect Physiol 64:74–80. https://doi.org/10.1016/j.jinsphys.2014.03.010

Detrain C, Deneubourg JL, Pasteels JM (1999) Information processing in social insects. Springer Science & Business Media

Detrain C, Verheggen FJ, Diez L, Wathelet B, Haubruge E (2010) Aphid-ant mutualism: how honeydew sugars influence the behaviour of ant scouts. Physiol Entomol 35:168–174. https://doi.org/10.1111/j.1365-3032.2010.00730.x

Detrain C, Fichaux M, Verheggen F (2017) Tuned protection of aphids by ants against a predatory hoverfly. Ecol Entomol 42:235–244. https://doi.org/10.1111/EEN.12378

Diamond SE, Sorger DM, Hulcr J, Pelini SL, Del Toro I, Hirsch C, Oberg E, Dunn RR (2012) Who likes it hot? A global analysis of the climatic, ecological, and evolutionary determinants of warming tolerance in ants. Glob Chang Biol 18:448–456. https://doi.org/10.1111/j.1365-2486.2011.02542.x

Diamond SE, Penick CA, Pelini SL, Ellison AM, Gotelli NJ, Sanders NJ, Dunn RR (2013) Using physiology to predict the responses of ants to climatic warming. Integr Comp Biol 53:965–974. https://doi.org/10.1093/ICB/ICT085

Docherty M, Wade FA, Hurst DK, Whittaker JB, Lea PJ (1997) Responses of tree sap-feeding herbivores to elevated CO2. Glob Chang Biol 3:51–59. https://doi.org/10.1046/j.1365-2486.1997.00096.x

Doebeli M, Knowlton N (1998) The evolution of interspecific mutualisms. Proc Natl Acad Sci 95:8676–8680. https://doi.org/10.1073/pnas.95.15.8676

Douglas AE (1998) Nutritional interactions in insect-microbial symbioses: aphids and their symbiotic bacteria *Buchnera*. Annu Rev Entomol 43:17–37. https://doi.org/10.1146/ANNUREV.ENTO.43.1.17

Douglas AE (2009) Endosymbionts and intracellular parasites. In: Schaechter M (ed) Desk encyclopedia of microbiology. Academic Press, pp 391–405

El-Ziady S, Kennedy J (1956) Beneficial effects of the common garden ant, *Lasius niger* on the black bean aphid *Aphis fabae* scopoli. Proc R Entomol Soc London 31:61–65

Endo S, Itino T (2012) The aphid-tending ant Lasius fuji exhibits reduced aggression toward aphids marked with ant cuticular hydrocarbons. Popul Ecol 54:405–410. https://doi.org/10.1007/S10144-012-0314-9

Endo S, Itino T (2013) Myrmecophilous aphids produce cuticular hydrocarbons that resemble those of their tending ants. Popul Ecol 55:27–34. https://doi.org/10.1007/S10144-012-0355-0

Engel V, Fischer MK, Wäckers FL, Völkl W (2001) Interactions between extrafloral nectaries, aphids and ants: are there competition effects between plant and homopteran sugar sources? Oecologia 129:577–584. https://doi.org/10.1007/S004420100765

Ferrari J, Vavre F (2011) Bacterial symbionts in insects or the story of communities affecting communities. Philos Trans R Soc B Biol Sci 366:1389–1400. https://doi.org/10.1098/RSTB.2010.0226

Fischer MK, Shingleton AW (2001) Host plant and ants influence the honeydew sugar composition of aphids. Funct Ecol 15:544–550. https://doi.org/10.1046/J.0269-8463.2001.00550.X

Fischer CY, Lognay GC, Detrain C, Heil M, Grigorescu A, Sabri A, Thonart P, Haubruge E, Verheggen FJ (2015) Bacteria may enhance species association in an ant–aphid mutualistic relationship. Chemoecology 25:223–232. https://doi.org/10.1007/s00049-015-0188-3

Fischer CY, Detrain C, Thonart P, Haubruge E, Francis F, Verheggen FJ, Lognay GC (2017) Bacteria may contribute to distant species recognition in ant–aphid mutualistic relationships. Insect Sci 24:278–284. https://doi.org/10.1111/1744-7917.12302

Freires ASA, Ferreira CAS, Nascimento G, Arnan X (2023) Carbohydrate-rich diet increases critical thermal maximum in ants. Entomol Exp Appl 171:240–247. https://doi.org/10.1111/EEA.13268

Gibbs A, Mousseau TA (1994) Thermal acclimation and genetic variation in cuticular lipids of the lesser migratory grasshopper (*Melanoplus sanguinipes*): effects of lipid composition on biophysical properties. Physiol Zool 67:1523–1543. https://doi.org/10.1086/physzool.67.6.30163910

Gibbs AG, Rajpurohit S (2010) Cuticular lipids and water balance. In: Blomquist GJ, Bagnères AG (eds) Insect hydrocarbons: biology, biochemistry, and chemical ecology. Cambridge University Press, Cambridge, pp 100–120

Greene MJ, Gordon DM (2003) Cuticular hydrocarbons inform task decisions. Nature 423:32–32. https://doi.org/10.1038/423032a

Hamann E, Blevins C, Franks SJ, Jameel MI, Anderson JT (2021) Climate change alters plant–herbivore interactions. New Phytol 229:1894–1910. https://doi.org/10.1111/NPH.17036

Hendrix DL, Wei YA, Leggett JE (1992) Homopteran honeydew sugar composition is determined by both the insect and plant species. Comp Biochem Physiol 101:23–27. https://doi.org/10.1016/0305-0491(92)90153-I

Hentley WT, Wade RN (2016) Global change, herbivores and their natural enemies. In: Johnson SN, Jones TH (eds) Global climate change and terrestrial invertebrates. John Wiley & Sons, Ltd, Chichester, pp 177–200

Hentley WT, Vanbergen AJ, Hails RS, Jones TH, Johnson SN (2014) Elevated atmospheric CO2 impairs aphid escape responses to predators and conspecific alarm signals. J Chem Ecol 40:1110–1114. https://doi.org/10.1007/s10886-014-0506-1

Herre EA, Knowlton N, Mueller UG, Rehner SA (1999) The evolution of mutualisms: exploring the paths between conflict and cooperation. Trends Ecol Evol 14:49–53. https://doi.org/10.1016/S0169-5347(98)01529-8

Heyworth ER, Smee MR, Ferrari J (2020) Aphid facultative symbionts aid recovery of their obligate symbiont and their host after heat stress. Front Ecol Evol 8:505867. https://doi.org/10.3389/FEVO.2020.00056

Holopainen JK, Blande JD, Sorvari J (2020) Functional role of extrafloral nectar in Boreal forest ecosystems under climate change. Forests 11:67. https://doi.org/10.3390/F11010067

Hullé M, Coeur d'Acier A, Bankhead-Dronnet S, Harrington R (2010) Aphids in the face of global changes. Comptes Rendus 333:497–503. https://doi.org/10.1016/j.crvi.2010.03.005

IPCC (2023) Climate change 2023: synthesis report. In: Shukla PR, Skea J, Calvo Buendía A et al. (eds) Contribution of working groups I, II, and III to the sixth assessment report of the Intergovernmental Panel on Climate Change. IPCC

Jerbi-Elayed M, Foray V, Tougeron K, Grissa-Lebdi K, Hance T (2021) Developmental temperature affects life-history traits and heat tolerance in the aphid parasitoid *Aphidius colemani*. Insects 12:852. https://doi.org/10.3390/INSECTS12100852

Jerbi-Elayed M, Tougeron K, Grissa-Lebdi K, Hance T (2022) Effect of developmental temperatures on Aphidius colemani host-foraging behavior at high temperature. J Therm Biol 103:103140. https://doi.org/10.1016/J.JTHERBIO.2021.103140

Jiggins C, Majerus M, Gough U (1993) Ant defence of colonies of *Aphis fabae* Scopoli (Hemiptera: Aphididae), against predation by ladybirds. Br J Entomol Nat Hist 6:129–137

Johnson SN, Staley JT, McLeod FAL, Hartley SE (2011) Plant-mediated effects of soil invertebrates and summer drought on above-ground multitrophic interactions. J Ecol 99:57–65. https://doi.org/10.1111/J.1365-2745.2010.01748.X

Jossart M, Hance T, Detrain C (2025) Infected connections: Unraveling the impact of a bacterial symbiont on ant-aphid partnership. Plos one, 20(6):e0326875. https://doi.org/10.1371/journal.pone.0326875

Kansman J, Nalam V, Nachappa P, Finke D (2020) Plant water stress intensity mediates aphid host choice and feeding behaviour. Ecol Entomol 45:1437–1444. https://doi.org/10.1111/een.12928

Kaszyca-Taszakowska N, Depa Ł (2022) Microbiome of the aphid genus *Dysaphis* Börner (Hemiptera: Aphidinae) and its relation to ant attendance. Insects 13:1089. https://doi.org/10.3390/INSECTS13121089

Katayama N, Suzuki N (2002) Cost and benefit of ant attendance for *Aphis craccivora* (Hemiptera: Aphididae) with reference to aphid colony size. Can Entomol 134:241–249. https://doi.org/10.4039/ENT134241-2

Katayama N, Hembry DH, Hojo MK, Suzuki N (2013) Why do ants shift their foraging from extrafloral nectar to aphid honeydew? Ecol Res 28:919–926. https://doi.org/10.1007/S11284-013-1074-5

Kessler D, Gase K, Baldwin IT (2008) Field experiments with transformed plants reveal the sense of floral scents. Science (80-) 321:1200–1202. https://doi.org/10.1126/SCIENCE.1160072

Kiers TE, Palmer TM, Ives AR, Bruno JF, Bronstein JL (2010) Mutualisms in a changing world: an evolutionary perspective. Ecol Lett 13:1459–1474. https://doi.org/10.1111/j.1461-0248.2010.01538.x

Kikuchi Y, Tada A, Musolin DL, Hari N, Hosokawa T, Fujisaki K, Fukatsu T (2016) Collapse of insect gut symbiosis under simulated climate change. MBio 7:1–8. https://doi.org/10.1128/mBio.01578-16

Kocmánková E, Trnka M, Eitzinger J, Formayer H, Dubrovský M, Semerádová D, Åalud Z, Juroch J, Možný M (2010) Estimating the impact of climate change on the occurrence of selected pests in the Central European region. Clim Res 44:95–105. https://doi.org/10.3354/CR00905

Kremer JMM, Nooten SS, Cook JM, Ryalls JMW, Barton CVM, Johnson SN (2018) Elevated atmospheric carbon dioxide concentrations promote ant tending of aphids. J Anim Ecol 87:1475–1483. https://doi.org/10.1111/1365-2656.12842

Lang C, Menzel F (2011) *Lasius niger* ants discriminate aphids based on their cuticular hydrocarbons. Anim Behav 82:1245–1254. https://doi.org/10.1016/j.anbehav.2011.08.020

Laubmeier AN, Tabassum N, Tenhumberg B (2023) Temperature fluctuation alters optimal predator community composition for anticipated biological control. Front Ecol Evol 10:998396. https://doi.org/10.3389/FEVO.2022.998396

Lee H, Calvin K, Dasgupta D, Krinner G, Mukherji A, Thorne P, Ruane A (2023) Climate change 2023 synthesis report summary for policymakers

Leroy PD, Sabri A, Heuskin S, Thonart P, Lognay G, Verheggen FJ, Francis F, Brostaux Y, Felton GW, Haubruge E (2011) Microorganisms from aphid honeydew attract and enhance the efficacy of natural enemies. Nat Commun 2:348. https://doi.org/10.1038/ncomms1347

Leybourne DJ, Preedy KF, Valentine TA, Bos JIB, Karley AJ (2021) Drought has negative consequences on aphid fitness and plant vigor: insights from a meta-analysis. Ecol Evol 11:11915–11929. https://doi.org/10.1002/ECE3.7957

Ma C, Sen, Hau B, Poehling HM (2004) The effect of heat stress on the survival of the rose grain aphid, *Metopolophium dirhodum* (Hemiptera: Aphididae). Eur J Entomol 101:327–331. https://doi.org/10.14411/eje.2004.042

Ma G, Rudolf VHW, Ma C (2015) Extreme temperature events alter demographic rates, relative fitness, and community structure. Glob Chang Biol 21:1794–1808. https://doi.org/10.1111/GCB.12654

Mailleux AC, Deneubourg JL, Detrain C (2003) Regulation of ants' foraging to resource productivity. Proc R Soc 270(2003):1609–1616. https://doi.org/10.1098/rspb.2003.2398

Martin S, Drijfhout F (2009) A review of ant cuticular hydrocarbons. J Chem Ecol 35:1151–1161. https://doi.org/10.1007/S10886-009-9695-4

Menzel F, Blaimer BB, Schmitt T (2017) How do cuticular hydrocarbons evolve? Physiological constraints and climatic and biotic selection pressures act on a complex functional trait. Proc R Soc B Biol Sci 284:20161727. https://doi.org/10.1098/RSPB.2016.1727

Menzel F, Zumbusch M, Feldmeyer B (2018) How ants acclimate: impact of climatic conditions on the cuticular hydrocarbon profile. Funct Ecol 32:657–666. https://doi.org/10.1111/1365-2435.13008

Mondor EB, Tremblay MN, Awmack CS, Lindroth RL (2004) Divergent pheromone-mediated insect behaviour under global atmospheric change. Glob Chang Biol 10:1820–1824. https://doi.org/10.1111/j.1365-2486.2004.00838.x

Montllor CB, Maxmen A, Purcell AH (2002) Facultative bacterial endosymbionts benefit pea aphids *Acyrthosiphon pisum* under heat stress. Ecol Entomol 27:189–195. https://doi.org/10.1046/J.1365-2311.2002.00393.X

Mooney E, Davidson B, Den Uyl J, Mullins M, Medina E, Nguyen P, Owens J (2019) Elevated temperatures alter an ant-aphid mutualism. Entomol Exp Appl 167:891–905. https://doi.org/10.1111/EEA.12839

Moran NA, McCutcheon JP, Nakabachi A (2008) Genomics and evolution of heritable bacterial symbionts. Annu Rev Genet 42:165–190. https://doi.org/10.1146/ANNUREV.GENET.41.110306.130119

Moreno-Delafuente A, Morales I, Garzo E, Fereres A, Viñuela E, Medina P (2021) Changes in melon plant phytochemistry impair *Aphis gossypii* growth and weight under elevated CO2. Sci Rep 11:1–12. https://doi.org/10.1038/s41598-021-81167-x

Müller CB, Godfray HCJ (1999) Indirect interactions in aphid–parasitoid communities. Popul Ecol 41:93–106. https://doi.org/10.1007/PL00011986

Nachappa P, Culkin CT, Saya PM, Han J, Nalam VJ (2016) Water stress modulates soybean aphid performance, feeding behavior, and virus transmission in soybean. Front Plant Sci 7:181744. https://doi.org/10.3389/fpls.2016.00552

Nascimento G, Câmara T, Arnan X (2022) Critical thermal limits in ants and their implications under climate change. Biol Rev 97:1287–1305. https://doi.org/10.1111/BRV.12843

Nault LR, Edwards LJ, Styer WE (1973) Aphid alarm pheromones: secretion and reception. Environ Entomol 2:101–105. https://doi.org/10.1093/EE/2.1.101

Nault L, Montgomery M, Bowers W (1976) Ant-aphid association: role of aphid alarm pheromone. Science (80-) 192:1349–1351. https://doi.org/10.1126/science.1273595

Ness J, Mooney K, Lach L (2010) Ants as mutualists. In: Ant ecology. Oxford University Press, pp 97–114

Newman JA, Gibson DJ, Hickam E, Lorenz M, Adams E, Bybee L, Thompson R (1999) Elevated carbon dioxide results in smaller populations of the bird cherry-oat aphid *Rhopalosiphum padi*. Ecol Entomol 24:486–489. https://doi.org/10.1046/j.1365-2311.1999.00210.x

Oehme V, Högy P, Zebitz CPW, Fangmeier A (2013) Effects of elevated atmospheric CO2 concentrations on phloem sap composition of spring crops and aphid performance. J Plant Interact 8:74–84. https://doi.org/10.1080/17429145.2012.736200

Oliver KM, Degnan PH, Burke GR, Moran NA (2010) Facultative symbionts in aphids and the horizontal transfer of ecologically important traits. Annu Rev Entomol 55:247–266. https://doi.org/10.1146/ANNUREV-ENTO-112408-085305

Ord TJ (2023) Drought-induced relocation of ant colonies and its consequences for the long-term spatial ecology of a population under stress. Funct Ecol 37:2231–2245. https://doi.org/10.1111/1365-2435.14383

Parr CL, Bishop TR (2022) The response of ants to climate change. Glob Chang Biol 28:3188–3205. https://doi.org/10.1111/GCB.16140

Pons I, Renoz F, Noël C, Hance T (2019) New insights into the nature of symbiotic associations in aphids: infection process, biological effects, and transmission mode of cultivable *Serratia symbiotica* bacteria. Appl Environ Microbiol 85. https://doi.org/10.1128/AEM.02445-18

Pringle EG, Akçay E, Raab TK, Dirzo R, Gordon DM (2013) Water stress strengthens mutualism among ants, trees, and scale insects. PLoS Biol 11. https://doi.org/10.1371/journal.pbio.1001705

Pringle EG, Novo A, Ableson I, Barbehenn RV, Vannette RL (2014) Plant-derived differences in the composition of aphid honeydew and their effects on colonies of aphid-tending ants. Ecol Evol 4:4065–4079. https://doi.org/10.1002/ECE3.1277

Rafferty NE, Caradonna PJ, Bronstein JL (2015) Phenological shifts and the fate of mutualisms. Oikos 124:14–21. https://doi.org/10.1111/OIK.01523

Rall BC, Vucic-Pestic O, Ehnes RB, Emmerson M, Brose U (2010) Temperature, predator–prey interaction strength and population stability. Glob Chang Biol 16:2145–2157. https://doi.org/10.1111/J.1365-2486.2009.02124.X

Renoz F, Pons I, Vanderpoorten A, Bataille G, Noël C, Foray V, Pierson V, Hance T (2019) Evidence for gut-associated *Serratia symbiotica* in wild aphids and ants provides new perspectives on the evolution of bacterial mutualism in insects. Microb Ecol 78:159–169. https://doi.org/10.1007/S00248-018-1265-2

Robinson A, Inouye DW, Ogilvie JE, Mooney EH (2017) Multitrophic interactions mediate the effects of climate change on herbivore abundance. Oecologia 185:181–190. https://doi.org/10.1007/s00442-017-3934-0

Rodrigues YK, Beldade P (2020) Thermal plasticity in insects' response to climate change and to multifactorial environments. Front Ecol Evol 8:271. https://doi.org/10.3389/fevo.2020.00271

Roeder KA, Roeder DV, Kaspari M (2018) The role of temperature in competition and persistence of an invaded ant assemblage. Ecol Entomol 43:774–781. https://doi.org/10.1111/EEN.12663

Roeder KA, Roeder DV, Bujan J (2021) Ant thermal tolerance: a review of methods, hypotheses, and sources of variation. Ann Entomol Soc Am 114:459–469. https://doi.org/10.1093/AESA/SAAB018

Römer D, Bollazzi M, Roces F (2018) Carbon dioxide sensing in the social context: leaf-cutting ants prefer elevated CO2 levels to tend their brood. J Insect Physiol 108:40–47. https://doi.org/10.1016/J.JINSPHYS.2018.05.007

Rosenblatt AE, Schmitz OJ (2016) Climate change, nutrition, and bottom-up and top-down food web processes. Trends Ecol Evol 31:965–975. https://doi.org/10.1016/j.tree.2016.09.009

Ryan GD, Sylvester EVA, Shelp BJ, Newman JA (2015) Towards an understanding of how phloem amino acid composition shapes elevated CO2-induced changes in aphid population dynamics. Ecol Entomol 40:247–257. https://doi.org/10.1111/een.12181

Sagata K, Gibb H (2016) The effect of temperature increases on an ant-hemiptera-plant interaction. PLoS One 11:1–14. https://doi.org/10.1371/journal.pone.0155131

Sakata H (1994) How an ant decides to prey on or to attend aphids. Res Popul Ecol 36:45–51. https://doi.org/10.1007/BF02515084

Sakata H, Hashimoto Y (2000) Should aphids attract or repel ants? Effect of rival aphids and extrafloral nectaries on ant-aphid interactions. Popul Ecol 42:171–178. https://doi.org/10.1007/PL00011996

Salt DT, Fenwick P, Whittaker JB (1996) Interspecific herbivore interactions in a high CO2 environment: root and shoot aphids feeding on cardamine. Oikos 77:326. https://doi.org/10.2307/3546072

Seeburger VC, Shaaban B, Schweikert K, Lohaus G, Schroeder A, Hasselmann M (2022) Environmental factors affect melezitose production in honeydew from aphids and scale insects of the order Hemiptera. J Apic Res 61:127–137. https://doi.org/10.1080/00218839.2021.1957350

Shingleton AW, Stern DL (2003) Molecular phylogenetic evidence for multiple gains or losses of ant mutualism within the aphid genus *Chaitophorus*. Mol Phylogenet Evol 26:26–35. https://doi.org/10.1016/S1055-7903(02)00328-7

Shingleton AW, Stern DL, Foster WA (2005) The origin of a mutualism: a morphological trait promoting the evolution of ant-aphid mutualisms. Evolution 59:921–926. https://doi.org/10.1554/04-584

Sprenger PP, Menzel F (2020) Cuticular hydrocarbons in ants (Hymenoptera: Formicidae) and other insects: how and why they differ among individuals, colonies, and species. Myrmecol News 30:1–26. https://doi.org/10.25849/myrmecol.news_030:001

Sprenger PP, Burkert LH, Abou B rengere, Federle W, Menzel F (2018) Coping with the climate: Cuticular hydrocarbon acclimation of ants under constant and fluctuating conditions. J Exp Biol 221. https://doi.org/10.1242/JEB.171488/262847/

Stadler B, Dixon AFG (1998) Costs of ant attendance for aphids. J Anim Ecol 67:454–459. https://doi.org/10.1046/J.1365-2656.1998.00209.X

Stadler B, Dixon AFG (2005) Ecology and evolution of aphid-ant interactions. Annu Rev Ecol Evol Syst 36:345–372. https://doi.org/10.1146/annurev.ecolsys.36.091704.175531

Stadler B, Kindlmann P, Šmilauer P, Fiedler K (2003) A comparative analysis of morphological and ecological characters of European aphids and lycaenids in relation to ant attendance. Oecologia 135:422–430. https://doi.org/10.1007/S00442-003-1193-8

Stiling P, Cornelissen T (2007) How does elevated carbon dioxide (CO2) affect plant-herbivore interactions? A field experiment and meta-analysis of CO2- mediated changes on plant chemistry and herbivore performance. Glob Chang Biol 13:1823–1842. https://doi.org/10.1111/j.1365-2486.2007.01392.x

Styrsky JD, Eubanks MD (2007) Ecological consequences of interactions between ants and honeydew-producing insects. Proc Biol Sci 274:151–164. https://doi.org/10.1098/rspb.2006.3701

Sun Y, Ge F (2011) How do aphids respond to elevated CO2? J Asia Pac Entomol 14:217–220. https://doi.org/10.1016/J.ASPEN.2010.08.001

Sun YC, Jing BB, Ge F (2009) Response of amino acid changes in *Aphis gossypii* (Glover) to elevated CO2 levels. J Appl Entomol 133:189–197. https://doi.org/10.1111/J.1439-0418.2008.01341.X

Sun Y, Su J, Ge F (2010) Elevated CO2 reduces the response of *Sitobion avenae* (Homoptera: Aphididae) to alarm pheromone. Agric Ecosyst Environ 135:140–147. https://doi.org/10.1016/j.agee.2009.09.011

Sun YC, Yin J, Chen FJ, Wu G, Ge F (2011) How does atmospheric elevated CO 2 affect crop pests and their natural enemies? Case histories from China. Insect Sci 18:393–400. https://doi.org/10.1111/j.1744-7917.2011.01434.x

Sun Y, Guo H, Ge F (2016) Plant–aphid interactions under elevated CO2: some cues from aphid feeding behavior. Front Plant Sci 7:180993. https://doi.org/10.3389/FPLS.2016.00502

Tougeron K, Iltis C (2022) Impact of heat stress on the fitness outcomes of symbiotic infection in aphids: a meta-analysis. Proc R Soc B 289. https://doi.org/10.1098/RSPB.2021.2660

Urban O (2003) Physiological impacts of elevated CO2 concentration ranging from molecular to whole plant responses. Photosynthetica 41:9–20. https://doi.org/10.1023/A:1025891825050

Van Baaren J, Lann C, Le, Jm Van Alphen J (2010) Consequences of climate change for aphid-based multi-trophic systems. In: Aphid Biodivers Under Environ Chang, pp 55–68. https://doi.org/10.1007/978-90-481-8601-3_4

Van Emden HF, Harrington R (2017) Aphids as crop pests. CABI Publishing, Wallingford

Van Oudenhove L, Billoir E, Boulay R, Bernstein C, Cerdá X (2011) Temperature limits trail following behaviour through pheromone decay in ants. Naturwissenschaften 98:1009–1017. https://doi.org/10.1007/s00114-011-0852-6

Van Oudenhove L, Boulay R, Lenoir A, Bernstein C, Cerda X (2012) Substrate temperature constrains recruitment and trail following behavior in ants. J Chem Ecol 38:802–809. https://doi.org/10.1007/S10886-012-0130-X

Vantaux A, Van den Ende W, Billen J, Wenseleers T (2011) Large interclone differences in melezitose secretion in the facultatively ant-tended black bean aphid Aphis fabae. J Insect Physiol 57:1614–1621. https://doi.org/10.1016/J.JINSPHYS.2011.08.014

Verheggen FJ, Diez L, Sablon L, Fischer C, Bartram S, Haubruge E, Detrain C (2012) Aphid alarm pheromone as a cue for ants to locate aphid partners. PLoS One 7:e41841. https://doi.org/10.1371/journal.pone.0041841

Villalta I, Rami L, Alvarez-Blanco P, Angulo E, Cerdá X, Boulay R (2020) Environmental and genetic constraints on cuticular hydrocarbon composition and nestmate recognition in ants. Anim Behav 159:105–119. https://doi.org/10.1016/J.ANBEHAV.2019.11.008

Völkl W, Woodring J, Fischer M, Lorenz MW, Hoffmann KH (1999) Ant-aphid mutualisms: the impact of honeydew production and honeydew sugar composition on ant preferences. Oecologia 118:483–491. https://doi.org/10.1007/S004420050751

Vorburger C, Gehrer L, Rodriguez P (2010) A strain of the bacterial symbiont *Regiella insecticola* protects aphids against parasitoids. Biol Lett 6:109–111. https://doi.org/10.1098/RSBL.2009.0642

Wagner D, Tissot M, Gordon D (2001) Task-related environment alters the cuticular hydrocarbon composition of harvester ants. J Chem Ecol 27:1805–1819. https://doi.org/10.1023/A:1010408725464

Way MJ (1963) Mutualism between ants and honeydew producing homopterans. Annu Rev Entomol 8:307–344. https://doi.org/10.1146/annurev.en.08.010163.001515

Wei J, Peng L, He Z, Lu Y, Wang F (2020) Potential distribution of two invasive pineapple pests under climate change. Pest Manag Sci 76:1652–1663. https://doi.org/10.1002/PS.5684

Wilkinson T, Ashford D, Pritchard J, Douglas A (1997) Honeydew sugars and osmoregulation in the pea aphid *Acyrthosiphon pisum*. J Exp Biol 200:2137–2143

Woodring J, Wiedemann R, Fischer MK, Hoffmann KH, Völkl W (2004) Honeydew amino acids in relation to sugars and their role in the establishment of ant-attendance hierarchy in eight species of aphids feeding on tansy (*Tanacetum vulgare*). Physiol Entomol 29:311–319. https://doi.org/10.1111/J.0307-6962.2004.00386.X

Yao I (2014) Costs and constraints in aphid-ant mutualism. Ecol Res 29:383–391. https://doi.org/10.1007/s11284-014-1151-4

Yao I, Akimoto SI (2002) Flexibility in the composition and concentration of amino acids in honeydew of the drepanosiphid aphid *Tuberculatus quercicola*. Ecol Entomol 27:745–752. https://doi.org/10.1046/j.1365-2311.2002.00455.x

Zhang B, Leonard SP, Li Y, Moran NA (2019) Obligate bacterial endosymbionts limit thermal tolerance of insect host species. Proc Natl Acad Sci USA 116:24712–24718. https://doi.org/10.1073/PNAS.1915307116

Zhou AM, Kuang BQ, Gao YR, Liang GW (2015) Sucrose triggers honeydew preference in the ghost ant, *Tapinoma melanocephalum* (Hymenoptera: Formicidae). Florida Entomol 98:1217–1222. https://doi.org/10.1653/024.098.0431

Zhou A, Qu X, Shan L, Wang X (2017) temperature warming strengthens the mutualism between ghost ants and invasive mealybugs. Sci Reports 71(7):1–10. https://doi.org/10.1038/s41598-017-01137-0

Chapter 18
Carbon and Nitrogen Fluxes in Leaf-Cutter Ant Sites in a Changing Climate

Angel S. Fernandez-Bou, Leticia M. Classen-Rodriguez, Philip W. Rundel, Thomas C. Harmon, Michael F. Allen, Carlos L. de la Rosa, and Luitgard Schwendenmann

18.1 Introduction

Leaf-cutter ants (species of *Acromyrmex* and *Atta*) are among the most ubiquitous and conspicuous animals in tropical forests of Central and South America. Their endless parades attract scientists and alarm farmers since they are industrious workers capable of harvesting large quantities of leaf and other plant material (Blanton and Ewel 1985; Wirth et al. 1997; Viana et al. 2004; Costa et al. 2008). They are major herbivores in the Neotropics, and each of their large colonies can harvest hundreds of kilograms of fresh vegetation per year (Hughes and Goulson 2002; Wirth et al. 2003),

A. S. Fernandez-Bou (✉)
SocioEnvironmental and Economic Development Strategies, LLC, Merced, California, USA

SocioEnvironmental and Education Network, Merced, California, USA

Sierra Nevada Research Institute, Department of Civil and Environmental Engineering, University of California, Merced, CA, USA

Union of Concerned Scientists, Berkeley, California, USA
e-mail: afernandezbou@ucmerced.edu

L. M. Classen-Rodriguez
SocioEnvironmental and Economic Development Strategies, Merced, CA, USA

SocioEnvironmental and Education Network, Merced, CA, USA

Sierra Nevada Research Institute, Department of Civil and Environmental Engineering, University of California, Merced, CA, USA

P. W. Rundel
Department of Ecology and Evolutionary Biology, University of California, Los Angeles, CA, USA
e-mail: rundel@ucla.edu

T. C. Harmon
Department of Civil and Environmental Engineering, University of California, Merced, CA, USA
e-mail: tharmon@ucmerced.edu

I. R. Leal et al. (eds.), *Ant Responses to Climate Change*, Entomology in Focus 10,
https://doi.org/10.1007/978-3-032-21127-9_18

representing up to 15% of leaves in their foraging ranges (Urbas da Sternberg et al. 2007, Fig. 18.1). Rather than directly eating the vegetative material they collect, leaf-cutter ants shred and masticate it, mixing it with their body fluids, to create a substrate to feed the symbiotic fungus *Leucoagaricus gongylophorus* (A. Möller) Heim, which produces hyphal nodules to feed the ants (Fig. 18.1). The fungus decomposes the substrate, breaking down toxins and complex carbon compounds, becoming a nutritious food for the colony. This fungus has been the fundamental diet for the leaf-cutter ants for more than 5 million years (Schultz and Brady 2008; Branstetter et al. 2017).

Because of their abundance and activity level, leaf-cutter ants are described as ecosystem engineers, organisms that create and modify habitats by changing the surrounding biotic or abiotic components of the forest ecosystem and regulating the availability of resources for other species (e.g., Jones et al. 1996; Jouquet et al. 2006). For example, leaf-cutter ants create gaps in the forest canopy by harvesting leaves, which changes the microclimate (e.g., light, temperature) and alters plant community composition (Corrêa et al. 2010; Meyer et al. 2011). In addition, the movement of substantial amounts of plant matter underground and the cultivation of fungi (Fig. 18.1) enhance the availability of nitrogen and other nutrients within nests and surrounding soils (Pinto-Tomás et al. 2009; Hudson et al. 2009; Meyer et al. 2013; Farji-Brener and Werenkraut 2015, 2017; Maurice-Lira et al. 2023; Baena et al. 2024). Ant nests are also considered biochemical hot spots due to high carbon and nutrient turnover rates relative to the surrounding soils (Swanson et al. 2019; Nascimento et al. 2024). High turnover of organic matter in leaf-cutter ant nests results in increased emissions of the greenhouse gases carbon dioxide (CO_2), methane (CH_4), and nitrous oxide (N_2O) (Fernandez-Bou et al. 2019).

This chapter examines the role of *Atta* and *Acromyrmex* leaf-cutter ants in the cycling of greenhouse gases in Central and South American forests, along with the potential impact of climate change on it. We begin with a description of nest structure, followed by a section on carbon and nutrient transformation and the role of leaf-cutter ants and associated fungi in biogeochemical cycling. In the last section, we discuss climate change's potential direct and indirect impacts on leaf-cutter ants and biogeochemical processes within nests and surrounding soil. Climate change has the potential to irreversibly alter the environmental conditions under which leaf-cutter ants thrive. This could lead to colony shrinkage or complete disappearance if one or more critical factors for the ants or their associated fungus exceed the tolerance level for their survival or metabolism.

M. F. Allen
Center for Conservation Biology, Department of Microbiology and Plant Pathology, University of California, Riverside, CA, USA
e-mail: mallen@ucr.edu

C. L. de la Rosa
de la Rosa Consultants, Escondido, CA, USA

L. Schwendenmann
Te Kura Mātai Taiao, School of Environment, Waipapa Taumata Rau, University of Auckland, Auckland, Aotearoa, New Zealand
e-mail: l.schwendenmann@auckland.ac.nz

Fig. 18.1 Leaf-cutter ant *Atta cephalotes* at different moments of its activity cycle at the La Selva Research Station, Costa Rica. (**a**) A whole nest of *Atta cephalotes*. (**b**) Leaf fragments dropped in *Atta cephalotes* nest entrances after a rain event. The ants do not pick up dropped leaves from the ground. Climate change is expected to increase the frequency and intensity of rainfall and storm events, resulting in changes in behavior, like foraging, or in environmental conditions, like soil moisture, potentially leading to a higher number of failed nests and colony migrations. (**c**) *Atta cephalotes* worker cutting a leaf fragment, using its jaws like scissors. (**d, e**) *Atta cephalotes* worker carrying a leaf fragment with a minimum (smallest cast) riding along and carrying a clump of dirt from the inside of the nest, part of their underground tunnel and chamber-making activities. (**f**) *Atta cephalotes* workers tending the fungus garden. Visible are the masticated leaf fragments and the white fungal hyphae

18.2 Leaf-Cutter Ant Ecosystems

18.2.1 Nest Structure

The nests of leaf-cutter ants can be massive, extending over hundreds of square meters above ground and up to 7 meters below ground, and they can contain hundreds to thousands of chambers to cultivate fungal gardens or dispose of refuse (Table 18.1, Figs. 18.1 and 18.2). Nest construction can result in a turnover of soil up to 10,000 kg ha^{-1} $year^{-1}$, which may surpass the impact on a global scale from termites (up to 5500 kg ha^{-1} $year^{-1}$) (Whitford and Eldridge 2013). Soil turnover from nest construction by leaf-cutter ants mixes organic-rich material from the upper soil layer with nutrient-poor soil, leading to soil heterogeneity. Nest construction changes physical soil properties such as porosity and bulk density, affecting soil aeration and water fluxes (e.g., Nascimento et al. 2024). The redistribution of organic-rich material and changes in soil physical properties potentially stimulate the breakdown of organic matter, increasing soil CO_2 emissions around nests.

The extensive network of nest tunnels and chambers actively regulates the nest environment to maintain optimal conditions for the growth of the fungus. Leaf-cutter ants exhibit sophisticated behavioral responses to internal microclimate conditions (Swanson 2017; Fernandez-Bou et al. 2020b; Nascimento et al. 2024). Nest architecture, including specialized turrets, plays a crucial role in nest ventilation.

Table 18.1 Surface area of the interior of the nest in *Atta* species

Species	Nest area (m^2)	Nest int. sf. area (m^2)	Maximum depth (m)	# chamber	# vent
A. laevigata[1,2]	26 to 67	84 to 563	4 to 7	1149 to 7864	60
A. capiguara[3,13,15]	19[3]; 65 to 221[15]	13[3]	1.7[3]; 3[13]; 3.35 to 5.85[15]	112[3]; 32[13]	21[3]
A. texana[4]	51	36	4.6	93	
A. bisphaerica[5,6,14]	19 to 74[5]; 42[14]	8 to 38[5]	2 to 2.5[5]; 2.9[6]; 3.3[14]	58 to 285[5]; 104 [14]	132[14]
A. sexdens[7,8,9]	56[7]	69	4[8,9]	390[7]	800[7]
A. vollenweideri[9]	50 to 78.5	127 to 170		250 to 3000	30 to 187
A. cephalotes[10,11,12]	30[11] to 67[12]		2.5[10]		

Nest internal surface area estimated from [1] Moreira et al. (2004); [2] Pereira-da-Silva (1975); [3] Mariconi et al. (1961); [4] Moser (2006); [5] Moreira et al. (2004); [6] Forti et al. (2018)); [7] Autuori in Gonçalves (1942); [8] Stahel and Geijskes (1941); [9] Jonkman (1980); [10] Stahel and Geijskes (1941); [11] Perfecto and Vandermeer (1993); [12] Fernandez-Bou et al. (2019); [13] Farias et al. (2020); [14] Farias et al. (2020); [15] Forti et al. (2017). To estimate the tunnel length, we assumed one chamber for every 25 cm of tunnels (based on observation of the images from the literature cited and personal experience with *Atta* nest excavation). *Nest int. sf. area* is the nest's internal surface area. The surface area of the chambers and tunnels was calculated using the measurements given or suggested by the authors of each work. *n* is the number of nests evaluated. Source: adapted from Fernandez-Bou (2019) and Swanson et al. (2019)

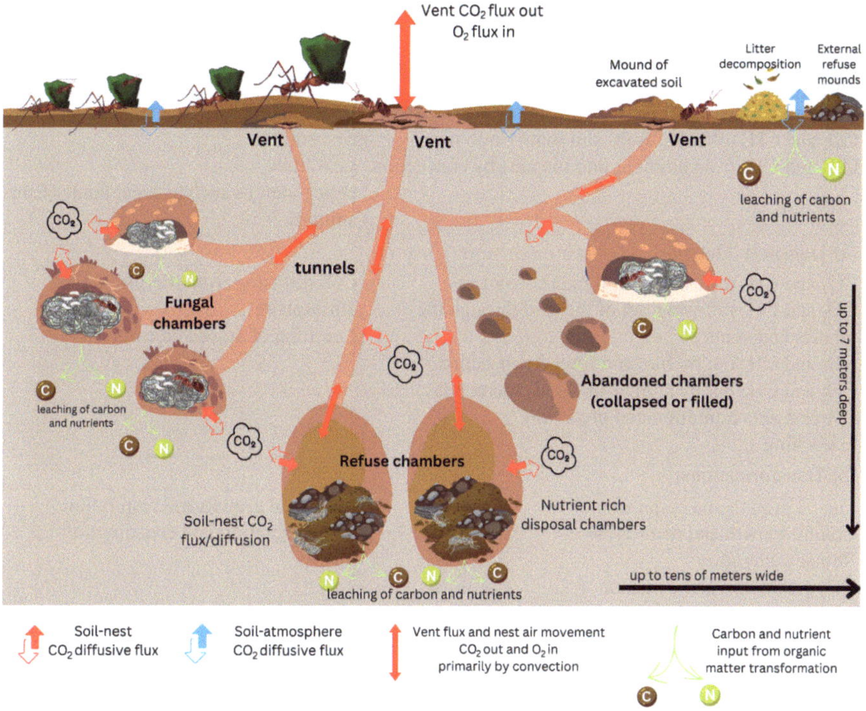

Fig. 18.2 Conceptual diagram of CO_2 exchange and transport pathways in *Atta cephalotes* nests, including (1) soil surface CO_2 efflux (diffusion), (2) nest vent CO_2 flux (convection and diffusion), where vent CO_2 stems from nest production (primarily fungal activity and refuse decay), and (3) soil-nest CO_2 and dissolved organic and inorganic carbon efflux. (Adapted from Fernandez-Bou et al. (2019) and Nascimento et al. (2024). Art by Leticia Classen-Rodriguez)

Wind-induced ventilation, primarily driven by surface winds, is a key mechanism for regulating the nest enviroment. This process is influenced by temperature gradients between the nest interior and the external environment, which are susceptible to alterations due to climate change.

18.2.2 *Carbon and Nitrogen Production and Transformation in Leaf-Cutter Ant Nests*

Understanding how leaf-cutter ants influence the soil around them requires investigating the carbon and nutrient cycles within their nests and nearby soil (Fig. 18.2 and Table 18.2). Leaf-cutter ants collect large quantities of nutrient-rich plant material, primarily leaves but may also include flowers and fruits, and deposit it in their nests (Hudson et al. 2009; Meyer et al. 2013; Farji-Brener and Werenkraut 2015; Maurice-Lira et al. 2023). Within leaf-cutter ant colonies, foraged plant material

Table 18.2 Carbon-related processes inside an *Atta* sp. nest and its surrounding soil

Inside the nest structure	In the soil around the nest
(a) Transport: Inputs	
Vegetative material CO_2 and CH_4 diffusing from soil matrix into nest CO_2 and CH_4 from ambient into the nest by convection	Root and hyphae growth toward the nest soil Leaf litter Other sources such as fecal matter from animals
(a) Transport: Outputs	
CO_2 and CH_4 diffusing to soil matrix from nest CO_2 and CH_4 diffusing out of the nest through the tunnels and vents CO_2 and CH_4 out by free and forced convection CO_2 and CH_4 out by animal-induced convection Off-nest ant respiration and predation C leaching	CO_2 diffusing from the soil to the atmosphere Leaching of dissolved carbon
(b) Transformations	
Fungal growth and respiration Colony growth and respiration Refuse generation CH_4 generation and consume	Microbial growth and respiration CH_4 generation and consumption
(c). Pools	
Fungus microbial biomass Ant biomass Vegetative material Refuse CO_2 and CH_4 in nest air	Root biomass, hyphae biomass Microbial biomass Dissolved carbon CO_2 and CH_4 in the soil air

Our results, estimations, and the literature suggest that the main transport mechanisms are the input of vegetation and the CO_2 out by convection. Source: adapted from Fernandez-Bou (2019)

serves as the primary substrate for the growth of the symbiotic fungus (Fig. 18.1) and hyphal production. Inside the nest, carbon is produced and transformed, driven by fungal growth, ant colony development, refuse decomposition, microbial activity in the nest and surrounding soil matrix, and CH_4 generation and consumption (Fig. 18.2). Carbon and nitrogen production and transformation within nests are also influenced by the preferential growth of plant roots and arbuscular mycorrhizal (AM) hyphae in the nutrient-rich nest environment. For example, at La Selva Research Station, part of the Organization for Tropical Studies, in Costa Rica, although the standing crop of fine roots and fungi is similar between nests and control samples, the turnover rates are much higher in nests due to enhanced oxygen levels and lower soil water content (Swanson et al. 2019). This high turnover likely contributes to greater microbial activity and carbon cycling in leaf-cutter ant nests compared to non-nest soils. Given that leaf-cutter ants provide continuous input of fresh leaves into fungal chambers, the relatively low solubility of CO_2, the CO_2 concentration in the nest chambers, tunnels, and vents can surpass 50,000 parts per million in volume (Table 18.3).

Table 18.3 CO_2 concentration in nests of *Atta* species

Species	CO_2 concentration in the nest and comments
Atta vollenweideri[1]	0.78% at 0.6 m; 1.9% at 2.1 m; 2.8% as CO_2 increases with depth
Atta laevigata[2]	>2% in the openings; ~4.5% in fungal chambers
Atta capiguara[2]	>2% in the openings; ~4.5% in fungal chambers
Atta cephalotes[3,4]	0.5% to 2% inside five different nest vents 0.31% to 1.02% at 10 to 20 cm inside vents
Atta sexdens[5]	1.40%, ranging from 1% to 2.5% (early colonies)

[1] Kleineidam and Roces (2000); [2] Bollazzi et al. (2012); [3] Harmon et al. (2015); [4] Fernandez-Bou et al. (2019); [5] Sousa et al. (2021)

Understanding the localized greenhouse gas emissions from leaf-cutter ant nests and scaling these observations to broader landscapes are crucial for accurately estimating their impact on regional and global carbon cycles. For example, in the tropical primary forest of Costa Rica at La Selva ResearchStation, the average nest surface area is 67 m^2, with 32 vents averaging 0.00021 m^2 each, resulting in a vent-to-nest area ratio of about 1:10,000. While vents emit substantially higher CO_2 concentrations, their overall impact on soil CO_2 emissions is relatively small when scaled by area. A reasonable range of soil CO_2 efflux rates is 4 kg CO_2 m^{-2} $year^{-1}$ to 7 kg CO_2 m^{-2} $year^{-1}$ (Schwendenmann and Veldkamp 2006). Nest surface areas range from 30 m^2 to 70 m^2 (Perfecto and Vandermeer 1993; Wirth et al. 2003), and the soil ground surface of a nest emits 120 kg CO_2 to 490 kg CO_2 annually. For vent efflux, based on an average vent efflux value of 2.1×10^4 kg CO_2 m^{-2} $year^{-1}$ and the frequency and geometry of the vents, the total vent CO_2 emissions are about 72 kg CO_2 per nest per year. This suggests that an average *A. cephalotes* nest area emits around 200 kg to 600 kg CO_2 per year, 15% to 60% more than an equivalent soil area away from nests in a lowland tropical forest. At La Selva, at least 1.2% of the forest surface is occupied by *A. cephalotes* nests. This translates to *A. cephalotes* contributing 0.2% to 0.7% of total CO_2 emissions in this Neotropical rainforest.

Other significant greenhouse gases are produced in forests besides CO_2, such as methane (CH_4, Mehring et al. 2021, Soper et al. 2019), which has a warming potential significantly greater than that of CO_2 (Dalal and Allen 2008). Methane is produced under low O_2 (anoxic) conditions through methanogenesis, where organisms respire CH_4 instead of CO_2 (Aronson et al. 2019). However, leaf-cutter ant nests are specially constructed to maintain oxygenated conditions, which support methanotrophic bacteria that consume CH_4 and release CO_2 instead (Aronson et al. 2019; Allen et al. 2023; Shulman et al. 2024). Even so, leaf-cutter ant nests have been identified as hotspots for methane (CH_4) emissions, further contributing to their greenhouse gas footprint.

Waste material (e.g., residue from the fungus garden, dead ants, plant refuse, and other colony detritus) is stored in separate underground chambers or external piles isolated from the fungus garden (Hart and Ratnieks 2002; Farji-Brener et al. 2016). Analysis of refuse dumps has shown high concentrations of carbon and nitrogen, high decomposition rates, and diverse microbial communities (da Sternberg et al. 2007; Soper et al. 2019). High nitrogen concentrations and anoxic conditions in refuse piles have been shown to provide ideal conditions for very high rates of N_2O production (Soper et al. 2019).

In summary, the activities of leaf-cutter ants significantly impact soil carbon and nitrogen dynamics. Harvesting vegetation and cultivating fungus introduce substantial organic material into the soil. The decomposition of this material, along with fungal and ant respiration, enhances CO_2 production within the nest. In addition, mixing soil layers during nest construction alters soil aeration and temperature, influencing nutrient turnover rates. These processes create biogeochemical hotspots that can affect broader ecosystem nutrient cycles.

18.2.3 Gas and Nutrient Flows

An important feature of leaf-cutter ant nests is the extensive network of tunnels and chambers maintained via continual excavation. This network actively regulates internal environmental conditions of relative humidity, temperature, and gas concentrations (CO_2, CH_4, N_2O, and water vapor). Soil pipes provide a gas or water transport pathway to or between nests and outside soils (Fig. 18.2).

In addition, vascular plant roots and fungal hyphae may provide functional gas transport pathways to and from nests. Vascular plant roots and fungal hyphal growth into leaf-cutter ant nests can channel more significant amounts of CO_2 and other gases toward well-ventilated nests, releasing CO_2 produced by the nest along with CO_2 transported laterally from adjacent soil. Oxygen diffuses in the reverse direction, supporting root and hyphal growth. Bottom-up canopy gaps are often found directly above leaf-cutter ant nests (Farji-Brener 2000), creating openings that increase light transmission, wind, and temperature (Meyer et al. 2013). By altering how nutrients, water, and gases are transported through the soil, leaf-cutter ants create patchiness in CO_2 fluxes that may affect estimates of ecosystem carbon balance.

18.2.3.1 Seasonality and Diel Patterns

The presence of leaf-cutter ant nests can influence the seasonality of soil CO_2 concentration. For example, at La Selva, measuring concentrations at monthly intervals over 2.5 years, we found that nest soils had significantly lower soil CO_2 concentrations during wet periods than non-nest soils (Fernandez-Bou et al. 2019). Nest soils exhibited lower CO_2 concentrations for identical precipitation amounts than non-nest soils. Based on soil CO_2 production estimates and soil surface CO_2 efflux measurements, we determined that the lower CO_2 concentrations in nest soils were not due to lower CO_2 production or higher soil surface CO_2 efflux rates (that would deplete the soil matrix of CO_2). Using flow-through chambers to measure nest vent CO_2 efflux, we found that their average emissions per unit area were four orders of magnitude greater than those from the soil surface. From this, we reasoned that the nest ventilation efflux and the large nest-soil interfacial surface area maintain low

CO_2 concentrations in the nest (relative to adjacent soils) to promote diffusive gas transfer, depleting the nearby soil CO_2 concentrations (Fernandez-Bou et al. 2019).

Nest soils also present different diel soil CO_2 concentration patterns compared with non-nest soils (Fig. 18.3, Fernandez-Bou 2019). The nest ventilation in *A. cephalotes* at La Selva is driven by free convection (when warm, moist air rises out of the nest due to its lower density relative to the ambient). This transport of CO_2 out of the nest decreases the CO_2 concentration in the nest air following a diel pattern driven by the external temperature (since nest temperature is relatively constant). Therefore, at night (when the outside temperature is lower), the CO_2 efflux from the nest vents is more prominent than during the day, following a diel pattern of CO_2 emissions. The evidence from La Selva suggests that the diel pattern shown by the soil CO_2 concentration near the nest is due to enhanced ventilation and gas exchange with the atmosphere through the nest's internal structure in the short term. In general, the diel pattern is more apparent in nest soils than in non-nest soils, which suggests additional soil ventilation provided by the adjacent nest structure. In addition, nest soils at La Selva were systematically around 1 °C warmer than non-nest soils, which can enhance convection and diffusion dynamics. The higher temperature in nest soils can be attributed to the metabolic activity of the fungal and refuse chambers in the nest. In 2017, during a nest excavation, we measured fungal temperature of 27.2 °C on average (three gardens, five measurements per garden), which was always warmer than the surrounding soil, which was 26.7 °C. This agrees with the measurements obtained by the sensors in the nest and explains the warmer temperatures in leaf-cutter ant nest soils. The concentration of O_2 was consistently higher in the nest than in non-nest soils at all the studied depths between 2 cm and 50 cm (Fernandez-Bou 2019).

18.2.3.2 Ventilation Through Free, Forced, and Animal-Driven Convection

The most important ventilation mechanisms in leaf-cutter ant nests are forced and free convection. Forced convection occurs when the wind causes changes in pressure at the mouth of nest vents, which can lead to the suction of the internal air toward the atmosphere if the pressure is negative (Kleineidam et al. 2001; Bollazzi et al. 2012) or can push atmospheric air into the nest if the pressure is positive.

Free convection occurs when there are differences in air density across the nest, leading to air movement. Changes in air density are primarily due to temperature differences (warmer air rises over cooler air). However, density differences can also occur due to differences in the molecular composition of the air. Water vapor is the gas that varies the most in the atmosphere, although humidity inside the nest is usually higher than in the atmosphere. Although free convection is a less efficient mechanism for nest ventilation than forced convection, it can be a significant mechanism for gas transport when cool, dense gas overlies warm, less dense gas (Fernandez-Bou et al. 2019). Relative humidity gradients can also impact air density within the nest since water molecules have lower molecular weights than dry air

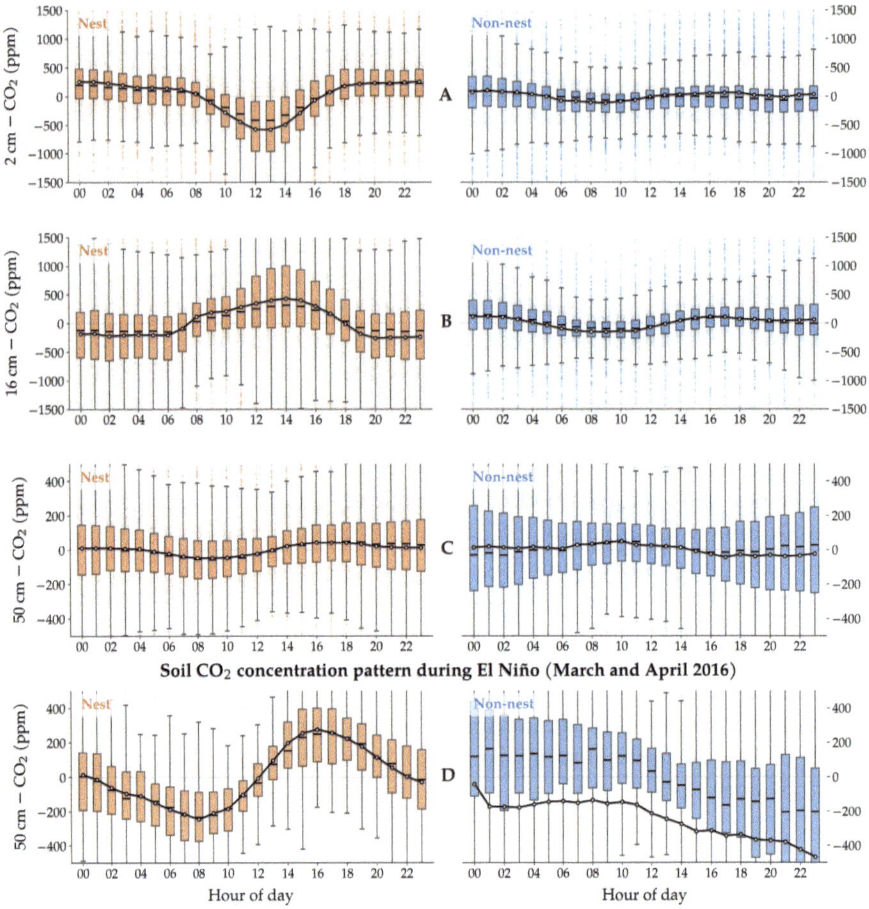

Fig. 18.3 Example from La Selva Research Station (Costa Rica) of the diel soil CO_2 concentration pattern in nest soils (left, orange) and non-nest soils (right, blue) calculated as the difference between the current soil CO_2 concentration and the average soil CO_2 concentration for the day. (**a**) represents the 2 cm depth, where the nest soil CO_2 concentration is lower during the warmer hours of the day. (**b**) depicts the 16 cm depth, with soil CO_2 concentration peaking after the warmest period of the day, when the soil is warmest, slightly lagging ambient temperature. (**c, d**) show soil CO_2 concentrations at 50 cm depth for the whole data series (**c**) and during El Niño conditions (**d**), respectively. **a**, **b**, and **c** were calculated from March 1, 2016, to January 23, 2018. **d** were calculated from March 1, 2016, to May 1, 2016, during El Niño (ENSO) conditions. (NOTE: We subtracted from each measured value the day's mean for both cases and plotted the resulting deviations with time. This way, we aimed to reduce seasonal bias when comparing daily patterns over time. Source: Fernandez-Bou 2019)

molecules (moist air rises over dry air). In confined spaces such as an *Atta* nest, less dense air of higher temperature and/or relative humidity will expand into regions of more dense air with a lower temperature and/or moisture by free convection transport until the density is uniform. Free and forced convection through nest vents and superficial emissions from the nest soil account for almost all nest emissions. Those emissions vary plenty among species, but for *A. cephalotes* at La Selva, they range from 200 kg $year^{-1}$ to 600 kg $year^{-1}$ (Fernandez-Bou et al. 2019).

Animal convection in leaf-cutter ant nests displaces an air volume equal to that of the ants and vegetative material that enter and exit the nest based on Archimedes' principle, which can be considered to be negligible (5 g to 15 g CO_2 per nest and year) (Fernandez-Bou 2019).

18.2.3.3 Ventilation Through Diffusion

Diffusion is the spontaneous and random motion of molecules from higher to lower concentrations until they are equalized (without a driving force), and higher temperatures enhance it. The contribution of diffusion to transport inside ant nests can be significant in the absence of convection. However, it is usually negligible compared to the effect of free and forced convection. In the ant nest context, CO_2 and other greenhouse gas concentrations are much higher near the refuse and fungal chambers due to metabolic activity (i.e., respiration). That creates a concentration gradient between the high-concentration regions of the nest and the surface outside the nest. Usually, if the airflow velocity due to convection is more than 0.0001 m s^{-1} (which is very likely to occur in *Atta* spp. nests), then diffusion can be negligible. For example, at the La Selva site in the primary forest of Costa Rica, CO_2 emissions from *A. cephalotes*' nests due to diffusion are three orders of magnitude less than emissions due to convection, and diffusion totals on average about 20 grams of CO_2 per nest and year (Fernandez-Bou et al. 2020b).

18.2.3.4 Leaching

Besides the movement of gases, leaching is another crucial pathway of carbon and nutrients in soils. Changes in physical soil properties (see Sect. 18.2.1) affect infiltration rates and the downward transport of organic matter, clay, dissolved organic matter (DOM), and nutrients. Dissolved organic matter is the organic matter fraction in solution that passes through a 0.45 μm filter and is the most mobile organic matter fraction (Bolan et al. 2011). Dissolved organic matter is a sensitive indicator for current and long-term soil biogeochemical processes (Kaiser and Kalbitz 2012). Despite a large body of knowledge on the influence of ant activities on soil physical properties and soil nutrients, the number of studies on the linkages between ants and the leaching of dissolved organic and inorganic matter across ecosystems is comparatively low (Stadler et al. 2006; Bierbaß et al. 2015; Ehrle et al. 2021), particularly in leaf-cutter ant nests.

A study conducted in leaf-cutter ant nests at La Selva showed that dissolved organic carbon (DOC), nitrogen (DON), and inorganic nitrogen (DIN) concentrations were higher in active nests than in non-nest soils (Table 18.4). Higher DON concentrations were also found in soil with ant nests in tropical forests (Wang et al.

Table 18.4 pH, organic carbon and nitrogen, and nitrate + nitrite concentrations in non-nests, active nests, and abandoned nests, La Selva ResearchStation, Costa Rica. Values are medians across soil, forest type, depth, and the 25th (P25) and 75th (P75) percentiles

Nest type	pH			DOC (mg L^{-1})			DON (mg L^{-1})		
	Median	P25	P75	Median	P25	P75	Median	P25	P75
Non-nest	5.02 a	4.69	5.30	0.888 a	0.372	1.660	0.042 a	0.019	0.095
Active nest	4.86 b	4.57	5.09	1.218 b	0.613	2.295	0.051 b	0.017	0.145
Abandoned nest	4.79 bc	4.50	5.02	0.953 ac	0.577	1.367	0.051 b	0.017	0.145
Nest type	**DOC:DON ratio**			**$SUVA_{254}$**			**Nitrate + nitrite (mg L^{-1})**		
	Median	**P25**	**P75**	**Median**	**P25**	**P75**	**Median**	**P25**	**P75**
Non-nest	14.7 a	4.9	37.2	1.55 a	0.69	4.91	0.262 a	0.128	0.507
Active nest	13.2 a	5.6	34.6	1.53 ab	0.73	3.44	0.354 bc	0.116	1.045
Abandoned nest	21.3 b	9.7	45.7	2.38 c	1.23	4.44	0.141 c	0.039	0.815

Different letters indicate significant differences between nest types for a given parameter ($p < 0.05$)

2017; Zuo et al. 2021). This is partially explained by the transformation and leaching of nutrient-rich organic matter in leaf-cutter ant nests (Farji-Brener and Werenkraut 2015, 2017; Maurice-Lira et al. 2023; Baena et al. 2024). Higher DIN concentrations may result from higher nitrogen mineralization rates in nest soils (Delgado-Baquerizo et al. 2019). Soil solution pH in leaf-cutter ant nests was more acidic than in non-nest soils. This is in line with findings from studies on underground-nesting ants in tropical forest soils (Zuo et al. 2021) and has been attributed to the decomposition of organic substances, resulting in higher DOC concentration. Changes in pH also affect the conversion of CO_2 to dissolved inorganic carbon (DIC), which can be lost through water flow through the soil profile or via soil pipes (Sollins et al. 1994). Estimates based on HYDRUS modeling indicate a loss of 8 g C m^{-2} $year^{-1}$ as DIC through the soil profile (unpublished data).

The DOC:DON ratio and Specific UV absorbance at 254 nm ($SUVA_{254}$) values are indicators of the chemical composition of DOC. The SUVA254 value has been used as a proxy for DOC aromaticity (Weishaar et al. 2003). Low DOC:DON ratios and $SUVA_{254}$ values in leachates from active nests suggest lower aromaticity and more readily assimilable organic carbon sources for microorganisms. The highest DOC: DON ratio and $SUVA_{254}$ values were found in abandoned nest soils. Studies have shown that the substrate is depleted in the more labile carbon sources when it ages (Maurice-Lira et al. 2023). Inorganic nitrogen concentration in abandoned nests was lower than in active and non-nest sites (Table 18.4). A rapid decline in inorganic nitrogen concentrations was also observed in *A. colombica* nests abandoned for more than one year (Hudson et al. 2009). In sum, the distinct soil chemistry of leaf-cutter ant nests, marked by elevated concentrations of dissolved organic carbon and nitrogen, lower pH, and unique DOC:DON ratios, underscores the significant impact of leaf-cutter ants on nutrient cycling and soil properties in the Neotropics.

In each plot (non-nest, active nest, and abandoned nest), three lysimeters (1905 L slim tube, Soilmoisture Equipment Corp., Santa Barbara, California, USA) were installed at depths of 20 cm, 60 cm, and 100 cm, each with a 500 ml collection bottle attached. Soil water samples were collected monthly from March 2015 to September 2017. At the beginning of each campaign, samplers at 20 cm, 60 cm, and 100 cm depth were prepared with vacuum of 50 kPa, 55 kPa, and 60k Pa respectively. The pH was measured with a pH 211 Hanna Instruments meter. Total nitrogen and carbon concentrations were measured using a TOC/TN analyzer (TOC-VCSN, TNM-1, Shimadzu, Kyoto, Japan). Nitrate and NH_4-N were measured by flow injection analysis (Alpkem Flow Solution IV Automated wet chemistry system, O.I. Analytical, College Station TX). The DON concentrations were estimated by subtracting inorganic N concentrations (NO_3 + NH_4) from total N concentrations. The DOC aromaticity was assessed for each soil solution sample by ultraviolet (UV) absorption spectrophotometry (UV-1700, Shimadzu Corporation, Kyoto, Japan). Spectra were acquired between 200 nm and 500 nm at a scan rate of 150 nm min^{-1} and with a uniform data point interval of 0.5 nm. Specific absorption at 254 nm (i.e., measured absorbance divided by the DOC concentration), referred to as Specific Ultraviolet Absorbance ($SUVA_{254}$), was used to assess the aromaticity of DOM.

18.3 Effects of Climate Change on Leaf-Cutter Ants and Associated Biogeochemical Processes

Climate models project significant warming across the Neotropics throughout the twenty-first century, with most scenarios indicating temperature increases of 2 °C to 4 °C by 2100 (Masson-Delmotte et al. 2021). Changes in rainfall patterns are a particular concern for the Neotropics (Masson-Delmotte et al. 2021). For lowland tropical forests, projections suggest an increased frequency of extreme weather events, including flooding and hurricanes, similar to those observed in recent years that have led to significant nest abandonment by leaf-cutter ants (Fernandez-Bou et al. 2019; Rundel et al. 2023). In contrast, a significant reduction in rainfall has been observed in Southeastern South America, Chile, Argentina, and northeastern Brazil, and drought conditions are predicted to increase (Vera et al. 2023). These climate alterations are likely to have cascading effects on ecosystem functioning through changes in ant distribution and behavior, soil structure, and biogeochemical processes.

18.3.1 Ant Distribution

Temperature strongly affects leaf-cutter ant distribution, abundance, metabolism, development, and activity patterns. Species distribution modeling suggests variable outcomes across different *Atta* species in places like Mexico. While *A. cephalotes* is predicted to increase its range, *Atta texana* and *A. mexicana* are expected to face range contractions due to rising temperatures, decreased rainfall, and increased drought frequency (Gómez-Díaz et al. 2023). While ants can regulate nest temperatures through architectural modifications and behavioral thermoregulation (see Chap. 2), their symbiotic fungus has relatively narrow thermal tolerance ranges that ultimately determine colony survival and distribution. For tropical leaf-cutter species, the ideal temperatures for fungal growth range from 20 °C to 30 °C, whereas species in cooler environments, such as *A. texana*, can tolerate fungal temperatures as low as 15 °C (Mueller et al. 2011). Some regions that are traditionally too cold for the fungus are experiencing higher temperatures, which may allow ants to expand their range (Nascimento et al. 2022).

18.3.2 Ant Activity and Behavior

The running speed of *Acromyrmex* and *Atta* species increases with temperature (Hurlbert et al. 2008), and warmer conditions can lead to increased foraging activity of the leaf-cutter ants (Bustamante et al. 2020). The high tolerance to severe micro-climatic conditions of some leaf-cutter ant species suggests that they may become highly prolific foragers as air and soil temperatures increase in a changing climate. Physiological and behavioral responses to increasing temperatures can alter colony function and ecological processes, such as prioritizing harvesting at cooler night-time hours, altering trail paths to maximize shade, and modifying nest architecture to enhance ventilation. These changes can affect colonies through increased energetic costs of nest cooling, reduced foraging efficiency during hotter periods, and potential mismatches in the development rates of different castes, all of which could have cascading effects on colony productivity and survival.

Climate change is increasingly disrupting the synchronization between leaf-cutter ant colony cycles and the seasonal availability of optimal plant resources, creating phenological mismatches with potential consequences for colony productivity and survival (Renner and Zohner 2018; Bizarria et al. 2021). The annual cycle of leaf-cutter colonies—including periods of intense growth, production of reproductive individuals, and the founding of new colonies—has historically been tightly coordinated with seasonal patterns of leaf production and plant nutritional quality.

Rainfall patterns heavily influence leaf-cutter ant foraging activity, and most species reduce or stop their activity outside the nest when it is raining (Fig. 18.1). In addition, fungus gardens need optimal humidity, generally around 90% of relative humidity. However, intense rainfall can cause the collapse of the chambers and parts of the nest, leading to the destruction or relocation of nests (Fernandez-Bou et al. 2019). As extreme climatic events become more intense and frequent due to climate change, some leaf-cutter ant nests may succumb to these altered conditions. For example, during one of our studies at La Selva (Costa Rica) from 2014 to 2018, there were multiple extreme climate events (several floods, a dry El Niño period, and a hurricane). Six of nine monitored leaf-cutter ant nests in that period were abandoned, and at least one moved to a different location around 50 meters from its original position. Most of the nests were abandoned after extreme floods (Fernandez-Bou et al. 2019; Rundel et al. 2023).

18.3.3 Soil Structure and Biogeochemical Processes

Modifications in soil temperature and moisture may lead to alterations in soil structure, affecting its porosity and permeability. These changes can influence the efficiency of gas diffusion between the nest and the surrounding soil, impacting the overall ventilation system of the nest (Bollazzi et al. 2012). They may also result in increased loss of carbon and nutrients from leaf-cutter ant nests and surrounding soil through leaching (Cleveland et al. 2007; Fernandez-Bou et al. 2019).

Changes in atmospheric processes through global warming alter wind patterns and intensities (Abatzoglou et al. 2018). This has the potential to alter forced convection mechanisms in leaf-cutter ant nests. If climate change increases local wind frequency and/or intensity, ventilation through forced convection will increase. The potential effects of climate change and global warming on free convection in leaf-cutter ant nests can include changes in the density and temperature gradients between the atmosphere and the nest tunnels that can modify ventilation efficiency by free convection.

Increasing temperatures and modifying rainfall patterns will affect ecosystem processes within leaf-cutter ant nests and surrounding soil through complex interactions between temperature, moisture, and biological factors that change the metabolic dynamics or Michaelis–Menten kinetics. Water and temperature are essential factors influencing CO_2 emissions (Fang and Moncrieff 2001; Updegraff et al. 2001), and leaf-cutter ants can modify those variables. Studies suggest that increasing temperature will result in higher decomposition rates, leading to higher CO_2 and N_2O emissions from tropical soils (Wood et al. 2012; McFarlane et al. 2024) and changing the CO_2 emission patterns from leaf-cutter ant nests (Fernandez-Bou et al. 2020a). However, in regions experiencing increased drought frequency, soil moisture limitations may temporarily reduce CO_2 production. At the same time, subsequent rewetting events could trigger pulses of enhanced microbial activity and greenhouse gas emissions from leaf-cutter ant nests (Fernandez-Bou et al. 2020a), potentially amplifying their role as biogeochemical hotspots in Neotropical ecosystems.

The effects of climate change on soil CO_2 dynamics will also vary with soil type. These impacts are expected to show complex interactions depending on their influence's temporal and spatial scales. For example, increased rainfall does not necessarily boost soil CO_2 production in clayey soils because very wet conditions lead to high soil tortuosity, making it more difficult for CO_2 to move in the soil matrix. Thus, increased tortuosity decreases soil CO_2 emissions by restricting the diffusive path of CO_2 to the atmosphere. Additionally, waterlogged soils reduce O_2 availability, which decreases CO_2 production. This indicates that water availability can be a limiting factor for CO_2 production in these Neotropical regions (Fernandez-Bou et al. 2020a). However, leaf-cutter ants can adapt to living in multiple types of soils that may respond differently to climate change, including changes in precipitation regimes and increased soil temperatures. This means that local conditions will depend on the local biogeochemistry of soils and ecosystems and the local effects of climate change.

The nitrogen transformation processes within leaf-cutter ant nests are particularly sensitive to temperature and moisture changes (Soper et al. 2019). Leaf-cutter ant nests serve as nitrogen enrichment zones through plant-derived nitrogen concentration and enhanced ammonification and nitrification rates mediated by specialized microbial communities (Pinto-Tomás et al. 2009). However, increasing temperatures accelerate volatile nitrogen losses through ammonia volatilization and denitrification, potentially reducing the nitrogen retention capacity of nest soils. Simultaneously, the more extreme wet-dry cycles projected under climate change scenarios can trigger pulses of nitrogen mineralization during rewetting events, creating temporary spikes in plant-available nitrogen that may be poorly synchronized with vegetation uptake periods (Erinle et al. 2021).

Elevated atmospheric CO_2 concentrations also represent a potential factor affecting biogeochemical cycling by leaf-cutter ants. While higher CO_2 concentrations can increase plant biomass, they also alter leaf biochemistry, generally decreasing nitrogen, increasing the concentration of carbon-based secondary metabolites, and changing the texture of the leaves. Altered temperature and precipitation regimes are shifting the timing of plant phenological events, including leaf flush and senescence, while simultaneously affecting the development rates of different ant castes. These changes may require colonies to harvest more significant amounts of vegetation to compensate for different nutrient compositions.

18.4 Conclusion

Changes in climatic conditions can affect the distribution and abundance of leaf-cutter ants, their behavior, and the biogeochemical processes within nests and surrounding soils. The direct effects of climate change include elevated temperatures, a modified rainfall regime, and an increasing frequency and severity of extreme weather events. Other abiotic changes include higher atmospheric CO_2 concentrations and ultraviolet radiation. As leaf-cutter ants thrive in disturbed environments, some species may be able to expand their distribution range to regions that are currently more temperate. Additionally, climate change can significantly impact the convection and diffusion venti processes in leaf-cutter ant nests by altering temperature gradients, soil moisture, vegetation availability, and soil structure. These changes can affect the ants' ability to regulate their nest environment and may have broader implications for their survival and ecological role. However, our current knowledge is insufficient to make confident predictions about how climate change will affect the distributions of leaf-cutter ants and their roles in geochemical cycling.

Acknowledgments We thank Diego Dierick and Michael Meredyth-Young for their help in fieldwork, especially setting up equipment, collecting data, and performing analyses. We also thank the staff of the Organization for Tropical Studies at the La Selva Research Station for their logistical support.

References

Abatzoglou JT, Dobrowski SZ, Parks SA, Hegewisch KC (2018) TerraClimate, a high-resolution global dataset of monthly climate and climatic water balance from 1958–2015. Scientific Data 5(1):170191. https://doi.org/10.1038/sdata.2017.191

Allen MF, Shulman H, Rundel PW et al (2023) Leaf-cutter ants–mycorrhizal fungi: observations and research questions from an unexpected mutualism. Front Fungal Biol 4

Aronson EL, Dierick D, Botthoff J et al (2019) ENSO-influenced drought drives methane flux dynamics in a tropical wet forest soil. J Geophys Res Biogeosci 124:2267–2276

Baena ML, de la Luz Avendaño-Yáñez M, González-Zamora A et al (2024) Atta mexicana waste dumps are hot spots of soil nutrients in an urban Neotropical cloud forest environment. Ecol Entomol 49:149–152. https://doi.org/10.1111/een.13288

Bierbaß P, Gutknecht JL, Michalzik B (2015) Nest-mounds of the yellow meadow ant (*Lasius flavus*) at the "Alter Gleisberg", Central Germany: hot or cold spots in nutrient cycling? Soil Biol Biochem 80:209–217

Bizarria R, Kooij PW, Rodrigues A (2021) Climate change influences Basidiome emergence of leaf-cutting ant cultivars. J Fungi 7:912. https://doi.org/10.3390/jof7110912

Blanton CM, Ewel JJ (1985) Leaf-cutting ant herbivory in successional and agricultural tropical ecosystems. Ecology 66:861–869

Bolan NS, Adriano DC, Kunhikrishnan A et al (2011) Chapter One - Dissolved organic matter: biogeochemistry, dynamics, and environmental significance in soils. In: Sparks DL (ed) Advances in agronomy. Academic Press, pp 1–75

Bollazzi M, Forti LC, Roces F (2012) Ventilation of the giant nests of Atta leaf-cutting ants: does underground circulating air enter the fungus chambers? Insect Soc 59:487–498. https://doi.org/10.1007/s00040-012-0243-9

Branstetter MG, Ješovnik A, Sosa-Calvo J et al (2017) Dry habitats were crucibles of domestication in the evolution of agriculture in ants. Proc R Soc B Biol Sci 284:20170095

Bustamante S, Amarillo-Suárez A, Wirth R (2020) Effects of pasture and forest microclimatic conditions on the foraging activity of leaf-cutting ants. Biotropica 52:697–708

Cleveland CC, Nemergut DR, Schmidt SK, Townsend AR (2007) Increases in soil respiration following labile carbon additions linked to rapid shifts in soil microbial community composition. Biogeochemistry, 82(3):229–240. https://doi.org/10.1007/s10533-006-9065-z

Corrêa MM, Silva PSD, Wirth R et al (2010) How leaf-cutting ants impact forests: drastic nest effects on light environment and plant assemblages. Oecologia 162:103–115. https://doi.org/10.1007/s00442-009-1436-4

Costa AN, Vasconcelos HL, Vieira-Neto EH, Bruna EM (2008) Do herbivores exert top-down effects in Neotropical savannas? Estimates of biomass consumption by leaf-cutter ants. J Veg Sci 19:849–854

da Sternberg LS, Pinzon MC, Moreira MZ et al (2007) Plants use macronutrients accumulated in leaf-cutting ant nests. Proc R Soc B Biol Sci 274:315–321

Dalal RC, Allen DE (2008) Greenhouse gas fluxes from natural ecosystems. Aust J Bot 56:369–407

Delgado-Baquerizo M, Eldridge DJ, Hamonts K, Singh BK (2019) Ant colonies promote the diversity of soil microbial communities. ISME J 13:1114–1118

Ehrle A, Kolle O, Tischer A et al (2021) Effects of mound building *Lasius flavus* on organic carbon and nutrient fluxes in soils of temperate grassland ecosystems. Pedobiologia 84:150701

Erinle KO, Bengtson P, Marschner P (2021) Rewetting intensity influences soil respiration and nitrogen availability. J Soil Sci Plant Nutr 21:2137–2144. https://doi.org/10.1007/s42729-021-00509-w

Fang C, Moncrieff J (2001) The dependence of soil CO2 efflux on temperature. Soil Biol Biochem 33:155–165

Farias AP, da Camargo RS, Andrade Sousa KK et al (2020) Nest architecture and colony growth of *Atta bisphaerica* grass-cutting ants. Insects 11:741. https://doi.org/10.3390/insects11110741

Farji-Brener I (2000) Do leaf-cutting ant nests make "bottom–up" gaps in neotropical rain forests?: a critical review of the evidence. Ecol Lett 3:219–227. https://doi.org/10.1046/j.1461-0248.2000.00134.x

Farji-Brener AG, Werenkraut V (2015) A meta-analysis of leaf-cutting ant nest effects on soil fertility and plant performance. Ecol Entomol 40:150–158. https://doi.org/10.1111/een.12169

Farji-Brener AG, Werenkraut V (2017) The effects of ant nests on soil fertility and plant performance: a meta-analysis. J Anim Ecol 86:866–877. https://doi.org/10.1111/1365-2656.12672

Farji-Brener AG, Elizalde L, Fernández-Marín H, Amador-Vargas S (2016) Social life and sanitary risks: evolutionary and current ecological conditions determine waste management in leaf-cutting ants. Proc R Soc B Biol Sci 283:20160625. https://doi.org/10.1098/rspb.2016.0625

Fernandez-Bou AS (2019) Leaf-cutter ant engineered Nest soil CO_2 dynamics in a neotropical rainforest. UC Merced

Fernandez-Bou AS, Dierick D, Swanson AC et al (2019) The role of the ecosystem engineer, the leaf-cutter ant *Atta cephalotes*, on soil CO_2 dynamics in a wet tropical rainforest. J Geophys Res Biogeosci 124:260–273

Fernandez-Bou AS, Dierick D, Allen MF, Harmon TC (2020a) Precipitation-drainage cycles lead to hot moments in soil carbon dioxide dynamics in a Neotropical wet forest. Glob Change Biol 26:5303–5319. https://doi.org/10.1111/gcb.15194

Fernandez-Bou AS, Dierick D, Harmon TC (2020b) Diel pattern driven by free convection controls leaf-cutter ant nest ventilation and greenhouse gas emissions in a Neotropical rain forest. Oecologia 192:591–601

Forti LC, Protti de Andrade AP, Camargo RDS et al (2017) Discovering the Giant Nest architecture of grass-cutting ants, *Atta capiguara* (Hymenoptera, Formicidae). Insects 8:39. https://doi.org/10.3390/insects8020039

Forti LC, Moreira AA, Camargo R d S et al (2018) Nest architecture development of grass-cutting ants. Rev Bras Entomol 62:46–50. https://doi.org/10.1016/j.rbe.2017.10.002

Gómez-Díaz JA, Baena ML, González-Zamora A, Delfín-Alfonso CA (2023) Potential present and future distributions of the genus *Atta* of Mexico. PLoS One 18:e0292072. https://doi.org/10.1371/journal.pone.0292072

Gonçalves C (1942) Contribuição para o conhecimento do gênero *Atta* Fabr., das formigas saúvas. Soc Bras Agron

Harmon TC, Dierick D, Trahan N et al (2015) Low-cost soil CO_2 efflux and point concentration sensing systems for terrestrial ecology applications. Methods Ecol Evol 6:1358–1362

Hart AG, Ratnieks FLW (2002) Waste management in the leaf-cutting ant *Atta colombica*. Behav Ecol 13:224–231. https://doi.org/10.1093/beheco/13.2.224

Hudson TM, Turner BL, Herz H, Robinson JS (2009) Temporal patterns of nutrient availability around nests of leaf-cutting ants (*Atta colombica*) in secondary moist tropical forest. Soil Biol Biochem 41:1088–1093. https://doi.org/10.1016/j.soilbio.2009.02.014

Hughes WO, Goulson D (2002) The use of alarm pheromones to enhance bait harvest by grass-cutting ants. Bull Entomol Res 92:213–218

Hurlbert AH, Ballantyne Iv F, Powell S (2008) Shaking a leg and hot to trot: the effects of body size and temperature on running speed in ants. Ecol Entomol 33:144–154. https://doi.org/10.1111/j.1365-2311.2007.00962.x

Jones CG, Lawton JH, Shachak M (1996) Organisms as ecosystem engineers. In: Ecosystem management: selected readings. Springer New York, New York, pp 130–147

Jonkman JCM (1980) Average vegetative requirement, colony size and estimated impact of *Atta vollenweideri* on cattle-raising in Paraguay. Z Für Angew Entomol 89:135–143. https://doi.org/10.1111/j.1439-0418.1980.tb03452.x

Jouquet P, Dauber J, Lagerlöf J et al (2006) Soil invertebrates as ecosystem engineers: intended and accidental effects on soil and feedback loops. Appl Soil Ecol 32:153–164

Kaiser K, Kalbitz K (2012) Cycling downwards–dissolved organic matter in soils. Soil Biol Biochem 52:29–32

Kleineidam C, Roces F (2000) Carbon dioxide concentrations and nest ventilation in nests of the leaf-cutting ant *Atta vollenweideri*. Insect Soc 47:241–248

Kleineidam C, Ernst R, Roces F (2001) Wind-induced ventilation of the giant nests of the leaf-cutting ant *Atta vollenweideri*. Naturwissenschaften 88:301–305. https://doi.org/10.1007/s001140100235

Mariconi FAM, Zamith APL, Castro UDP (1961) Contribuição para. o conhecimento da "saúva parda" *Atta capiguara* Gonçalves, 1944. An Esc Super Agric Luiz Queiroz 18:301–312. https://doi.org/10.1590/S0071-12761961000100020

Masson-Delmotte V, Zhai P, Pirani A et al (eds) (2021) Climate change 2021: the physical science basis. Contribution of Working Group I to the Sixth Assessment Report of the Intergovernmental Panel on Climate Change. Cambridge University Press, Cambridge/New York

Maurice-Lira J, Romero-Napoles J, Lara-Herrera A, et al (2023) Effect of *Atta mexicana* wastes on the nutritional and microbiological chaacteristics of soil. Appl Ecol Environ Res 21

McFarlane KJ, Cusack DF, Dietterich LH, Hedgpeth AL, Finstad KM, Nottingham AT (2024) Experimental warming and drying increase older carbon contributions to soil respiration in lowland tropical forests. Nature Communications 15(1):7084

Mehring AS, Martin RM, Delavaux CS et al (2021) Leaf-cutting ant (*Atta cephalotes*) nests may be hotspots of methane and carbon dioxide emissions in tropical forests. Pedobiologia 87–88:150754. https://doi.org/10.1016/j.pedobi.2021.150754

Meyer ST, Leal IR, Tabarelli M, Wirth R (2011) Ecosystem engineering by leaf-cutting ants: nests of *Atta cephalotes* drastically alter forest structure and microclimate. Ecol Entomol 36:14

Meyer ST, Neubauer M, Sayer EJ et al (2013) Leaf-cutting ants as ecosystem engineers: topsoil and litter perturbations around Atta cephalotes nests reduce nutrient availability. Ecol Entomol 38:497–504

Moreira A, Forti LC, Andrade AP et al (2004) Nest architecture of *Atta laevigata* (F. Smith, 1858) (Hymenoptera: Formicidae). Stud Neotropical Fauna Environ 39:109–116. https://doi.org/10.1080/01650520412331333756

Moser JC (2006) Complete excavation and mapping of a Texas leafcutting ant nest. Ann Entomol Soc Am 99:891–897. https://doi.org/10.1603/0013-8746(2006)99[891:CEAMOA]2.0.CO;2

Mueller UG, Mikheyev AS, Hong E et al (2011) Evolution of cold-tolerant fungal symbionts permits winter fungiculture by leafcutter ants at the northern frontier of a tropical ant–fungus symbiosis. Proc Natl Acad Sci 108:4053–4056. https://doi.org/10.1073/pnas.1015806108

Nascimento G, Câmara T, Arnan X (2022) Critical thermal limits in ants and their implications under climate change. Biol Rev 97:1287–1305. https://doi.org/10.1111/brv.12843

Nascimento DL, Chiapini M, Vidal-Torrado P et al (2024) The underestimated role of leaf-cutting ants in soil and geomorphological development in neotropical America. Earth-Sci Rev 248:104650. https://doi.org/10.1016/j.earscirev.2023.104650

Pereira-da-Silva V (1975) Contribuição ao estudo das populações de *Atta sexdens rubropilosa* Forel, e *Atta laevigata* (Fr. Smith) no Estado de São Paulo (Hym.: Formicidae). Stud Entomol 18:201–250

Perfecto I, Vandermeer J (1993) Distribution and turnover rate of a population of Atta cephalotes in a tropical rain Forest in Costa Rica. Biotropica 25:316–321. https://doi.org/10.2307/2388789

Pinto-Tomás AA, Anderson MA, Suen G et al (2009) Symbiotic nitrogen fixation in the fungus gardens of leaf-cutter ants. Science 326:1120–1123. https://doi.org/10.1126/science.1173036

Renner SS, Zohner CM (2018) Climate change and phenological mismatch in trophic interactions among plants, insects, and vertebrates. Annu Rev Ecol Evol Syst 49:165–182. https://doi.org/10.1146/annurev-ecolsys-110617-062535

Rundel PW, Harmon TC, Fernandez-Bou AS, Allen MF (2023) Collaborative use of sensor networks and cyberinfrastructure to understand complex ecosystem interactions in a tropical rainforest: challenges and lessons learned. Sensors 23. https://doi.org/10.3390/s23229081

Schultz TR, Brady SG (2008) Major evolutionary transitions in ant agriculture. Proc Natl Acad Sci 105:5435–5440

Schwendenmann L, Veldkamp E (2006) Long-term CO_2 production from deeply weathered soils of a tropical rain forest: evidence for a potential positive feedback to climate warming. Glob Change Biol 12:1878–1893
Shulman HB, Aronson EL, Dierick D et al (2024) Leafcutter ants enhance microbial drought resilience in tropical forest soil. Environ Microbiol Rep 16:e13251. https://doi.org/10.1111/1758-2229.13251
Sollins P, Sancho F, Mata R, Sanford RL (1994) Soils and soil process research. Selva Ecol Nat Hist Neotropical Rain For:34–53
Soper FM, Sullivan BW, Osborne BB et al (2019) Leaf-cutter ants engineer large nitrous oxide hot spots in tropical forests. Proc R Soc B Biol Sci 286:20182504. https://doi.org/10.1098/rspb.2018.2504
Sousa KKA, Camargo RS, Caldato N et al (2021) Carbon dioxide levels in initial nests of the leaf-cutting ant Atta sexdens (Hymenoptera: Formicidae). Sci Rep 11:20562. https://doi.org/10.1038/s41598-021-00099-8
Stadler B, Schramm A, Kalbitz K (2006) Ant-mediated effects on spruce litter decomposition, solution chemistry, and microbial activity. Soil Biol Biochem 38:561–572
Stahel G, Geijskes D (1941) Weitere Untersuchungen über Nestbau und Gartenpilz von *Atta cephalotes* L. und *Atta sexdens* L. Hym. Formicidae. Verl Nicht Ermittelbar
Swanson AC (2017) Disturbance, restoration, and soil carbon dynamics in desert and tropical ecosystems. University of California, Riverside
Swanson AC, Schwendenmann L, Allen MF et al (2019) Welcome to the Atta world: a framework for understanding the effects of leaf-cutter ants on ecosystem functions. Funct Ecol 33:1386–1399. https://doi.org/10.1111/1365-2435.13319
Updegraff K, Bridgham SD, Pastor J et al (2001) Response of CO2 and CH4 emissions from peatlands to warming and water table manipulation. Ecol Appl 11:311–326
Vera C, Skansi M, González M (2023) Características de las sequías en sur de Sudamérica [SISSA-ART-001-2023]
Viana LR, Santos JC, Arruda LJ et al (2004) Foraging patterns of the leaf-cutter ant *Atta laevigata* (Smith)(Myrmicinae: Attini) in an area of cerrado vegetation. Neotrop Entomol 33:391–393
Wang S, Wang H, Li J, Zhang Z (2017) Ants can exert a diverse effect on soil carbon and nitrogen pools in a Xishuangbanna tropical forest. Soil Biol Biochem 113:45–52
Weishaar JL, Aiken GR, Bergamaschi BA et al (2003) Evaluation of specific ultraviolet absorbance as an indicator of the chemical composition and reactivity of dissolved organic carbon. Environ Sci Technol 37:4702–4708. https://doi.org/10.1021/es030360x
Whitford WG, Eldridge DJ (2013) 12.19 effects of ants and termites on soil and geomorphological processes. In: Shroder JF (ed) Treatise on geomorphology. Academic Press, San Diego, pp 281–292
Wirth R, Beyschlag W, Ryel RJ, Hölldobler B (1997) Annual foraging of the leaf-cutting ant *Atta colombica* in a semideciduous rain forest in Panama. J Trop Ecol 13:741–757
Wirth R, Herz H, Ryel RJ et al (2003) Herbivory of leaf cutting ants. A case study in the tropical rainforest of Panama
Wood TE, Cavaleri MA, Reed SC (2012) Tropical forest carbon balance in a warmer world: A critical review spanning microbial-to ecosystem-scale processes. Biological Reviews 87(4):912–927
Zuo Q, Wang S, Wang P et al (2021) Ant-mediated effects on soil nitrogen mineralization vary with species in a tropical forest. Catena 203:105352

Part VI
Conclusion

Chapter 19
Ant Responses to Climate Change: A Synthesis

Inara R. Leal, Marcelo Tabarelli, Fernanda M. P. Oliveira, Xavier Arnan, and Alan N. Andersen

19.1 Understanding the Impacts of Climate Change

Anthropogenic climate change is a primary threat to biodiversity and ecosystem function worldwide (Mooney et al. 2009; Frishkoff et al. 2016; Lapola et al. 2023; Keck et al. 2025). It is therefore important to understand the likely impacts of climate change on individuals, on species, on interactions between species, and on the provision of ecosystem services. Ants are a dominant faunal group globally in terms of biomass, the flow of energy and nutrients, and the provision of ecosystem services (Folgarait 1998; Del Toro et al. 2012; Elizalde et al. 2020). This makes them valuable bioindicators of ecological change in land management (Andersen and Majer 2004). Ants are therefore an important focal group for understanding the likely impacts of climate change on biodiversity and ecosystem function.

An understanding of the importance of climatic variation is foundational to inferring likely impacts of climate change. This book has described how climatic variation influences ants at all levels of biological organisation, from individuals to communities, and how this has important implications for interactions between ants and other species and for the ecological services that ants provide (Fig. 19.1). In

I. R. Leal (✉) · M. Tabarelli · F. M. P. Oliveira
Departamento de Botânica, Universidade Federal de Pernambuco, Recife, Pernambuco, Brazil
e-mail: inara.leal@ufpe.br; marcelo.tabarelli@ufpe.br; fmp.oliveira@ufpe.br

X. Arnan
Universidade de Pernambuco – Campus Garanhuns, Garanhuns, Pernambuco, Brazil
e-mail: xavier.arnan@upe.br

A. N. Andersen
Research Institute for the Environment and Livelihoods, Charles Darwin University, Brinkin, Northern Territory, NT, Australia
e-mail: alan.andersen@cdu.edu.au

I. R. Leal et al. (eds.), *Ant Responses to Climate Change*, Entomology in Focus 10,
https://doi.org/10.1007/978-3-032-21127-9_19

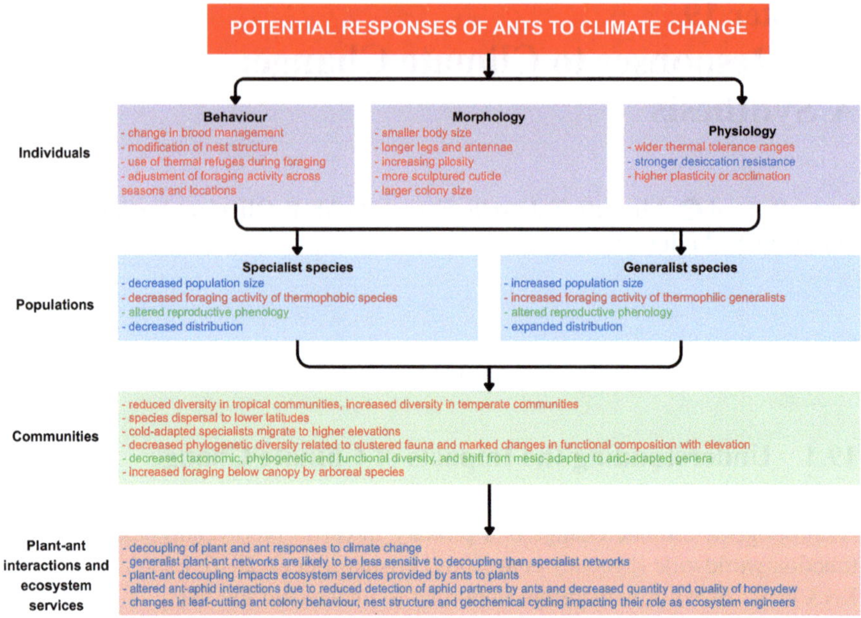

Fig. 19.1 Potential responses of ants to climate change across different levels of biological organisation, as reported in previous chapters. Climate change includes increased temperature (with responses shown in red), altered rainfall patterns (green), and both factors combined (blue). At the individual level, potential responses involve alterations in behaviour (Cerdá et al., Chap. 2), morphology (Gibb and Riskas, Chap. 3), and physiology (Nascimento et al., Chap. 4; Bujan and Roeder, Chap. 5). At the population level, predictions relate to the dynamics and distribution of both native (Centeno et al., Chap. 6) and invasive species (Berstelmeier and Bates, Chap. 7). At the community level, macroecological patterns are observed across latitudinal (Andersen et al., Chap. 8), elevational (Warren II, Chap. 9; Bishop and Zammit, Chap. 10) and rainfall (Arnan et al., Chap. 11) gradients, as well as along local vertical (ground-to-canopy) gradients in forests (Leahy, Chap. 12). In the context of plant-ant interactions, potential responses include changes in mutualistic networks (Câmara et al., Chap. 13) and in interactions among extrafloral nectary-bearing plants, herbivores and ants (Leal et al., Chap. 14). Finally, potential responses to climate change extend to a range of ecosystem services provided by ants, including general service provision (Elizalde et al., Chap. 15), seed dispersal (Oliveira et al., Chap. 16), plant protection via aphid tending (Blanchard and Detrain, Chap. 17) and the ecosystem engineering role of leaf-cutting ants (Fernandez-Bou et al., Chap. 18)

particular, temperature and rainfall are primary drivers of ant behaviour, morphology and physiology, of ant species distributions and of community dynamics (Andersen 1995; Dunn et al. 2009; Gibb et al. 2015). Accordingly, the effects of climate change on ants and their ecosystem functions are potentially far-reaching and profound.

We identify rising temperature as the primary agent of climate-change impacts on ants, through its effects on behavioural adaptations (Cerdá et al., Chap. 2), morphology (Gibb and Riskas, Chap. 3), thermal tolerance (Nascimento et al., Chap. 4), and desiccation resistance (Bujan and Roeder, Chap. 5). Behaviour often provides the first and most flexible response, influencing foraging dynamics, brood care and the efficiency of chemical communication under heat stress. It allows individuals to adjust activity patterns, exploit thermal refuges or modify nest architecture or location to avoid harmful temperatures (Cerdá et al., Chap. 2). Shifts in foraging activity patterns are especially likely as thermal limitations constrain foraging opportunities. Foraging flexibility in space and time reduces the vulnerability of species to global warming (Cerdá et al., Chap. 2), suggesting that distributional modelling based on broad-scale climatic niches might overestimate range shifts under future climates.

When behavioural strategies are insufficient, morphological and physiological mechanisms determine survival, activity, and reproduction under rising temperatures. Body size is the most critical morphological trait, given that it is a driver of metabolic rate, thermal tolerance, starvation resistance and desiccation resistance (Gibb and Riskas, Chap. 3). Warming favours small body size in ants, which may have significant consequences for species interactions and provision of ecosystem services. Responses of traits such as appendage length and colour may depend on changes in habitat structure that result from climate change. However, the impacts of climate change on morphological traits will be regulated by other aspects of ant biology, such as physiology (Nascimento et al., Chap. 4 and Bujan and Roeder, Chap. 5), and the interaction between selection on workers and the colony.

Physiological mechanisms such as critical thermal limits and thermal performance curves (Nascimento et al., Chap. 4) exhibit phylogenetic constraints and varying plasticity. Cold tolerance is often more labile than heat tolerance, and some evidence indicates limited potential for acclimation or rapid adaptation in many species. At the same time, increasing drought frequency and intensity highlight desiccation resistance as a key trait, particularly in species from dry or variable environments (Bujan and Roeder, Chap. 5). Desiccation resistance is shaped by body size, cuticle traits, hydrocarbon composition, and behavioural adjustments, yet remains comparatively understudied despite its critical role in species persistence under drier climates. Together, these physiological responses mediate how ant workers and colonies withstand environmental stress, ultimately influencing population dynamics, species distributions and ecosystem function in a warming and drying world.

Demographic responses to climatic variation vary across species. Overall, increasing temperature and aridity tend to influence ant population size and distribution by favouring generalist species over specialists (Centeno-Alvarado et al., Chap. 6). Generalists, which tend to have broader ecological niches and greater thermal tolerance, often maintain or increase their populations under warming and prolonged drought scenarios (Almeida et al. 2024; Kwon et al. 2024). In contrast, thermally sensitive or microhabitat-specialist species typically experience declines (Almeida et al. 2024, Kwon et al. 2024). Range shifts toward higher latitudes and elevations and disruptions in interspecific interactions collectively reflect the metapopulation processes driving species' responses to climate change (Centeno-Alvarado et al., Chap. 6). Phenological changes include an earlier reproduction, increased nuptial flight frequency in response to rainfall, and reduced male production at higher temperatures (Centeno-Alvarado et al., Chap. 6). Finally, population changes related to climate also include genetic level, such as genome expansion in colder, wetter areas and altered gene expression when faced with heat and drought, indicating divergent evolutionary trajectories along climate differences (Centeno-Alvarado et al., Chap. 6).

There is particular concern over potential range expansions of invasive ant species under climate change. Ants are among the world's worst invasive species, having significant economic as well as environmental impacts, and in some cases posing serious risks to human health (Holway et al. 2002). The impacts of climate change on invasive ant species are likely to be highly idiosyncratic, including potential range contractions as well as expansions (Berstelmeier and Bates, Chap. 7). The modelling of future distributions is especially problematic for invasive species outside their native ranges because they often show substantial shifts in their climatic niches following invasion (Berstelmeier and Bates, Chap. 7).

Changes in species distributions and demographic dynamics across gradients of temperature and aridity (Centeno-Alvarado et al., Chap. 6) result in the reorganisation of communities along macro-scale gradients such as latitude and elevation. For example, under global warming, it is likely that many species will move to lower latitudes (Andersen et al., Chap. 8) and cold-adapted specialists will shift to higher elevations (Warren II, Chap. 9). Distributional shifts will potentially lead to marked changes in the functional composition of ant communities, in relation to both temperature (Bishop and Zammit, Chap. 10) and rainfall (Arnan et al., Chap. 11). However, in forested ecosystems the steepest thermal gradients occur vertically within a site, from the ground to the canopy (Leahy, Chap. 12). Tree canopies are hotter, drier, and more variable than on the forest floor below, and such variation provides opportunities for behavioural adaptation (Cerdá et al., Chap. 2) in situ without the need for geographic range shifts (Leahy, Chap. 12). This is especially the case for arboreal species whose foraging ranges from the tree canopy down to the shrub layer, which can represent an ambient thermal gradient spanning several degrees Celsius. Moreover, the temperature of canopy structures can vary by more than 15 °C depending on their direct exposure to sunlight (Stark et al. 2017). Further variation in the thermal landscape is provided by the diel cycle, where temperature varies markedly both within the day and between day and night. Such dynamic thermal variation within a site, along vertical and temporal dimensions, is far greater than is variation in mean temperature at substantial macro-scales (Scheffers et al. 2013).

Ants and plants are differentially sensitive to climatic variation and so ants and their current vegetation associations are unlikely all to shift together under climate change. Based on current climate associations along a latitudinal gradient in southeastern Australia, the responses of ants to climate change are predicted to be extensively decoupled from those of plants (Caddy-Retalic et al. 2019). This means that sites within a species' climatic niche might no longer support its preferred vegetation type. Such decoupling suggests that impacts on the ranges of ant species will be more marked than predicted by models based on climatic niches alone. This is illustrated by the incorporation of habitat suitability into niche modelling of rainforest ants in the Australian Wet Tropics, which resulted in >90% of communities in the region being predicted to lose up to half their similar habitats compared with a much lower 50–60% when considering only direct climate-change impacts (Nowrouzi et al. 2019).

A decoupling of ant and plant responses to climate change has important implications for ant-plant mutualism networks more generally (Câmara et al., Chap. 13). Generalist networks, where species interact with many partners, are likely to be more resilient to decoupling than are specialist networks, which allow for less adaptability (Câmara et al., Chap. 13). This is the case for facultative mutualistic interactions between extrafloral nectaries (EFN)-bearing plants and ants in the Brazilian Caatinga dry forest (Oliveira et al. 2021). Increased aridity reduces phylogenetic and functional diversity of EFN-bearing plants (Arnan et al. 2022), and it reduces the quality of plant-protection provided by ants (Oliveira et al. 2021). This is associated with decreases in EFN size (Câmara et al. 2022). However, a global meta-analysis indicates that the reproductive outcomes of EFN-bearing plants remain consistent across different levels of mean annual precipitation worldwide and therefore will be little affected by an increasingly arid future climate (Leal et al., Chap. 14).

Climate change also has important implications for the full range of ecological services provided by ants, including soil bioturbation, nutrient cycling, pest control, seed dispersal and pollination (Elizalde et al., Chap. 15). For example, increasing aridity in Caatinga reduces the quality of seed dispersal services provided by ants as measured by the rate of seed removed and dispersal distance (Oliveira et al. 2019). Additionally, distribution modelling under future climate scenarios of the ant *Dinoponera quadriceps* (Ponerinae), a key disperser of the myrmecochorous seeds, indicates a significant reduction in overlap between this ant and all the myrmecochorous species modelled (Oliveira et al., Chap. 16). Many plant populations will therefore no longer co-occur with their key seed disperser, and they will therefore receive inferior dispersal services.

Climate change can be expected to affect other ecosystem services provided by ants. For example, disruptions to chemical communication may impede the detection of aphid partners (Blanchard and Detrain, Chap. 17), potentially compromising the protection services ants provide to plants through aphid tending. Moreover, drought combined with high temperatures markedly affects both the quantity and quality of aphid honeydew (Blanchard and Detrain, Chap. 17), which could reduce the nutritional rewards that sustain these mutualisms and, in turn, diminish ants' effectiveness in pest control or plant defence. Changes in temperature, rainfall and extreme weather events also influence colony behaviour, nest structure and geochemical cycling by leaf-cutter ants, key ecosystem engineers, throughout the Neotropics (Fernandez-Bou et al., Chap. 18). Additionally, climate change can significantly impact the convection and diffusion processes in leaf-cutting ant nests by altering temperature gradients, soil moisture, vegetation availability and soil structure (Fernandez-Bou et al., Chap. 18). These changes can affect the ants' ability to regulate their nest environment and may have broader implications for their survival and ecological role, especially in human-modified landscapes.

The above synthesis shows that climate change has the potential to affect virtually all aspects of ant biology and ecology (Fig. 19.1). However, the likelihood of such impacts is highly uncertain. This is partly because the behaviour of many ant species is highly flexible. It is also because the climatic relationships described above tend to be weak and so have low predictive power. Moreover, the relationships are typically macroecological rather than necessarily across the scale of projected climate change (Andersen et al., Chap. 8). For example, a trend that occurs across a gradient of 20 °C mean temperature will not necessarily be expressed over the rise of 2–3 °C projected under climate change. Much therefore remains to be known about the impacts of climate change on ants.

19.2 Identifying Climate-Sensitive Zones

Macroecological relationships with climatic variables can be highly non-linear (Kreyling et al. 2014). This highlights the importance of identifying climate-sensitive zones—where rates of biodiversity change are particularly high and therefore biodiversity change under future climates is likely to be most pronounced. These represent priority areas for ongoing monitoring and potentially management (Williams et al. 1995).

Such climate-sensitive zones have been identified for ant species composition in monsoonal Australia, where highest rates of species turnover occur at the southern transition to the central arid zone, and in the transition between semi-arid and mesic savanna to the north (Andersen, Chap. 8). In mountains of Australia's Wet Tropics, there is a pronounced discontinuity in ant species composition between 600 m and 800 m above sea level, which is the height of orographic cloud; this height represents a strong environmental threshold of seasonal stability in moisture associated with cloud 'stripping' (Nowrouzi et al. 2016). The height of orographic cloud is expected to rise substantially under global warming, and so ant communities occurring at 600–800 m are likely to experience especially high rates of change in species composition.

The identification of climate-sensitive zones is especially relevant when high rates of species turnover involve major changes in functional composition. For example, ant faunas of deserts have highly distinctive functional composition, featuring a prominence of specialized granivores, high levels of thermophily and characteristic morphology (Arnan et al., Chap. 11). However, it is unclear if thresholds occur along aridity gradients where there is a rapid switch from 'mesic' to 'arid' communities, or whether there is continuous variation in arid-adapted features. Similarly, although there is clear change in ant functional composition along elevation gradients in relation to temperature (Bishop and Zammit, Chap. 10), it is unclear if such change is continuous or involves tipping points where there is rapid change from one state to another.

Differential climate sensitivity also occurs at the biogeographical scale. The wet tropics of eastern Amazon, Africa and Indo-Malaya have been identified as being especially sensitive, with the projected higher temperatures and decreased rainfall likely to rearrange current patterns of ant diversity and composition in by the end of twenty-first century (Parr and Bishop 2022).

19.3 Interactive Effects of Climate Change and Anthropogenic Disturbance

As humans are everywhere, climate change is not expected to operate as an isolated driver but rather to interact with local disturbances as drivers of ecological reorganisation. Over recent decades, local anthropogenic disturbances—such as land use change, habitat loss, overharvesting of natural resources, and wildfires—have exerted more immediate and pronounced impacts on biodiversity and ecosystems than has climate change alone (Sala et al. 2000; Northrup et al. 2019; Jaureguiberry et al. 2022). However, evidence continues to accumulate suggesting that local disturbances and climate change interact in complex and synergistic ways, resulting in shifts in ecological attributes from the population to the ecosystem level (Ghazoul et al. 2015; Lapola et al. 2023). For instance, recent findings in solitary bees show that warmer winter temperatures—driven by climate change—exacerbate the effects of pesticide exposure, drastically reducing longevity and potentially compromising reproductive success (Albacete et al. 2023). This highlights that not only habitat alteration but also other human-derived stressors, such as chemical contamination, can interact with climate warming in unpredictable and severe ways.

While each of these drivers individually poses substantial threats to biodiversity, ecosystem functioning and resilience, their interactive effects are often more disruptive than the sum of their parts (Mooney et al. 2009; Frishkoff et al. 2016; Malhi et al. 2020; Lapola et al. 2023; Keck et al. 2025). One example is deforestation and its consequent habitat loss and fragmentation, which alters microclimatic conditions by increasing edge effects and habitat openness in general (e.g. Ewers and Banks-Leite 2013; Lapola et al. 2023; Slater et al. 2024). This structural simplification leads to higher solar radiation exposure, elevated surface temperatures, reduced soil and air moisture, and stronger winds (Ewers and Banks-Leite 2013; Slater et al. 2024). This makes human-modified landscapes particularly vulnerable during the extreme weather conditions that are becoming more frequent and intense under climate change (Frishkoff et al. 2016; Malhi et al. 2020; Lapola et al. 2023). These altered conditions may cross ecological thresholds, shifting systems into new stable states. A prime example of this is the 'savannisation' of rainforests, in which deforestation and degradation can act as tipping points in the conversion of forests into more simplified, open, and desiccated vegetation types that facilitate the invasion of grasses and increase the likelihood of fire (Malhi et al. 2014, 2020; Nobre et al. 2016). Such savannisation is a widespread occurrence in the southern Amazon basin and results in a comprehensive transformation of plant and animal communities involving the replacement of forest specialists with open-adapted generalists (Brando et al. 2014; Nobre et al. 2016), including for ants (Paolucci et al. 2017).

Ant species with narrow physiological tolerances or specific dietary and microhabitat requirements are particularly vulnerable to the effects of human disturbance and climate change, while generalist or disturbance-tolerant species are often favoured (e.g. Leal et al. 2017; Almeida et al. 2024; Kwon et al. 2024). For instance, specialist predators, tree specialists and climate specialists are especially sensitive to forest fragmentation in rain forests (Leal et al. 2012), as well as to extensive livestock and exploitation of forest products in dry forests (Oliveira et al. 2017). Such sensitivity is likely to increase with global warming and increasing aridity (Leal et al. 2017). As a result of generalist-specialist winner-loser replacement, ecological communities exposed to ongoing anthropogenic disturbance and climate change will likely result in increasing biotic homogenisation, detected across taxonomic, phylogenetic and functional dimensions (Gibb et al. 2015; Arnan et al. 2018, 2020; Ribeiro-Neto et al. 2023). Particularly concerning are the cascading impacts on plant-animal interactions and the ecosystem services they sustain. For example, interactive effects of increasing aridity and chronic disturbance (i.e. extensive livestock and exploitation of forest products) in Caatinga dry forest decreases the specialisation (Câmara et al. 2018) and temporal stabilisation (Câmara et al. 2019) of mutualistic interactions between EFN-bearing plants and ants. This more generalist and temporally instable interactions is associated with a reduction in EFN size as aridity increases (Câmara et al. 2022), which result in a reduction in the quality of plant protection services provided by ants (Oliveira et al. 2021). Similar results have been observed for ant-mediated seed dispersal, in which the quality of the seed dispersal service declines as chronic disturbance (Leal et al. 2014) and aridity (Oliveira et al. 2019) increase. Notably, rising temperatures in Caatinga may increase the vulnerability of plant-ant mutualisms to anthropogenic disturbances (Câmara et al. 2025).

19.4 Future Directions

Climate change has the potential to affect all aspects of ant biology, from individual, behavioural, morphological and physiological traits to the structure of communities, and will likely lead to substantial decoupling of ant-plant associations. This has direct implications for ecological interactions and key ecosystem services such as seed dispersal, herbivore control, and nutrient cycling.

Changes to ant behaviour, species composition and ant-mediated ecological services will challenge traditional conservation and management frameworks. Anticipating these trajectories is essential for safeguarding both ant diversity and the broader ecological processes they underpin. With this in mind, we identify the following ten priority research questions:

1. What traits or combinations of traits best predict vulnerability or resilience to climate change in different biomes?
2. To what extent can behavioural or physiological plasticity buffer ant populations and communities against rapid climate change?
3. What is the potential for evolutionary responses in ants to keep pace with the speed of climate change, and under what conditions might these responses be promoted or constrained?
4. Can we develop predictive frameworks that integrate ant thermal physiology, species interactions, and landscape context to forecast ecosystem service dynamics under climate change?
5. What role will extremes (rather than changes in means) of temperature and rainfall play in the impact of climate change on ants and their ecosystem functions?
6. To what extent can macroecological relationships reliably predict impacts from the much smaller scale of projected climate change?
7. How does climate sensitivity vary among biomes and biogeographical regions?
8. Within regions, where are the 'climate-sensitive zones' where monitoring ant communities can serve as early indicators of biodiversity shifts?
9. To what extent will the responses of plants and ants be decoupled, producing novel associations under a future climate?
10. How will climate change interact with anthropogenic disturbance to drive biodiversity change?

Addressing these questions will help fill major knowledge gaps and build a more mechanistic understanding of how climate change is reshaping ant ecology and the broader ecosystems in which they operate.

We conclude with a positive story to tell with ants. Despite the likely impacts of climate change, ants are an ecologically resilient group, with at least some species in nearly every region capable of adapting—and even thriving—under novel environmental conditions (Andersen 2019). These disturbance-tolerant or climate-resilient species are likely to become increasingly prominent components of future ecosystems. We should recognise the potential of these species as agents of recovery, making them valuable allies in ecosystem restoration—resilient ants can play key functional roles in degraded habitats, promoting processes such as soil turnover, nutrient cycling, seed dispersal, and the maintenance of biotic interactions (Almeida et al. 2024; Elizalde et al., Chap. 15). Identifying the traits that underpin their success, and strategically harnessing their ecological functions, offers a promising, sustainable, cost-effective, and scalable pathway to foster resilience and functionality in transitioning ecosystems. In this sense, ants could become not just indicators of change (Tiede et al. 2017), but tools for building the next generation of biodiverse, multifunctional and adaptive ecosystems.

References

Albacete S, Sancho G, Azpiazu C, Rodrigo A, Molowny-Horas R, Sgolastra F, Bosch J (2023) Bees exposed to climate change are more sensitive to pesticides. Glob Chang Biol 29:6248–6260. https://doi.org/10.1111/gcb.17482

Almeida T, Arnan X, Capowiez Y, Hedde M, Mesléard F, Dutoit T, Blight O (2024) Ants in restoration ecology: why, what's and the way forward. Land Degrad Dev 35:1284–1295. https://doi.org/10.1002/ldr.5006

Andersen AN (1995) A classification of Australian ant communities, based on functional groups which parallel plant life-forms in relation to stress and disturbance. J Biogeogr 22:15–29. https://doi.org/10.2307/2846070

Andersen AN (2019) Responses of ant communities to disturbance: five principles for understanding the disturbance dynamics of a globally dominant faunal group. J Anim Ecol 88(3):350–362. https://doi.org/10.1111/1365-2656.12907

Andersen AN, Majer JD (2004) Ants show the way down-under: invertebrates as bioindicators in land management. Front Ecol Environ 2:291–298. https://doi.org/10.1890/1540-9295(2004)002[0292:ASTWDU]2.0.CO;2

Arnan X, Arcoverde GB, Pie MR, Ribeiro-Neto JD, Leal IR (2018) Increased anthropogenic disturbance and aridity reduce phylogenetic and functional diversity of ant communities in Caatinga dry forest. Sci Total Environ 631-632:429–438. https://doi.org/10.1016/j.scitotenv.2018.03.037

Arnan X, Cerdá X, Rodrigo A (2020) Do forest fires make biotic communities homogeneous or heterogeneous? Patterns of taxonomic, functional, and phylogenetic ant beta diversity at local and regional landscape scales. Front Forests Glob Chang 3. https://doi.org/10.3389/ffgc.2020.00067

Arnan X, Silva CHF, Reis DQA, Oliveira FMP, Câmara T, Ribeiro EMS, Andersen AN, Leal IR (2022) Individual and interactive effects of chronic anthropogenic disturbance and rainfall on taxonomic, functional and phylogenetic composition and diversity of extrafloral nectary-bearing plants in Brazilian Caatinga. Oecologia 198:267–277. https://doi.org/10.1007/s00442-021-05074-8

Brando PM, Balch JK, Nepstad DC, Morton DC, Putz FE, Coe MT, Silvério D, Macedo MN, Davidson EA, Nóbrega CC, Alencar A, Soares-Filho BS (2014) Abrupt increases in Amazonian tree mortality due to drought–fire interactions. Proc Natl Acad Sci USA 111(17):6347–6352. https://doi.org/10.1073/pnas.1305499111

Caddy-Retalic S, Hoffmann B, Guerin G, Andersen A, Wardle G, McInerney C, Lowe A (2019) Plant and ant assemblages predicted to decouple under climate change. Divers Distrib 25:551–567. https://doi.org/10.1111/ddi.12858

Câmara T, Leal IR, Blüthgen N, Oliveira FMP, Queiroz RT, Arnan X (2018) Effects of chronic anthropogenic disturbance and rainfall on the specialization of ant-plant mutualistic networks in the Caatinga, a Brazilian dry forest. J Anim Ecol 87:1022–1033. https://doi.org/10.1111/1365-2656.12820

Câmara T, Leal IR, Blüthgen N, Oliveira FMP, Arnan X (2019) Anthropogenic disturbance and rainfall variation threaten the stability of plant-ant interactions in the Brazilian Caatinga. Ecography 42:1960–1972. https://doi.org/10.1111/ecog.04531

Câmara T, Reis DQA, Arnan X, Oliveira FMP, Arruda ECP, Leal IR (2022) Drought-induced reductions in plant defenses: insights from extrafloral nectaries in the Caatinga dry forest. Biotropica 54:12–17. https://doi.org/10.1111/btp.13041

Câmara T, Cavalcante NT, Falcão HM, Santana E, Silva-Teixeira GS, Arnan X (2025) Increased temperatures could heighten vulnerability of an ant–plant mutualism. Oecologia 207:8. https://doi.org/10.1007/s00442-024-05646-4

Del Toro I, Ribbons RR, Pelini SL (2012) The little things that run the world revisited: a review of ant-mediated ecosystem services and disservices (hymenoptera: Formicidae). Myrmecol News 17:133–146

Dunn RR, Agosti D, Andersen AN, Arnan X, Bruhl CA, Cerdá X, Ellison AM, Fisher BL, Fitzpatrick MC, Gibb H, Gotelli NJ, Gove AD, Guenard B, Janda M, Kaspari M, Laurent EJ, Lessard JP, Longino JT, Majer JD, Menke SB, McGlynn TP, Parr CL, Philpott SM, Pfeiffer M, Retana J, Suarez AV, Vasconcelos HL, Weiser MD, Sanders NJ (2009) Climatic drivers of hemispheric asymmetry in global patterns of ant species richness. Ecol Lett 12:324–333. https://doi.org/10.1111/j.1461-0248.2009.01291.x

Elizalde L, Arbetman M, Arnan X, Eggleton P, Leal IR, Lescano MN, Saez A, Werenkraut V, Pirk GI (2020) The ecosystem services provided by social insects: traits, management tools and knowledge gaps. Biol Rev 95:1418–1441. https://doi.org/10.1016/j.ecoser.2018.12.002

Ewers RM, Banks-Leite C (2013) Fragmentation impairs the microclimate buffering effect of tropical forests. PLoS One 8(3):e58093. https://doi.org/10.1371/journal.pone.0058093

Folgarait PJ (1998) Ant biodiversity and its relationship to ecosystem functioning: a review. Biodivers Conserv 7:1221–1244. https://doi.org/10.1023/A:1008891901953

Frishkoff LO, Karp DS, Flanders JR, Zook J, Hadly EA, Daily GC, M'Gonigle LK (2016) Climate change and habitat conversion favour the same species. Ecol Lett 19(9):1081–1090. https://doi.org/10.1111/ele.12645. Epub 2016 Jul 11. PMID: 27396714

Ghazoul J, Burivalova Z, Garcia-Ulloa J, King LA (2015) Conceptualizing forest degradation. Trends Ecol Evol 30:622–632. https://doi.org/10.1016/j.tree.2015.08.001

Gibb H, Sanders NJ, Dunn RR, Watson S, Photakis M, Abril S, Andersen AN, Angulo E, Armbrecht I, Arnan X, Baccaro FB, Bishop TR, Boulay R, Castracani C, Del Toro I, Delsinne T, Diaz M, Donoso DA, Enriquez ML, Fayle TM, Feener DH, Fitzpatrick MC, Gomez C, Grasso DA, Groc S, Heterick B, Hoffmann BD, Lach L, Lattke J, Leponce M, Lessard JP, Longino J, Lucky A, Majer J, Menke SB, Mezger D, Mori A, Munyai TC, Paknia O, Pearce-Duvet J, Pfeiffer M, Philpott SM, Souza JLP, Tista M, Vasconcelos HL, Vonshak M, Parr CL (2015) Climate mediates the effects of disturbance on antassemblage structure. Proc Roy Soc London, Ser B, Biol Sci 282:8. https://doi.org/10.1098/rspb.2015.0418

Holway DA, Lach L, Suarez AV, Tsutsui ND, Case TJ (2002) The causes and consequences of ant invasions. Annu Rev Ecol Syst 33:181–233. https://doi.org/10.1146/annurev.ecolsys.33.010802.150444

Jaureguiberry P, Titeux N, Wiemers M, Bowler DE, Coscieme L, Golden AS, Guerra CA, Jacob U, Takahashi Y, Settele J, Díaz S, Molnár Z, Purvis A (2022) The direct drivers of recent global anthropogenic biodiversity loss. Sci Adv 8(45):eabm9982. https://doi.org/10.1126/sciadv.abm9982

Keck F, Peller T, Alther R, Barouillet C, Blackman R, Capo E, Chonova T, Couton M, Fehlinger L, Kirschner D, Knüsel M, Muneret L, Oester R, Tapolczai K, Zhang H, Altermatt F (2025) The global human impact on biodiversity. Nature 641:395–400. https://doi.org/10.1038/s41586-025-08752-2

Kreyling J, Jentsch A, Beier C (2014) Beyond realism in climate change experiments: gradient approaches identify thresholds and tipping points. Ecol Lett 17:125. https://doi.org/10.1111/ele.12193

Kwon TS, Lee CM, Kim TW et al (2024) Selection of climate variables in ant species distribution models: case study in South Korea. Int J Biometeorol 68(2):263–277. https://doi.org/10.1007/s00484-023-02588-z

Lapola DM, Pinho P, Barlow J, Aragão LEOC, Berenguer E, Carmenta R, Liddy HM, Seixas H, Silva CVJ, Silva-Junior CHL, Alencar AAC, Anderson LO, Armenteras D, Brovkin V, Calders K, Chambers J, Chini L, Costa MH, Faria BL, Fearnside PM, Ferreira J, Gatti L, Gutierrez-Velez VH, Han Z, Hibbard K, Koven C, Lawrence P, Pongratz J, Portela BTT, Rounsevell M, Ruane AC, Schaldach R, da Silva SS, von Randow C, Walker WS (2023) The drivers and impacts of Amazon forest degradation. Science 379:eabp8622. https://doi.org/10.1126/science.abp8622

Leal IR, Filgueiras BKC, Gomes JP, Andersen AN (2012) Effects of habitat fragmentation on ant richness and functional composition in Atlantic Forest of northeastern Brazil. Biodivers Conserv 21:1687–1701. https://doi.org/10.1007/s10531-012-0271-9

Leal LC, Andersen AN, Leal IR (2014) Anthropogenic disturbance reduces seed-dispersal services for myrmecochorous plants in the Brazilian Caatinga. Oecologia 174:173–181. https://doi.org/10.1007/s00442-013-2740-6

Leal IR, Ribeiro-Neto JD, Arnan X, Oliveira FMP, Arcoverde GB, Feitosa RM, Andersen AN (2017) Ants of the Caatinga: diversity, biogeography, and functional responses to anthropogenic disturbance and climate change. In: Silva JMC, Leal IR, Tabarelli M (eds) Caatinga: the largest tropical dry forest region in South America. Springer International Publishing, Cham, pp 65–95

Malhi Y, Gardner TA, Goldsmith GR, Silman MR, Zelazowski P (2014) Tropical forests in the Anthropocene. Annu Rev Environ Resour 39:125–159. https://doi.org/10.1146/annurev-environ-030713-155141

Malhi Y, Franklin J, Seddon N, Solan M, Turner MG, Field CB, Knowlton N (2020) Climate change and ecosystems: threats, opportunities and solutions. Philos Trans R Soc B 375:20190104. https://doi.org/10.1098/rstb.2019.0104

Mooney H, Larigauderie A, Cesario M, Elmquist T, Hoegh-Guldberg O, Lavorel S, Mace GM, Palmer M, Scholes R, Yahara T (2009) Biodiversity, climate change, and ecosystem services. Curr Opin Environ Sustain 1:46–54. https://doi.org/10.1016/j.cosust.2009.07.006

Nobre CA, Sampaio G, Borma LSB, Castilla-Rubio JC, Silva JS, Cardoso M (2016) Land-use and climate change risks in the Amazon and the need of a novel sustainable development paradigm. Proc Natl Acad Sci USA 113(39):10759–10768. https://doi.org/10.1073/pnas.1605516113

Northrup JM, Rivers JW, Yang Z, Betts MG (2019) Synergistic effects of climate and land-use change influence broad-scale avian population declines. Glob Chang Biol 25:1561–1575. https://doi.org/10.1111/gcb.14571

Nowrouzi S, Andersen AN, Macfadyen S, Staunton KM, VanDerWal J, Robson SKA (2016) Ant diversity and distribution in relation to latitude and elevation in the Australian Wet Tropics: the importance of seasonal moisture stability. PLoS One 11:e0153420. https://doi.org/10.1371/journal.pone.0153420

Nowrouzi S, Bush A, Harwood T, Staunton KM, Robson SKA, Andersen AN (2019) Incorporating habitat suitability into community projections: ant responses to climate change in the Australian Wet Tropics. Divers Distrib 25:1273–1288. https://doi.org/10.1111/ddi.12935

Oliveira FMP, Ribeiro-Neto JD, Andersen AN, Leal IR (2017) Chronic anthropogenic disturbance as a secondary driver of ant community structure: interactions with soil type in Brazilian Caatinga. Environ Conserv 44:115–123. https://doi.org/10.1017/S0376892916000291

Oliveira FMP, Andersen AN, Arnan X, Ribeiro-Neto JD, Arcoverde GB, Leal IR (2019) Effects of increasing aridity and chronic anthropogenic disturbance on seed dispersal by ants in Brazilian Caatinga. J Anim Ecol 88:870–880. https://doi.org/10.1111/1365-2656.12979

Oliveira FMP, Câmara T, Durval JIF, Oliveira CLS, Arnan X, Andersen AN, Ribeiro EMS, Leal IR (2021) Plant protection services mediated by extrafloral nectaries decline with aridity but are not influenced by chronic anthropogenic disturbance in Brazilian Caatinga. J Ecol 109:260–272. https://doi.org/10.1111/1365-2745.13469

Paolucci LN, Schoereder JH, Brando PM, Andersen AN (2017) Fire-induced forest transition to derived savannas: cascading effects on ant communities. Biol Conserv 214:295–302. https://doi.org/10.1016/j.biocon.2017.08.020

Parr CL, Bishop TR (2022) The response of ants to climate change. Glob Change Biol 28(10):3188–3205. https://doi.org/10.1111/gcb.16140

Ribeiro-Neto JD, Oliveira FMP, Arcoverde GB, Tabarelli M, Leal IR, Arnan X (2023) Aridity and chronic anthropogenic disturbances cause a taxonomic, functional, and phylogenetic homogenization of ant communities in a Caatinga dry forest. Biol Conserv 284:110151. https://doi.org/10.1016/j.biocon.2023.110151

Sala OE, Chapin FS III, Armesto JJ, Berlow E, Bloomfield J, Dirzo R, Huber-Sanwald E, Huenneke LF, Jackson RB, Kinzig A, Leemans R, Lodge DM, Mooney HA, Oesterheld M, Poff NLR, Sykes MT, Walker BH, Walker M, Wall DH (2000) Global biodiversity scenarios for the year 2100. Science 287:1170–1174. https://doi.org/10.1126/science.287.5459.1770

Scheffers BR, Phillips BL, Laurance WF, Sodhi NS, Diesmos A, Williams SE (2013) Increasing arboreality with altitude: a novel biogeographic dimension. Proc R Soc B 280:20131581. https://doi.org/10.1098/rspb.2013.1581

Slater HD, Gillingham PK, Pratt V, Eaton B, Fletcher S, Abdullah A, Supriadi KAH (2024) Living on the edge: forest edge effects on microclimate and terrestrial mammal activity in disturbed lowland forest in Sumatra, Indonesia. Oryx 58:228–239. https://doi.org/10.1017/S0030605323000212

Stark AY, Adams BJ, Fredley JL, Yanoviak SP (2017) Out on a limb: thermal microenvironments in the tropical forest canopy and their relevance to ants. J Therm Biol 69:32–38. https://doi.org/10.1016/j.jtherbio.2017.06.002

Tiede Y, Wallis CI, Bendix J, Brandl R, Farwig N (2017) Ants as indicators of environmental change and ecosystem processes. Ecol Indic 83:527–537. https://doi.org/10.1016/j.ecolind.2017.01.029

Williams RJ, Cook GD, Braithwaite RW, Andersen AN, Corbett LK (1995) Australia's wet-dry tropics: identifying the sensitive zones. In: Pernatta J, Leemans R, Elder D, Humphrey S (eds) The impact of climate change on ecosystems and species: terrestrial biomes. IUCN, Gland, pp 39–65

MIX
Papier aus verantwortungsvollen Quellen
Paper from responsible sources
FSC® C105338

FSC
www.fsc.org

If you have any concerns about our products,
you can contact us on
ProductSafety@springernature.com

In case Publisher is established outside the EU,
the EU authorized representative is:
Springer Nature Customer Service Center GmbH
Europaplatz 3, 69115 Heidelberg, Germany

Printed by Libri Plureos GmbH
in Hamburg, Germany